Mathematical Sciences Research Institute Publications 26

Mathematical Sciences Research Institute Publications

Volume 1	Freed and Uhlenbeck: Instantons and Four-Manifolds Second Edition
Volume 2	Chern (ed.): Seminar on Nonlinear Partial Differential Equations
Volume 3	Lepowsky, Mandelstam, and Singer (eds.): Vertex Operators in Mathematics and Physics
Volume 4	Kac (ed.): Infinite Dimensional Groups with Applications
Volume 5	Blackadar: K-Theory for Operator Algebras
Volume 6	Moore (ed.): Group Representations, Ergodic Theory, Operator Algebras, and Mathematical Physics
Volume 7	Chorin and Majda (eds.): Wave Motion: Theory, Modelling, and Computation
Volume 8	Gersten (ed.): Essays in Group Theory
Volume 9	Moore and Schochet: Global Analysis on Foliated Spaces
Volume 10	Drasin, Earle, Gehring, Kra, and Marden (eds.): Holomorphic Functions and Moduli I
Volume 11	Drasin, Earle, Gehring, Kra, and Marden (eds.): Holomorphic Functions and Moduli II
Volume 12	Ni, Peletier, and Serrin (eds.): Nonlinear Diffusion Equations and their Equilibrium States I
Volume 13	Ni, Peletier, and Serrin (eds.): Nonlinear Diffusion Equations and their Equilibrium States II
Volume 14	Goodman, de la Harpe, and Jones: Coxeter Graphs and Towers of Algebras
Volume 15	Hochster, Huneke, and Sally (eds.): Commutative Algebra
Volume 16	Ihara, Ribet, and Serre (eds.): Galois Groups over Q
Volume 17	Concus, Finn, Hoffman (eds): Geometric Analysis and Computer Graphics
Volume 18	Bryant, Chern, Gardner, Goldschmidt, Griffiths: Exterior Differential Systems
Volume 19	Alperin (ed.): Arboreal Group Therapy

(continued after page 416)

H. Judah W. Just H. Woodin
Editors

Set Theory of the Continuum

Springer-Verlag
New York Berlin Heidelberg London Paris
Tokyo Hong Kong Barcelona Budapest

Haim Judah
Department of Mathematics
and Computer Science
Bar Ilan University
52900 Ramatgan, Israel

Winfried Just
Department of Mathematics
Ohio University
Athens, OH 45701-2979
USA

Hugh Woodin
Department of Mathematics
University of California
Berkeley, CA 94720
USA

Mathematical Sciences
Research Institute
1000 Centennial Drive
Berkeley, CA 94720
USA

The Mathematical Sciences Research Institute wishes to acknowledge support by the National Science Foundation.

AMS Subject Classifications: 04-06, 03Exx

Library of Congress Cataloging-in-Publication Data
Set theory of the continuum / Haim Judah, Winfried Just, Hugh Woodin, editors.
p. cm. – (Mathematical Sciences Research Institute publications ; 26)
Papers presented at an MSRI workshop Oct. 16–20, 1989.
Includes bibliographical references.
ISBN 0-387-97874-7 (U.S.)
1. Set theory—Congresses. I. Judah, Haim. II. Just, Winfried. III. Woodin, Hugh. IV. Mathematical Sciences Research Institute (Berkeley, Calif.). V. Series.
QA248.S414 1992
511.3′22—dc20 92-28316

Printed on acid-free paper.

Production managed by Hal Henglein; manufacturing supervised by Jacqui Ashri.
Photocomposed copy prepared by the Mathematical Sciences Research Institute using TeX.
Printed and bound by Edwards Brothers, Inc., Ann Arbor, MI.
Printed in the United States of America.

9 8 7 6 5 4 3 2 1

ISBN 0-387-97874-7 Springer-Verlag New York Berlin Heidelberg
ISBN 3-540-97874-7 Springer-Verlag Berlin Heidelberg New York

PREFACE

During the academic year 1989–90 MSRI organized a "Logic Year." As part of this Logic Year several activities took place in set theory.

In the week of October 16–20 1989, a workshop was held. The workshop focused on the many set theoretical aspects of the continuum and was entitled, "Set Theory and the Continuum." The workshop was organized (with extensive and greatly appreciated help from the staff at MSRI) by one of us (Woodin).

A year-long seminar on set theory was organized by H. Judah in the autumn and continued, after Judah's departure, by D. A. Martin in the spring. Other seminars and series of talks in set theory that lasted for periods of several months were given by M. Magidor and J. Steel/P. Welch. There were also many talks on set theory given on a more informal basis, or in seminars that tried to foster interaction between the subdisciplines of Mathematical Logic, and between Mathematical Logic and the rest of Mathematics.

This volume is primarily an account of the talks presented at the meeting, but is also intended to reflect the whole spectrum of activities in set theory during the entire year. It has been divided into two sections. The first is the "talks" section and for the most part includes survey papers by invited speakers derived from their talks given during the workshop. There are three exceptions however: The paper by P. Dehornoy gives account of his approach to results by Richard Laver, one of the speakers invited to the workshop who could not attend. The other two exceptions are papers by Mac Lane and Mathias. They are not based on the workshop (although Mac Lane was an invited speaker), but on a series of polemic talks on the role of set theory as a foundation of mathematics that Mac Lane and Mathias gave alternatingly over the Logic Year. Their short contributions to this volume reflect some of the flavour of their controversy, and highlight the major points each of them was making.

The second section includes the research papers. Those have been subject to refereeing, with the same criteria applied as for publication in leading journals.

Here is a list of speakers at the workshop:

J. Baumgartner : $\mathfrak{c}^{++}$
H. Becker : Descriptive set theoretic phenomena in analysis and topology
M. Foreman : Amenable group actions on the integers, an independence result
M. Gitik : The singular cardinals problem revisited again
S. Jackson : Admissible Souslin cardinals in $L(\mathbb{R})$
H. Judah : Measure and category
A. Kechris : Descriptive dynamics
A. Louveau : Classifying Borel structures
S. Mac Lane : Topos-theoretic versions of the continuum
M. Magidor : The singular cardinals problem revisited
S. Shelah : Is cardinal arithmetic interesting?
R. Shore : Degrees of constructibility
S. Simpson : Reverse mathematics and dynamical systems
T. Slaman : Global properties of degree structures
R. Soare : Continuity properties of Turing degrees and games applied to recursion theory
J. Steel : Fine structure and inner models of Woodin cardinals
S. Todorcevic : Forcing axioms
B. Velickovic : OCA and automorphisms of $P(\omega)$/finite

The organizer of the workshop would like to thank all of the participants for they are in essence responsible for its success. We, the editors of this volume, thank all those who contributed; their work is evident. We particularly would like to express our sincere appreciation to all the referees without whom this volume would not really have been possible. As is usually the case, the magnitude of their contribution is not evident.

The workshop would not have occurred without the help of the MSRI staff, in particular without Irving Kaplansky. This volume would not exist were it not for the technical assistance of Arlene Baxter, David Mostardi, Margaret Pattison, and Sean Brennan.

Though uninvited, Nature also decided to give a talk. Midway through the workshop at 5:04 pm on Tuesday, October 17 the Loma Prieta earthquake occurred, measuring approximately 7.1 on the Richter scale.

Haim Judah
Winfried Just
Hugh Woodin

CONTENTS

Research Papers

DESCRIPTIVE SET THEORETIC PHENOMENA IN ANALYSIS AND TOPOLOGY

Howard Becker

1. Introduction

This paper is concerned with a portion of descriptive set theory, namely the theory of (boldface) $\boldsymbol{\Sigma}^1_n$, $\boldsymbol{\Pi}^1_n$ and $\boldsymbol{\Delta}^1_n$ sets in Polish spaces, for $n \leq 3$. We assume the reader has some familiarity with this subject, in fact with the logicians' version of this subject. Moschovakis [37] is the basic reference and we generally follow his notation and terminology. A *Polish space* is a topological space homeomorphic to a separable complete metric space. All uncountable Polish spaces are Borel isomorphic, and a Borel isomorphism preserves $\boldsymbol{\Sigma}^1_n$ sets, so as far as the abstract theory is concerned, there is only one space [37,1G]. But particular examples happen to live in particular spaces, so in this paper we will consider many different spaces, all Polish.

The theorems of descriptive set theory show that there are pointsets exhibiting various phenomena, e.g., that there are universal sets. While we know that such things exist, it remains an interesting problem to find examples of the given phenomenon which arise naturally in some context in analysis, topology, algebra, logic, etc. The principal purpose of this paper is to give some natural examples of three types of descriptive set theoretic phenomena, examples which occur in analysis and topology. The three phenomena are: true $\boldsymbol{\Sigma}^1_n$ sets (§2), universal sets (§3), and inseparable pairs (§4). This paper is mainly a list of such examples, both ancient and modern, along with references, definitions, questions, remarks, comments, asides, and occasionally even a few hints at proofs. This is a survey; we make no claim to completeness. The word "natural" is not a technical term — it just reflects the author's personal esthetic judgment. It is undoubtedly true that naturalness is in the eye of the beholder. It is also undoubtedly true that some natural examples are more natural than others.

Descriptive set theory has its historical origins in analysis, but it has moved a long way from its origins. The subject has been studied largely for its own sake, or for its connections with logic. In recent years, a number of mathematicians — both logicians and analysts — have been studying situations in analysis where ideas and results of descriptive set theory may be relevant. This paper is a part of that trend, but only a small part. There is a lot more to the subject which might be called "connections between descriptive set theory and analysis," than finding natural examples of descriptive set theoretic phenomena.

This paper is a revised version of a talk given at the Workshop on Set Theory and the Continuum, held at MSRI in October 1989, a workshop at which, incidentally, several of the talks illustrated the trend mentioned in the previous paragraph. I thank the organizers of the workshop, and MSRI, for enabling me to participate. I thank Robert Lubarsky, Frank Tall and Hugh Woodin for their comments after the talk, comments which have led to some revisions. This paper has been heavily influenced by about six years of conversations with Alexander Kechris, whom I also thank.

2. Classification of Pointsets in the Projective Hierarchy

One of the most important theorems about $\underset{\sim}{\Sigma}^1_n$ ($\underset{\sim}{\Pi}^1_n$) sets is that these sets actually exist — that is, there exist pointsets which are $\underset{\sim}{\Sigma}^1_n$ ($\underset{\sim}{\Pi}^1_n$) but not $\underset{\sim}{\Delta}^1_n$. We call such a pointset *true* $\underset{\sim}{\Sigma}^1_n$ (*true* $\underset{\sim}{\Pi}^1_n$). In §1 we give some natural examples of such pointsets, or in other words, for certain natural pointsets, we give the set's exact classification in the projective hierarchy.

The following is one of the oldest (1936) and best known natural examples.

Example 1. (Mazurkiewicz [36]).

Space: $C[0,1]$.
Pointset: $E_1 = \{ f \ : \ f \text{ is differentiable} \}$.
Classification: True $\underset{\sim}{\Pi}^1_1$.

To be precise, let us take *differentiable* to mean having a finite derivative everywhere, and at endpoints we consider the one-sided derivative (although the classification true $\underset{\sim}{\Pi}^1_1$ would still be valid under any other reasonable definition.) That E_1 is $\underset{\sim}{\Pi}^1_1$ is trivial — it is defined by applying a universal quantifier to a Borel matrix: $(\forall x \in [0,1])(f'(x) \text{ exists })$. The content of Mazurkiewicz's Theorem is that it is no simpler than $\underset{\sim}{\Pi}^1_1$, that is, that there is no way to define E_1 without using a universal quantifier.

We next consider four more examples in the same space.

Example 2. (Mauldin [35]).

Space: $C[0,1]$.
Pointset: $E_2 = \{f : f \text{ is nowhere differentiable} \}$.
Classification: True $\underset{\sim}{\Pi}^1_1$.

Example 3. (Woodin).

Space: $C[0,1]$.
Pointset: $E_3 = \{f : f$ satisfies Rolle's Theorem $\}$.
Classification: True $\underset{\sim}{\Sigma}^1_1$.

Example 4. (Woodin).

Space: $C[0,1]$.
Pointset: $E_4 = \{\, f \,:\, f$ satisfies the Mean Value Theorem $\}$.
Classification: True $\underset{\sim}{\Pi}^1_2$.

Example 5. (Humke-Laczkovich [17]).

Space: $C[0,1]$.
Pointset: $E_5 = \{\, f \,:\, (\exists g \in C\,[0,1])\,(f = g \circ g)\,\}$.
Classification: True $\underset{\sim}{\Sigma}^1_1$.

Saying that f *satisfies Rolle's Theorem* means, of course, that f satisfies the conclusion of Rolle's Theorem, that is:

For all $a, b \in [0,1]$, if $a < b$ and $f(a) = f(b)$, then there exists a c in (a,b) such that f is differentiable at c and $f'(c) = 0$.

Thus all differentiable functions are in E_3, but also some nondifferentiable functions. Similarly, f *satisfies the Mean Value Theorem* means:

For all $a, b \in [0,1]$, if $a < b$ then there exists a c in (a,b) such that f is differentiable at c and $f'(c) = \dfrac{f(b) - f(a)}{b - a}$.

The upper bounds on the complexity are obvious, except for Example 3. To see that E_3 is $\underset{\sim}{\Sigma}^1_1$ note that for continuous f, f satisfies Rolle's Theorem iff:

For all rational a, b, d, if $0 \le a < d < b \le 1$ and either $f(d) > \max(f(a), f(b))$ or $f(d) < \min(f(a), f(b))$, then there exists a c in (a,b) such that $f'(c) = 0$.

It is perhaps surprising that E_3 and E_4 have different complexity, since the Mean Value Theorem is usually thought of as being pretty much the same thing as Rolle's Theorem — to prove the Mean Value Theorem, you add a linear function and apply Rolle's Theorem. But this actually explains the difference in complexity, that is, explains where the extra universal quantifier comes from: f satisfies the Mean Value Theorem iff *for every* linear function L, $f+L$ satisfies Rolle's Theorem.

A pointset B in a Polish space Y is called *complete* $\underset{\sim}{\Sigma}_n^1$ (*complete* $\underset{\sim}{\Pi}_n^1$) if B is $\underset{\sim}{\Sigma}_n^1$ ($\underset{\sim}{\Pi}_n^1$) and for every Polish space X and every $\underset{\sim}{\Sigma}_n^1$ ($\underset{\sim}{\Pi}_n^1$) pointset $A \subset X$, there is a Borel measurable function $H : X \to Y$ such that $A = H^{-1}[B]$. Such an H is said to *reduce* A to B.

Clearly any complete $\underset{\sim}{\Sigma}_n^1$ ($\underset{\sim}{\Pi}_n^1$) set is true $\underset{\sim}{\Sigma}_n^1$ (or $\underset{\sim}{\Pi}_n^1$), and indeed this is the most common method of proving a given set is true $\underset{\sim}{\Sigma}_n^1$ (or $\underset{\sim}{\Pi}_n^1$). It was essentially the method used in the original proofs of the lower bounds for Examples 1, 2 and 4. But it is not the only method. The original proof that E_3 is not $\underset{\sim}{\Pi}_1^1$ is as follows: If E_3 was $\underset{\sim}{\Pi}_1^1$, then E_4 would also be $\underset{\sim}{\Pi}_1^1$, since $f \in E_4$ iff for every linear function L, $f+L$ is in E_3; but Woodin proved that E_4 is not $\underset{\sim}{\Pi}_1^1$. This argument does not show that E_3 is complete $\underset{\sim}{\Sigma}_1^1$. The only known proof that E_5 is not Borel is the proof in [17], where it is proved that every Borel set is reducible to E_5 by a continuous function. This argument also does not show completeness.

Assuming $\underset{\sim}{\Pi}_1^1$-determinacy (equivalently, assuming $\forall x \subset \omega$, $x^\sharp$ exists), every true $\underset{\sim}{\Sigma}_1^1$ set is complete $\underset{\sim}{\Sigma}_1^1$. This is a theorem of Wadge —see [44]. So assuming strong axioms (actually all that is needed is $0^\sharp$), E_3 and E_5 are complete $\underset{\sim}{\Sigma}_1^1$. At the time of the Set Theory Workshop, it was not known whether the completeness of E_3 was provable in ZFC, but Woodin has subsequently shown that it is. It is still open whether the completeness of E_5 is provable in ZFC.

For any Polish space X, denote by X^ω the topological product of countably many copies of X. The space X^ω is also Polish. We next consider some examples from the space $(C[0,1])^\omega$. The points in this space are sequences of functions. Note that the topology on $C[0,1]$ will always be the same topology considered above, that is, the topology of uniform convergence, and pointclasses such as $\underset{\sim}{\Sigma}_n^1$ refer to this topology. We will be considering pointwise convergence of sequences in $C[0,1]$, but we will never consider the topology of pointwise convergence on $C[0,1]$ (which, incidentally, is not a Polish topology).

Example 6.

Space: $(C[0,1])^\omega$.
Pointset: $E_6 = \{ \langle f_i \rangle : \langle f_i \rangle \text{ converges pointwise } \}$.
Classification: True $\underset{\sim}{\Pi}_1^1$.

Example 7.

Space: $(C[0,1])^\omega$.

Pointset: $E_7 = \{ \langle f_i \rangle : \langle f_i \rangle$ converges pointwise to a continuous limit $\}$.

Classification: True $\underset{\sim}{\Pi}^1_1$.

One might also consider the set of uniformly convergent sequences, but this is a Borel set in $(C[0,1])^\omega$, and Borel sets are unworthy of inclusion in the list of examples.

It is not obvious that E_7 is $\underset{\sim}{\Pi}^1_1$, since at first glance, defining E_7 appears to require an existential quantifier — one must say $\exists g \in C[0,1]$ such that g is the limit of the sequence. The following theorem is very useful in pointclass computations via quantifier-counting — it handles Example 7, as well as some other computations that appear later in this paper. If x and y are points in recursively presented Polish spaces X and Y, respectively, then $x \leq_h y$ means that x is hyperarithmetic-in-y, or equivalently, x is $\Delta^1_1(y)$. (This is defined in Moschovakis [37,3D and 3E] for points in arbitrary recursively presented Polish spaces. The reader who prefers doing recursion theory in ω^ω can view points in X and Y as being encoded by elements of ω^ω, and then work with the codes.)

Theorem 2.1. (Kleene - see [37, 4D.3]).

The pointclass $\underset{\sim}{\Pi}^1_1$ is closed under quantification of the form: $\exists x \leq_h y$. That is, if $P \subset X \times Y \times Z$ is $\underset{\sim}{\Pi}^1_1$ and $Q \subset Y \times Z$ is defined by

$$Q(y,z) \iff (\exists x \leq_h y) P(x,y,z),$$

then Q is also $\underset{\sim}{\Pi}^1_1$.

Returning to Example 7, the continuous limit (if it exists) is clearly hyperarithmetic in the sequence. Hence $\langle f_i \rangle \in E_7$ iff:

$$(\exists g \leq_h \langle f_i \rangle)(g \in C[0,1] \text{ and } (\forall x \in [0,1])(\lim_{i \to \infty} f_i(x) = g(x)))\cdot$$

So by Kleene's Theorem, E_7 is $\underset{\sim}{\Pi}^1_1$. This use of recursion theoretic methods is not necessary; there are very classical ways to prove that E_7 is $\underset{\sim}{\Pi}^1_1$.

Regarding lower bounds on complexity, the fact that E_6 and E_7 are not Borel is a very elementary reduction argument. It is the sort of theorem that was probably known to classical descriptive set theorists of the 1930's — and if it wasn't, it should have been. However the earliest explicit statements of these results that I have been able to find appeared in the late 1980's: [7] for E_6 and [8] for E_7. Assani [3], [4] contain some theorems which are similar to these results, and the proof given in Assani's papers does indeed show that E_6 and E_7 are complete $\underset{\sim}{\Pi}^1_1$ — this is apparently the first published proof. (Assani [3], [4] is concerned with weakly Cauchy and weakly convergent sequences in various Banach spaces, including $C[0,1]$. For more information on this topic, see Becker [9].)

Example 8.

Space: $(C[0,1])^\omega$.
Pointset: $E_8 = \{\langle f_i\rangle : \text{ Some subsequence of } \langle f_i\rangle \text{ converges uniformly }\}$.
Classification: True $\underset{\sim}{\Sigma}^1_1$.

Example 8 is another folklore result. It has perhaps the easiest proof of any natural example — so easy, that we put it in this paper. Example 8 is a special case of Example 9, below. Since uniform convergence is convergence in the topology of the space $C[0,1]$, the pointset E_8 has an analog for any space. That is, for any space X, we can consider the pointset in X^ω of all sequences which have a convergent subsequence (with respect to the topology of X).

Example 9.

Space: X^ω, X a fixed Polish space which is not σ-compact.
Pointset: $E_9 = \{\ \langle x_i\rangle\ :\ \text{ Some subsequence of } \langle x_i\rangle \text{ converges }\ \}$.
Classification: True $\underset{\sim}{\Sigma}^1_1$.

Proof. A Polish space which is not σ-compact contains a closed copy of ω^ω. So it will suffice to prove that E_9 is complete $\underset{\sim}{\Sigma}^1_1$ for the space $X = \omega^\omega$. The set of nonwellfounded trees on $\omega\backslash\{0\}$ (i.e., those trees which have an infinite branch) is a complete $\underset{\sim}{\Sigma}^1_1$ set in the Polish space Tr of all trees on $\omega\backslash\{0\}$ — see [26] for details. We construct a continuous function $H : Tr \to (\omega^\omega)^\omega$ which reduces the nonwellfounded trees to E_9, and thereby complete the proof. For $\sigma \in \omega^{<\omega}$, let $x_\sigma \in \omega^\omega$ be σ followed by an infinite string of 0's. For T a tree, let $H(T) = \langle y_i^T\rangle \in (\omega^\omega)^\omega$ be such that

$$\{y_i^T : i \in \omega\} \ = \ \{x_\sigma : \sigma \in T \text{ or length } (\sigma) = 1\},$$

and such that $y_j^T \neq y_i^T$ for $i \neq j$; such a sequence $\langle y_i^T\rangle$ can be constructed from T in a continuous way. Note that $\langle y_i^T\rangle$ has a convergent subsequence iff T has an infinite branch.

Returning to the space $(C[0,1])^\omega$, we give three more examples. In all three, the obvious upper bound obtained by quantifier-counting, is, in fact, the best upper bound.

Example 10. (Becker [8]).

Space: $(C[0,1])^\omega$.
Pointset: $E_{10} = \{\langle f_i \rangle :$ Some subsequence of $\langle f_i \rangle$ converges pointwise $\}$.
Classification: True $\underset{\sim}{\Sigma}^1_2$.

Example 11. (Becker [8]).

Space: $(C[0,1])^\omega$.
Pointset: $E_{11} = \{\langle f_i \rangle :$ Some subsequence of $\langle f_i \rangle$ converges pointwise to a continuous limit $\}$.
Classification: True $\underset{\sim}{\Sigma}^1_2$.

Example 12. (Becker [8]).

Space: $(C[0,1])^\omega$.
Pointset: $E_{12} = \{ \langle f_i \rangle : (\forall g \in C[0,1])$ (Some subsequence of $\langle f_i \rangle$ converges pointwise to g) $\}$.
Classification: True $\underset{\sim}{\Pi}^1_3$.

Intuitively, E_{12} is the set of sequences which generate the whole space $C[0,1]$, in a particular way. For example, if $\langle p_i \rangle$ is an enumeration of all polynomials with rational coefficients, then by Weierstrass's Theorem, $\langle p_i \rangle \in E_{12}$; but this example is somewhat atypical, since in this case we can always get uniform convergence, whereas the definition of E_{12} only requires pointwise convergence. Example 12 is an extreme point of this paper — we will not go beyond the third level of the projective hierarchy. There is an open problem related to this example. A *Baire*-1 function is a pointwise limit of continuous functions. Let

$$\hat{E}_{12} = \{ \langle f_i \rangle \in (C[0,1])^\omega : (\forall \text{ Baire-1 function } g) \\ (\text{Some subsequence of } \langle f_i \rangle \text{ converges pointwise to } g)\}.$$

$\hat{E}_{12}$ is a subset of E_{12}, and perhaps a more natural set. It can be shown that $\hat{E}_{12} \neq E_{12}$. $\hat{E}_{12}$ is also clearly a $\underset{\sim}{\Pi}^1_3$ set. But it is an open question whether it is true $\underset{\sim}{\Pi}^1_3$ — it may be simpler.

Since we are considering convergence of sequences of functions, it is only appropriate that we look at Fourier series.

Example 13. (Ajtai-Kechris [2]).

Space: $C[0,2\pi]$ or $L^p[0,2\pi]$, for fixed $p \geq 1$.
Pointset: $E_{13} = \{ f :$ The Fourier series of f converges everywhere $\}$.
Classification: True $\underset{\sim}{\Pi}^1_1$.

The fact that E_{13} is $\underset{\sim}{\Pi}_1^1$ was published in 1931 by Kuratowski [28], where it is attributed to Banach. This fact may appear to be a triviality, a straightforward exercise in quantifier-counting; that appearance is correct, but the reader should keep in mind that Kuratowski invented quantifier-counting (the "Tarski-Kuratowski algorithm" — see [41]), and this was one of its first applications. The problem of whether E_{13} is true $\underset{\sim}{\Pi}_1^1$ was posed in that paper, and solved 56 years later by Ajtai and Kechris.

In Examples 1 and 2, we considered everywhere differentiable and nowhere differentiable functions. By analogy, after Example 13, we should consider functions whose Fourier series diverges everywhere. For most of the spaces of Example 13, this is the empty set; by a famous theorem of Carleson and Hunt, for $p > 1$, for any L^p function f (hence for any continuous f), the Fourier series of f converges almost everywhere. But Kolmogorov proved that there exists an L^1 function with everywhere divergent Fourier series. The next example is a strengthening of Kolmogorov's Theorem.

Example 14. (Kechris [22]).

Space: $L^1[0, 2\pi]$.
Pointset: $E_{14} = \{ f \ : \text{The Fourier series of } f \text{ diverges everywhere } \}$.
Classification: True $\underset{\sim}{\Pi}_1^1$.

For any space X, let $\mathcal{K}(X)$ denote the space of all nonempty compact subsets of X, with the Hausdorff metric δ:

$$\delta(K, K') = \sup\{ d(x, K), d(y, K') \ : \ x \in K', y \in K \}.$$

If X is Polish, so is $\mathcal{K}(X)$.

Example 15. (Hurewicz [18]).

Space: $\mathcal{K}(X)$, X a fixed uncountable Polish space.
Pointset: $E_{15} = \{ K \ : \ K \text{ is countable } \}$.
Classification: True $\underset{\sim}{\Pi}_1^1$.

To see that E_{15} is $\underset{\sim}{\Pi}_1^1$, note that $K \in E_{15}$ iff $\forall$ perfect set $P, P \not\subset K$. Example 15 is the oldest natural example (1930). The fact that E_{15} is not Borel has an interesting application to Banach space theory, due to Bourgain — see Rosenthal [42].

We next consider some natural examples from topology. We are interested in determining the complexity of notions such as connectedness, path-connectedness and simple connectedness, for compact subsets of $\mathbb{R}^n$, i.e., in the space $\mathcal{K}(\mathbb{R}^n)$. (A topological space is *simply connected* if it is path-connected and it has no holes. A topological space Y *has no holes* if every map from the unit circle into Y is homotopic to a constant map; equivalently, every map from the unit circle into Y can be extended to a map from the closed unit disc into Y.) For any X, $\{ K \in \mathcal{K}(X) : K \text{ is connected } \}$ is closed, hence uninteresting. For subsets of the line, connected = path-connected = simply connected. Thus the only interesting cases are path-connectedness and simple connectedness in $\mathcal{K}(\mathbb{R}^n)$ for $n \geq 2$. For certain dimensions n the answer is known. That constitutes the next three examples.

Example 16. (Ajtai, Becker [10]).

Space: $\mathcal{K}(\mathbb{R}^n)$, for fixed $n \geq 3$.
Pointset: $E_{16} = \{ K : K \text{ is path-connected} \}$.
Classification: True $\underset{\sim}{\Pi}^1_2$.

Example 17. (Becker [10]).

Space: $\mathcal{K}(\mathbb{R}^2)$.
Pointset: $E_{17} = \{ K : K \text{ is simply connected} \}$.
Classification: True $\underset{\sim}{\Pi}^1_1$.

Example 18. (Becker [10]).

Space: $\mathcal{K}(\mathbb{R}^n)$, for fixed $n \geq 4$.
Pointset: $E_{18} = \{ K : K \text{ is simply connected } \}$.
Classification: True $\underset{\sim}{\Pi}^1_2$.

In $\mathcal{K}(\mathbb{R}^2)$, we have upper and lower bounds on path-connectedness.

Theorem 2.2. (Ajtai, Becker [10]).
In the space $\mathcal{K}(\mathbb{R}^2)$, the pointset $\{ K : K \text{ is path-connected} \}$ is $\underset{\sim}{\Pi}^1_2$ and it is not $\underset{\sim}{\Pi}^1_1$.

The *exact* classification of path-connectedness in $\mathcal{K}(\mathbb{R}^2)$ is not known — it may be complete $\underset{\sim}{\Pi}^1_2$, it may be complete $\underset{\sim}{\Sigma}^1_1$, or it may be somewhere in between. (All $\underset{\sim}{\Sigma}^1_1$ sets are reducible to it.) Simple connectedness in three dimensions is also only partly classified: It is known to be $\underset{\sim}{\Pi}^1_2$ and it is known that it is neither $\underset{\sim}{\Pi}^1_1$ nor $\underset{\sim}{\Sigma}^1_1$ [10].

Example 16 has an application. Let $\mathcal{C} \subset \mathcal{K}(\mathbb{R}^n)$ be a collection of pointsets which is closed under continuous image. Say that $F \in \mathcal{K}(\mathbb{R}^n)$ *generates* $\mathcal{C}$ if $\mathcal{C}$ is the set of all continuous images of F (in $\mathbb{R}^n$). What we have in mind here is a well-known theorem of Hahn and Mazurkiewicz which characterizes metric spaces which are the continuous image of the closed unit interval, $[0, 1]$: A metric space Y is the continuous image of $[0, 1]$ iff

Y is compact, connected and locally connected. In our terminology, $[0,1]$ generates $\{ K \in \mathcal{K}(\mathbb{R}^n) : K \text{ is connected and locally connected} \}$. Similarly, the Cantor set generates all of $\mathcal{K}(\mathbb{R}^n)$. The application of Example 16 is that there is no Hahn-Mazurkiewicz Theorem for path-connectedness.

Theorem 2.3. *For $n \geq 3$, there is no set $F \in \mathcal{K}(\mathbb{R}^n)$ which generates*

$$\mathcal{C} = \{ K \in \mathcal{K}(\mathbb{R}^n) \ : \ K \textit{ is path-connected} \}.$$

Proof. For any F, the set

$$\{ K \in \mathcal{K}(\mathbb{R}^n) \ : \ K \text{ is the continuous image of } F \}$$

is $\underset{\sim}{\Sigma}^1_1$. By Example 16, $\mathcal{C}$ is not $\underset{\sim}{\Sigma}^1_1$.

For $\mathbb{R}^2$, this question seems to be open.

In one respect, Example 17 is very strange. Normally, in classifying a natural example in the projective hierarchy, getting the upper bound is either trivial or easy; the difficult part is getting the lower bound. Example 17 is an exception. The hard part is proving that E_{17} is $\underset{\sim}{\Pi}^1_1$. We give an outline of the proof, below.

Theorem 2.4. *In the space $\mathcal{K}(\mathbb{R}^2)$, the pointset $NH = \{K : K$ has no holes$\}$ is $\underset{\sim}{\Pi}^1_1$.*

The analog of 2.4 for 3 or more dimensions is false. The difference between 2 dimensions and ≥ 3 dimensions, is that in $\mathbb{R}^2$ we have the Jordan Curve Theorem. A *Jordan curve* is a one-to-one map of the circle into the plane. By the Jordan Curve Theorem, any Jordan curve has a well defined *inside* and *outside*. We need the following topological theorem. For any set $A \subset \mathbb{R}^2$, the following are equivalent:

(a) A has no holes.

(b) For any Jordan curve J, if $J \subset A$ then Inside$(J) \subset A$.

Now (b) is a $\underset{\sim}{\Pi}^1_1$ condition, which proves 1.4. The set E_{17} is, by definition, the intersection of the $\underset{\sim}{\Pi}^1_1$ set NH and the set

$$PC = \{ K \in \mathcal{K}(\mathbb{R}^2) \ : \ K \text{ is path-connected} \}.$$

But PC is *not* a $\underset{\sim}{\Pi}^1_1$ set (by Theorem 2.2). So we have not yet succeeded in showing that E_{17} is $\underset{\sim}{\Pi}^1_1$.

Theorem 2.5. (Becker[11]). *Let $K \subset \mathbb{R}^2$ be compact and simply connected. Let $p, q \in K$. There is a path γ from p to q, lying in K, such that $\gamma \leq_h (K, p, q)$.*

By 2.5, we have that for $K \in \mathcal{K}(\mathbb{R}^2)$, K is simply connected iff:

$$K \in NH \text{ and } (\forall p, q \in K)(\exists \gamma \leq_h (K, p, q))$$
$$(\gamma \text{ is a path from } p \text{ to } q \text{ lying in } K).$$

By 2.1 and 2.4, the above formula shows that E_{17} is $\underset{\sim}{\Pi}^1_1$.

We have three remarks about 2.5. First, this theorem is false for arbitrary path-connected, as opposed to simply connected, sets; for if it was true, the above argument would show that path-connectedness in $\mathbb{R}^2$ is $\underset{\sim}{\Pi}^1_1$, which, by 2.2, is not so. Second, "hyperarithmetic" is best possible. Our third remark is that this is an example of the use of effective methods to prove a noneffective theorem. The fact that E_{17} is (boldface) $\underset{\sim}{\Pi}^1_1$ is a statement of classical descriptive set theory, a statement which does not in any way involve recursion theoretic concepts. (I am tempted to call it a "classical theorem," but that could be misunderstood.) The only known proof, the one given above, uses recursion theory.

We now leave topology and return to analysis. A set $A \subset [0, 2\pi]$ is called a *set of uniqueness* if no nonzero trigonometric series converges to 0 at every point of $[0, 2\pi] \setminus A$. Thus the sets of uniqueness are a type of exceptional set, or notion of smallness, which comes up in harmonic analysis.

Example 19. (Kaufman [19], Solovay).

Space: $\mathcal{K}([0, 2\pi])$.
Pointset: $E_{19} = \{ K : K \text{ is a set of uniqueness} \}$.
Classification: True $\underset{\sim}{\Pi}^1_1$.

Kechris and Louveau have written a book [23] about connections between descriptive set theory and various types of exceptional sets, including sets of uniqueness. So we do not pursue the subject here, but instead refer the reader to [23]. We do, however, wish to point out that the Kaufman-Solovay Theorem was used by Debs-Saint Raymond [13] in their solution of an old problem in the theory of sets of uniqueness; specifically, they proved that every set of uniqueness (which has the property of Baire) is first category. But later Kechris-Louveau [23, VIII, §3] came up with a different proof, not involving descriptive set theory.

3. Representation Theorems and Universal Sets

If $A \subset X \times Y$ is a pointset in a product space, then for all $x \in X$, A_x denotes the vertical section of A above x: $A_x = \{ y \in Y : (x,y) \in A \}$. A set $U \subset X \times Y$ is called a *universal set* for $\underset{\sim}{\Sigma}^1_n \restriction Y$ if U is $\underset{\sim}{\Sigma}^1_n$ and every $\underset{\sim}{\Sigma}^1_n$ subset of Y is equal to U_x for some $x \in X$. For all n, and for all uncountable Polish spaces X and Y, there exists a universal set $U \subset X \times Y$ for $\underset{\sim}{\Sigma}^1_n \restriction Y$ (and similarly for $\underset{\sim}{\Pi}^1_n$ and for the Borel classes $\underset{\sim}{\Sigma}^0_n$ and $\underset{\sim}{\Pi}^0_n$); see [37,1D.2 and 1E.3]. In §3 we give some examples of universal sets which occur in nature. These results are representation theorems — they state that every $\underset{\sim}{\Sigma}^1_n$ subset of Y can be represented in a particular manner — and the representation gives us the universal set.

For any $f \in C[0,1]$, let R_f be the range of the derivative of f:

$$R_f = \{ y \in \mathbb{R} : (\exists x \in [0,1])(f \text{ is differentiable at } x \text{ and } f'(x) = y) \}.$$

Clearly for any f, R_f is a $\underset{\sim}{\Sigma}^1_1$ set of real numbers. The converse is also true — every $\underset{\sim}{\Sigma}^1_1$ set of real numbers can be represented in this manner.

Theorem 3.1. (Poprougénko [39]). *Let $S \subset \mathbb{R}$ be any $\underset{\sim}{\Sigma}^1_1$ set. There exists an $f \in C[0,1]$ such that $S = R_f$.*

Define $R \subset C[0,1] \times \mathbb{R}$ as follows:

$$R = \{ (f,y) : (\exists x \in [0,1])(f \text{ is differentiable at } x \text{ and } f'(x) = y) \}.$$

Then R is $\underset{\sim}{\Sigma}^1_1$, and the vertical sections of R are the R_f's. Thus by Poprougénko's Theorem, R is a universal set for $\underset{\sim}{\Sigma}^1_1 \restriction \mathbb{R}$.

A theorem of calculus, due to Darboux, states that if f is differentiable everywhere, then f' satisfies the intermediate value property. Hence if f is differentiable everywhere, R_f is an interval.

It is possible to prove a stronger version of the previous theorem: 3.1 holds *uniformly*. There are several ways to make this precise; we choose to use the following *S-m-n* style formulation. Let $A \subset \omega^\omega \times \mathbb{R}$ be any $\underset{\sim}{\Sigma}^1_1$ set. There exists a continuous function $H : \omega^\omega \to C[0,1]$ such that for all $z \in \omega^\omega$, $A_z = R_{H(z)}$. In other words, 3.1 says that there exists an f such that $A_z = R_f$; the uniform version of 3.1 gives us a continuous H which actually computes such an f, i.e., H computes an index for A_z with respect to the universal set R. (If A is lightface Σ^1_1, e.g., A is the canonical universal set, then H can be taken to be a recursive function.)

Suppose $B \subset \omega^\omega$ is an arbitrary $\underset{\sim}{\Sigma}^1_1$ set. Then there is a $\underset{\sim}{\Sigma}^1_1$ (in fact, G_δ) set $A \subset \omega^\omega \times \mathbb{R}$ such that B is the projection of A, that is, $B(z) \iff$

$\exists y\, A(z,y)$. Note that the function H given by the uniform version of 3.1 reduces B to the set $\{ f : \exists y\, (y \in R_f) \} \subset C[0,1]$. That is, H reduces B to the set $C[0,1] \setminus E_2$. This proves that $C[0,1] \setminus E_2$ is complete $\underset{\sim}{\Sigma}^1_1$, hence E_2 is complete $\underset{\sim}{\Pi}^1_1$. This is Mauldin's Theorem (Example 2), although it is not Mauldin's proof.

Similarly, suppose $C \subset \omega^\omega$ is an arbitrary $\underset{\sim}{\Pi}^1_2$ set. Then there is a $\underset{\sim}{\Sigma}^1_1$ set $A \subset \omega^\omega \times \mathbb{R}$ such that $C(z) \iff \forall y\, A(z,y)$. Then H reduces C to the set $\{ f : \forall y (y \in R_f) \}$, so this set is complete $\underset{\sim}{\Pi}^1_2$, which gives us another example.

Example 20.

Space: $C[0,1]$.
Pointset: $E_{20} = \{ f : (\forall y \in \mathbb{R})(y \text{ is in the range of } f') \}$.
Classification: True $\underset{\sim}{\Pi}^1_2$.

This method of proof is very general. If U is any universal set satisfying the above uniformity property, then by putting a universal or existential quantifier in front of the universal set, one obtains a complete set in the appropriate pointclass. In practice, representation theorems tend to hold uniformly. Hence taking such a theorem and putting quantifiers in front of the universal set, generates examples of pointsets of a particular complexity. (The naturalness of the examples is another matter.)

This type of proof is fairly new. The abstract idea of completeness (e.g., that complete $\underset{\sim}{\Pi}^1_2$ sets are true $\underset{\sim}{\Pi}^1_2$), and the use of completeness arguments for natural examples was known classically. The abstract notion of universal set was also known classically, as were various specific representation theorems, including 3.1. But the idea of uniformity, of calculating an index, was not considered classically; it seems to be a recursion theoretic idea, even though it can be formalized in a boldface setting with no mention of recursion theory. This method of pointclass computation, via universal sets and uniformity, was first applied to natural examples in descriptive set theory by Kechris, around 1984.

Consider the complex Banach space c_0. Let $T : c_0 \to c_0$ be a bounded linear operator. Let

$$\begin{aligned} E_T &= \text{ the set of eigenvalues of } T \\ &= \{ \lambda \in \mathbb{C} : (\exists \bar{v} \in c_0)(\bar{v} \neq \bar{0} \text{ and } T(\bar{v}) = \lambda \bar{v}) \}. \end{aligned}$$

Clearly for any T, E_T is a $\underset{\sim}{\Sigma}^1_1$ set of complex numbers. Since T is a *bounded* operator, the set of eigenvalues must be bounded.

Theorem 3.2. (Kaufman [20]). *Let S be any bounded $\underset{\sim}{\Sigma}_1^1$ set of complex numbers. There exists a bounded linear operator $T : c_0 \to c_0$ such that $S = E_T$.*

The set E_T is sometimes called the *point spectrum* of T. A related, and more important, concept is that of the *spectrum* of T, which is

$$\{ \lambda \in \mathbb{C} : T - \lambda I \text{ is not invertible } \}.$$

In contrast to 3.2, for any Banach space V and any bounded linear $T : V \to V$, the spectrum of T is always a nonempty compact set.

Several other representation theorems for $\underset{\sim}{\Sigma}_1^1$ sets have appeared in the literature, for example Bagemihl-McMillan [5], Lorentz-Zeller [30], and Nishiura [38]. There is only one basic type of representation theorem for $\underset{\sim}{\Sigma}_2^1$ sets which is known, although there are a number of variations on the same theme. I know of no representation theorem for $\underset{\sim}{\Sigma}_n^1$ sets, for $n \geq 3$.

We now take up the $\underset{\sim}{\Sigma}_2^1$ case. For any $\langle f_i \rangle \in (C[0,1])^\omega$, let $A_{\langle f_i \rangle}$ be the following subset of $C[0,1]$:

$$\{ g \in C[0,1] : \text{Some subsequence of } \langle f_i \rangle \text{ converges pointwise to } g \}.$$

For any $\langle f_i \rangle$, $A_{\langle f_i \rangle}$ is a $\underset{\sim}{\Sigma}_2^1$ set.

Theorem 3.3. (Becker[8]). *Let $S \subset C[0,1]$ be any $\underset{\sim}{\Sigma}_2^1$ set. There exists an $\langle f_i \rangle \in (C[0,1])^\omega$ such that $S = A_{\langle f_i \rangle}$.*

For example, the set of differentiable functions, E_1, can be represented as an $A_{\langle f_i \rangle}$, but the set of functions satisfying the Mean Value Theorem, E_4, cannot be.

Theorem 3.3 also holds uniformly. Putting quantifiers in front of the universal set gives the proof for Examples 11 and 12.

Representation theorems of this sort have a lot of corollaries. Theorem 3.3, for example, enables one to take any theorem about $\underset{\sim}{\Sigma}_2^1$ sets and translate it into a theorem about the $A_{\langle f_i \rangle}$'s. In this paper we give only one such corollary. Let $B \subset (C[0,1])^\omega \times C[0,1]$ be the set $\{ (\langle f_i \rangle, g) : g \notin A_{\langle f_i \rangle} \}$. A *uniformization* for B is a choice function which assigns to each $\langle f_i \rangle$ in the domain of B, a g such that $(\langle f_i \rangle, g) \in B$.

Theorem 3.4. *If ZFC is consistent, then so are each of the following two theories.*

(a) *$ZF + DC+$ There does not exist a uniformization for B.*
(b) *$ZFC+$ There is no uniformization for B which is ordinal-definable from a real parameter.*

The consistency of (a) follows from that of (b) by going into the model $L(\mathbb{R})$. Toward proving the consistency of (b), consider the following proposition.

($\star$) *There is a* $\underset{\sim}{\Pi}^1_2$ *relation which has no uniformization ordinal-definable from a real.*

It follows from ($\star$) and (the uniform version of) Theorem 3.3 that B has no uniformization ordinal-definable from a real. And ($\star$) is known to be consistent. It holds in the model obtained by adding $\aleph_1$ Cohen reals to L; the nonuniformizable $\underset{\sim}{\Pi}^1_2$ relation is $\{ (x,y) : y \notin L(x) \}$. These consistency results are variants of a theorem of Levy [29].

Two open questions were posed in [8]. The first question involved an analog of 3.3 for Baire-1 functions, the second question (due to Kechris) involved an analog of 3.3 for weak convergence. For any $\langle f_i \rangle \in (C[0,1])^\omega$, let $\hat{A}_{\langle f_i \rangle}$ be the following set of Baire-1 functions:

$$\{ g : \text{ Some subsequence of } \langle f_i \rangle \text{ converges pointwise to } g \}.$$

The Baire-1 functions do not form a Polish space in any natural way. But the Baire-1 functions can be encoded by elements of $(C[0,1])^\omega$, each sequence in $(C[0,1])^\omega$ encoding its pointwise limit, if it exists. The set of codes is $\underset{\sim}{\Pi}^1_1$ and the induced equivalence relation on codes is also $\underset{\sim}{\Pi}^1_1$. We say that a set of Baire-1 functions is $\underset{\sim}{\Sigma}^1_2$ if its set of codes is $\underset{\sim}{\Sigma}^1_2$. It is not hard to see that for any $\langle f_i \rangle \in (C[0,1])^\omega$, $\hat{A}_{\langle f_i \rangle}$ is a $\underset{\sim}{\Sigma}^1_2$ set of Baire-1 functions. The first question was: Is it true that for any $\underset{\sim}{\Sigma}^1_2$ set S of Baire-1 functions, there is an $\langle f_i \rangle \in (C[0,1])^\omega$ such that $S = \hat{A}_{\langle f_i \rangle}$? This is still open. It is not even known whether the set of discontinuous Baire-1 functions can be represented as an $\hat{A}_{\langle f_i \rangle}$. A positive answer to this open question would provide a positive answer to the open question about $\hat{E}_{12}$ following Example 12.

For any $\langle f_i \rangle \in (C[0,1])^\omega$, let $K_{\langle f_i \rangle}$ be the following subset of $C[0,1]$:

$$\{ g : \text{ Some subsequence of } \langle f_i \rangle \text{ converges weakly to } g \}.$$

Weak convergence means convergence in the weak topology of the Banach space $C[0,1]$; in more concrete terms, $\langle f_i \rangle$ *converges weakly* to g means $\langle f_i \rangle$ is uniformly bounded and $\langle f_i \rangle$ converges pointwise to g. Again, $K_{\langle f_i \rangle}$ is $\underset{\sim}{\Sigma}^1_2$. The second question was: Is it true that for any $\underset{\sim}{\Sigma}^1_2$ set $S \subset C[0,1]$, there is an $\langle f_i \rangle \in (C[0,1])^\omega$ such that $S = K_{\langle f_i \rangle}$? This question was recently answered by Kaufman, who proved the following strong version of 3.3.

Theorem 3.5. (Kaufman [21]). *Let $S \subset C[0,1]$ be any $\underset{\sim}{\Sigma}^1_2$ set. There exists an $\langle f_i \rangle \in (C[0,1])^\omega$ such that $S = A_{\langle f_i \rangle} = K_{\langle f_i \rangle}$.*

The remarks and corollaries mentioned above for 3.3, are also valid for 3.5.

Kaufman's original proof [21] of 3.5 involved some deep results from harmonic analysis, specifically a version of Ivashev-Musatov's Theorem (see [23,p. 294]); this is in contrast to the proof of 3.3 in [8], in which the analysis used is all fairly elementary. Later, Freiling and Louveau, independently, found a way to eliminate the harmonic analysis, so there now exists a proof of 3.5 suitable for a set theory workshop.

The Borel classes $\underset{\sim}{\Sigma}^0_n$ and $\underset{\sim}{\Pi}^0_n$ also have universal sets, and a number of representation theorems (hence natural universal sets) have appeared in the literature. Even if one has no interest in Borel sets, only in projective sets, the subject of representation theorems for Borel sets would still be worth looking at, since putting a universal or existential quantifier in front of a universal set for some (large enough) level of the Borel hierarchy, would give a complete $\underset{\sim}{\Pi}^1_1$ or complete $\underset{\sim}{\Sigma}^1_1$ set.

Consider a power series $\sum_{i=0}^{\infty} c_i z^i$ $(c_i \in \mathbb{C})$ which has radius of convergence 1. Let T be the unit circle, and let $B_{\langle c_i \rangle}$ be the subset of T consisting of those points at which the power series converges:

$$B_{\langle c_i \rangle} = \{\, z \in T \;:\; \textstyle\sum_{i=0}^{\infty} c_i z^i \text{ converges } \}.$$

T is a Polish space, and for any $\langle c_i \rangle \in \mathbb{C}^\omega$, the pointset $B_{\langle c_i \rangle}$ is $F_{\sigma\delta}$ ($\underset{\sim}{\Pi}^0_3$). Can every $\underset{\sim}{\Pi}^0_3$ subset of T be represented as a set of the form $B_{\langle c_i \rangle}$, for some $\langle c_i \rangle$? If so, we would have a nice example of a universal set for $\underset{\sim}{\Pi}^0_3$.

This question appeared in print in some papers in the 1940's and 50's, papers in which weak versions of a positive answer were proved. The problem was solved in 1978 by Lukašenko [31], who showed that there exists a G_δ subset of T which cannot be represented as a $B_{\langle c_i \rangle}$ —see Körner [27], for more information. One of the positive partial results proved is the following.

Theorem 3.6. (Herzog-Piranian [16]). *Let $S \subset T$ be any F_σ ($\underset{\sim}{\Sigma}^0_2$) set. There exists a power series $\sum_{i=0}^{\infty} c_i z^i$ with radius of convergence 1 such that $S = B_{\langle c_i \rangle}$.*

This is a representation theorem, but it does not give a universal set — it gives a $\underset{\sim}{\Pi}^0_3$ set in the plane such that every $\underset{\sim}{\Sigma}^0_2$ set occurs as a vertical section.

We now consider two more natural examples. We identify power series with the space $\mathbb{C}^\omega$.

Example 21.

Space: $\mathbb{C}^\omega$.
Pointset: $E_{21} = \{\langle c_i \rangle$: The power series $\sum_{i=0}^{\infty} c_i z^i$ converges everywhere on $T\}$.
Classification: True $\underset{\sim}{\Pi}_1^1$.

Example 22.

Space: $\mathbb{C}^\omega$.
Pointset: $E_{22} = \{\langle c_i \rangle$ The power series $\sum_{i=0}^{\infty} c_i z^i$ diverges everywhere on $T\}$.
Classification: True $\underset{\sim}{\Pi}_1^1$.

Note that the set

$$\{ (\langle c_i \rangle, r) \in \mathbb{C}^\omega \times [0,\infty] \ : \ r \text{ is the radius of convergence of } \textstyle\sum_{i=0}^{\infty} c_i z^i \}$$

is a Borel set in $\mathbb{C}^\omega \times [0,\infty]$. Hence intersecting the set of power series with radius of convergence 1, with either E_{21} or E_{22} , gives a set which is also true $\underset{\sim}{\Pi}_1^1$.

Even though Theorem 3.6 does not give us an honest universal set, (the uniform version of) it is still sufficient to prove that E_{21} is complete $\underset{\sim}{\Pi}_1^1$. If $P \subset \omega^\omega$ is an arbitrary $\underset{\sim}{\Pi}_1^1$ set, there is an F_σ set $A \subset \omega^\omega \times T$ such that $P(x) \Leftrightarrow \forall z A(x,z)$, so the completeness proof works. But if we try to prove that E_{22} is complete $\underset{\sim}{\Pi}_1^1$ by this method, we run into problems. It is not true that any $\underset{\sim}{\Pi}_1^1$ set in ω^ω is obtained by applying a universal quantifier to a G_δ set in $\omega^\omega \times T$; in fact, since T is compact, the subset of ω^ω obtained by applying a universal quantifier to a G_δ set in $\omega^\omega \times T$ will itself be a G_δ. So the fact that E_{22} is complete $\underset{\sim}{\Pi}_1^1$ does *not* follow from 3.6. It does follow, however, from (the uniform version of) Theorem 3.7, below, a theorem which is another positive partial result in the direction of the conjectured (but false) $F_{\sigma\delta}$ representation theorem.

A set $E \subset T$ has *logarithmic measure* 0 if for each $\varepsilon > 0$, there is a sequence $\langle I_n \rangle$ of open intervals of T such that $E \subset \cup I_n$, length $(I_n) < 1$ and

$$\sum_{n=0}^{\infty} \frac{-1}{\log(\text{ length } (I_n))} < \varepsilon.$$

Theorem 3.7. (Erdös-Herzog-Piranian [15]). *Let $S \subset T$ be any $G_\delta(\underset{\sim}{\Pi}_2^0)$ set whose closure has logarithmic measure 0. There exists a power series $\sum_{i=0}^{\infty} c_i z^i$ with radius of convergence 1 such that $S = B_{\langle c_i \rangle}$.*

4. Inseparable Pairs

Given three pointsets A, B and C in the same space, we say that C *separates* A and B if $A \subset C$ and $B \cap C = \emptyset$. It is well known that any pair of disjoint $\mathbf{\Sigma}_1^1$ sets can be separated by a Borel set. It is also well known that there is a pair of disjoint $\mathbf{\Pi}_1^1$ sets which cannot be separated by any Borel set [37,4B.12]; we call such a pair *Borel-inseparable.* In §4 we give some natural examples of Borel-inseparable pairs of $\mathbf{\Pi}_1^1$ sets, as well as of the analogous phenomenon at a higher level of the projective hierarchy. This entire section is based on Becker [7].

We describe below a general procedure for taking a natural example of a true $\mathbf{\Pi}_1^1$ set and turning it into a natural example of a Borel-inseparable pair of $\mathbf{\Pi}_1^1$ sets. This procedure will thus generate a large number of examples, all of which in some sense look alike. For a genuinely different natural example, see Dellacherie-Meyer [14].

Consider the space $C[0,1]$, and in this space consider the two pointsets:

$$A_0 = E_1 = \{\, f \,:\, f \text{ is differentiable } \}.$$
$$A_1 = \{\, f \,:\, \text{There is exactly one } x \in [0,1] \text{ such that } f'(x) \text{ does not exist} \,\}.$$

Note that for functions f in A_1, since the point where f is not differentiable is unique, that point is hyperarithmetic-in-f; hence by Theorem 2.1, A_1 is a $\mathbf{\Pi}_1^1$ set. Thus A_0 and A_1 are a pair of $\mathbf{\Pi}_1^1$ sets in the same space, and they are clearly disjoint.

Theorem 4.1. *A_0 and A_1 are a Borel-inseparable pair of $\mathbf{\Pi}_1^1$ sets.*

Another way of looking at Theorem 4.1 is as an *overspill* theorem: Any Borel property true of all differentiable functions, also holds for some function with exactly one point of nondifferentiability. This implies, of course, that A_1 itself is not a Borel set; hence A_1 is another natural example of a true $\mathbf{\Pi}_1^1$ set.

There is nothing special about the numbers 0 and 1 — any other numbers would work just as well. Let

$$A_n = \{f : \text{There are exactly } n \text{ points in } [0,1] \text{ at which } f' \text{ does not exist}\}.$$

For any m and n, if $0 \leq m < n \leq \aleph_0$, then A_m and A_n are a Borel-inseparable pair of $\mathbf{\Pi}_1^1$ sets.

The procedure used for going from Example 1 to Theorem 4.1 is very general. It takes a $\mathbf{\Pi}_1^1$ set of the form "points with no singularities," and creates a Borel-inseparable pair of $\mathbf{\Pi}_1^1$ sets, the above set and the set of "points with exactly one singularity." Using this procedure, one can mindlessly and mechanically convert natural examples of true $\underset{\sim}{\Pi}^1_1$ sets into natural examples of Borel-inseparable pairs (and also convert a proof of the former into a proof of the latter — see [7]).

We now mindlessly and mechanically apply this procedure to Example 6. Consider the space $(C[0,1])^\omega$, and in this space consider the two pointsets:

$$B_0 = E_6 = \{\, \langle f_i \rangle \;:\; \langle f_i \rangle \text{ converges pointwise } \}.$$
$$B_1 = \{\, \langle f_i \rangle \;:\; \text{There is exactly one } x \in [0,1] \text{ such that } \langle f_i(x) \rangle \text{ diverges} \,\}.$$

Theorem 4.2. *B_0 and B_1 are a Borel-inseparable pair of $\underset{\sim}{\Pi}^1_1$ sets.*

Similarly, Examples 2, 13, 14, 21 and 22 can be converted into natural examples of a Borel-inseparable pair of $\underset{\sim}{\Pi}^1_1$ sets.

For Example 17, another true $\underset{\sim}{\Pi}^1_1$ set, the situation is more interesting. Consider the space $\mathcal{K}(\mathbb{R}^2)$. Let

$$\begin{aligned} C_0 = E_{17} &= \{\, K \;:\; K \text{ is simply connected } \} \\ &= \{\, K \;:\; K \text{ is path-connected and } K \text{ has no holes} \,\}. \end{aligned}$$

Then let

$$C_1 = \{\, K \;:\; K \text{ is path-connected and } K \text{ has exactly one hole} \,\}.$$

Theorem 4.3. *C_0 and C_1 are Borel-inseparable.*

There are many ways to give a precise definition of C_1 ; it is not clear to me that any of these definitions is a $\underset{\sim}{\Pi}^1_1$ definition. But regardless of whether or not C_1 is $\underset{\sim}{\Pi}^1_1$, C_0 and C_1 are still Borel-inseparable.

This procedure obviously does not work on *every* natural example of a true $\underset{\sim}{\Pi}^1_1$ set. It clearly cannot work for E_{15} or for $(C[0,1])^\omega \setminus E_8$, since it is not possible for a compact set to have exactly one perfect subset, or for a sequence to have exactly one convergent subsequence.

ZFC is sufficient to answer almost every question about the first level of the projective hierarchy, and some questions about the second level, but virtually no questions about the third or higher levels. Beginning in 1968 with Addison-Moschovakis [1] and Martin [32], determinacy axioms have been brought into the subject to answer these questions. Assuming determinacy, a fairly complete theory of projective sets has emerged; one could almost (but not quite) say that we understand $\underset{\sim}{\Pi}^1_n$, for arbitrary n, as well as we understand $\underset{\sim}{\Pi}^1_1$. For an account of this theory, and of determinacy axioms, see Moschovakis [37]. Assuming determinacy, the theory of projective sets exhibits a periodicity of order 2; that is, the pointclasses $\underset{\sim}{\Pi}^1_1$, $\underset{\sim}{\Pi}^1_3$, $\underset{\sim}{\Pi}^1_5$, etc., have similar structural properties, as do the pointclasses $\underset{\sim}{\Pi}^1_2$, $\underset{\sim}{\Pi}^1_4$, $\underset{\sim}{\Pi}^1_6$, etc.

Since the pointclasses $\underset{\sim}{\Pi}^1_1$ and $\underset{\sim}{\Pi}^1_3$ have the same properties, one would expect this procedure for converting true $\underset{\sim}{\Pi}^1_1$ sets into a Borel-inseparable pair of $\underset{\sim}{\Pi}^1_1$ sets, would also work two levels up. This expectation turns out to be correct. We apply the procedure to a natural example of a true $\underset{\sim}{\Pi}^1_3$ set: Example 12. Consider the space $(C[0,1])^\omega$, and in this space consider the two pointsets:

$$D_0 = E_{12} = \{ \langle f_i \rangle \ : (\forall g \in C[0,1]) \text{ (Some subsequence of } \langle f_i \rangle \text{ converges pointwise to } g) \}.$$

$$D_1 = \{ \langle f_i \rangle \ : \text{ There is exactly one } g \in C[0,1] \text{ such that no subsequence of } \langle f_i \rangle \text{ converges pointwise to } g \}.$$

(Incidentally, D_1 is nonempty. This follows from Theorem 3.3.)

Theorem 4.4. *Assume $\underset{\sim}{\Delta}^1_2$-determinacy.*
(a) D_0 and D_1 are $\underset{\sim}{\Pi}^1_3$ sets.
(b) D_0 and D_1 cannot be separated by any $\underset{\sim}{\Delta}^1_3$ set.

Martin-Steel [33] showed that $\underset{\sim}{\Delta}^1_2$-determinacy follows from the existence of a Woodin cardinal with a measurable cardinal above it (actually from a little less).

Theorem 4.4 is definitely not provable in ZFC. It is false in L, as shown by the following theorem.

Theorem 4.5. *Assume that there exists a $\underset{\sim}{\Delta}^1_3$-good wellordering of the reals. Then every pair of disjoint $\underset{\sim}{\Pi}^1_3$ sets can be separated by a $\underset{\sim}{\Delta}^1_3$ set.*

See [37,Ch. 5] for a proof, as well as for the definition of a *good* wellordering. The Axiom of Constructibility, $V = L$, implies that there is a $\underset{\sim}{\Delta}^1_2$-good, hence $\underset{\sim}{\Delta}^1_3$-good, wellordering of the reals [37,8F.7]; therefore it implies that 4.4 is false. In fact, if a Woodin cardinal exists, then there is an inner model with a Woodin cardinal, in which there is a $\underset{\sim}{\Delta}^1_3$-good wellordering of the reals (Martin-Steel [34]). This essentially means that Theorem 4.4 cannot be proved from any large cardinal or determinacy axiom weaker than $\underset{\sim}{\Delta}_2^1$-determinacy.

I know nothing about the strength of 4.4, either in terms of what it implies, or in terms of relative consistency . It may be equiconsistent with ZFC. Or it may outright imply $\underset{\sim}{\Delta}_2^1$-determinacy. This is related to the fourth Victoria Delfino problem [25, p. 281].

Determinacy is used twice in the proof of 4.4 — once to prove (a), and once to prove (b). Recall that to prove that A_1, B_1 and other "exactly one singularity" sets are $\underset{\sim}{\Pi}_1^1$, we used Theorem 2.1. In the very first paper in which determinacy was applied to descriptive set theory [1], a $\underset{\sim}{\Pi}_3^1$-analog of Theorem 2.1 was proved, assuming $\underset{\sim}{\Delta}_2^1$-determinacy (see [37, 6B.2 and 4D.3]); this can then be applied to prove 4.4 (a). The proof of Borel-inseparability in Theorems 4.1, 4.2 and 4.3 (and in all the other examples), uses the theorem of Lusin that every $\underset{\sim}{\Delta}_1^1$ set is the one-to-one continuous image of a closed set. The $\underset{\sim}{\Delta}_3^1$-analog of this theorem was proved somewhat later by Moschovakis (see [37, 6E.14]), again assuming $\underset{\sim}{\Delta}_2^1$-determinacy; using this result, we prove 4.4 (b).

We thus have two more open problems. While parts (a) and (b) of Theorem 4.4 cannot both be provable in ZFC, it is possible that (a) is provable in ZFC, and it is also possible that (b) is provable in ZFC.

We now return to ZFC, and close §4 by pointing out that some natural examples of disjoint $\underset{\sim}{\Pi}_1^1$ sets can be separated by a (not necessarily natural) Borel set. E_1 and E_2, the sets of differentiable and nowhere differentiable functions, respectively, are separated by $\{ f : f'(p) \text{ exists } \}$, where p is a fixed point in $[0,1]$. Similarly, the pairs (E_{13}, E_{14}) and (E_{21}, E_{22}) can be separated. $(C[0,1] \setminus E_3)$, the set of functions which fail to satisfy Rolle's Theorem, is a $\underset{\sim}{\Pi}_1^1$ set, clearly disjoint from E_1. We describe a Borel set $B \subset C[0,1]$ which contains E_1 and is disjoint from $(C[0,1] \setminus E_3)$. For any $f \in C[0,1]$, and any a, b such that $0 \le a < b \le 1$, let $M(f,a,b)$ denote

$$\{ x \in [a,b] : x \text{ is a maximum of } f \restriction [a,b] \}.$$

Note that $M(f,a,b)$ is a nonempty closed set. Similarly, let $m(f,a,b)$ be the set of minima. Define $B \subset C[0,1]$ as follows. $f \in B$ iff:

For all rational a and b, if $0 \le a < b \le 1$, then

$\{x \in M(f,a,b) : f'(x) = 0 \text{ or } x = a \text{ or } x = b\}$ is comeager in $M(f,a,b)$

and

$$\{x \in m(f,a,b) : f'(x) = 0 \text{ or } x = a \text{ or } x = b\} \text{ is comeager in } m(f,a,b).$$

Then B separates. That B is Borel follows from the fact that the pointclass of Borel sets is closed under quantification of the form: "For a comeager set of x's" [37, 4F.19].

5. Remarks on Other Descriptive Set Theoretic Phenomena

There are two parts of the subject of natural examples of descriptive set theoretic phenomena that we have so far ignored. First, we do wish to point out for the record, that there are other types of phenomena, besides the three types considered in the previous three sections of this paper, for which natural examples exist. One of these other phenomena is that of norms. Every $\underset{\sim}{\Pi}^1_1$ set P admits a $\underset{\sim}{\Pi}^1_1$-norm, and if P is true $\underset{\sim}{\Pi}^1_1$, the norm will have length ω_1. Such norms do occur in nature. The countable compact sets (Example 15) form a pointset which is true $\underset{\sim}{\Pi}^1_1$, and the Cantor-Bendixson rank is a natural $\underset{\sim}{\Pi}^1_1$-norm on this pointset. For other natural norms, see Ajtai-Kechris [2], Bourgain [12], Kechris-Louveau [23], [24], Kechris-Woodin [26], and Ramsamujh [40].

Second, there are several types of descriptive set theoretic phenomena which, as far as I know, are not exhibited by any *natural* example. Whenever this situation occurs it presents a challenge: either find a natural example, or explain why there are none. There exists a pair of disjoint $\underset{\sim}{\Sigma}^1_2$ sets which cannot be separated by a $\underset{\sim}{\Delta}^1_2$ set; I know of no examples of this that are at all natural (or even of candidates for such an example, for which the proof is missing). I know of no natural examples of pointsets which have been proved to be $\underset{\sim}{\Delta}^1_2$ but not in the σ-algebra generated by $\underset{\sim}{\Pi}^1_1$ (although I do have some candidates).

There are many natural examples of Borel relations R in a product space $X \times Y$ which have no Borel uniformization. In fact, whenever the projection of R, $\{x : \exists y\, R(x,y)\}$, is true $\underset{\sim}{\Sigma}^1_1$, R is such a relation; for if R did have a Borel uniformization, then by Theorem 2.1, the projection of R would be Borel. For example, let

$$X = \mathcal{K}(\mathbb{R}^2) \times \mathbb{R}^2 \times \mathbb{R}^2,$$

let Y be the space of paths in $\mathbb{R}^2$, that is,

$$Y = C[0,1] \times C[0,1],$$

and let $R \subset X \times Y$ be the following relation:

$$R = \{((K,p,q),\gamma) : p,q \in K \text{ and } \gamma \text{ is a path from } p \text{ to } q \text{ lying in } K\}.$$

Then R is a closed set in $X \times Y$, and by Theorem 2.2 its projection is not Borel, so R has no Borel uniformization (cf. Theorem 2.5).

However it is a theorem that there are Borel relations R in $X \times Y$ such that R has no Borel uniformization *and* $\forall x \exists y R(x, y)$; there are no known natural examples of such an R in analysis or topology. There are examples in logic, such as:

$$R(x, y) \Leftrightarrow y \text{ encodes a countable } \omega\text{-model of } ZFC^- \text{ containing } x.$$

It would be interesting to find an example of this in analysis or topology.

This last problem is related to "reverse mathematics" (see Simpson [43]). It is more or less equivalent to the problem of finding a natural proposition in analysis or topology, $\phi = \forall x \exists y R(x, y)$, such that ϕ is true but ϕ is not provable in KP. (KP is the Kripke-Platek axioms for set theory — see [6]. The true statement ϕ would presumably be provable in ZFC.)
This situation is difficult to explain — why are there no *natural* examples? One intriguing possibility is that such natural examples exist but they have not yet been found, because the world just does not know how to use axioms stronger than KP to prove natural theorems in analysis and topology. If this is the case, logicians could conceivably make a positive contribution to analysis or topology by figuring out how to use stronger axioms.

I also know of no natural examples of pointsets which occur in the fourth level (or higher levels) of the projective hierarchy. In this case, a plausible explanation for the lack of examples has been proposed. According to Rogers [41, p. 322], "... the human mind seems limited in its ability to understand and visualize beyond four or five alternations of quantifier."

REFERENCES

1. J. W. Addison and Y. N. Moschovakis, *Some consequences of the axiom of definable determinateness*, Proc. Nat. Acad. Sci., U.S.A. **59** (1968), 708–712.
2. M. Ajtai and A. S. Kechris, *The set of continuous functions with everywhere convergent Fourier series*, Trans. Amer. Math. Soc. **302** (1987), 207–221.
3. I. Assani, *Une caractérisation des Banach réticulés faiblement séquentiellement complets*, C. R. Acad. Sci. Paris Sér. I Math. **298** (1984), 445–448.
4. I. Assani, *Quelques propriétés mesurables de diverses suites d'un espace de Banach séparable E dans E^N*, Math. Scand. **58** (1986), 301–310.
5. F. Bagemihl and J. E. McMillan, *Characterization of the sets of angular and global convergence, and of the sets of angular and global limits, of functions in a half-plane*, Fund. Math. **59** (1966), 177-187.
6. J. Barwise, *Admissible Sets and Structures: An Approach to Definability Theory*, Springer-Verlag, Berlin, 1975.
7. H. Becker, *Some examples of Borel-inseparable pairs of coanalytic sets*, Mathematika **33** (1986), 72–79.
8. H. Becker, *Pointwise limits of subsequences and $\underset{\sim}{\Sigma}^1_2$ sets*, Fund. Math. **128** (1987), 159–170.
9. H. Becker, *The descriptive set theory of sequences in separable Banach spaces*, in preparation.
10. H. Becker, *Path-connectedness, simple connectedness and the projective hierarchy*, in preparation.
11. H. Becker, *Simply connected sets and hyperarithmetic paths*, in preparation.
12. J. Bourgain, *On separable Banach spaces, universal for all separable reflexive spaces*, Proc. Amer. Math. Soc. **79** (1980), 241–246.
13. G. Debs and J. Saint Raymond, *Ensembles boréliens d'unicité et d'unicité au sens large*, Ann. Inst. Fourier (Grenoble) **37** (1987), 217–239.
14. C. Dellacherie and P. A. Meyer, *Ensembles analytiques et temps d'arrêt*, Séminaire de Probabilités IX, Lecture Notes in Mathematics, vol. 465, ed. P. A. Meyer, Springer-Verlag, Berlin, 1975, pp. 373–389.
15. P.Erdös, F. Herzog and G. Piranian, *Sets of divergence of Taylor series and of trigonometric series*, Math. Scand. **2** (1954), 262–266.
16. F. Herzog and G. Piranian, *Sets of convergence of Taylor series,I*, Duke Math. J. **16** (1949), 529–534.
17. P.D. Humke and M.Laczkovich, *The Borel structure of iterates of continuous functions*, Proc. Edinburgh Math. Soc. **32** (1989), 483–494.
18. W. Hurewicz, *Zur Theorie der analytischen Mengen*, Fund. Math. **15** (1930), 4–17.
19. R. Kaufman, *Fourier transforms and descriptive set theory*, Mathematika **31** (1984), 336–339.
20. R. Kaufman, *On some operators in c_0*, Israel J. Math. **50** (1985), 353–356.
21. R. Kaufman, *Topics on analytic sets*, preprint.
22. A. S. Kechris, *Sets of everywhere singular functions*, Recursion Theory Week, Lecture Notes in Mathematics, vol. 1141, ed. H.D. Ebbinghaus, G.H. Müller and G.E. Sacks, Springer-Verlag, Berlin, 1985, pp. 233–244.
23. A. S. Kechris and A. Louveau, *Descriptive Set Theory and the Structure of Sets of Uniqueness*, London Math. Soc. Lecture Notes, vol. 128, Cambridge University Press, Cambridge, 1987.
24. A. S. Kechris and A. Louveau, *A classification of Baire class 1 functions*, Trans. Amer. Math. Soc. (to appear).
25. A. S. Kechris and Y. N. Moschovakis, Eds., *Cabal Seminar 76-77*, Lecture Notes In Mathematics **689** (1978), Springer-Verlag, Berlin.

26. A. S. Kechris and W. H. Woodin, *Ranks of differentiable functions*, Mathematika **33** (1986), 252–278.
27. T. W. Körner, *The behavior of power series on their circle of convergence*, Banach Spaces, Harmonic Analysis and Probability Theory, Lecture Notes in Mathematics, vol. 995, ed. R.C. Blei and S.J. Sidney, Springer-Verlag, Berlin, 1983, pp. 56–94.
28. K. Kuratowski, *Evaluation de la classe borélienne ou projective d'un ensemble de points a l'aide des symboles logiques*, Fund. Math. **17** (1931), 249–272.
29. A. Levy, *Definability in axiomatic set theory,I*, Logic, Methodology and Philosophy of Science, ed. Y. Bar-Hillel, North-Holland, Amsterdam, 1965, pp. 127–151.
30. G. G. Lorentz and K. Zeller, *Series rearrangements and analytic sets*, Acta Mathematica **100** (1958), 149–169.
31. S. Ju. Lukašenko, *Sets of divergence and nonsummability for trigonometric series* (in Russian), Vestnik Moskov. Univ. Ser I Mat. Meh. no. 2 (1978), 65–70; (English translation: Moscow Univ. Math. Bull. **33** ((1978) no. 2), 53–57).
32. D. A. Martin, *The axiom of determinateness and reduction principles in the analytical hierarchy*, Bull. Amer. Math. Soc **74** (1968), 687–689.
33. D. A. Martin and J. R. Steel, *A proof of projective determinacy*, Jour. Amer. Math. Soc. **2** (1989), 71–125.
34. D. A. Martin and J. R. Steel, *Iteration trees*, Xeroxed notes (1989).
35. D. Mauldin, *The set of continuous nowhere differentiable functions*, Pacific J. Math. **83** (1979), 199–205.
36. S.Mazurkiewicz, *Über die Menge der differenzierbaren Funktionen*, Fund. Math. **27** (1936), 244–249.
37. Y.N. Moschovakis, *Descriptive Set Theory*, North-Holland, Amsterdam, 1980.
38. T. Nishiura, *Uncountable-order sets for radial-limit functions*, Real Analysis Exchange **10** (1984-85), 50–53.
39. G. Poprougénko, *Sur l'analyticité des ensembles (A)*, Fund. Math **18** (1932), 77–84.
40. T. I. Ramsamujh, *A comparison of the Dini and Jordan tests*, Real Analysis Exchange **12** (1986-87), 510–515.
41. H. Rogers, *Theory of Recursive Functions and Effective Computability*, McGraw Hill, New York, 1967.
42. H. P. Rosenthal, *On applications of the boundedness principle to Banach space theory, according to J. Bourgain*, Séminaire d'Initiation á l'Analyse, ed. G. Choquet, M. Rogalski, J. Saint-Raymond. 18e Année: 1978/1979 vol. 29, Publications Mathématique de l'Université Pierre et Marie Curie, Université de Paris VI, 1979.
43. S.G. Simpson, *Reverse mathematics*, Recursion Theory, Procedings of Symposia In Pure Mathematics, vol. 42, ed. A. Nerode and R. A. Shore, American Mathematical Society, Providence, 1985, pp. 461–471.
44. R. Van Wesep, *Wadge degrees and descriptive set theory*, Cabal Seminar 76-77, Lecture Notes in Mathematics, vol. 689, ed. A. S. Kechris and Y.N. Moschovakis, Springer-Verlag, Berlin, 1978, pp. 151–170.

DEPARTMENT OF MATHEMATICS,
UNIVERSITY OF SOUTH CAROLINA, COLUMBIA SC 29208

RESEARCH SUPPORTED IN PART BY NSF GRANT DMS-8601731

AN ALTERNATIVE PROOF OF LAVER'S RESULTS ON THE ALGEBRA GENERATED BY AN ELEMENTARY EMBEDDING

PATRICK DEHORNOY

ABSTRACT. Richard Laver recently gave an achieved proof for some natural conjectures about the algebraic structure generated by the iteration of an elementary embedding of a rank into itself. The aim of this paper is to give an alternative proof for these results.

The freeness of the algebra generated by an elementary embedding into itself had been conjectured by many set theorists, and has been proved by Richard Laver recently. For λ a limit ordinal let $\mathcal{E}_\lambda$ be the collection of all $j : V_\lambda \to V_\lambda$ such that j is an elementary embedding of $(V_\lambda, \in)$ into itself distinct from identity. For k,ℓ in $\mathcal{E}_\lambda$, we write $j[\ell]$, or simply $j\ell$, for $\bigcup_{\alpha<\lambda} j(\ell|_{V_\alpha})$, and denote by $\mathfrak{a}_j$ the closure of $\{j\}$ under this operation. Then $\mathfrak{a}_j$ (together with the operation above) is a (monogenic) left distributive structure, *i.e.* it satisfies the identity

$$x(yz) = (xy)(xz).$$

Now introduce $\mathfrak{f}_1$ to be the monogenic *free* left distributive structure: $\mathfrak{f}_1$ is easily represented as the quotient of the set $\mathcal{W}$ made by all wellformed terms using a single variable say 'a' and a fixed binary operator under the least congruence $\equiv$ that forces the distributivity condition to hold. Then Laver proves in [La] the following results about the structure $\mathfrak{a}_j$

Theorem. *Assume that $\mathcal{E}_\lambda$ is nonempty for some λ;*

i) *(for every j in $\mathcal{E}_\lambda$,) $\mathfrak{a}_j$ is isomorphic to $\mathfrak{f}_1$ (i.e. is free);*
ii) *there exists a linear ordering $<$ on $\mathfrak{f}_1$ such that the left translations are strictly increasing mappings of $(\mathfrak{f}_1, <)$ into itself;*
iii) *the word problem for $\mathcal{W}/\equiv$ is decidable.*

We give here an alternative proof of this theorem. For the history, it happens that all the results used in this proof were already available some time ago (the material of Proposition 1 below is presented in [De 2]), but there cannot be any doubt about the priority for the theorem above because the author was hopelessly unable to complete a proof by himself, and he only understood his stupidity when seeing at the beginning of Laver's paper that the missing piece (Proposition 2 below) was known for several years (yet unpublished) and resulted from elegant, but short and basic, computations on critical points and not from some difficult analysis.

Laver's proof and the present one are in some sense complementary: so to speak, R. Laver starts from an intensive study of the elementary embeddings and establishes enough properties of $\mathfrak{a}_j$ to prove that $\mathfrak{a}_j$ must resemble $\mathfrak{f}_1$, while our proof starts from a purely algebraic analysis of $\mathfrak{f}_1$ and establishes that $\mathfrak{f}_1$ must resemble $\mathfrak{a}_j$. Technically, this means that these proofs use the specific properties of the elementary embeddings captured in Proposition 2 in different places: at the beginning of the construction in Laver's proof, at the end in order to conclude in the present one. This discrepancy leads to different developments of the basic result: thanks to the critical point, Laver's method does not only prove the decidability of the word problem, but it also provides a unique normal form result that is not included in the present proof and has a great intrinsic interest; on the other hand, a deepening of the methods below suggests an effective approach for solving the word problem independently from any set theoretical hypothesis. One can hope for future common developments.

We turn to the proof of the theorem. It happens that the notations in [La] and [De] are mostly compatible or, at least, isomorphic. We shall use the following ones in the sequel. First, to be short, any left distributive set endowed with a left distributive binary operation will be called a *LD-magma.* If $\mathfrak{g}$ is an LD-magma, and x, y are elements of $\mathfrak{g}$, we write $x <^{\mathfrak{g}} y$ if for some positive p there exist $z_1, \ldots, z_p$ in $\mathfrak{g}$ such that y is $(\ldots((xz_1)z_2)\ldots)z_p$. The result we prove is the following one:

Proposition 1. *Assume that $\mathfrak{g}$ is a monogenic LD-magma such that $<^{\mathfrak{g}}$ is irreflexive;*

i) *$\mathfrak{g}$ is isomorphic to $\mathfrak{f}_1$ (i.e. is free)*
ii) *there exists a linear ordering $<$ on $\mathfrak{f}_1$ such that the left translations are strictly increasing mappings of $(\mathfrak{f}_1, <)$ into itself*
iii) *the word problem for $\mathcal{W}/\!\equiv$ is decidable.*

Laver's theorem follows from this proposition and the following one established in the beginning of [La]:

Proposition 2. *(Laver) The relation* $<^{\mathfrak{a}_j}$ *is irreflexive.*

The proof of Proposition 1 uses two ingredients, one trick and one more structural result. The trick is

Lemma 3. *Let x, y be arbitrary members of a monogenic LD-magma $\mathfrak{g}$; then there exists z in $\mathfrak{g}$ such that xz and yz are equal.*

The structural result is

Lemma 4. *Assume that xz and yz are equal members of $\mathfrak{f}_1$; then at least one of $x<^{\mathfrak{f}_1}y$, $x = y$, $y<^{\mathfrak{f}_1}x$ must hold.*

Proof of Proposition 1. (from Lemmas 3 and 4). Assume that $\mathfrak{g}$ is a monogenic LD-magma and let π be the canonical projection of $\mathfrak{f}_1$ onto $\mathfrak{g}$ mapping a to a generator of $\mathfrak{g}$. We wish to show that π is injective. Let x, y be distinct members of $\mathfrak{f}_1$; by Lemmas 3 and 4, at least one of $x<^{\mathfrak{f}_1}y$, $y<^{\mathfrak{f}_1}x$ must hold; assume the first. Clearly π respects $<$, so $\pi x<^{\mathfrak{g}}\pi y$ holds as well. If $<^{\mathfrak{g}}$ is supposed to be irreflexive, πx and πy cannot be equal, π is injective and $\mathfrak{g}$ is free.

Now notice that $<^{\mathfrak{f}_1}$ is from now on irreflexive since it is isomorphic to $<^{\mathfrak{g}}$; it immediately follows that $<^{\mathfrak{f}_1}$ is a linear ordering on $\mathfrak{f}_1$. Then left distributivity shows that $x<^{\mathfrak{f}_1}y$ implies $zx<^{\mathfrak{f}_1}zy$ for every z, so point ii) is straightforward.

We turn to the word problem for $\mathcal{W}/\equiv$. We shall go into a more precise analysis of $\equiv$ later, but, for the moment it is enough to notice that there always exists an algorithm that, when a word S is given, enumerates all words that are equivalent to S. The problem of course is in recognizing that inequivalent words are inequivalent. But start with arbitrary words S, T and enumerate (using a convenient ordering) all pairs (S', T') such that $S' \equiv S$ and $T' \equiv T$ hold. If we denote by $[U]$ the class of the word U in $\mathfrak{f}_1$, it follows once again from Lemmas 3 and 4 that at least one of $[S]<^{\mathfrak{f}_1}[T]$, $[S] = [T]$, $[T]<^{\mathfrak{f}_1}[S]$ must hold, and this means that, in the preceding enumeration, a pair (S', T') will eventually appear such that, respectively, S' is, with obvious meaning, a left factor of T', or S' is equal to T', or T' is a left factor of S'. In the second case, we conclude that S and T are equivalent; otherwise we conclude that they are not because $<^{\mathfrak{f}_1}$ is irreflexive. So the algorithm solves the word problem correctly. □

So the nontrivial content of the proof is in the lemmas.

Proof of Lemma 3. Assume that a is a generator of $\mathfrak{g}$; we define inductively an element $a^{(n)}$ in $\mathfrak{g}$ for every integer n by $a^{(0)} = a$ and $a^{(n+1)} = aa^{(n)}$. We claim first that for every x in $\mathfrak{g}$ the equality $xa^{(n)} = a^{(n+1)}$ holds for n large enough. This is proved inductively on the complexity of x when expressed as a term constructed from a. If x is a, the equality holds for every n by definition. Now if x is yz and the property holds for y and z, we obtain for n large enough

$$xa^{(n)} = (xy)a^{(n)} = (xy)(xa^{(n-1)}) = x(ya^{(n-1)}) = xa^{(n)} = a^{(n+1)}$$

and the property holds for x as well. It follows that if x, y are arbitrary elements of $\mathfrak{g}$ then $xa^{(n)}$ and $ya^{(n)}$ are equal for n large enough. □

Proof of Lemma 4. This is more crucial. The property follows from the general analysis of the relation $\equiv$ which is initiated in [De 1]. We shall extract the basic arguments that are used in the present case. First of all, in order to understand what happens in $\mathcal{W}$, we need a convenient representation of the terms. This is done when seeing these terms as binary trees in the usual way. We address the nodes of a binary tree using finite sequences of 0 (for 'left') and 1 (for 'right'). The set of all such sequences will be denoted by $\mathbb{S}$, and the empty sequence (address for the root of the tree) by Λ . For S in $\mathcal{W}$ and u in $\mathbb{S}$, we denote by $S_{/u}$ the subterm of S corresponding to the subtree with root in u (if defined). For instance if S is the term $a((aa)(aa))$, $S_{/0}$ is a, $S_{/1}$ is $(aa)(aa)$, $S_{/110}$ is a, while $S_{/L}$ is S itself and $S_{/1100}$ is not defined.

In order to describe the equivalence $\equiv$, we *polarize* it as follows. First we denote by Δ the partial mapping of $\mathcal{W}$ into itself that maps every term that can be written as $S(TU)$ on the corresponding term $ST(SU)$; then we introduce for every u in $\mathbb{S}$ the mapping $\Delta^{(u)}$ that is similar to Δ but acts below u: the action of $\Delta^{(u)}$ on S consists in distributing the subterm $S_{/u0}$ to each of the subterms $S_{/u10}$ and $S_{/u11}$ (when these subterms are defined). If S is the term in the example above, then S lies both in the domains of Δ and of $\Delta^{(1)}$, and the respective images are the terms

$(a(aa))(a(aa))$ and $a(((aa)a)((aa)a))$. Now, for S, T in $\mathcal{W}$, let us write $S \longrightarrow T$ if there is a finite composition of $\Delta^{(u)}$'s that maps S to T. It should be clear that $\equiv$ is the equivalence relation generated by the relation $\longrightarrow$. Two claims are needed to prove Lemma 4.

Claim 1. *Assume $S \longrightarrow T$; then if $S_{/0^p}$ is defined, there exists an integer $q \geq p$ such that $S_{/0^p} \longrightarrow T_{/0^q}$ holds.*

Proof. Easy. Using induction, it suffices to prove the result for T being the image of S under $\Delta^{(u)}$. Three cases can occur. If u is 0^i with $i < p$, then $T_{/0^{i+k+1}}$ is $S_{/0^{i+k}}$ for every $k \geq 1$, so $S_{/0^p} = T_{/0^{p+1}}$, and therefore $S_{/0^p} \longrightarrow T_{/0^{p+1}}$, hold. If u is $0^p u'$ for some u', then $T_{/0^p}$ is the image of

$S_{/0^p}$ under $\Delta^{(u')}$, and therefore $S_{/0^p} \longrightarrow T_{/0^p}$ holds. If u is $0^i1u'$ for some u', then $T_{/0^p}$ is $S_{/0^p}$, and therefore $S_{/0^p} \longrightarrow T_{/0^p}$ holds. □

Claim 2. *Assume $S \equiv T$; then there exists some U in $\mathcal{W}$ such that $S \longrightarrow U$ and $T \longrightarrow U$ both hold.*

Proof. This is the hard core. However, if only the result of claim 2 is needed, the details are rather easily and quickly checked. We shall only sketch the arguments since the details appear in [De 1]. The point is getting a convenient notion of *derivation* for the words. For every S in $\mathcal{W}$ there exists another word called ∂S that is a kind of 'lower common extension' for all the possible images of S under some $\Delta^{(u)}$ (there is only a finite number of u's such that a given term belongs to the domain of $\Delta^{(u)}$): if T is the image of S under $\Delta^{(u)}$, then $T \longrightarrow \partial S$ holds (and ∂S is nearly minimal with that property). The main property is that the mapping ∂ is compatible with $\longrightarrow$: $S \longrightarrow T$ implies $\partial S \longrightarrow \partial T$, and it follows that, if T is the image of S under the composition of k successive mappings $\Delta^{(u_1)}, \ldots, \Delta^{(u_k)}$, then $T \longrightarrow \partial^k S$ holds. Claim 2 easily follows, by showing that $T \equiv S$ implies that, for k large enough, $T \longrightarrow \partial^k S$ and, trivially, $S \longrightarrow \partial^k S$ hold. □

Remark. In the proof above, the mapping ∂ is effective (∂ has a simple inductive definition), while the integer 'k' arising at the end is not, and therefore claim 2 is not sufficient to solve the word problem for $\mathcal{W}/\equiv$.

We can now easily prove Lemma 4. Assume that $xz = yz$ holds in $\mathfrak{f}_1$; choose words S, T, U in $\mathcal{W}$ that represent respectively x, y, z. Then $SU \equiv TU$ holds, henceforth by claim 2 there exists V such that $SU \longrightarrow V$ and $TU \longrightarrow V$ hold. Now S is $SU_{/0}$, so, by claim 1, there exists an integer $q \geq 1$ such that $S \longrightarrow V_{/0^q}$ holds, and, symmetrically, there exists an integer $r \geq 1$ such that $T \longrightarrow V_{/0^r}$ holds. It follows that x is the equivalence class of $V_{/0^q}$, while y is the class of $V_{/0^r}$. Now $q = r$ implies $x = y$, $q > r$ implies $x <^{\mathfrak{f}_1} y$ and $q < r$ implies $y <^{\mathfrak{f}_1} x$, and we are done. □

The proof of Laver's theorem is complete. We shall briefly discuss some further points. Let (IH) be the hypothesis 'the relation $<^{\mathfrak{f}_1}$ is irreflexive'. We first notice that, if, for x in $\mathfrak{f}_1$, γ_x is the left translation map associated to x, then γ_x is an homomorphism of $(\mathfrak{f}_1, \bullet, <^{\mathfrak{f}_1})$ into the sub-LD-magma of $\mathfrak{f}_1$ generated by xa, and, if (IH) holds, this homomorphism is an isomorphism; moreover the mapping $x \mapsto \gamma_x$ is injective. It follows that, if (IH) is true, then $\mathfrak{f}_1$ admits left cancellation and that the mapping $x \mapsto \gamma_x$ gives a realization of $\mathfrak{f}_1$ as a LD-magma with the following properties:

its elements are increasing injections of some linear ordering into itself;

if f, g belong to this LD-magma and z in the image of f, $(fg)(z)$ is $f(g(f^{-1}(z)))$.

This indicates some kind of resemblance between $\mathfrak{f}_1$ and $\mathfrak{a}_j$ (whose elements are elementary embeddings and induce increasing injections of the ordinals into themselves). However one easily verifies that the linear ordering $<^{\mathfrak{f}_1}$ is not wellfounded in any case.

About conjecture (IH), a natural approach is to try to construct directly monogenic LD-magmas. Let us write $x<_1^{\mathfrak{g}}y$ if x, y are in some LD-magma $\mathfrak{g}$ and, for some z in $\mathfrak{g}$, y is xz. Conjecture (IH) claims that the relation $<_1^{\mathfrak{f}_1}$ has no cycle. Laver's proof of Proposition 2 shows inductively that $<_1^{\mathfrak{a}_j}$ has no k-cycle for any integer k, and rests upon the properties of the critical point mapping. Partial results can be obtained by capturing some of these properties. For instance [De 3] describes a monogenic LD-magma $\mathfrak{d}$ endowed with a mapping

$$\text{crit} : \mathfrak{d} \longrightarrow \mathbb{N}$$

such that, for every x, y in $\mathfrak{d}$, $\text{crit}(xy) \neq \text{crit}(x)$. This is enough to prove that 1-cycles can exist for $<_1$ neither in $\mathfrak{d}$, nor in $\mathfrak{f}_1$. Likewise it can be easily shown that if an LD-magma $\mathfrak{g}$ is endowed with a mapping

$$\text{crit} : \mathfrak{g} \longrightarrow Ord$$

such that the three rules $\text{crit}(x) \leq \text{crit}(y) \Rightarrow \text{crit}(xy) > \text{crit}(y)$, $\text{crit}(x) > \text{crit}(y) \Rightarrow \text{crit}(xy) = \text{crit}(y)$ and $\text{crit}(x) < \text{crit}(y) \Rightarrow \text{crit}(zx) < \text{crit}(zy)$ are obeyed (this is trivially the case for $\mathfrak{a}_j$), then no 2-cycle can exist for $<_1$ in $\mathfrak{g}$. But these rules don't seem to be sufficient for going further, and, on the other hand, no example of an LD-magma satisfying them (unless $\mathfrak{a}_j$) is known to the author.

However it *seems* to be possible to give a direct proof of the irreflexivity of $<^{\mathfrak{f}_1}$ without using the existence of an auxiliary LD-magma like $\mathfrak{a}_j$. To do that, one has to strengthen considerably the arguments used in claim 2 (and developed in [De 1]). These developments deeply involve the structure of the monoid $\mathcal{M}(\Delta)$ generated by all the mappings $\Delta^{(u)}$'s using composition, and in particular the presentation of this monoid from its generators. It happens that $\mathcal{M}(\Delta)$ extends in some sense the infinite braid group B_∞. Then the point is getting for the members of $\mathcal{M}(\Delta)$ convenient normal form, and this question is closely connected with similar questions for braid groups. The results are still far from complete.

Finally we can notice that the description of a congruence in terms of a monoid of basic rewriting rules can be performed as well for other identities like commutativity or associativity. Some groups appear in this way, but, due to the fact that a variable is repeated twice in the distributivity identity but not in the latter ones, the problems in these cases are very easily compared with the corresponding ones in the distributivity case ([De 4]). Another natural extension of the present questions consists in introducing the composition as a second operation for elementary embeddings, as in [La]. It happens that most of the results in the 'two operations' case can be deduced from the results in the present 'one operation' case ([De 5]).

Note. (December 1991) A direct proof of the irreflexivity hypothesis has been completed recently along the lines above. It uses algebraic methods which are related with Garside's calculus on braid groups. As an application one obtains a new example of distributive operation by defining on the braid group B_∞ an operation $*$ by

$$x * y = x\tau(y)\sigma_1\tau(x)^{-1},$$

where $\sigma_1, \sigma_2 \ldots$ are the generators of B_∞ and τ is the endomorphism which maps σ_i to σ_{i+1}. Let $\mathfrak{b}$ be the closure of 1 under $*$: $\mathfrak{b}$ is a monogenic LD-magma, and the relation $<_L^{\mathfrak{b}}$ has no 3-cycle (it is conjectured that $\mathfrak{b}$ is free). The LD-magma $\mathfrak{d}$ above is a quotient of $\mathfrak{b}$.

References

[De 1] P. Dehornoy, *Free distributive groupoids*, Journal of Pure and Applied Algebra **61** (1989), 123–146.

[De 2] P. Dehornoy, *Sur la structure des gerbes libres*, C.R. Acad. Sci. Paris **309-I** (1989), 143–149; (with an Abridged English Version).

[De 3] P. Dehornoy, *Algebraic properties of the shift mapping*, Proc. AMS **106-3** (1989), 617–623.

[De 4] P. Dehornoy, *Structural monoids associated to equational varieties*, Proc. AMS; (to appear).

[De 5] P. Dehornoy, *The adjoint representation of left distributive structures*, Communications in Algebra; (to appear).

[La] R. Laver, *The left distributive law and the freeness of an algebra of elementary embeddings*, Advances in Mathematics; (to appear).

Département de Mathématiques, Université de Caen, 14032 Caen, FRANCE
E-mail address: frcaen51.bitnet

SOME OTHER PROBLEMS IN SET THEORY

MATTHEW FOREMAN

The intention of this short note is to publicize some opportunities for set theorists to work on some analytical problems of a somewhat unusual flavor to the set-theorists palate. No claim of authorship or originality of these problems is made (rather the contrary!). I present the definitions necessary to formally understand the problems. The content of my talk at the workshop is summarized in a research announcement in the Bulletin of the AMS [F1]. I will not reproduce it here though it may be useful as motivation for these problems and as a "point of view". The problems are arranged to allow natural discussion. The reader will be trusted to make his own ranking of importance.

Definition. A (discrete) group G is *amenable* iff there is a finitely additive probability measure (fapm) $\mu : \mathcal{P}(G) \to [0,1]$ such that μ is G-invariant. (For all $X \subseteq G$, $g \in G$, $\mu(X) = \mu(gX)$). G is *locally finite* iff any finitely generated subgroup of G is finite.

Locally finite groups are amenable, and amenable groups have a vestige of local finiteness in that every amenable group has the *Følner Property*: If $X \subseteq G$ is finite and $\varepsilon > 0$ then there is a subset $F \subseteq G$, with $X \subseteq F$ and for all $g \in X$,

$$\frac{|gF \Delta F|}{|F|} < \varepsilon.$$

(This is equivalent to amenability.)

If a group G acts on a set X, then one can study the G-invariant finitely additive probability measures on X, the *invariant means*. More generally, if $B \subseteq \mathcal{P}(X)$ is a G-invariant Boolean algebra, one can study the G-invariant means on B.

To illustrate, let $\mathbb{Z}$ act on $\mathbb{Z}$ by addition. Then a $\mathbb{Z}$-invariant mean is "equivalent" to a classical *Banach limit* on $\ell^\infty(\mathbb{Z})$.

The crudest property of a collection of invariant means is cardinality. In particular, with a given G-action on X, one can ask if there is *more than one invariant mean*, or even if an invariant mean exists. Banach [B] showed that for each of $\mathbb{R}$, S^1 and $\mathbb{R}^2$ there is more than one translation invariant finitely additive measure (giving the unit interval measure one). More recently, Margulis [M] and Sullivan [S] showed that for $n \geq 4$ there is a unique isometry invariant finitely additive probability measure on the Lebesgue measurable subsets of S^n. Drinfeld [D] proved the analogous result for $n = 2, 3$. A striking feature of these proofs is that they heavily use the structure of the group. In particular, Banach used amenability (admittedly before it was defined) and the others used representation-theoretic properties of $SO(n)$ that imply non-amenability. □

We begin with:

I. Rosenblatt's Question: Can there be an amenable group G acting on a set X that induces a *unique* invariant mean on X?

Remarks. a) Standard amenability considerations imply that at least one invariant mean exists.
b) A series of papers of Rosenblatt, Rosenblatt-Talagrand [R-T] and Krasa [K] culminated in the result of Krasa that solvable groups do not induce unique invariant means. □

The most concrete case is $X = \mathbb{N}$. Here the results are as follows:

If G is "analytic" (as a subset of $\mathbb{N}^{\mathbb{N}}$) then the answer is no [F1]. Yang [Y] proved under C. H. that there is a locally finite group G acting on $\mathbb{N}$ with a unique invariant mean. This was improved in [F1] to show that, assuming M.A., *every* free ultrafilter on $\mathbb{N}$ is the unique invariant mean with respect to some locally finite group. The proofs of these two results are quite different. One difference is that Yang's mean maps $\mathcal{P}(\mathbb{N})$ *onto* $[0, 1]$ and doesn't readily generalize under M.A.

II. Problem: Assume M.A. Is there a locally finite group acting on $\mathbb{N}$ inducing a unique invariant mean $\mu : \mathcal{P}(\mathbb{N}) \to [0, 1]$ that maps *onto* $[0, 1]$?
□

Using the construction of [F1], given ultrafilters $U_1 \cdots U_n$ on ω it is easy to build a locally finite group acting on $\mathbb{N}$ with the property that every invariant mean is an affine combination of $U_1, \ldots, U_n$. I conjecture that there is a notion of dimension such that given an amenable group G, $\{\mu : \mu$ is a G-invariant mean$\}$ either has finite dimension or cardinality $2^{2^{\omega}}$. A weak version of this is:

III. Problem. Is there an amenable group of permutations of $\mathbb{N}$, such that $2^{\aleph_0} < |\{\text{invariant means}\}| < 2^{2^{\aleph_0}}$? □

In [F1], it is shown that adding $\kappa \geq \aleph_2$ Cohen reals to a model of C. H. yields a model where every locally finite group of permutations of $\mathbb{N}$ has at least two invariant means. Unfortunately, the proof doesn't settle:

IV. Problem. Is it consistent with ZFC that every amenable group of permutations of $\mathbb{N}$ has at least two invariant means? □

Perhaps closer to the original motivating questions, one can ask:

V. Problem. Is there an amenable (locally finite?) group of measure preserving transformations of the unit interval that uniquely determines Lebesgue measure as a finitely additive probability measure on the measurable subsets of $[0, 1]$?

As far as I know, nothing is known about this problem. □

Returning to the results of Drinfeld, Margulis and Sullivan about $SO(n+1)$ invariant means on S^n, we note that in order to get the representation–theoretic machinery going, they needed that every $SO(n+1)$-invariant mean on the Lebesgue measurable subsets of $S^n (n \geq 2)$ gives each Lebesgue-null set measure zero. This was accomplished by remarking that every Lebesgue measure zero set X is contained in a Lebesgue measure zero set Y that has a measurable paradoxical decomposition. This clearly uses completeness of Lebesgue-measure. In particular, to my knowledge, the following is open:

VI. Problem. Let $n \geq 2$. Is Lebesgue measure the unique $SO(n + 1)$-invariant finitely additive probability measure on the *Borel* subsets of S^n?

It was widely conjectured that this was false and that there was a rotation-invariant finitely additive probability measure μ on the Borel subsets of S^2 that gave meager sets measure zero. The existence of such a measure was disproved in [D-F].

In [D-F], we showed that if X is a Polish space with a group G of homomorphism that acts freely on a comeager subset of X and contains a subgroup isomorphic to the free group on 2-generators, then there is no G-invariant finitely additive probability measure on the Borel subsets of X that gives meager sets measure zero. My final problem asks whether it suffices that G be non-amenable.

VII. Problem. Suppose X is a Polish space and G is a non-amenable group of homeomorphisms of X that acts freely on a comeager subset of X. Can there be a G-invariant finitely additive probability measure on the Borel subsets of X giving meager sets measure zero?

References

[B] S. Banach, *Sur le problème de la mesure*, Fund. Math. **4** (1923), 7–33.

[D] V. G. Drinfeld, *Solution of the Banach-Ruziewicz problem on S^2 and S^3*, Functional Anal. Appl. **18** (1984), 77–78.

[D-F] R. Dougherty and M. Foreman, *The Banach-Tarski Paradox with "nice" pieces*, In preparation.

[F1] M. Foreman, *Amenable Group Actions on the Integers: An Independence Result*, Bulletin of the AMS **21(2)** (October, 1989).

[F2] M. Foreman, *Locally finite groups of permutations of* $\mathbb{N}$ *acting on* ℓ^∞, A Tribute to Paul Erdös (Baker, Bollabás and Hajnal, eds.), Cambridge University Press, 1990.

[K] S. Krasa, *Non-uniqueness of invariant means for amenable group actions*, Monatshefte fur Mathematik **100** (1985), 21–125.

[M] G. Margulis, *Finitely additive invariant measures on Euclidean spaces*, J. Ergodic Theory and Dyn. systems **2(3)** (1982).

[R-T] J. Rosenblatt and M. Talagrand, *Different types of invariant means*, J. London Math. Soc. **24** (1981), 525–532.

[S] D. Sullivan, *For $n > 3$, there is only one finitely additive rotationally invariant measure on the n-sphere defined on all Lebesgue measurable sets*, Bull. Amer. Math. Soc. (N.S.) **4** (1981), 121–123.

Department of Mathematics, Ohio State University, Columbus OH 43210
Partially supported by NSF Grant DMS 7521991

GAMES IN RECURSION THEORY AND CONTINUITY PROPERTIES OF CAPPING DEGREES

LEO HARRINGTON AND ROBERT I. SOARE

ABSTRACT. It is shown here that there are no maximal minimal pairs of recursively enumerable (r.e.) degrees. Combining this with a dual theorem by Ambos-Spies, Lachlan and Soare for r.e. degrees cupping to $\mathbf{0}'$ it follows that any open formula $F(x, y)$ of two free variables in the language of the r.e. degrees, $\mathbf{R}$, which holds for r.e. degrees $\mathbf{a} \neq \mathbf{b}$, where $\mathbf{a}, \mathbf{b} \neq \mathbf{0}, \mathbf{0}'$, holds continuously in a neighborhood about $\mathbf{a}$ and $\mathbf{b}$. This is the best possible continuity result for formulas in general, because it fails for open formulas of three or more variables and also for formulas with quantifiers.

1. INTRODUCTION

The fundamental problems concerning structures in recursion theory such as the Turing degrees, the recursively enumerable (r.e.) degrees, or the lattice of r.e. sets, are questions of definable properties, decidability of the first order theory, classification of algebraic properties, and classification of automorphisms. The attempt to resolve these questions may be thought of abstractly as a game between two players. Roughly, the first player called "RED" attempts to produce *definable* properties, to code into the structure some undecidable theory (perhaps even true arithmetic) in order to prove undecidability, and to prove that the structure is *rigid*, namely has no nontrivial automorphisms. The second player called "BLUE" attempts to prove *nondefinability* of various elements or subclasses of the structure, and to generate as many automorphisms of the structure as possible, since automorphisms can be used to prove that a property is not definable.

For the structure which is the lattice, $\mathcal{E}$, of r.e. sets under inclusion, the struggle between RED and BLUE has been rather equal. In the direction of RED Harrington (unpublished) and independently Herrmann [7] have shown undecidability of the elementary theory of $\mathcal{E}$. In the BLUE direction Soare [13] produced a method for generating automorphisms of $\mathcal{E}$, which has been used by him and others to generate many automorphism results demonstrating uniformity of structure of $\mathcal{E}$. Recently, Harrington and Soare [6] have strengthened the results of RED by producing several new definable properties of $\mathcal{E}$, the most striking of which [5] solves a fundamental problem stemming from Post's Program [9]. In the direction of BLUE, Harrington, Lachlan, and Soare have also recently developed a powerful method for generating many new automorphisms of $\mathcal{E}$. Harrington and Soare have pressed both of these new methods (developing new definable properties and new automorphisms) to close the gap on certain important questions by either producing a definable property for RED, or by producing an automorphism for BLUE which proves that no such property can exist.

However, for the case of the Turing degrees $(\mathbf{D}, <)$ in general, or for the r.e. degrees, $(\mathbf{R}, <)$ in particular, the situation is very unbalanced, since there is an impressive array of results for RED and virtually none for BLUE. For example, Cooper [2] has recently proved that the jump operation is definable in $(\mathbf{D}, <)$, that the subclass of r.e. degrees $(\mathbf{R}, <)$ is definable there, and that all of the jump classes, $\mathbf{H}_n$ and $\mathbf{L}_n$ for $0 \leq n$, are definable in $(\mathbf{D}, <)$. Slaman and Woodin [12] have proved a number of results about definability with parameters for $(\mathbf{D}, <)$. In particular, they have shown that $(\mathbf{D}, <)$ has at most countably many automorphisms and that if $(\mathbf{R}, <)$ is rigid, then so is $(\mathbf{D}, <)$. No nontrivial automorphisms of $(\mathbf{R}, <)$ have been produced in spite of the vigorous and determined efforts by several very good recursion theorists.

The main result of this paper, Theorem 2.2, and its dual, Theorem 2.3, can be viewed as results for the BLUE player for the r.e. degrees $\mathbf{R}$. These theorems imply that any open formula $F(x, y)$ which holds for r.e. degrees $\mathbf{a} \neq \mathbf{b}$ holds continuously in a neighborhood about $\mathbf{a}$ and $\mathbf{b}$. Since an element $\mathbf{a}$ in $\mathbf{R}$ is *definable* by a formula $F(x)$ in $\mathbf{R}$ just if $\mathbf{a}$ is the unique element of $\mathbf{R}$ satisfying $F(x)$ in $\mathbf{R}$, a continuity result may be viewed as a kind of nondefinability result.

We use the standard definitions and notation in recursion theory as found in Soare [14]. From now on all sets and degrees will be r.e. even if not specified as such, although for emphasis we may also explicitly refer to them as being r.e.

2. The Continuity Results

In this section we state the two main results on continuity for cupping and capping, and we derive as a corollary the continuity result for open formulas of two variables.

Definition 2.1. *A pair* $\mathbf{a}$ *and* $\mathbf{b}$ *of r.e. degrees form* a minimal pair *if* $\mathbf{a}$ *and* $\mathbf{b}$ *are nonzero and* $\mathbf{a} \cap \mathbf{b} = \mathbf{0}$.

We say that $\mathbf{0} < \mathbf{a} < \mathbf{0}'$ is *capping* if $\mathbf{a}$ is half of a minimal pair, and *cupping* if $\mathbf{a} \cup \mathbf{b} = \mathbf{0}'$ for some $\mathbf{b} < \mathbf{0}'$. The next theorem is the main result of this paper. It asserts that there is no minimal pair of r.e. degrees which is maximal with respect to the property of being a minimal pair.

Theorem 2.2. *(Harrington and Soare). If r.e. degrees* $\mathbf{a}$ *and* $\mathbf{b}$ *form a minimal pair then there is an r.e. degree* $\mathbf{c} > \mathbf{a}$ *such that* $\mathbf{c}$ *and* $\mathbf{b}$ *form a minimal pair.*

For each statement about $\mathbf{R}$ there is a dual statement where $\mathbf{0}$, $\mathbf{0}'$, $<$, $\cup$, and $\cap$ are replaced by $\mathbf{0}'$, $\mathbf{0}$, $>$, $\cap$, and $\cup$, respectively. The next theorem is the dual of Theorem 2.2, and will not be proved here, but will appear in [1].

Theorem 2.3. *(Ambos-Spies, Lachlan, and Soare [1]). Given r.e. degrees* $\mathbf{a}$ *and* $\mathbf{b}$ *such that* $\mathbf{0} < \mathbf{a} < \mathbf{0}'$ *and* $\mathbf{a} \cup \mathbf{b} = \mathbf{0}'$ *there exists an r.e. degree* $\mathbf{c} < \mathbf{a}$ *such that* $\mathbf{c} \cup \mathbf{b} = \mathbf{0}'$.

From now on we fix the language $\mathcal{L} = L(<, \cup, \cap, \mathbf{0}, \mathbf{0}')$ of the r.e. degrees $(\mathbf{R}, <, \cup, \cap, \mathbf{0}, \mathbf{0}')$ with partial order, supremum (cupping), infimum (capping), least element $\mathbf{0}$ and greatest element $\mathbf{0}'$. Note that for $\mathbf{a}, \mathbf{b} \in \mathbf{R}$ $\mathbf{a} \cup \mathbf{b}$ always exists but $\mathbf{a} \cap \mathbf{b}$ does not always exist because $(\mathbf{R}, <, \cup, \cap, \mathbf{0}, \mathbf{0}')$ forms an upper semi-lattice but not a lattice. Hence, we view $\cap$ in $\mathcal{L}$ as a 3-place relation symbol rather than as a binary function symbol. Note also that the other operations and constants $\cup, \cap, \mathbf{0}, \mathbf{0}'$ can be defined in $(\mathbf{R}, <)$ using quantifiers, but we prefer the language $\mathcal{L}$ to $L(<)$ because we will be considering open (i.e. quantifier free) formulas of $\mathcal{L}$.

Definition 2.4. *Let* $F(x, y)$ *be any formula in the language* $\mathcal{L}$. *For* $\mathbf{a}, \mathbf{b} \in \mathbf{R}$, *we say that* F holds in a neighborhood of $(\mathbf{a}, \mathbf{b})$ *if there exist r.e. degrees* $\mathbf{a}_0 < \mathbf{a}$, $\mathbf{a}_1 > \mathbf{a}$, $\mathbf{b}_0 < \mathbf{b}$, *and* $\mathbf{b}_1 > \mathbf{b}$ *such that* $F(\mathbf{x}, \mathbf{y})$ *holds for all r.e. degrees* $\mathbf{x} \in (\mathbf{a}_0, \mathbf{a}_1)$, *and* $\mathbf{y} \in (\mathbf{b}_0, \mathbf{b}_1)$.

Similar definitions could be made for a formula F of n variables for any $n \geq 1$. If $F(x)$ is a open formula of $\mathcal{L}$ of one variable which holds at $\mathbf{a}$, $\mathbf{0} < \mathbf{a} < \mathbf{0}'$ then using the Sacks Density Theorem (see [14, p. 142]) it is easy to see that $F(x)$ holds in a neighborhood of $\mathbf{a}$. As a corollary of Theorems 2.2 and 2.3 we now extend this result to open formulas $F(x, y)$ of $\mathcal{L}$ of two variables.

Corollary 2.5. *Let $F(x,y)$ be any open formula of $\mathcal{L}$ with two variables, such that $F(\mathbf{a},\mathbf{b})$ holds for r.e. degrees $\mathbf{a} \neq \mathbf{b}$, $\mathbf{a},\mathbf{b} \neq \mathbf{0},\mathbf{0}'$. Then $F(x,y)$ holds in a neighborhood of $(\mathbf{a},\mathbf{b})$.*

Proof. Without loss of generality we may assume that the formula $F(\mathbf{a},\mathbf{b})$ specifies the complete (atomic) diagram of $\mathbf{a}$, $\mathbf{b}$, $\mathbf{0}$ and $\mathbf{0}'$. It is easy to see that $F(\mathbf{a},\mathbf{b})$ must be logically equivalent to one of the following cases.

Case 1. Assume that $F(\mathbf{a},\mathbf{b})$ asserts that $\mathbf{0} < \mathbf{a} < \mathbf{b} < \mathbf{0}'$. Then use the Sacks Density Theorem to construct the necessary degrees $\mathbf{a}_0$, $\mathbf{a}_1$, $\mathbf{b}_0$, and $\mathbf{b}_1$. The case of $\mathbf{0} < \mathbf{b} < \mathbf{a} < \mathbf{0}'$ is the same.

Case 2. Assume that $F(\mathbf{a},\mathbf{b})$ implies that

$$\mathbf{a}|\mathbf{b} \;\;\&\;\; \mathbf{a} \cap \mathbf{b} = \mathbf{0}.$$

Then by the Lachlan Nondiamond Theorem [14, p. 162] $F(\mathbf{a},\mathbf{b})$ must also imply that $\mathbf{a} \cup \mathbf{b} \neq \mathbf{0}'$. Now apply Theorem 2.2 to $\mathbf{a}$ and $\mathbf{b}$ to produce $\mathbf{a}_1 > \mathbf{a}$ such that $\mathbf{a}_1$ and $\mathbf{b}$ form a minimal pair. Next apply Theorem 2.2 to $\mathbf{b}$ and $\mathbf{a}_1$ to produce $\mathbf{b}_1 > \mathbf{b}$ such that $\mathbf{a}_1$ and $\mathbf{b}_1$ form a minimal pair. Using the Sacks Density Theorem choose any $\mathbf{a}_0$ such that $\mathbf{0} < \mathbf{a}_0 < \mathbf{a}$, and similarly choose $\mathbf{b}_0$. Then $F(\mathbf{x},\mathbf{y})$ holds for all $\mathbf{x} \in (\mathbf{a}_0,\mathbf{a}_1)$, and $\mathbf{y} \in (\mathbf{b}_0,\mathbf{b}_1)$.

Case 3. Assume that $F(\mathbf{a},\mathbf{b})$ implies that

$$\mathbf{a}|\mathbf{b} \;\&\; \mathbf{a} \cup \mathbf{b} = \mathbf{0}'.$$

The proof is entirely dual to the preceding paragraph except with Theorem 2.3 in place of Theorem 2.2.

Case 4. Assume that $F(\mathbf{a},\mathbf{b})$ implies that

$$\mathbf{a}|\mathbf{b} \;\&\; \mathbf{a} \cap \mathbf{b} \neq \mathbf{0} \;\&\; \mathbf{a} \cup \mathbf{b} \neq \mathbf{0}'.$$

Then there are degrees $\mathbf{c}$ and $\mathbf{d}$ such that

$$\mathbf{a} \cup \mathbf{b} = \mathbf{c} \;\&\; \mathbf{0} < \mathbf{d} < \mathbf{a},\mathbf{b}.$$

Without loss of generality we may assume that $\mathbf{d}$ is low. By the Robinson low splitting theorem [14, p.224] there are incomparable r.e. degrees $\mathbf{e}_0$ and $\mathbf{e}_1$ such that $\mathbf{a} = \mathbf{e}_0 \cup \mathbf{e}_1$ and $\mathbf{d} < \mathbf{e}_i$ for $i = 0,1$. Now either $\mathbf{e}_1 \not\leq \mathbf{b}$ or $\mathbf{e}_0 \not\leq \mathbf{b}$ (since otherwise $\mathbf{a} \leq \mathbf{b}$). Let $\mathbf{a}_0$ be whichever of $\mathbf{e}_0$ and $\mathbf{e}_1$ satisfies $\mathbf{e}_i \not\leq \mathbf{b}$. Similarly, choose $\mathbf{b}_0$ such that $\mathbf{d} < \mathbf{b}_0 < \mathbf{b}$ and $\mathbf{b}_0 \not\leq \mathbf{a}$. Hence, $\mathbf{d} \leq \mathbf{a}_0$, $\mathbf{b}_0$ so $\mathbf{a}_0$ and $\mathbf{b}_0$ do not form a minimal pair, and $\mathbf{a}_0|\mathbf{b}_0$.

Using a theorem of Robinson [14, VIII.4.7, p. 146] choose $\mathbf{a}_1$ such that $\mathbf{a} < \mathbf{a}_1 < \mathbf{c}$ and $\mathbf{b}_0 \not< \mathbf{a}_1$. Similarly, choose $\mathbf{b}_1$ such that $\mathbf{b} < \mathbf{b}_1 < \mathbf{c}$ and

$\mathbf{a}_0 \not< \mathbf{b}_1$. Note that $\mathbf{a}_1|\mathbf{b}_1$ and $\mathbf{a}_1 \cup \mathbf{b}_1 = \mathbf{c} < \mathbf{0}'$. Then $F(\mathbf{x}, \mathbf{y})$ holds for all $\mathbf{x} \in (\mathbf{a}_0, \mathbf{a}_1)$, and $\mathbf{y} \in (\mathbf{b}_0, \mathbf{b}_1)$. $\square$

Note that Corollary 2.5 cannot be extended either to open formulas of $\mathcal{L}$ of three or more variables or to formulas of $\mathcal{L}$ with quantifiers, and hence Corollary 2.5 is the best continuity result for general formulas of $\mathcal{L}$ which we can obtain. To see this recall [14, IX.2.3, p. 160] that there are nonzero r.e. degrees $\mathbf{a}$, $\mathbf{b}$ and $\mathbf{c}$ satisfying the formula

$$G(\mathbf{x}, \mathbf{y}, \mathbf{z}) : \mathbf{x} > \mathbf{z} \ \& \ \mathbf{y} > \mathbf{z} \ \& \ \mathbf{x} \cap \mathbf{y} = \mathbf{z},$$

namely $\mathbf{z}$ is a *branching* degree with branches $\mathbf{x}$ and $\mathbf{y}$. Note that the open formula G cannot hold in any neighborhood of $\mathbf{a}$, $\mathbf{b}$ and $\mathbf{c}$. Furthermore, the formula with quantifiers

$$H(z) : (\exists x)(\exists y)G(x, y, z)$$

cannot hold in any neighborhood of $\mathbf{c}$, because Fejer [3] has proved the density of the nonbranching degrees.

3. The Requirements

The next few sections will be devoted to a proof of Theorem 2.2. Fix r.e. degrees $\mathbf{a}$ and $\mathbf{b}$ such that $\mathbf{0} < \mathbf{a} < \mathbf{0}'$ and $\mathbf{a} \cap \mathbf{b} = \mathbf{0}$, and fix r.e. sets $A \in \mathbf{a}$ and $B \in \mathbf{b}$. We will construct an r.e. set C such that $\mathbf{c} = \deg(A \oplus C)$ satisfies $\mathbf{a} < \mathbf{c}$, and $\mathbf{c} \cap \mathbf{b} = \mathbf{0}$.

In what follows upper case Greek letters, Φ, Ψ, Γ, Δ, Ξ, and Λ, will represent Turing reductions with oracles (i.e. recursive functionals) as defined in [14, Chapter III], and the corresponding lower case Greek letters φ, ψ, γ, δ, ξ, and λ, will represent their use functions. Other lower case Greek letters, θ, and η without corresponding upper case Greek letters will represent partial recursive functions. Upper case Roman letters A, B, C, D, and E will represent r.e. sets. All these symbols may have subscripts or even double subscripts.

To illustrate how the proof was discovered we view it as a game in the style of Lachlan [8], where the two players RED and BLUE have roles analogous to those in §1. For emphasis we may sometimes put a "hat" over the sets or functionals such as $\hat{C}$ played by us (the BLUE player), and leave "unhatted" the sets and functionals played by the opponent (the RED player), as in the automorphism notation [14, Chapter XV].

To prove Theorem 2.2, it suffices to meet for all $i \in \omega$ the following requirements.

$$\mathcal{P}_i: \qquad \Xi_i^A = \hat{C} \implies (\exists \hat{\Lambda}_i)[\hat{\Lambda}_i^{A\oplus\hat{C}} = K]$$

$$\mathcal{R}_i: \qquad D_i = {\Phi_i}^{A\oplus\hat{C}} = \Psi_i^B \implies (\exists \hat{\Gamma}_i, \hat{\Delta}_i, \hat{E}_i)[\hat{E}_i = \hat{\Gamma}_i^A = \hat{\Delta}_i^B \\ \& \ [D_i \text{ nonrecursive } \implies \hat{E}_i \text{ nonrecursive }]].$$

Here we assume that $\{\Xi_i\}_{i\in\omega}$ is a listing of all recursive functionals, and $\{(\Phi_i, \Psi_i, D_i)\}_{i\in\omega}$ is a listing of all triples (Φ, Ψ, D) such that Φ and Ψ are recursive functionals and D is an r.e. set.

Since $\mathcal{R}_i$ is a complicated requirement we break it into the simpler subrequirements $\mathcal{T}_i$ and $\mathcal{S}_{i,j}$ as follows:

$$\mathcal{T}_i: \qquad D_i = {\Phi_i}^{A\oplus\hat{C}} = \Psi_i^B \implies (\exists \hat{\Gamma}_i, \hat{\Delta}_i, \hat{E}_i)[\hat{E}_i = \hat{\Gamma}_i^A = \hat{\Delta}_i^B],$$

$$\mathcal{S}_{i,j}: \qquad \hat{E}_i = \theta_{i,j} \implies (\exists \hat{\eta}_{i,j})[D_i = \hat{\eta}_{i,j}].$$

Here we assume that for each i, $\{\theta_{i,j}\}_{j\in\omega}$ is a listing of all partial recursive functions.

4. The Tree of Strategies

To meet the requirements we need a tree argument like that used to build a minimal pair of high r.e. degrees (see [14, p. 310]). The present argument also has features of the theorem on promptly simple degrees [14, p. 289]. We assume that the reader is familiar with the notation and use of the tree method as presented in [14, Chapter XIV].

In the next section we define the atomic strategies to meet each kind of requirement, the *σ-module* to meet a $\mathcal{P}_i$-requirement, the *τ-module* to meet a $\mathcal{T}_i$-requirement, and the *α-module* to meet an $\mathcal{S}_{i,j}$-requirement, and we will describe there the *outcomes* of each module which we now denote as follows.

(i) The outcomes of the σ-module are denoted by

$$\{d_0, d_1, \cdots\} \bigcup \{w\}.$$

(ii) The outcomes of the τ-module are denoted by

$$\{d_0, e_0, d_1, e_1, \cdots\} \bigcup \{w\}.$$

(iii) The outcomes of the α-module are denoted by $\{s, g, w\}$.

Consider the set of outcomes

$$S = \{d_0, e_0, d_1, e_1, \cdots\} \bigcup \{s, g, w\}$$

considered as a set ordered from left to right as listed. The *tree of outcomes* T is a subset of finite sequences of S defined as follows.

We define T and the strategy assigned to each $\rho \in T$ by induction on $|\rho|$ as follows, where $|\rho|$ denotes the length of ρ.

(i) If $\rho \in T$ and $|\rho| = 3i$ assign to ρ the τ-module strategy for $\mathcal{T}_i$ and put $\rho^\frown\langle a\rangle$ in T for each $a \in \{d_0, e_0, d_1, e_1, \cdots\} \bigcup \{w\}$.

(ii) If $\rho \in T$ and $|\rho| = 3i + 1$ put $\rho^\frown\langle a\rangle$ in T for $a \in \{s, g, w\}$, and assign to ρ the α-module strategy for $\mathcal{S}_{i,j}$ if $\rho = \tau^\frown\langle e_j\rangle$. If $\rho = \tau^\frown\langle w\rangle$ or $\rho = \tau^\frown\langle d_j\rangle$ assign no strategy to ρ.

(iii) If $\rho \in T$ and $|\rho| = 3i + 2$ assign to ρ the σ-module strategy for $\mathcal{P}_i$ and put $\rho^\frown\langle a\rangle$ in T for each $a \in \{d_0, d_1, \cdots\} \bigcup \{w\}$. Lower case Greek letters α, β, ρ, σ, τ represent nodes on T.

5. The Basic Modules for Each Requirement

We now define for each type of requirement $\mathcal{P}_i$, $\mathcal{T}_i$, $\mathcal{S}_{i,j}$ an atomic strategy (*module*) for that requirement. We partition ω into the disjoint union of infinite recursive sets

$$\omega = \bigcup \left\{\omega^{[\beta]} : \beta \in T\right\},$$

where $\omega^{[\beta]}$ denotes $\omega^{[n]}$ for n the code number of β in some effective coding of T.

If node $\beta \in T$ is assigned to requirement $\mathcal{P}_i$, $\mathcal{T}_i$, or $\mathcal{S}_{i,j}$, we may write Φ_β, Ψ_β,D_β, Ξ_β, θ_β, for the functionals and sets Φ_i, Ψ_i,D_i, Ξ_i, $\theta_{i,j}$, played by RED and write $\hat{\Gamma}_\beta$, $\hat{\Psi}_\beta$, $\hat{E}_\beta$, $\hat{\Lambda}_\beta$, $\hat{\eta}_\beta$, for the versions constructed by β as candidates for the corresponding functionals and sets $\hat{\Gamma}_i$, $\hat{\Psi}_i$, $\hat{E}_i$, $\hat{\Lambda}_i$, $\hat{\eta}_{i,j}$, played by BLUE.

In describing the construction we regard all sets and functionals as being in a state of formation and we will use A, B, $\hat{C}$, Φ, Ψ, $\hat{\Gamma}$, $\hat{\Delta}$, and so on to denote the current approximations to these objects during a given stage. Thus, A denotes the finite set of elements which have been enumerated in A, Φ denotes the functional determined by the finite set of instructions which have already been enumerated in Φ, p denotes the current value of parameter p, and similarly for all sets and functionals played by either RED or BLUE. When necessary to avoid confusion we append $[s]$ and write $A[s]$, $\Phi[s]$ or p$[s]$ to denote the result by the end of stage s as in [14, p. 315]. We also append $[s]$ to a whole expression to denote the current value of all the symbols mentioned there. Finally, to improve readability we let A_s, B_s, and C_s denote $A[s]$, $B[s]$, and $C[s]$ for the sets A, B, and C only.

In what follows we view the use functions, for example $\hat{\lambda}^{A\oplus\hat{C}}(x)$ for $\hat{\Lambda}^{A\oplus\hat{C}}(x)$, as movable markers whose position at the end of stage s is denoted by $\hat{\lambda}(x)[s]$, and such that once defined $\hat{\lambda}(x)$ can become redefined only if some $z \leq \hat{\lambda}(x)$ is enumerated in $A \oplus \hat{C}$. We will ensure that as a function of two variables $\hat{\lambda}(x)[s]$ (when defined) will be nondecreasing in s and strictly increasing in x. We may assume that for each s each use function $\hat{\lambda}(x)[s]$ is defined for at most finitely many arguments x, and that whenever marker $\hat{\lambda}(x)$ is newly placed on a value z then z is *fresh*, i.e. z exceeds all previous values $\hat{\lambda}(y)[s]$ for all y and s (and therefore $z \notin \hat{C}_s$). We also assume that all functionals played by RED satisfy the "hat condition" [14, p. 131], namely

(1) $[\varphi^A(x)[s]\downarrow = u \;\&\; (\exists z \leq u)[z \in A_{s+1} - A_s\;]] \implies \Phi^A(x)[s+1]\uparrow .$

5.1. The σ-module for $\mathcal{P}_i$. (This is the Sacks coding strategy as in the Sacks Density Theorem [14, p.142].) If $\sigma \in T$ satisfies $|\sigma| = 3i+2$ we assign to σ the following strategy for $\mathcal{P}_i$ called the *σ-module.* Define

$$\ell(\Xi_i, \hat{C})[s] = \max\left\{x : (\forall y < x)[\Xi^A_{i,s}(y)[s]\downarrow = \hat{C}(y)[s]]\right\}.$$

For convenience in §5.1 we drop the subscript i from Ξ_i, $\hat{\Lambda}_i$, ξ_i, λ_i, and other sets and functions.

If at stage $s+1$, $\xi(x)[s+1]\downarrow$, and $x < \ell(\Xi, \hat{C})[s]$ but $\hat{\lambda}(x)[s]\uparrow$ then we define $\hat{\Lambda}^{A\oplus\hat{C}}(x) = K(x)$ and $\hat{\lambda}^{A\oplus\hat{C}}(x) = z \in \omega^{[\sigma]}$, $z > \xi^A(x)$. If later some $y \leq \xi^A(x)$ enters A causing $\xi^A(x)$ to become undefined then y allows BLUE to make $\hat{\lambda}^{A\oplus\hat{C}}(x)$ undefined also and $\hat{\lambda}(x)$ may be later redefined as above. If $x \in K[s]$, $x < \ell(\Xi, \hat{C})[s]$, and $\hat{\Lambda}(x)[s]\downarrow = 0$, then at stage $s+1$ BLUE enumerates $\hat{\lambda}(x)[s]$ in $\hat{C}$, redefines $\hat{\Lambda}(x)[s+1] = K(x)[s]$ and redefines $\hat{\lambda}(x)$, as above.

Now suppose that Ξ^A is total and $= \hat{C}$. Then clearly $\hat{\Lambda}^{A\oplus\hat{C}}$ is total and $= K$, so $K \leq_T A$ contrary to hypothesis. Hence, we can choose

$$x = (\mu y)[\Xi^A(y) \neq \hat{C}(y)].$$

It may be that either $\Xi^A(x)\downarrow \neq \hat{C}(x)$ or that $\xi(x)[s]\downarrow$ for finitely many s, so that $\lim_s \ell(\Xi, \hat{C})[s] < \infty$. This is the σ-outcome w, in which case σ acts finitely often and hence contributes at most finitely many elements to $\hat{C}$.

The second possibility is that $\xi(x)[s]\downarrow$ for infinitely many s but

$$\lim_s \xi(x)[s] = \infty. \tag{2}$$

This is the σ-outcome d_x (for *divergence* of $\xi(x)$). In this case σ may act infinitely often and may put infinitely many elements into $\hat{C}$. Note however that

(3) $(\forall s)(\forall y)[\xi(y)[s] \leq \hat{\lambda}(y)[s]$ if both are defined]; and

(4) $(\exists t)(\forall s \geq t)[\ \hat{\lambda}(y)[s] \in \hat{C}_{s+1} - \hat{C}_s \Rightarrow y > x\ ]$

Now, (2), (3), and (4) imply that the set of elements z enumerated in $\hat{C}$ by σ is recursive because after stage t any such z satisfies $z \geq \xi(x)$, but $\lim_s \xi(x)[s] = \infty$.

Furthermore, we will arrange that every node α such that $\sigma\widehat{\ }\langle d_x\rangle \subseteq \alpha$ will not act on z (for example to restrain z from $\hat{C}$) until a stage s such that $z < \xi_\sigma(x)[s]$ since thereafter σ will not want to enumerate z in $\hat{C}$. Hence, (except for finite injury by $\hat{\lambda}_\sigma(y)$ for the finitely many $y < x$) the action of α will not be interfered with by the higher priority strategy σ.

Note that for the σ-module to succeed we need only that

(5) $$\liminf_s R_\sigma[s] < \infty$$

where we define

(6) $$R_\beta[s] = max\,\{\, r_\alpha[s] : \alpha \subset \beta\ \vee\ \alpha <_L \beta \,\},$$

and where $r_\alpha[s]$ is the $\hat{C}$ restraint imposed by node α at the end of stage s.

5.2. The τ-module for $\mathcal{T}_i$. If $\tau \in \mathcal{T}$ and $|\tau| = 3i$ we assign to τ the following strategy called the *τ-module*. Define

$$\ell(i,s) = \max\left\{x : (\forall y < x)[D_i(y)[s] = \Phi_i^{A\oplus\hat{C}}(y)[s] = \Psi_i^B(y)[s]]\right\}, \text{ and}$$

$$m(i,s) = \max\,\{\ell(i,t) : t \leq s\}\,.$$

A stage s is *i-expansionary* if $s = 0$ or if $\ell(i,s) > m(i,s-1)$. For convenience we drop the subscript i from Φ_i, Ψ_i, $\hat{E}_i$, $\hat{\Gamma}_i$, as before.

If s is i-expansionary, $x < \ell(i,s)$ and $\hat{\gamma}^A(x)[s]\uparrow$ but $\varphi^{A\oplus\hat{C}}(x)\downarrow$ then at stage $s+1$ we define $\hat{\Gamma}^A(x)[s+1] = \hat{E}(x)$ and define $\hat{\gamma}^A(x)$ such that

(7) $$\hat{\gamma}^A(x) \geq \varphi^{A\oplus\hat{C}}(x).$$

If at some stage s either (7) holds for x or $\hat{\gamma}(x)$ is undefined then we say that x is *honest* at s, and *dishonest* otherwise. If $\hat{\gamma}^A(x)$ is defined and some $y \leq \hat{\gamma}(x)$ enters A then we allow $\hat{\gamma}(x)$ and $\hat{\Gamma}(x)$ to become undefined, and may later redefine them as above.

Similarly, if s is an i-expansionary stage, $x < \ell(i,s)$, $\psi(x)[s]\downarrow$ and $\hat{\delta}(x)[s]\uparrow$ then define $\hat{\Delta}^B(x)[s+1] = \hat{E}(x)$ and define $\hat{\delta}^B(x)[s+1]$ such that

$$\hat{\delta}^B(x) \geq \psi^B(x). \tag{8}$$

(Note that since $\hat{C} \not\leq_T A$ it will be impossible to guarantee (7) for all x because after $\hat{\gamma}^A(x)$ is defined and honest, a change in $\hat{C} \restriction \varphi(x)$ will allow $\varphi^{A\oplus\hat{C}}(x)$ to be redefined so that x is no longer honest. However for *certain* x, BLUE will keep x honest by imposing $\hat{C}$-restraint on all $y \leq \varphi(x)$ whenever $\varphi(x)\downarrow$. For an honest such x, $\varphi(x)$ can then only be redefined by an $A \restriction z$ change for $z \leq \varphi(x)+1$ which allows $\hat{\gamma}(x)$ to be redefined also because x is honest. Keeping at least one honest x is the key to the α-module below. Note that unlike (7), we can guarantee (8) for all x and all stages s because any later B change which allows $\psi^B(x)$ to become undefined also allows BLUE to redefine $\hat{\delta}^B(x)$.)

The τ-module has outcome w in case there are at most finitely many i-expansionary stages, and it has outcome d_j if there are infinitely many i-expansionary stages, and j is minimal such that

$$\lim_s \varphi_i(j)[s] = \infty \;\vee\; \lim_s \psi_i(j)[s] = \infty. \tag{9}$$

Otherwise it follows that $\Phi_i^{A\oplus\hat{C}} = \Psi_i^B = D_i$, and the action of the τ-module ensures that $\hat{\Gamma}_\tau^A = \hat{\Delta}_\tau^B = \hat{E}_\tau$. This could be called outcome e of the τ-module. However, in this case to complete the action to meet requirement $\mathcal{R}_i$ we must satisfy $\mathcal{S}_{i,j}$ for all j. To give the construction a chance to do this we split outcome e into infinitely many outcomes $\{e_0, e_1, \cdots\}$ and attach to node $\alpha = \tau\widehat{\ }\langle e_j\rangle$ the following α-module for $\mathcal{S}_{i,j}$.

5.3. The α-module for $\mathcal{S}_{i,j}$. If $|\tau| = 3i$ and $\alpha = \tau\widehat{\ }\langle e_j\rangle$ then we assign to α the following strategy for $\mathcal{S}_{i,j}$ called the *α-module*, which is the key part of the entire proof. (For a more intuitive but less formal description the reader should now read §8 before proceeding.)

For convenience we drop the subscript i from various sets and functions as above and we also write θ_α and $\hat{\eta}_\alpha$ in place of $\theta_{i,j}$ and $\hat{\eta}_{i,j}$.

The α-module will require various parameters such as x_α, y_α, r_α whose values at the end of stage s will be denoted by $x_\alpha[s]$, $y_\alpha[s]$, $r_\alpha[s]$. During a given stage we let x_α, y_α, r_α denote the current value of these parameters.

Let $\ell(\hat{E}, \theta_\alpha)[s]$ denote the first disagreement of $\hat{E}$ and θ_α at the end of stage s (which must exist because $\theta_\alpha[s]$ is finite).

The α-module consists of the following steps.

Step 1. At step $v+1$ if there exists $z \in \omega^{[\alpha]}$ such that $z < \ell(i,v)$, $z < \ell(\hat{E},\theta_\alpha)$, $\theta_\alpha(z)\downarrow= 0$, and $z > x_\alpha[v]$, $y_\alpha[v]$ if either of the latter is defined, then let $y_\alpha[v+1]$ be the maximum such z.

Step 2. (Open α-gap). If $A_{s+1} \restriction z \neq A_s \restriction z$ for $z = y_\alpha[s]$ or $z = \varphi(x_\alpha)[s]+1$ (and the α-module is not currently in an open gap) then *open* an α-gap at stage $s+1$.

Step 2a. Set the $\hat{C}$-restraint $r_\alpha[s+1] = 0$.

Step 2b. If $x_\alpha[s] < y_\alpha[s]$ then define $x_\alpha[s+1] = y_\alpha[s]$, let $y_\alpha[s+1]$ be undefined, and define $\hat{\eta}_\alpha(z) = D(z)[s]$ for all $z \leq x_\alpha[s+1]$, z not yet in $dom(\hat{\eta}_\alpha)$.

Step 3. (Close α-gap). If an α-gap was last opened at stage $s+1$ and t is the next i-expansionary stage $> s+1$ then we *close* the α-gap at stage $t+1$. The stages v such that $s+1 \leq v \leq t$ are the *gap* stages and non gap stages are called *cogap* stages.

Step 3a. (Successful close). Suppose $\hat{\delta}^B(x_\alpha)[v]\uparrow$ for some v, $s+1 \leq v \leq t$. Enumerate x_α in $\hat{E}_\alpha[t+1]$, define $r_\alpha[t+1] = 0$, and take no action for the α-module at any stage $t' > t+1$.

Step 3b. (Unsuccessful close). Otherwise. Define $\hat{C}$ restraint $r_\alpha[t+1] = \hat{\gamma}(x_\alpha)[t+1]$. (Note that since $t+1$ is i-expansionary we may assume that at stage $t+1$ the τ-module has already defined $\hat{\gamma}(x_\alpha) > \varphi(x_\alpha)$ as in §5.2.)

We now describe the possible outcomes of the α-module and for each outcome the progress made on requirement $\mathcal{R}_i$ or $\mathcal{S}_{i,j}$.

Outcome s. There is some stage $t+1$ at which α successfully closes an α-gap.

In this case $x = x_\alpha[t]$ is enumerated in $\hat{E}$ at stage $t+1$. Note that at some stage $v < t$, $x = y_\alpha[v]$ and $\theta_\alpha(x)[v]\downarrow= 0$ by Step 1. Hence, x witnesses that $\theta_\alpha \neq \hat{E}$ so the requirement $\mathcal{S}_{i,j}$ is satisfied at all stages $w \geq t+1$.

Notice that if Step 3a applies to x then we still have

$$\hat{\Gamma}^A(x) = \hat{\Delta}^B(x) = \hat{E}(x)[t+1].$$

This is because $x = x_\alpha[s+1] = x_\alpha[t]$, $\hat{\delta}(x)[v]\uparrow$ for some v, $s+1 \leq v \leq t$, and $\hat{\gamma}(x)[s+1]\uparrow$. (The latter follows because if Step 2 applies at $s+1$ for $z = y_\alpha[s]$ then $x_\alpha[s+1] = y_\alpha[s] = z$, and $\varphi(z)[s+1]\uparrow$ by (1), but if Step 2 applies at $s+1$ with $z = x_\alpha[s]$ then $A_{s+1} \restriction u \neq A_s \restriction u$ for $u = \varphi(z)[s]+1$ so $\Phi(z)[s+1]\uparrow$ by (1) and $\hat{\gamma}(x)\uparrow$ by (7) and the honesty of x_α.)

Hence, $\hat{\delta}(x)[t]\uparrow$ and $\hat{\gamma}(x)[t]\uparrow$ because after $s+1$ if either $\hat{\delta}(x)$ or $\hat{\gamma}(x)$ becomes undefined then the τ-module does not redefine them until $t+1$ because t is the next i-expansionary stage $> s+1$. Now at stage $t+1$ we arrange that the α-module *first* enumerates $x = x_\alpha[t]$ in $\hat{E}[t+1]$ and *then* the τ-module redefines $\hat{\Gamma}^A(x) = \hat{\Delta}(x) = \hat{E}(x)[t+1]$.

Outcome w. In this outcome the α-module opens at most finitely many gaps and never closes one successfully.

We may suppose the each α-gap once opened is eventually closed, else there are at most finitely many i-expansionary stages so the correct outcome of the τ-module is w and not e_j (contrary to our assumption that we are in the case of τ-outcome $\alpha = \tau\widehat{\ }\langle e_j\rangle$.) Therefore, α *opens* at most finitely many gaps.

Thus, $\lim_s y_\alpha[s] < \infty$ because otherwise A nonrecursive implies that there exist infinitely many z and s such that $z \in A[s+1] - A[s]$ and $z \leq y_\alpha[s]$ so that α opens infinitely many gaps by Step 2, for $z = y_\alpha[s]$.

However, by Step 1 if $\lim_s y_\alpha[s] < \infty$ then either $\lim_s \ell(\hat{E}, \theta_\alpha)[s] < \infty$ (in which case $\mathcal{S}_{i,j}$ is satisfied) or $\limsup_s \ell(i,s) < \infty$ (in which case the hypotheses of $\mathcal{R}_i$ are not satisfied) so requirement $\mathcal{R}_i$ is met.

Outcome g. In this outcome α opens infinitely many gaps and closes each unsuccessfully.

Case 1. $x = \lim_s x_\alpha[s] < \infty$. Then $y = \lim_s y_\alpha[s] < \infty$ also, since $x_\alpha[s+1] = y_\alpha[s]$ by Step 2 if α opens a gap at stage $s+1$. Hence, for almost every s if α opens a gap at stage $s+1$ then $z \in A[s+1] - A[s]$ for some $z \leq \varphi(x)[s]$. Hence, $\lim_s \varphi(x)[s] = \infty$, so $\Phi(x)\uparrow$ and requirement $\mathcal{R}_i$ is met.

Case 2. $\lim_s x_\alpha[s] = \infty$. Then $\lim_s y_\alpha[s] = \infty$ also, since every value $x_\alpha[s]$ was $y_\alpha[v]$ for some $v < s$. Hence, $\theta_\alpha = \hat{E}$, and $\limsup_s \ell(i,s) = \infty$ by Step 1. But also $\hat{\eta}_\alpha$ is total because $x_\alpha[s]$ is nondecreasing in s, and whenever α opens a gap at $s+1$ then $\hat{\eta}_\alpha(z)$ is defined for all $z \leq x_\alpha[s+1]$. In this case we prove the following Proposition and hence conclude that requirement $\mathcal{R}_i$ is met.

Proposition 5.1. *$\hat{\eta}_\alpha = D$ and hence D is recursive.*

Proof. Suppose $\hat{\eta}(p) = D(p)[s]$ is first defined at stage $s+1$. Let $x = x_\alpha[s+1]$, and note that $p \leq x$. We must show $\hat{\eta}(p) = D(p)[s']$ for all $s' \geq s$.

Now α opens a gap at $s+1$ for $x = x_\alpha$ and hence $\varphi(x)[s+1]\uparrow$, and $\hat{\gamma}(x)[s+1]\uparrow$. Hence, $\hat{\delta}^B(x)[s+1]\downarrow$ else this gap is later closed successfully by Step 3a. Thus, $\psi^B(x)[s+1]\downarrow$ by (8) and

$$(\forall z \le x)[\hat{\eta}(z) = D(z)[s+1]], \tag{10}$$

$$(\forall z \le x)[D(z)[s+1] = \Psi^B(z)[s+1]], \tag{11}$$

because (11) held at the stage $v \le s+1$ when the τ-module last defined $\hat{\delta}(x)$ to satisfy (8) and because any $B \upharpoonright \hat{\delta}(x)$ change at w, $v < w \le s$, would have caused $\hat{\delta}(x)$ to become undefined at stage $w+1$ by (1).

Suppose this α-gap is closed at stage $t+1$. Since this is an unsuccessful close $B_t \upharpoonright u = B_{s+1} \upharpoonright u$ where $u = \psi^B(x)[s+1]$. But since t is i-expansionary, (10) and (11) imply

$$(\forall z \le x)[\hat{\eta}(z) = D(z)[t] = \Phi^{A\oplus\hat{C}}(z)[t] = \Psi^B(z)[t]]. \tag{12}$$

Hence, at stage $t+1$, *first* the τ-module defines $\hat{\Gamma}^A(x) = \hat{E}(x)$ and

$$\hat{\gamma}^A(x) > \varphi^{A\oplus\hat{C}}(x), \tag{13}$$

and *then* the α-module defines $\hat{C}$-restraint $r_\alpha[s+1] = \hat{\gamma}(x)$ which remains in force and hence ensures (13) during the *cogap*, i.e. during those stages v, $t+1 \le v \le s'$, where $s'+1$ is the least stage $> t+1$ at which α opens a gap. If $u = \varphi^{A\oplus\hat{C}}(x)[t+1]$ then this restraint ensures that no $z \le u$ enters $\hat{C}$ at any stage v, $t+1 \le v \le s'$ and the condition in Step 2 for opening a gap ensures that no $z \le u$ enters A at such a stage v because $s'+1$ is the next gap opening stage $> t$. Hence,

$$(\forall z \le x)[\Phi(z)[s'] = \Phi(z)[t] = D(z)[t] = \hat{\eta}(z)] \tag{14}$$

Now repeat the argument with $x' = x_\alpha[s'+1]$ in place of x. Repeating the argument for each gap opening stage $s'' > s'$ we see that

$$(\forall v \ge s)(\forall z \le x)[\Phi(z)[v] = \hat{\eta}(z) \vee \Psi(z)[v] = \hat{\eta}(z)]$$

and hence $\hat{\eta}(z) = D(z)$ since $\liminf_s \ell(i,s) = \infty$. $\square$

6. The Construction

We now combine the strategies assigned to each node $\beta \in T$ to give the full construction. The following conventions and notation closely follow those of the nonbounding construction in [14, pp. 327-330].

In addition to the above symbols if $\beta \in T$ is assigned the α-module then β will also have associated parameters x_β, y_β, r_β as previously discussed. A parameter p once assigned a value retains that value until redefined, the current value of p is denoted simply by p, and $p[s]$ denotes the value at the end of stage s.

To *initialize* node β at a given stage means to let all the parameters x_β, y_β, r_β and all functionals $\hat{\Gamma}_\beta$, $\hat{\Delta}_\beta$, $\hat{\Lambda}_\beta$, $\hat{\eta}_\beta$, become undefined on all values and to let $\hat{E}_\beta[s] = \emptyset$. (Later new β action may redefine them.)

At the end of the construction we will define the *true path* $f \in [T]$ of the construction. We will approximate f by defining during each stage s of the construction a string $\pi[s] \in T$ such that

$$f = \liminf_s \pi[s].$$

We say that s is a *β-stage* if $s = 0$ or $\beta \subseteq \pi[s]$.

For any $\beta \in T$ define

$$(15) \quad P_\beta[s] = \min\left\{\xi_\sigma(k)[v] : \sigma\widehat{\ }\langle d_k\rangle \subset \beta \ \&\ |\sigma| \equiv 2 \bmod 3 \ \&\ v \leq s\right\}.$$

Note that $P_\beta[s]$ is nondecreasing in s. Note also that, except for finitely many s, if $\sigma \subset \beta$ contributes an element z to $\hat{C}$ at stage $s+1$ then $z > P_\beta[s]$.

The construction is as follows.

Stage $s = 0$. Initialize all nodes $\beta \in T$. Define $\pi[0] = \emptyset$, the empty node of T.

Stage $s+1$. The construction will proceed by substages $t \leq s+1$. We refer to substage t of stage $s+1$ as *stage* $(s+1, t)$. The value of a parameter p (such as π) at the end of substage t will be denoted by p_t. We will arrange that $|\pi_t| = t$ and $\pi_t \subset \pi_{t+1}$. Only node π_t can act at substage $t+1$. After substage $t+1 = s+1$ we will define $\pi[s+1] = \pi_{t+1}$.

Substage $t = 0$. Define $\pi_t = \emptyset$. Go to substage 1.

Substage $t+1 \leq s+1$. Given π_t.

Case 1. $|\pi_t| = 3i$ for some i. Let $\tau = \pi_t$. Let v_0 be the maximum τ-stage $< s$. We define $\ell(i,s)$ as in the τ-module in §5.2 except that now we only consider $\Phi_i(x)$ and $\Psi_i(x)$ computations such that

$$\varphi_i(x)[s], \quad \psi_i(x)[s] < P_\tau[s].$$

Let $m(\tau)$ be the maximum of $\ell(i,w)$ for all τ-stages $w \leq v_0$. If $m(\tau) < \ell(i,s)$ then s is a *τ-expansionary stage.*

Subcase 1a. Assume s is not τ-expansionary. Define $\pi_{t+2} = \tau\widehat{\ }\langle w\rangle\widehat{\ }\langle w\rangle$ and go to substage $t+3$.

Subcase 1b. Assume s is τ-expansionary. Choose the least j such that either:

$$\varphi_i(j)[v_0] \neq \varphi_i(j)[s] \vee \psi_i(j)[v_0] \neq \psi_i(j)[s], \quad \text{or} \tag{16}$$

$$\tau\hat{\ }\langle e_j \rangle \text{ requires attention,} \tag{17}$$

where we say that $\alpha = \tau\hat{\ }\langle e_j \rangle$ *requires attention* if the α-module is now ready to perform some action according to Step 1, 2, or 3 of the α-module in §5.3. (Note that j exists because s is τ-expansionary so (16) must hold for some j.)

If j satisfies (16) then let $\pi_{t+2} = \tau\hat{\ }\langle d_j \rangle\hat{\ }\langle w \rangle$, define $\hat{\Gamma}^A_\tau(y) = \hat{\Delta}^A_\tau(y) = \hat{E}_\tau(y)[s]$ for all $y < \ell(i,s)$, define $\hat{\gamma}_\tau(y) > \varphi(y)$ and $\hat{\delta}_\tau(y) > \psi(y)$ for all $y < \ell(i,s)$ with $\hat{\Gamma}_\tau(y)$ or $\hat{\Delta}_\tau(y)$ defined, and go to substage $t+3$.

If j satisfies (17) then let $\pi_{t+1} = \alpha = \tau\hat{\ }\langle e_j \rangle$. If the α-module wants to successfully close a gap according to Step 3a then define $\hat{\Gamma}^A_\tau(x) = \hat{\Delta}^A_\tau(x) = 1$ for $x = x_\alpha[s]$ (in anticipation that the α-module will enumerate x in $\hat{E}_\tau$ at substage $t+2$), and otherwise define $\hat{\Gamma}^A_\tau(x) = \hat{\Delta}^A_\tau(x) = \hat{E}_\tau(x)[s]$. In addition, define $\hat{\Gamma}^A_\tau(y) = \hat{\Delta}^A_\tau(y) = \hat{E}_\tau(y)[s]$ for all $y < \ell(i,s)$, $y \neq x_\alpha[s]$. Also define $\hat{\gamma}_\tau(y) > \varphi(y)$ and $\hat{\delta}_\tau(y) > \psi(y)$ for all $y < \ell(i,s)$ with $\hat{\Gamma}_\tau(y)$ or $\hat{\Delta}_\tau(y)$ defined. Go to stage $t+2$ and let α act in Case 2 as follows.

Case 2. $|\pi_t| = 3i+1$ for some i. We may assume that $\pi_t = \tau\hat{\ }\langle e_j \rangle$ for some j and $\alpha = \pi_t$ wants to perform some step in the α-module since otherwise the action in Case 1 caused us to go to substage $t+3$ moving directly from some τ-node to some σ-node.

The action for $\alpha = \pi_t$ is just the same as in the α-module in §5.3 except that: in Step 1 we also require that $z < P_\alpha[v]$ before we define $y_\alpha[v+1] = z$; in Step 2 we replace the condition $A_{s+1} \restriction z \neq A_s \restriction z$ by $A_{s+1} \restriction z \neq A_{v_0} \restriction z$; where v_0 is the greatest α-stage $\leq s$; and in Step 3 we replace "i-expansionary stage" by "τ-expansionary stage".

Define

$$\pi_{t+1} = \begin{cases} \pi_t\hat{\ }\langle s \rangle & \text{if the } \alpha\text{-module acts under Step 3a,} \\ \pi_t\hat{\ }\langle w \rangle & \text{if the } \alpha\text{-module acts under Step 1 or Step 3b,} \\ \pi_t\hat{\ }\langle g \rangle & \text{if the } \alpha\text{-module acts under Step 2,} \end{cases}$$

Go to Substage $t+2$.

Case 3. $|\pi_t| = 3i+2$ for some i. Let $\sigma = \pi_t$. Let R_σ denote the current value at the end of stage (s,t) of that function $R_\sigma[v]$ as defined in (6).

Step 1. If $x \in K[s]$, $\hat{\Lambda}^{A\oplus\hat{C}}_\sigma(x)\downarrow = 0$, $x < \ell(\Xi_i, \hat{C})[s]$, and $R_\sigma < \hat{\lambda}_\sigma(x)[s]$ then enumerate $\hat{\lambda}_\sigma(x)[s]$ in $\hat{C}$, let $\hat{\lambda}_\sigma(y)$ become undefined for all

$y \geq x$, and redefine $\hat{\Lambda}_\sigma(y)$ and $\hat{\lambda}_\sigma(y)$ as in Step 2. Initialize all nodes β such that $\sigma\char`^\langle d_x\rangle \subseteq \beta$ or $\sigma\char`^\langle d_x\rangle <_L \beta$.

Step 2. If $x < \ell(\Xi_i, \hat{C})[s]$ and $\hat{\Lambda}_\sigma(x)$ is currently undefined, then define $\hat{\Lambda}_\sigma(x) = K(x)[s]$, and define $\hat{\lambda}_\sigma(x) = z \in \omega^{[\sigma]}$, $z > \xi_i(x)$, and z such that $\hat{\lambda}_\sigma(x) = z$ will satisfy the use function conventions stated just before §5.1.

Define $\pi_{t+1} = \sigma\char`^\langle d_j\rangle$ if $j < \ell(\Xi_\sigma, \hat{C})[s]$ is minimal such that $\xi_\sigma(j)$ is currently undefined or has changed in value since the last $\sigma$-stage. If there is no such $j$ define $\pi_{t+1} = \sigma\char`^\langle w\rangle$. This completes substage $t+1$.

At the end of substage $t+1 = s+1$, define $\pi[s+1] = \pi_{t+1}$ and initialize all nodes β such that $\pi[s+1] <_L \beta$.

7. The Verification

To complete the proof of Theorem 2.2 we need to prove that for every i requirements $\mathcal{P}_i$ and $\mathcal{R}_i$ are satisfied.

Define the *true path* $f \in [T]$ of the construction by

$$f = \liminf_s \pi[s], \tag{18}$$

namely $f \restriction n = \liminf_s \pi[s] \restriction n$, for all n. (Since the tree T is infinitely branching it is not obvious that this lim inf exists, but we will establish it by proving (19) by induction on n for $\beta = f \restriction n$.)

We will show that each requirement is satisfied by the unique node $\beta \subset f$ which is assigned to that requirement.

Fix $\beta \subset f$. By the definition of f we know $\pi[s] <_L \beta$ for at most finitely many s. Also each $\sigma \subset \beta$ initializes β at most finitely often according to Case 3 Step 1 of the construction in §6. Hence, there is a stage s_β such that β is never initialized and $\pi[s] \not<_L \beta$ at any stage $s \geq s_\beta$.

In addition we will assume the following inductive hypotheses for β,

$$(\exists^\infty s)[s \text{ a } \beta\text{-stage}], \tag{19}$$

$$\lim_s P_\beta[s] = \infty, \quad \text{and} \tag{20}$$

$$R_\beta = \liminf_s \{\, R_\beta[s] : s \text{ a } \beta\text{-stage}\,\} < \infty, \tag{21}$$

where $P_\beta[s]$ was defined in (15) and $R_\beta[s]$ was defined in (6).

We now examine the case $\beta = f \restriction 3i$ and we show in Lemmas 7.1 and 7.2 that the modules for β and $\beta^+ = f \restriction (3i+1)$ satisfy $\mathcal{R}_i$, and that (19), (20), and (21) hold for $\beta^{++} = f \restriction (3i+2)$. (Later we do the analogous verification for $\beta = f \restriction (3i+2)$ and $\mathcal{P}_i$ and β^+ in Lemmas 7.3 and 7.4.) Let $\beta = f \restriction 3i$ and $\tau = \beta$.

Lemma 7.1. *Requirement $\mathcal{R}_i$ is satisfied.*

Proof. Assume $\Phi_i^{A\oplus\hat{C}} = \Psi_i^B = D_i$. Hence,

(22) $$\lim_s \ell(i,s) = \infty,$$

there are infinitely many τ-expansionary stages, and $f(3i) \neq d_j$ because Φ_i and Ψ_i are total.

Hence, the τ-module constructs $\hat{\Gamma}_\tau$, $\hat{\Delta}_\tau$, and $\hat{E}_\tau$ such that $\hat{\Gamma}_\tau^A = \hat{\Delta}_\tau^B = \hat{E}_\tau$. (This uses the remark in Outcome s of §5.3 about why $\hat{\Gamma}_\tau(x)$ and $\hat{\Delta}_\tau(x)$ remain correct when x enters $\hat{E}_\tau$.) However, by hypothesis $\mathbf{a} \cap \mathbf{b} = \mathbf{0}$, so $\hat{E}_\tau$ must be recursive. Choose the least j such that $\theta_{i,j} = \hat{E}_\tau$. Hence,

(23) $$\lim_s \ell(\hat{E}_\tau, \theta_{i,j})[s] = \infty, \quad \text{and}$$

(24) $$(\forall k < j)[\lim_s \ell(\hat{E}_\tau, \theta_{i,k})[s] < \infty].$$

Fix $\alpha = \tau\hat{\ }\langle e_j\rangle$. Now by (24) and the totality of Φ_i and Ψ_i we know that $\pi[s] <_L \alpha$ for finitely many s, in particular never at stage $s \geq s_\alpha$ defined above.

From (22), (23) and Step 1 of the α-module it follows that $\lim_s y_\alpha[s] = \infty$. Hence, A nonrecursive implies that there are infinitely many s such that at stage $s+1$ the α-module is ready to open a gap via Step 1 where $A_{s+1} \upharpoonright z \neq A_{v_0} \upharpoonright z$ for $z = y_\alpha[s]$. By (22) each α-gap must later be closed. Therefore, α opens and closes infinitely many gaps and $f(3i) = e_j$ so $\alpha \subset f$.

Hence, $D_i = \hat{\eta}_\alpha$ and D_i is recursive by the same proof as in Proposition 5.1. (Here we are using (20), the definition of s_α and the fact that $\varphi(x_\alpha)[s] < P_\alpha[s]$ to see that no element $z \leq r_\alpha[s]$ is enumerated in $\hat{C}$ at a stage $s \geq s_\alpha$.) □

Lemma 7.2. *Node $\beta^{++} = f \upharpoonright (3i+2)$ satisfies the inductive hypotheses (19), (20), and (21).*

Proof. Note that if β^{++} exists (i.e. satisfies (19)) then β^{++} clearly satisfies (20) because $P_{\beta^{++}}[s] = P_\beta[s]$. We now show that β^{++} exists and satisfies (21). First note that $\beta = \tau$ satisfies (19) by inductive hypothesis.

Case 1. Suppose there are finitely many τ-expansionary stages. Then $\beta^{++} = \beta\hat{\ }\langle w\rangle\hat{\ }\langle w\rangle$ and β^{++} satisfies (19), (20), and (21).

Case 2. Suppose there are infinitely many τ-expansionary stages, but $\Phi_i(j)\uparrow$ or $\Psi_i(j)\uparrow$ with j minimal. Then $\beta^+ = \tau\hat{\ }\langle d_j\rangle$ or $\beta^+ = \tau\hat{\ }\langle e_k\rangle$ for some $k < j$ because the tree T below τ is finitely branching to the left of outcome d_j. If $\beta^+ = \tau\hat{\ }\langle d_j\rangle$ then $\beta^{++} = \tau\hat{\ }\langle d_j\rangle\hat{\ }\langle w\rangle$ and β^+ clearly satisfies (19) and (21).

If $\beta^+ = \alpha = \tau\hat{\ }\langle e_k\rangle$ then $\beta^+ = \alpha\hat{\ }\langle a\rangle$ for some $a \in \{s, g, w\}$. If $a \in \{s, w\}$ then $r_0 = \lim_s r_\alpha[s] < \infty$. If $a = g$ then $r_\alpha[s+1] = 0$ for each of the infinitely many stages $s+1$ at which α opens a gap. In either case (21) holds for $\beta^{++} = \beta^{+\frown}\langle a\rangle$.

Case 3. Otherwise, RED constructs $\Phi_i = \Psi_i = D_i$, so the τ-module constructs $\hat{\Gamma}_\tau = \hat{\Delta}_\tau = \hat{E}_\tau$. But then RED must make $\hat{E}_\tau$ recursive, say $\hat{E}_\tau = \theta_{i,k}$ with k minimal. Then $\beta^+ = \alpha = \tau^\frown\langle e_k\rangle$, $\beta^{++} = \alpha^\frown\langle g\rangle$ and the argument is the same as in the second paragraph of Case 2 above.

Note that since (19) holds for β^{++} we have that $\beta^{++} \subset f$ because it is obvious in each case that $\pi[s] <_L \beta^{++}$ for at most finitely many s, and likewise in the proof of Lemma 7.4, although we will not mention it there explicitly. □

Lemma 7.3. *Requirement $\mathcal{P}_i$ is satisfied.*

Proof. Assume $\Xi_i^A = \hat{C}$. Then $\lim_s \ell(\Xi_i, \hat{C})[s] = \infty$. Hence, the σ-module constructs $\hat{\Lambda}_\sigma^{A\oplus\hat{C}}$ total such that $\hat{\Lambda}_\sigma^{A\oplus\hat{C}}(x) = K(x)$ for all x satisfying $R_\sigma < \hat{\lambda}_\sigma(x)$ by (21) and Case 3 of the construction. Thus, $K \leq_T A \oplus \hat{C} \leq_T A$ contrary to the hypothesis on A. □

Lemma 7.4. *If $\beta = f \restriction (3i+2)$ then $\beta^+ = f \restriction (3i+3)$ satisfies (19), (20) and (21).*

Proof. Clearly, if β^+ exists then β^+ satisfies (21) because $r_\beta[s]$ is never defined so $R_{\beta^+}[s] = R_\beta[s]$ for all s. Let $\sigma = \beta = f \restriction (3i+2)$. By Lemma 7.3 choose

$$x = (\mu y)\neg[\Xi_i^A(y)\downarrow = \hat{C}(y)].$$

If $\Xi_i^A(x)\downarrow \neq \hat{C}(x)$ or $\xi(x)[s]\downarrow$ for at most finitely many s then

$$\lim_s \ell(\Xi_i, \hat{C})[s] < \infty,$$

the σ-module performs finitely much action, and $\beta^+ = \beta^\frown\langle w\rangle$. In this case $P_{\beta^+}[s] = P_\beta[s]$ for all s, and almost every β-stage is also a β^+-stage, so (19) and(20) hold for β^+.

Otherwise we have

$$\lim_s \xi_i(x)[s] = \infty, \tag{25}$$

in which case (19) holds for β^+ by the construction, and (20) follows for β^+ from (15) and (20) for β. □

This completes the proof of Theorem 2.2. □

8. Intuition about the α-module

The following intuition may help to explain the α-module. For α and τ as in §5.3, the α-module assumes that if

$$\Phi^{A\oplus\hat{C}} = \Psi^B = D \tag{26}$$

then the τ-module constructs

$$\hat{\Gamma}^A = \hat{\Delta}^B = \hat{E} \tag{27}$$

as in §5.2. The α-module must then show that if

$$\hat{E} = \theta_\alpha \tag{28}$$

then the function $\hat{\eta}_\alpha$ constructed by α satisfies

$$\hat{\eta}_\alpha = D. \tag{29}$$

Roughly, α looks for a stage v at which there is a sufficiently large and honest element $y \in \omega^{[\alpha]}$, $x_\alpha < y$, y not yet in $\hat{E}$ such that $\theta_\alpha(y)[v]\downarrow$. The α-module then defines $x_\alpha[v+1] = y$ and $\hat{\eta}(z) = D(z)$ for all $z \leq x_\alpha$. Let $x_\alpha = x_\alpha[v+1]$. Now since $\theta_\alpha(x_\alpha)[v]\downarrow$ RED must impose sufficiently much restraint on A and B to ensure that for all $w \geq v$ either

$$\hat{\Gamma}^A(x_\alpha)[w]\downarrow = \theta(x_\alpha)[w], \text{ or} \tag{30}$$

$$\hat{\Delta}^B(x_\alpha)[w]\downarrow = \theta(x_\alpha)[w], \tag{31}$$

because otherwise α enumerates x_α in $\hat{E}$ refuting (28) forever. Hence, α will likewise use (30) and (31) to show that for all $w \geq v$ either

$$\Phi^{A\oplus\hat{C}}(x_\alpha)[w]\downarrow = \hat{\eta}(x_\alpha), \text{ or} \tag{32}$$

$$\Psi^B(x_\alpha)[w]\downarrow = \hat{\eta}(x_\alpha), \tag{33}$$

so that (29) follows from (26).

Now x_α is honest when first appointed, and α imposes $\hat{C}$ restraint $r_\alpha = \hat{\gamma}(x_\alpha)$ whenever α is not in a gap for x_α (i.e. is in a cogap) to ensure that x_α remains honest. If $\varphi(x_\alpha)\downarrow = u$ and $z \in A_{s+1} - A_s$ for some $z \leq u$ then $\varphi(x_\alpha)[s+1]\uparrow$, and $\hat{\gamma}(x_\alpha)[s+1]\uparrow$ so α opens a gap at stage $s+1$ and α defines $\hat{C}$ restraint $r_\alpha[s+1] = 0$, because α knows that RED must hold the computation (33) during this gap until the next τ-expansionary stage $t+1$, which is when the τ-module next redefines $\hat{\gamma}(x_\alpha)$, and x is of course honest at $t+1$ by (7). At stage $t+1$ α reimposes $\hat{C}$ restraint $r_\alpha[t+1] = \hat{\gamma}(x_\alpha)[t+1]$ to keep x_α honest until the next gap is opened.

Thus, (29) follows because (33) holds in the gaps and (32) holds in the cogaps. Notice also that this is true not merely for x_α itself but also for all

$z \leq x_\alpha$, because $\hat{\gamma}(z)[s]$ and $\hat{\delta}(z)[s]$ are nondecreasing in z, so $\hat{\gamma}(z)[s]\downarrow$ if $\hat{\gamma}(x_\alpha)[s]\downarrow$, and likewise for $\hat{\delta}(z)$.

If RED ensures (26) and (28) then α must make $\hat{\eta}$ total and hence must arrange that $\lim_s(x_\alpha)[s] = \infty$ because $x_\alpha[s]$ is the maximum element in $\mathrm{dom}(\hat{\eta}[s])$. The role of y_α in §5.3 is merely as a placeholder for a future value of x_α. We arrange that $y_\alpha < \ell(i, v), y < \ell(\hat{E}, \theta_\alpha)$ (so $\theta_\alpha(y_\alpha)\downarrow$) and $x_\alpha[s] < y_\alpha[s]$. If (26) and (28) hold then $\lim_s y_\alpha[s] = \infty$. Now since A is nonrecursive there is some s such that $A_{s+1} \upharpoonright z \neq A_s \upharpoonright z$ for $z = y_\alpha[s]$. But then $\varphi(z)[s+1]\uparrow$ by (1), so $\hat{\gamma}(z)[s+1]\uparrow$, and z is honest. Hence, α defines $x_\alpha[s+1] = z = y_\alpha[s]$. Repeating this for new values of y_α it follows that $\lim_s x_\alpha[s] = \infty$ because $\lim_s y_\alpha[s] = \infty$.

(In the α-module presented in §5.3 we have incorporated a suggestion made to us by David Seetapun. In our original α-module, at Step 1a we appointed $y_\alpha \in \omega^{[\alpha]}$ to be a fresh element such that $\hat{\gamma}(y_\alpha)\uparrow$. In Step 1b we waited for a stage when $\hat{\gamma}_\alpha(y_\alpha)\downarrow$ (and therefore y_α honest) and we established $\hat{C}$ restraint $r_\alpha > \varphi(y_\alpha)$ to keep y_α honest whenever $\varphi(y_\alpha)$ was defined. Finally, in Step 1c (rather than in Step 2 in §5.3) when $y_\alpha < \ell(\hat{E}, \theta_\alpha)$ and $y_\alpha < \ell(i, s)$ we defined $x_\alpha[s+1] = y_\alpha[s]$ and $\hat{\eta}_\alpha(z) = D_i(z)$ for all $z \leq x_\alpha[s+1]$. In Step 2 we opened a gap only for an $A \upharpoonright u$ change where $u = \varphi(x_\alpha) + 1$. We made sure r_α exceeded both $\varphi(x_\alpha)$ and $\varphi(y_\alpha)$ whenever either was defined. The outcomes $\{w, s, g\}$ and their progress on the requirements were similar with the following exception. If α opens finitely many gaps and closes none successfully (so the outcome is w) then it may happen that $\lim_s x_\alpha[s] < \infty$, $y = \lim_s y_\alpha[s] < \infty$, and $\lim_s \varphi(y)[s] = \infty$. In this case $\mathcal{R}_i$ is satisfied by divergence of $\Phi_i(y)$, but we may have $\limsup_s r_\alpha[s] = \infty$ and $\liminf_s r_\alpha[s] = 0$ which we handle in as the case of α-outcome g. The main difference is that the α-module in §5.3 has the more conventional property that if the outcome of α is w then α acts finitely often.)

9. Historical Remarks and Related Results

The idea of studying continuity properties of r.e. degrees arose from a question of Lachlan posed in 1967. Lachlan asked whether every r.e. degree $\mathbf{0} < \mathbf{a} < \mathbf{0}'$ has a *major subdegree* namely whether

$$(\forall \mathbf{a})_{\mathbf{0}<\mathbf{a}<\mathbf{0}'}(\exists \mathbf{c})_{\mathbf{c}<\mathbf{a}}(\forall \mathbf{b})[\mathbf{a} \cup \mathbf{b} = \mathbf{0}' \implies \mathbf{c} \cup \mathbf{b} = \mathbf{0}']. \tag{34}$$

This question generated much effort but few results. Stob [15] used contiguous degrees and *wtt*-reducibility to show that there are incomparable r.e. degrees $\mathbf{a}$ and $\mathbf{b}$ such that $\mathbf{b}$ is the unique complement to $\mathbf{a}$ in the interval $[\mathbf{0}, \mathbf{a} \cup \mathbf{b}]$, and Ambos-Spies proved a dual result.

Attention then turned to the dual of (34). In 1987 two recursion theorists announced proofs of the negation of Theorem 2.2, and one presented his result at a meeting in October, 1987. After receiving a written version of the proof, M. Lerman discovered the error and pointed it out to us. We then realized that the author was making a fundamental mistake in trying to combine two methods from [14] and we began to try to refute his claim. In December, 1987 we proved Theorem 2.2, and during the fall of 1988, Ambos-Spies, Lachlan and Soare proved the dual, Theorem 2.3, which is the nonuniform version of (34). Both were presented by Soare at the Recursion Theory meeting in Oberwolfach, Germany in March, 1989. After hearing that talk, C. G. Jockusch and M. Stob made some helpful observations which led to the present formulation of Corollary 2.5 in place of an earlier version presented in the lecture. Very recently Sui [16] has suggested another method of doing the α-module for Theorem 2.2 in the style of the promptly simple degree theorem [14, p. 284].

During the fall of 1990 David Seetapun [10] used a very interesting $0'''$-priority argument to prove that every r.e. degree $\mathbf{0} < \mathbf{a} < \mathbf{0}'$ is *locally noncappable* namely

$$(35) \qquad (\forall \mathbf{a})_{\mathbf{0}<\mathbf{a}<\mathbf{0}'}(\exists \mathbf{c})_{\mathbf{a}<\mathbf{c}}(\forall \mathbf{b})_{\mathbf{b}<\mathbf{c}}[\mathbf{a}\cap\mathbf{b}=\mathbf{0} \implies \mathbf{b}=\mathbf{0}].$$

From this it follows that Theorem 2.2 holds uniformly in $\mathbf{a}$, namely $\mathbf{c}$ can be required to depend only on $\mathbf{a}$ and not on $\mathbf{b}$. (This can also be obtained by converting the proof in §4–§7 into a $\mathbf{0}'''$-priority argument and putting the α-modules for $\mathcal{S}_{i,j}$ at infinitely many levels of the tree T in stead of letting all be immediate successors of the node τ for $\mathcal{T}_i$. However, the technical difficulties in carrying this out are the same as those for proving (35) which Seetapun had to overcome.) From (35) Seetapun also concluded that there are no maximal nonbounding degrees. All these are further results in the direction of the BLUE player.

The general question (34) of the major subdegree remains open. (A negative solution of the general major subdegree problem and some partial positive cases had been announced by Cooper and Slaman, but these have all been subsequently withdrawn.) Seetapun [11] has given a positive answer for the case where $\mathbf{a}$ is low_2.

10. Modified Sacks Coding and One Point Extension of Embeddings

Notice that in the conclusion of requirement $\mathcal{P}_i$ in §3 we used $\hat{\Lambda}_i^{A\oplus\hat{C}} = K$ rather than $\hat{\Lambda}_i^{\hat{C}} = K$, which is usually used in the Sacks coding strategy. The difference is that now in our σ-strategy in §5.1 to meet $\mathcal{P}_i$ BLUE enumerates $\hat{\lambda}(x)$ in $\hat{C}$ only when x enters K, and not because of an A change. In the conventional strategy where BLUE is building only $\hat{\Lambda}^{\hat{C}} = K$, an $A \restriction y$ change for some $y \leq \xi^A(x)$ causes $\xi^A(x)$ to move and then BLUE must enumerate $\hat{\lambda}^{\hat{C}}(x)$ into $\hat{C}$ to ensure

$$\xi(x) < \hat{\lambda}(x). \tag{36}$$

This action causes much more enumeration into $\hat{C}$ than with our σ-module as presented in §5.1. In our case such an $A \restriction y$ changes automatically allows $\hat{\lambda}^{A\oplus\hat{C}}$ (x) to be redefined, namely if (36) holds before the A change then it holds after the A change without enumerating $\hat{\lambda}(x)$ into $\hat{C}$.

The significance of this is that combining this new σ-strategy with the rest of Sacks Density Theorem method (see [13, p. 142]) it is easy to prove:

Theorem 10.1. *If C, D, F, and G are r.e. sets such that $D <_T C$, $F \not\leq_T D$, and $C \not\leq_T G$ then there exists an r.e. set A such that $D <_T A <_T C$, $F \not\leq_T A$, and $A \not\leq_T G$.*

This improves a theorem of R. W. Robinson [13, Exer. VIII.4.7, p. 146] since his theorem required the added hypothesis "$G \leq_T C$ " because both G and C were required to compute uniformly in i the infinite recursive set contributed to A for the sake of the positive requirement $A \neq \{i\}^G$, whereas in our new strategy only C is required. Note also that Theorem 10.1 is clearly an additional continuity result for the BLUE player in the spirit of this paper.

Harrington and Shelah [4] proved that the elementary theory of $(\mathbf{R}, <)$ is undecidable, but it is unknown which *fragments* of this theory, such as the $\forall\exists$-sentences valid in $(\mathbf{R}, <)$, form a decidable class. In particular, attention has focused on the subclass of $\forall\exists$-sentences of the form,

$$\begin{aligned}(\forall x_1)\dots(\forall x_n)[D(x_1,\dots,x_n)\\ \implies (\exists y_1)\dots(\exists y_m)D_1(x_1,\dots,x_n,y_1,\dots,y_m)],\end{aligned} \tag{37}$$

where D and D_1 are open diagrams in $L(<)$ such that D_1 extends D. This has been also called the *extension of embeddings* question, because we wish to decide whether for all r.e. degrees $\mathbf{a}_1, \mathbf{a}_2, \dots, \mathbf{a}_n$ such that

$$(\mathbf{R}, <) \models D(\mathbf{a}_1, \dots, \mathbf{a}_n)$$

there exist r.e. degrees $\mathbf{b}_1, \mathbf{b}_2, \dots, \mathbf{b}_m$ such that

$$(\mathbf{R}, <) \models D_1(\mathbf{a}_1, \dots, \mathbf{a}_n, \mathbf{b}_1, \dots, \mathbf{b}_m).$$

Slaman and Soare have noted that Theorem 10.1 (suitably extended for finitely many sets C_i, D_i, F_i, G_i, $i \leq k$, and combined with standard results from [13] such as the minimal pair theorem, existence of branching degrees, and embedding posets in $(\mathbf{R}, <)$), gives a solution for the *one point* extension of embeddings, namely the case where $m = 1$.

11. Acknowledgments

The first author was supported by National Science Foundation Grant DMS 89-10312, and the second author by National Science Foundation Grant DMS 88-07389. The second author presented the results in this paper and those in the companion paper [1] by Ambos-Spies, Lachlan, and Soare, at the *Workshop on Set Theory and the Continuum*, October 16–20, 1989, while the authors were visiting the Mathematical Sciences Research Institute in Berkeley, California, during the Special Year in Mathematical Logic, from September 1, 1989 through August 24, 1990, where they were partially supported by National Science Foundation Grant DMS 85-05550.

References

1. K. Ambos-Spies, A. H. Lachlan, and R. I. Soare, *The continuity of cupping to* $\mathbf{0}'$, Ann. Pure Appl. Logic, to appear.
2. S. B. Cooper, *The jump is definable in the structure of the degrees of unsolvability*, Bull. Amer. Math. Soc. **25** (1990), 151–158.
3. P. A. Fejer, *The density of the nonbranching degrees*, Ann. Pure Appl. Logic *24* (1983), 113-130.
4. L. Harrington and S. Shelah, *The undecidability of the recursively enumerable degrees* (research announcement), Bull. Amer. Math. Soc. (N. S.) **6** (1982) 79–90.
5. L. Harrington and R. I. Soare, *Post's Program and incomplete recursively enumerable sets*, Proceedings National Academy of Sci. USA, to appear.
6. L. Harrington and R. I. Soare, *Definable classes and automorphisms of recursively enumerable sets*, to appear.
7. E. Herrmann, *The undecidability of the elementary theory of the lattice of recursively enumerable sets* (abstract), In: Frege Conference 1984, Proceedings of the International Conference at Schwerin, GDR, Akademie-Verlag, Berlin, GDR, 66–72.
8. A. H. Lachlan, *On some games which are relevant to the theory of recursively enumerable sets*, Ann. of Math. (2) **91** (1970), 291-310.
9. E. L. Post, *Recursively enumerable sets of positive integers and their decision problems*, Bull. Amer. Math. Soc. **50** (1944), 284–316.
10. D. Seetapun, *Every recursively enumerable degree is locally noncappable*, to appear.
11. D. Seetapun, *Every* low_2 *recursively enumerable degree is locally noncuppable*, to appear.
12. T. Slaman and W. H. Woodin, *Definability in degree structures*, to appear.
13. R. I. Soare, *Automorphisms of the recursively enumerable sets, Part I: Maximal sets*, Ann. of Math. (2), **100** (1974), 80–120.
14. R. I. Soare, *Recursively Enumerable Sets and Degrees: A Study of Computable Functions and Computably Generated Sets*, Springer-Verlag, Heidelberg, 1987.
15. M. Stob, *wtt-degrees and T-degrees of recursively enumerable sets*, J. Symbolic Logic, **48** (1983), 921–930.
16. Y. Sui, *On the problem of the critical bound*, Acta Mathematica Sinica, to appear.

Department of Mathematics, University of California, Berkeley CA 94720

Department of Mathematics, University of Chicago, Chicago IL 60637

ADMISSIBILITY AND MAHLONESS IN $L(\mathbb{R})$

Steve Jackson

1. Introduction

Our purpose here is to present some recent results about the structural theory of $L(\mathbb{R})$ assuming the axiom of determinacy. We will focus our attention "high-up" in the $L(\mathbb{R})$ hierarchy, in a sense to be made precise momentarily. In particular, we will be considering cardinals κ corresponding to highly closed pointclasses. We will therefore be far beyond the projective sets. The results presented here will appear shortly in [2] with complete proofs. We will therefore omit some proofs or merely sketch an outline, although we provide enough details in the case of our main result (Theorem 1) so that the reader may reconstruct the proof.

There are several reasons for considering these problems. To understand the first, we review briefly the problem of developing the so-called "very-fine" structure theory for $L(\mathbb{R})$. The well-known axiom of determinacy, introduced by Mycielsky and Steinhaus in the 60's, asserts that every two player integer valued game is determined. Beginning with the work of Martin and Moschovakis in the late 60's, and continuing with the work of Solovay, Kechris, Steel and others, a reasonable theory of the projective sets was developed assuming AD. This theory described the properties of the projective sets in terms of the so-called projective ordinals, $\underset{\sim}{\delta}^1_n$. (We refer the reader to [5] and [8] for their definitions and basic properties.)

However, the theory did not compute the values of these ordinals, nor establish all their properties. In 1985, the author, building on some ideas of Martin, was able to complete a program originally conceived of by Kunen for computing the $\underset{\sim}{\delta}^1_n$. The main result proved was that $\underset{\sim}{\delta}^1_{2n+1} = [\aleph_{\left.\omega^{\omega^{\cdot^{\cdot^{\omega}}}}\right\}2n+1}]^+$ (also, $\underset{\sim}{\delta}^1_{2n+2} = (\underset{\sim}{\delta}^1_{2n+1})^+$ was previously known). The theory developed for this analysis also described in detail the cardinal structure for cardinals $\kappa < \aleph_{\epsilon_0} = \sup_n \underset{\sim}{\delta}^1_n$. For example, it can be verified that all regular cardinals $< \aleph_{\epsilon_0}$ are measurable (we refer the reader to [3] for the main part of the argument for the computation of all the projective ordinals, and to the forthcoming [4] for the special case of $\underset{\sim}{\delta}^1_5$). We refer to this analysis as the "very-fine" structure theory for the projective sets.

Extending work of Martin-Steel, and building upon ideas of Moschovakis, John Steel developed a "fine-structure" theory for the model $L(\mathbb{R})$ with AD. It is so called as it is in analogy with Jensen's fine-structure theory of L (see [9]). This theory extends the earlier theory of the projective sets to the collection of all sets of reals in $L(\mathbb{R})$. As with the earlier theory of the projective sets, however, this theory is not sufficiently detailed to answer some questions. For example, the question of whether every regular cardinal $\kappa < \Theta$ (= the supremum of the lengths of prewellorderings of the continuum) is measurable remains open. One goal is to extend the "very-fine" structure theory for the projective sets throughout all of $L(\mathbb{R})$ to answer these questions. This has in fact been done for the lower levels of the $L(\mathbb{R})$ hierarchy, although this has not appeared yet. Exactly how far one can proceed using current methods is not clear, although the theory has been extended past the least inaccessible cardinal in $L(\mathbb{R})$. The analysis is inductive, and becomes progressively more detailed the higher one goes in $L(\mathbb{R})$. One idea, then, is to leap-frog this inductive analysis and consider directly cardinals κ "high-up" in $L(\mathbb{R})$. The hope is that ideas developed in this context might suggest ways to proceed in extending the full theory, and might also lead to a more unified, simpler presentation.

A second, related, reason is that one might be able to isolate some of the obstacles that arise in extending the theory through all of $L(\mathbb{R})$, and thus may see the limitations of the current methods.

2. Background and Preliminaries

We work throughout in the theory $AD + V = L(\mathbb{R})$. We will assume the reader is familiar with the basic aspects of determinacy theory. We will also need some facts from the theory of $L(\mathbb{R})$ as in [9], although we will summarize these for the reader below.

The reader not familiar with these facts may either take them on faith, or interpret our statement into a more specific context (such as $\kappa =$ the ordinal of the inductive sets) for which they are more apparent.

For consistency and technical convenience, we use now the Jensen hierarchy $J_\alpha(\mathbb{R})$ for the model $L(\mathbb{R})$. Throughout, we will be considering cardinals which we refer to as "admissible Suslin cardinals". By this, we mean a Suslin cardinal κ such that $J_\kappa(\mathbb{R})$ is Σ_1-admissible. This is equivalent to saying that the pointclass $\undertilde{\Sigma}_1 - J_\kappa(\mathbb{R})$ (i.e. the sets definable in $J_\kappa(\mathbb{R})$ from Σ_1 formulas with real parameters) is closed under real quantification and that there is a $\undertilde{\Sigma}_1 - J_\kappa(\mathbb{R}$ definable partial map from $\mathbb{R}$ onto $J_\kappa(\mathbb{R})$ (c.f. [9] lemma 2.5). Such a cardinal must be much larger than the projective ordinals, for example, but exactly how much larger? This question will guide our discussion here.

Our main theorem (stated in the next section) extends and generalizes a result of Kechris-Woodin [6] that Θ is Mahlo, and also a result of Moschovakis that the pointclass of sets semi-recursive in ${}^3E$ lies strictly within the inductive sets. Our ideas also borrow from some ideas of Harrington [1] where the first recursively Mahlo ordinal is studied. Our main result will be that an admissible Suslin cardinal κ must be very large in the Mahlo hierarchy. Our methods will also allow us to pinpoint a potential obstacle to extending the very-fine structure theory further.

We collect now some facts about admissible Suslin cardinals we will require.

(F1) The set of Suslin cardinals $\kappa' < \kappa$ is c.u.b. in κ. In fact, the Suslin cardinals $\kappa' < \kappa$ for which for which $S(\kappa') = \underset{\sim}{\Sigma}_1 - J_\alpha(\mathbb{R})$ for some $\alpha < \kappa$ is c.u.b. in κ (c.f. [9] theorem 4.3 and corollary 4.4). Here $S(\kappa)$ denotes the pointclass of κ-Suslin sets.

(F2) $\underset{\sim}{\Sigma}_1 - J_\kappa(\mathbb{R})$ has the prewellordering (in fact, scale) property. In fact, there is a $\underset{\sim}{\Sigma}_1$ formula φ with real parameters such that $\varphi^{J_\kappa(\mathbb{R})}$ defines a prewellordering of length κ. and such that for c.u.b. many $\kappa' < \kappa$, $\varphi^{J_{\kappa'}(\mathbb{R})}$ defines the restriction of $\varphi^{J_\kappa(\mathbb{R})}$ to those reals of rank $< \kappa'$.

(F3) κ is (weakly) inaccessible and has the strong partition property $\kappa \to (\kappa)^\kappa$ (we refer the reader to [7] for a proof).

We first give a precise definition, due to Kleinberg, for the generalized Mahlo order of κ, which we denote by $o(\kappa)$, valid whenever $\kappa \to (\kappa)^\kappa$. If $S \subseteq \kappa$ is stationary and consists of limit ordinals of uncountable cofinality, we say S is thin if $\forall \alpha \in S (S \cap \alpha$ is not stationary in $\alpha)$. We say S is thick if $\forall \alpha \in S (S \cap \alpha$ stationary in $\alpha \to \alpha \in S)$. It is easy to see that if S is stationary and S' is the set of thin points of S, i.e., $S' = \{\alpha \in S : S \cap \alpha$ is not stationary in $\alpha\}$, then S' is thin and still stationary (given any c.u.b. $C \subseteq \kappa$, the least limit point of C in S is in S').

Suppose now that $S_1, S_2 \subseteq \kappa$ are stationary, thin, and consist of ordinals of uncountable cofinality. We say $S_1 < S_2$ iff $\exists C \subseteq \kappa$ C is c.u.b. and $\forall \alpha \in C \cap S_2 (S_1 \cap \alpha$ is stationary in $\alpha)$. We say S_1, S_2 are equivalent if there is a c.u.b. set on which they agree (alternatively, one could work throughout considering the ordering $\leq$ on thick stationary sets defined by $S_1 \leq S_2 \longleftrightarrow S_1 \subseteq S_2$ on a c.u.b. subset of κ. A variant of the following claim shows that the strict part *of* $\leq$ *is* a wellordering).

The following claim is due to John Steel:

Claim. *$<$ is a wellordering on the equivalence classes of thin stationary sets.*

Proof (sketch). Given $S_1, S_2 \subseteq \kappa$, we partition increasing functions $f : \kappa \to \kappa$ according to whether $\alpha_{S_1}(f) < \alpha_{S_2}(f)$, $\alpha_{S_1}(f) = \alpha_{S_2}(f)$, or $\alpha_{S_1}(f) > \alpha_{S_2}(f)$, where $\alpha_S(f) =$ the least limit point of the range of f in S. By the strong partition property, let $A \subseteq \kappa$ be a homogeneous set of size κ. Let $C \subseteq \kappa$ be the set of limit points of κ. If the first case of the partition holds, it is not difficult to check that for $\alpha \in C \cap S_2$, $S_1 \cap \alpha$ is stationary in α. Similarly, in the third case we get $S_2 < S_1$, and in the second case a c.u.b. set on which S_1, S_2 agree. Also, this argument shows that $<$ *is* wellfounded as otherwise we get an infinite descending set of ordinals $\alpha_{S_1}(f) > \alpha_{S_2}(f) >$. □

To make the definition of $o(\kappa)$ coincide with the usual definition of Mahloness in the small cases, we consider the relation $<$ *restricted* to the inaccessible cardinals:

Definition. $o(\kappa) =$ the rank of $<$ restricted to thin stationary $S \subseteq$ (inaccessible Suslin cardinals less than κ).

(Note: restricting S to the inaccessibles has the effect of deleting the first κ many equivalence classes of the original relation, namely those corresponding to the various cofinalities below κ).

Following again Kleinberg, we define for thin, stationary $S \subseteq \kappa$ a normal measure ν_S which we call the corresponding atomic normal measure. Namely, $A \subseteq \kappa$ has measure one with respect to ν_S iff $\exists C \subseteq \kappa(C$ is c.u.b. and $\forall \alpha \in C \cap S(\alpha \in A))$. It follows from the strong partition relation on κ that ν_S is a normal measure on κ.

We will explore the size of $o(\kappa)$ in the next sections.

3. The Main Result

We define in this section the notion of a *local* well-founded relation on $\mathbb{R}$. Our main theorem will be that for κ an admissible Suslin cardinal, $o(\kappa) \geq \delta =$ the supremum of the lengths of the local well-founded relations at κ. In the following section, we investigate the nature of δ. In particular, we show that $\mathrm{cof}(\delta) \geq \kappa^+$, δ is "closed under ultrapowers" in a sense to be made precise, and we give a result which rules out many regular cardinals $\geq \kappa^+$ as candidates for $\mathrm{cof}(\delta)$. We believe that in the presence of the complete very fine- structure theory below κ, this last result should generalize to "δ is regular".

First, however, we would like to mention some results which help to place our results in perspective, but will not be needed for the proofs. Corresponding to κ we have an inductive-like pointclass $\utilde{\Gamma} = \utilde{\Sigma}_1 - J_\kappa(\mathbb{R})$. We may also define a projective-like hierarchy above $\utilde{\Gamma}$ by: $\utilde{\Sigma}_1^* = \exists^{\mathbb{R}}(\utilde{\Gamma} \wedge \breve{\utilde{\Gamma}})$, $\utilde{\Pi}_2^* = \forall^{\mathbb{R}}\Sigma_1^*$, etc. Also, we let $\utilde{\delta}_n^* (= \utilde{\delta}_n^*(\kappa))$ = the supremum of the lengths of the $\utilde{\Delta}_n^*$ well-founded relations on $\mathbb{R}$ (where $\utilde{\Delta}_n^* = \utilde{\Sigma}_n^* \cap \utilde{\Pi}_n^*$ as usual). It is not difficult to see that $\kappa^{++} \leq \utilde{\delta}_1^* \leq j_\omega(\kappa)$ for all such κ, where $j_\omega(\kappa)$ denotes the ultrapower of κ by the ω-cofinal normal measure on κ. Also, a straightforward computation shows that any proper initial segment of the prewellordering of stationary subsets of κ is $\utilde{\Sigma}_2^*$. Hence, $o(\kappa) \leq \utilde{\delta}_2^*$. By the theorems presented here, it follows that $o(\kappa) > \utilde{\delta}_1^*$. Hence, $\utilde{\delta}_1^* < o(\kappa) \leq \utilde{\delta}_2^*$ for all admissible Suslin cardinals. Finally, it is a theorem (unpublished) of Woodin that for such κ, $o(\kappa) = \utilde{\delta}_2^*$ iff κ is $\utilde{\Pi}_2^*$-reflecting (i.e. whenever φ is of the form $\forall^{\mathbb{R}}\exists^{\mathbb{R}}(\psi_1 \wedge \neg\psi_2)$ where ψ_1, ψ_2 are Σ_1 formulas with real parameters, then $J_\kappa(\mathbb{R}) \models \varphi \Rightarrow \exists\alpha < \kappa(J_\alpha(\mathbb{R}) \models \varphi)$. $\utilde{\Pi}_2^*$-reflecting is stronger than admissibility as it implies that the admissibles below κ are stationary in κ. Thus, for κ = the first admissible Suslin cardinal = the closure ordinal of the inductive sets, we have $\utilde{\delta}_1^* < o(\kappa) < \utilde{\delta}_2^*$.

We assume for the remainder of this section that κ denotes a fixed admissible Suslin cardinal.

We define now the notion of a local wellfounded relation. We say the transitive wellfounded relation $\prec$ *on* reals is *local* if it satisfies the following:

(1) For each inaccessible Suslin cardinal $\alpha < \kappa$ there is a uniquely defined well-founded relation $\prec_\alpha$ on reals. Also, there is a function $F : \kappa \to \kappa$ and a formula φ (with real parameters) from the language of set theory such that for all inaccessible Suslin cardinals $\alpha < \kappa(x \prec_\alpha y) \Longleftrightarrow \varphi^{J_{F(\alpha)}(\mathbb{R})}(x, y, \alpha)$, for all x, y.

(2) There is a c.u.b. $C \subseteq \kappa$ such that if $\alpha \in C \cup \{\kappa\}$ and α is an inaccessible Suslin cardinal, then for $x, y \in$ field $(\prec_\alpha)$ if $x \prec_\alpha y$ holds then there is a c.u.b. $\bar{C} \subseteq \alpha$ such that for all inaccessible Suslin cardinals $\beta \in \bar{C}$, if $y \in$ field $(\prec_\beta)$ then x is also, and $x \prec_\beta y$ holds.

We continue with the definition in a moment. We define first for $\alpha \leq \kappa$ an inaccessible Suslin cardinal what it means for α to be x-Mahlo, for $x \in$ field $(\prec_\alpha)$. Namely, we say that α is x-Mahlo if for all $z \in$ field $(\prec_\alpha)$ with $z \prec_\alpha x$, the set $S_z^\alpha = \{\beta < \alpha : \beta$ is an inaccessible Suslin cardinal, $z \in$ field $(\prec_\beta)$, and β is z-Mahlo$\}$ is stationary in α. If x *is* $\prec_\alpha$-minimal we say α is x-Mahlo if α is Mahlo is the ordinary sense. Also, we say α is $|\prec_\alpha|$-Mahlo if for all $x \in$ field $(\prec_\alpha)$, α is x-Mahlo. It follows readily that for $\alpha \leq \kappa$ in C

as above for property 2 that if α is x-Mahlo for some $x \in$ field $(\prec_\alpha)$, then $o(\alpha) \geq |x|_{\prec_\alpha} - 1$ (where $|x|_{\prec_\alpha}$ denotes the rank of x *in* $\prec_\alpha$, and $\lambda - 1 = \lambda$ for limit λ). This follows since for all $z \prec_\alpha x$, S_z^α is stationary in α, and if $z_1 \prec_\alpha z_2 \prec_\alpha x$, then $(S_{z_1}^\alpha)' < (S_{z_2}^\alpha)'$ in the ordering on stationary sets, where $(S_{z_1}^\alpha)'$, $(S_{z_2}^\alpha)'$ denote the thin points of these sets. To see this, let $C \subseteq \alpha$ be as in (2) for z_1 and z_2. Then for $\beta \in C \cap S_{z_2}^\alpha$, $z_1 \in$ field $(\prec_\alpha)$ and $z_1 \prec_\alpha z_2$ from (2). Hence $S_{z_1}^\alpha$ is stationary in α, and therefore so is $(S_{z_1}^\alpha)'$. Also from property 2 it follows that if $\alpha \in C \cup \{\kappa\}$ for this C and α is not x-Mahlo for some $x \in$ field $(\prec_\alpha)$, then there is a c.u.b. $\bar{C} \subseteq \alpha$ such that for $\beta \in \bar{C}$, β is not x-Mahlo.

We now continue with the definition of local:

(3) For each $x \in$ field $(\prec)$ (where $\prec$ abbreviates $\prec_\kappa$), there is a function $f_x : \kappa \to \kappa$, a c.u.b. $C_x \subseteq \kappa$, and formulas $\psi_x^1, \ldots, \psi_x^n$ (here n may depend on x) in the language of set theory with real parameters, each of the form $\psi_x^i(w_1, w_2, w_3) \longleftrightarrow \forall z \in w_3 \bar{\psi}_x^i(w_1, w_2, z)$ for some $\bar{\psi}_x^i$, such that for all inaccessible Suslin cardinals $\alpha \in C_x \cup \{\kappa\}$, α closed under f_x, if $x \in$ field $(\prec_\alpha)$ then for all $y \in \mathbb{R}$, ($y \in$ field $(\prec_\alpha)$ and $y \prec_\alpha x) \Longleftrightarrow \exists$ a c.u.b. $D \subseteq \alpha[\forall \beta \in D\ \psi_x^{1^{J_{f_x(\beta)}(\mathbb{R})}}(y, \beta, D \cap \beta) \vee \ldots \vee \forall \beta \in D\ \psi_x^{n^{J_{f_x(\beta)}(\mathbb{R})}}(y, \beta, D \cap \beta)]$.

(4) For each $x \in$ field $(\prec)$ there is a c.u.b. $D_x \subseteq \kappa$ such that for all inaccessible Suslin cardinals $\alpha \in D_x$, if $x \notin$ field $(\prec_\alpha)$, then there is a c.u.b. $D \subseteq \alpha$ such that for all inaccessible Suslin cardinals $\beta \in D$, $x \notin$ field $(\prec_\beta)$.

(5) For each $x \in$ field $(\prec)$, there is a c.u.b. $E_x \subseteq \kappa$ such that for all inaccessible Suslin cardinals $\alpha \in E_x$, if $x \notin$ field $(\prec_\alpha)$, then there is a $y \in$ field $(\prec_\alpha)$ such that α is not y-Mahlo.

(6) For each $x \in$ field $(\prec)$, if κ is x-Mahlo, then the set $\bar{S}_x = \{\beta < \alpha : \beta$ is an inaccessible Suslin cardinal and $x \in$ field $(\prec_\alpha)\}$ is stationary in κ.

We will reference clauses in the definition of local by their numbering here, *e.g.* " ... property (3) ... ".

We now state the main theorem of this section:

Theorem 1. $(AD + V = L(\mathbb{R}))$ *Let κ be an admissible Suslin cardinal, and let $\prec$ be a local wellfounded relation at κ. Then κ is $|\prec|$-Mahlo.*

We present an outline of the proof of this theorem. It is convenient to start with the fact (Kechris-Woodin [6], see our introductory remarks) that κ is Mahlo. Alternatively, one may reword our proof here slightly (essentially by eliminating the main "case 4" in our arguments) to reprove this result.

We suppose the theorem fails, and fix a real $\bar{\bar{x}} \in$ field $(\prec)$ such that κ is not $\bar{\bar{x}}$- Mahlo, and we assume $\bar{\bar{x}}$ is chosen with $|\bar{\bar{x}}|_{\prec}$ minimal. It follows that there is an $\bar{x} \prec \bar{\bar{x}}$ in the field of $\prec$ such that κ is $\bar{x}$-Mahlo and $N = \{\alpha < \kappa : \alpha$ is an inaccessible Suslin cardinal, $\bar{x} \in$ field $(\prec_\alpha)$, and α is $\bar{x}$-Mahlo$\}$ is not stationary.

We fix $f_{\bar{x}}$, $C_{\bar{x}}$ as in (3) in the definition of local, and fix a c.u.b. $C \subseteq$ the Suslin cardinals below κ satisfying:

(a) For all inaccessible Suslin cardinals $\alpha \in C$, either $\bar{x} \notin$ field $(\prec_\alpha)$, or $\bar{x} \in$ field $(\prec_\alpha)$ but α is not $\bar{x}$-Mahlo.
(b) For all inaccessible Suslin cardinals $\alpha \in C$, if $\bar{x} \notin$ field $(\prec_\alpha)$, then there is an x' in field $(\prec_\alpha)$ such that α is not x'-Mahlo. We are using (5) here.
(c) For all inaccessible Suslin cardinals $\alpha \in C$, if $\bar{x} \notin$ field $(\prec_\alpha)$, then there is a c.u.b. $D \subseteq \alpha$ such that for β an inaccessible Suslin cardinal in D, $\bar{x} \notin$ field $(\prec_\beta)$. We are using (4) here.
(d) C is closed under $f_{\bar{x}} : \kappa \to \kappa$ and consists of limit points of $C_{\bar{x}}$.

We define now for each $\alpha \in C$ two sets of reals A_α, B_α. We think of A_α as a set of codes for α, and B_α as a "sufficiently complete set at the αth level". We define B_α to be the canonical universal $\underset{\sim}{\Sigma}_1 - J_\alpha(\mathbb{R})$ set, i.e., $B_\alpha = \{(x, y) : x = \langle n, x' \rangle$ where n codes a Σ_1-formula φ, and $J_\alpha(\mathbb{R}) \models \varphi(x', y)\}$, where $\langle\rangle$ denotes a standard coding of $\omega \times \mathbb{R}$ into $\mathbb{R}$. We define A_α inductively through the following cases:

(0) $\alpha =$ the least element of C. We set $A_\alpha = \{\bar{0}\}$, where $\bar{0} =$ the constant real 0.
(1) $\alpha =$ the $(\beta+1)^{th}$ element of C for some β. We set $A_\alpha = \{\langle 1, x\rangle : x \in A_\beta\}$.
(2) α a limit and $\exists \beta < \alpha$ and a set A wadge reducible to B_β such that $A \subseteq_{\alpha'} \bigcup_{<\alpha} A'_\alpha$, and A is "unbounded" in α in the obvious sense (i.e. $\forall \alpha_1 < \alpha \exists \alpha_2 < \alpha (\alpha_2 > \alpha_1$ and $\exists z \in A \cap A_{\alpha_2})$. In this case, we set $A_\alpha = \{\langle 2, x, y\rangle : x \in A_\beta$ for some $\beta < \alpha$ and if $A = \{z : y(z) \in B_\beta$, then $A \subseteq_{\alpha'} \bigcup_{<\alpha} A'_\alpha$ and A is "unbounded" in $\alpha\}$. Here, $y(z)$ is the result of applying the continuous function coded by y to z.
(3) Case 2 fails, and α is inaccessible but not Mahlo. We set $A_\alpha = \{\langle 3, \sigma\rangle : \forall x \in_{\alpha'} \bigcup_{<\alpha} A'_\alpha \;\; \sigma(x) \in_{\alpha'} \bigcup_{<\alpha} A'_\alpha$, and $D = \{\beta < \alpha : \forall x \in_{\beta'} \bigcup_{<\beta} A'_\beta(\sigma(x) \in_{\beta'} \bigcup_{<\beta} A'_\beta)\}$ is c.u.b. in $\alpha \cap C$, and for all limit points β of D, $\beta < \alpha$, β is not inaccessible $\}$.

(4) $\alpha \in C$, α Mahlo but not $|\prec_\alpha|$-Mahlo. We set $A_\alpha = \{\langle 4, x, \sigma\rangle : x \in$ field $(\prec_\alpha)$, α is x-Mahlo, $S_x^\alpha = \{\beta < \alpha : \beta$ is an inaccessible Suslin cardinal, $x \in$ field $(\prec_\beta)$, and β is x-Mahlo$\}$ is not stationary in α, and σ defines a c.u.b. $D \subseteq \alpha \cap C$ (as in case 3) such that for all inaccessible Suslin cardinals $\beta \in D$, either $x \notin$ field $(\prec_\beta)$ or β is not x-Mahlo and one of the following holds:
   (i) for $\beta \in D$, $\bar{x} \notin$ field $(\prec_\beta)$.
   (ii) for some $m \le n(= n(\bar{x}))$ and all $\beta \in D$, $\psi_{\frac{m}{x}}^{J_{f_{\bar{x}}(\beta)}(\mathbb{R})}(x, \beta, D \cap \beta)$ holds$\}$.

This completes the definition of the A_α.

One may easily check that $A_\alpha \cap A_\beta = \emptyset$ for $\alpha < \beta$ in C.

Next, one may verify that each $\alpha \in C$ gets a code, (i.e. each $A_\alpha \neq \emptyset$ for $\alpha \in C$). By case 1, we may assume a is a limit point of C. By case 2 and the coding lemma we may assume that α is regular. By case 3, we may assume α is Mahlo. Since $\alpha \in C$, either $\bar{x} \notin$ field $(\prec_\alpha)$ or $\bar{x} \in$ field $(\prec_\alpha)$ and α is not x-Mahlo. In either case, it can be shown from the definition of C that α gets a code from case 4.

Also, an inspection of the definition of the A_α shows that each A_α, $B_\alpha \in \underset{\sim}{\Delta} = \underset{\sim}{\Gamma} \cap \check{\underset{\sim}{\Gamma}}$ where $\underset{\sim}{\Gamma} = \underset{\sim}{\Sigma}_1 - J_\kappa(\mathbb{R})$. It follows (say from the coding lemma) that $A =_\alpha \bigcup_{<\kappa} A_\alpha \in \underset{\sim}{\Gamma}$. By admissibility, we must have $A \in \underset{\sim}{\Gamma} - \check{\underset{\sim}{\Gamma}}$ as the induction defining the A_α has size κ.

Finally, we obtain a contradiction by showing that $\neg A \in \underset{\sim}{\Gamma}$. We write out a formula which computes this. Intuitively, the formula asserts that x is not a code if there is a real z which codes countably many reals $z_0 = x$, $z_1, \dots, z_n, \dots$, and z_{n+1} by not being a code witnesses that z_n is not a code.

We claim that $x \in \neg A \longleftrightarrow \Omega(x)$, where $\Omega(x)$ is the statement: $\exists z[z$ codes countably many reals $z_0 = x, z_1, \dots, z_n, \dots$, such that for all n, (z_n, z_{n+1}) satisfy one of the following:

(0) z_n is not of the correct syntactical form to be a code.
(1) $z_n = \langle 1, x\rangle$ and $z_{n+1} = x$.
(2) $z_n = \langle 2, x, y\rangle$ for some x, y and either $z_{n+1} = x$ or ($x \in A$ and $y(z_{n+1}) \in B_{|x|}$).
(3) $z_n = \langle 3, \sigma\rangle$ for some σ and $\exists z[z \in A$, $\sigma(z) = z_{n+1}$, and "z_n is not the code of any ordinal $\le |z|$"].
(4) $z_n = \langle 4, x, \sigma\rangle$ for some x, σ and one of the following holds:
   (i) $\exists z \in A[\sigma(z) = z_{n+1}$ and "z_n does not code any ordinal $\le |z|$"].
   (ii) $\exists z \in A[$"$|z| \in C_\sigma$, the c.u.b. set coded by σ" and $\exists \alpha_0 \le |z|(\alpha_0$ is an inaccessible Suslin cardinal, $\bar{x} \in$ field $(\prec_{\alpha_0})$, and $\alpha_0 \in C_\sigma)$

& $\exists\alpha_1 \le |z|(\psi\frac{1}{x}^{J_{f_{\bar{x}}(\alpha_n)}(\mathbb{R}}(x,\alpha_1,C_\sigma\cap\alpha_1)$ fails, α_1 is an inaccessible Suslin cardinal, and $\alpha_1 \in C_\sigma)$

& $\exists\alpha_n \le |z|(\psi\frac{n}{x}^{J_{f_{\bar{x}}(\alpha_n)}(\mathbb{R}}(x,\alpha_n,C_\sigma\cap\alpha_n)$ fails, α_n is an inaccessible Suslin cardinal, and $\alpha_n \in C_\sigma)$

& $\forall\beta \le |z|$("z_n does not code any ordinal $\le \beta$")]&$z_{n+1} = z_n$.

It is easy to check, using the fact that $\underset{\sim}{\Gamma}$ is closed under real quantifiers that Ω defines a $\underset{\sim}{\Gamma}$ set.

We now claim that $x \in \neg A \longleftrightarrow \Omega(x)$.

First assume that $x \in \neg A$. One shows then that for any z_n, if $z_n \in \neg A$ then we may find a $z_{n+1} \in \neg A$ such that (z_n, z_{n+1}) satisfy one of the cases in Ω. By case 0, we may assume that z_n is of the correct syntactical form to be a code. Cases 1,2,3 are relatively easy. For case 4, we may find such a z_{n+1} unless $\sigma = \sigma(z_n)$ codes a c.u.b. set C_σ which we assume to be the case. We may further assume subcase ii does not hold as otherwise we may take $z_{n+1} = z_n$.*It* follows that for some $m \le n(= n(x)$, where now $z_n = \langle 4, x, \sigma\rangle)$ and all inaccessible Suslin cardinals $\beta \in C_\sigma$ that $\psi\frac{m}{x}^{J_{f_{\bar{x}}(\beta)}(\mathbb{R})}(x,\beta,C_\sigma\cap\beta)$ holds. From (3) in the definition of local it now follows that $x \in$ field $(\prec)$ and $x \prec \bar{x}$. However, from the definition of Mahloness it follows that we may find α a limit point of C_σ such that $x \in$ field $(\prec_\alpha)$ and α is x-Mahlo.

Taking the least such α it follows that z_n codes α, a contradiction.

For the other direction, assume $\Omega(x)$ holds and show $x \in \neg A$. We assume not, so $x \in A$. We let $z_0 = x$, $z_1, \dots, z_n, \dots$, witness Ω. One then shows by induction on n that each $z_n \in A$ and $|z_{n+1}| < |z_n|$, a contradiction. For fixed z_n, we consider cases on the syntactical form of z_n. The argument in all cases is straightforward.

Hence $x \in \neg A \longleftrightarrow \Omega(x)$, so $\neg A \in \underset{\sim}{\Gamma}$, a contradiction.

4. Results about δ

Definition. For κ an admissible Suslin cardinal, we define $\delta(= \delta(\kappa))$ to the supremum of the lengths of the local well-founded relations at κ.

Thus, the results of the previous section give that for κ an inaccessible Suslin cardinal, $o(\kappa) \ge \delta$. We state now some results concerning the size of δ. We will present only a rough outline of the proofs, referring the reader to [2] for details.

Theorem 2. *δ is a limit ordinal.*

Proof. One checks that *if* $\prec$ *is* a local well-founded relation of length $\gamma+1$, then we may find a local relation $\prec'$ *of* length $\gamma + 2$. We may assume $\bar{0} \notin$

field $(\prec)$ nor in the field of any of *the* $\prec_\alpha$, and we let $x_m \in$ field $(\prec)$ be such that $|x_m| = \gamma$. We define $\prec'$ *by* adjoining *to* $\prec$ the relations $y \prec' \bar{0}$ for all $y \prec x_m$, or $y = x_m$. We define $\prec'_\alpha$ for $\alpha < \kappa$ by similarly adjoining these relations whenever $x_m \in$ field $(\prec_\alpha)$. It is not difficult to check that $\prec'$ is local. □

We note that it is not clear in general whether or not δ is attained as the length of a single local well-founded relation.

Theorem 3. *δ is "closed under ultrapowers". That is, if $\gamma < \delta$ and S is a representative for the γth equivalence class for stationary subsets of κ, and if ν_γ denotes the corresponding atomic normal measure on κ, then $j_{\nu_\gamma}(\kappa) < \delta$, where j_{ν_γ} refers to the embedding corresponding to ultrapower by the measure ν_γ.*

Proof. Fix $\gamma < \delta$, and fix a local well-founded relation $\prec$ *at* κ of length $\gamma+2$, and reals $x_m, x'_m \in$ field $(\prec)$ with $|x_m|_\prec = \gamma$, $|x'_m|_\prec = \gamma+1$, and $x_m \prec x'_m$. From theorem 1 it follows that $S_m = \{\alpha < \kappa : \alpha$ is an inaccessible Suslin cardinal, $x_m \in$ field $(\prec_\alpha)$, and α is x_m-Mahlo$\}$ is stationary. Also, from theorem 1 and (2) in the definition of local, it follows that $S'_m = \{\alpha < \kappa : \alpha$ is an inaccessible Suslin cardinal, $S_m \cap \alpha$ is stationary in α, and α is x'_m-Mahlo$\}$ is stationary. Further, the rank of S_m in the ordering on stationary sets is at least γ. It suffices,therefore, to show that $j_{S_m}(\kappa) < \delta$, where j_{S_m} refers to the embedding from the atomic normal measure corresponding to S_m (this follows since an easy argument shows that if $S_1 < S_2$ in the ordering on stationary sets then $j_{S_1}(\kappa) < j_{S_2}(\kappa)$).

We define another local well-founded relation $\prec'$ of length $j_{S_m}(\kappa)$ as follows:

(1) The field *of* $\prec'$ consists of reals x such that $x \prec x_m$ or $x = x_m$ together with pairs (x_m, y) where y codes via the uniform coding lemma a function $f_y : \kappa \to \kappa$.

(2) For the reals x such that $x \prec x'_m$ or $x = x'_m$, the ordering $\prec'$ agrees with $\prec$. Also, all such reals x are *set* $\prec'$ to all pairs (x_m, y). Further, we set $(x_m, y) \prec' (x_m, y')$ iff $[f_y]_{S_m} < [f'_y]_{S_m}$.

(3) For $\alpha < \kappa$, we define $\prec'_\alpha = \prec_\alpha$ if $x'_m \notin$ field $(\prec_\alpha)$ or $S_m \cap \alpha$ is not stationary in α.

If $x'_m \in$ field $(\prec_\alpha)$ and $S_m \cap \alpha$ is stationary in α, we define $\prec'_\alpha$ similarly *to* $\prec$ (using the same x'_m).

One may then check that $\prec'$ is local, which finishes the proof. □

It follows from theorems 2 and 3 that δ must be quite large.For example, $\delta > \kappa^+, \kappa^{++}$, etc. We now state without proof two further theorems which have the flavor of saying "δ is regular". In fact, as we mentioned earlier, we believe that in the presence of the complete very-fine structure theory for $L(\mathbb{R})$ below κ that the proof of theorem 5 should generalize to show this.

Theorem 4. $\operatorname{cof}(\delta) > \kappa$.

Theorem 5. *Let $\prec$ be a local well-founded relation at κ with $z_m \in$ field $(\prec)$ and so (by theorem 1) $S = \{\alpha < \kappa : \alpha$ is an inaccessible Suslin cardinal and α is z_m-mahlo$\}$ is stationary in κ. Let V denote the corresponding atomic normal measure. Then* $\operatorname{cof}(\delta) \neq j_V(\kappa)$.

In fact, using the argument in the proof of theorem 5, we can show that $\operatorname{cof}(\delta) \neq \kappa^+, \kappa^{++}, \ldots, \kappa^{+n}, \ldots$ Theorem 5, then, is just ruling out as possibilities for $\operatorname{cof}(\delta)$ certain regular cardinals which are easily presented (one can show that for κ having the strong partition relation, the ultrapower of κ by any semi-normal measure is regular, where a measure is semi-normal if it gives every c.u.b. set measure one- a definition and result of Kleinberg).

5. Conclusion

For κ an admissible Suslin cardinal, we have shown that $o(\kappa) \geq$ *some* δ for which $\operatorname{cof}(\delta) > \kappa$ and δ is "closed under ultrapowers". We have also stated a result which suggest that δ should be regular. We state explicitly:

Conjecture. For κ a Suslin cardinal, $o(\kappa)$ being regular and closed under ultrapowers implies that κ is admissible (i.e. $\underset{\sim}{\Sigma}_1 - J_\kappa(\mathbb{R})$ is closed under real quantification).

Finally, we remark that using methods similar to the proof of theorem 5 we can show that δ carries a κ^+- additive measure, which in turn induces a non-atomic normal measure V on κ with $j_V(k) > \delta$. This seems to parallel some results of Woodin "high up" at $\kappa = \underset{\sim}{\delta}^2_1$ on the existence of normal measures with strength [10]. This may be important for extending the $L(\mathbb{R})$ theory. We have not been able to a corresponding version of theorem 5 for this measure on δ (i.e. rule out the various $j_v(\kappa)$ as candidates for the least cardinal where additivity of the measure fails). Thus, it is not clear whether or not δ is (or should be) measurable.

References

1. Harrington, L.A., *The Superjump and the First Recursively Mahlo Ordinal*, Generalized Recursion Theory, Studies in Logic and the Foundations of Mathematics, vol. 79, North-Holland, Amsterdam, 1974, pp. 43-52.
2. Jackson, S., *Admissible Suslin Cardinals in* $L(\mathbb{R})$, Journal of Symbolic Logic (to appear).
3. Jackson, S., *AD and the Projective Ordinals, Cabal Seminar 81-85*, Lecture Notes in Mathematics **1333** (1988), Springer-Verlag, 117-220.
4. Jackson, S., *A Computation of* $\underset{\sim}{\delta}^1_5$, in preparation.
5. Kechris, A.S., *AD and Projective Ordinals, Cabal Seminar 76-77*, Lecture Notes in Mathematics **689** (1978), Springer-Verlag, 91-132.
6. Kechris, A.S., *Determinacy and the Structure of* $L(\mathbb{R})$, Proceedings of Symposia in Pure Mathematics **42** (1985), 271-283.
7. Kechris, A.S., Kleinberg. E.M., Moschovakis, Y.N., and Woodin, W.H., *The Axiom of Determinacy, Strong Partition Properties, and Non-Singular Measures, Cabal Seminar 77-79*, Lecture Notes in Mathematics **839** (1981), Springer-Verlag, 75-100.
8. Moschovakis, Y.N., *Descriptive Set Theory*, North-Holland, Amsterdam, 1980.
9. Steel, J.R., *Scales in* $L(\mathbb{R})$, *Cabal Seminar 79-81*, Lecture Notes in Mathematics **1019** (1983), Springer-Verlag, 107-156.
10. Woodin, W.H., *Large Cardinals and Determinacy*, in preparation.

Department of Mathematics, University of North Texas, Denton TX 76203

SET THEORY OF REALS: MEASURE AND CATEGORY

HAIM JUDAH

1. INTRODUCTION

The study of the Lebesgue measurability and of the Baire property of sets of reals is a natural and old domain of mathematical research.

Although its main paradigm is searching for the existence of pathological sets of reals, this research has produced a well-supported mathematical theory about the non-pathological sets.

At the end of the last century, it was known that the number of the perfect subsets of the real line is equal to the number of the reals. Using this, Bernstein constructed a set $A \subseteq \mathbf{R}$ such that for every perfect set B, both $B \cap A$ and $B \cap \mathbf{R} \backslash A$ are not empty. Sets with this property are nowadays called Bernstein sets. They are not measurable, and they do not have the property of Baire.

On the non-pathological side, Sierpiński and Łuzin showed that the analytic sets are measurable and have the property of Baire.

In 1912, E. Borel introduced the concept of strong measure zero sets and conjectured that the strong measure zero sets are exactly the countable sets of reals. At the same time, using the Continuum Hypothesis, Łuzin built an uncountable set that has countable intersection with every meager set. Sierpiński proved that this Łuzin set also has strong measure zero. More sophisticated strong measure zero sets were studied by Rothberger during the 40's and 50's.

With the work of P. Cohen, a new era began: the study of properties of measure and category in various models of set theory.

2. THE KUNEN-MILLER CHART

Let us start with the following definitions.

$A(m) \equiv$ the union of less than continuum many measure zero sets has measure zero,

$B(m) \equiv$ the real line is not the union of less than continuum many measure zero sets,

$U(m) \equiv$ every set of reals of cardinality less than the continuum has measure zero,

$C(m) \equiv$ there does not exist a family F of measure zero sets, of cardinality less than the continuum, and such that every measure zero set is covered by some member of F.

$A(c)$, $B(c)$, $U(c)$, and $C(c)$ are defined similarly, with "first category" replacing "measure zero."

In the pre-forcing era of set theory, the following implications were known (see [**31**]):

$$\begin{array}{lcl} A(m) \Rightarrow B(m) \Rightarrow C(m) & ; & A(m) \Rightarrow U(m) \Rightarrow C(m) \\ A(c) \Rightarrow B(c) \Rightarrow C(c) & ; & A(c) \Rightarrow U(c) \Rightarrow C(c) \\ B(m) \Rightarrow U(c) & ; & B(c) \Rightarrow U(m). \end{array}$$

The non-trivial implications are due to Rothberger [**35**]. After the invention of forcing, a number of models were constructed to demonstrate that most of the other conceivable implications between these properties do not hold.

In Miller [**28**], a chart—later called the "Kunen-Miller chart"—diagramming these implications was published. Most of these implications were already ruled out by constructions of models due to Martin-Solovay [**27**], Kunen [**25**], and most notably, Miller himself, in trying to show that there are no other implications.

The chart left a few questions open, and folk wisdom said that these questions would not be provable in ZFC, since their "measure-category mirror images" were already known to be not provable in ZFC.

The main advance was given by Bartoszyński [**3**], where he proved that

$$A(m) \Rightarrow A(c) \qquad ; \qquad C(c) \Rightarrow C(m).$$

In this way we have, in ZFC, the following diagram of implications:

$$\begin{array}{ccccc} & & A(m) & & \\ & & \downarrow & & \\ \swarrow & & A(c) & & \searrow \\ & \swarrow & & \searrow & \\ B(m) \rightarrow U(c) & & & & B(c) \rightarrow U(m) \\ & \searrow & & \swarrow & \\ \searrow & & C(c) & & \swarrow \\ & & \downarrow & & \\ & & C(m) & & \end{array}$$

In Judah-Shelah [**14**], it was proved that no more implications can be proven in ZFC.

The main technical advance presented in [**14**] was that a countably supported iteration of forcing notions, each preserving outer measure one, also preserves outer measure one. Other preservation theorems for finitely supported iterations are proved in the same paper. The most remarkable proof is that "not adding generic filter for amoeba forcing" is preserved under finitely supported iterations. A weaker theorem was proved by A. Kamburelis [**24**].

3. The Cichoń diagram

Let me introduce two new properties:

$$\begin{aligned} wD &\equiv \forall F \in [\omega^\omega]^{<\mathfrak{c}}\, \exists g \in \omega^\omega\ \forall f \in F\, \exists^\infty n\, f(n) < g(n), \\ D &\equiv \forall F \in [\omega^\omega]^{<\mathfrak{c}}\, \exists g \in \omega^\omega\ \forall f \in F\, \forall^\infty n\, f(n) < g(n). \end{aligned}$$

It was well known that $B(c) \Rightarrow wD$, and it is not hard to show that $A(c) \Rightarrow D$. Cichon displayed all these properties in "Cichon's diagram":

$$\begin{array}{ccccccc} B(m) & \longrightarrow & U(c) & \longrightarrow & C(c) & \longrightarrow & C(m) \\ \uparrow & & \uparrow & & \uparrow & & \uparrow \\ | & & D & & wD & & | \\ | & & \uparrow & & \uparrow & & | \\ A(m) & \longrightarrow & A(c) & \longrightarrow & B(c) & \longrightarrow & U(m) \end{array}$$

In addition,

$$\begin{aligned} A(c) &\equiv B(c)\ \&\ D \\ C(c) &\equiv U(c) \vee wD. \end{aligned}$$

In the context of this diagram, like before, a natural question arises: are these the only implications between these sentences that are provable in ZFC? It turns out that the answer to this question is positive: every combination of those sentences which does not contradict the implications in the diagram is consistent with ZFC.

This is proved "step-by-step", i. e., by giving a model for each implication. The last five models are given in Bartoszyński-Judah-Shelah [**8**]. Although our paradigm is to look for asymmetries between measure and category, in the construction of the models we can recognize some kind of symmetry.

Let $\mathcal{W}$ be the set of sentences obtained from the sentences A, B, U, C,

D, and wD using logical connectives. Define $^*: \mathcal{W} \to \mathcal{W}$ inductively by

$$\phi^* = \begin{cases} \neg\psi^* & \text{if} \quad \phi = \neg\psi \\ {\psi_1}^* \vee {\psi_2}^* & \text{if} \quad \phi = \psi_1 \vee \psi_2 \\ \neg C & \text{if} \quad \phi = A \\ \neg U & \text{if} \quad \phi = B \\ \neg B & \text{if} \quad \phi = U \\ \neg A & \text{if} \quad \phi = C \\ \neg wD & \text{if} \quad \phi = D \\ \neg D & \text{if} \quad \phi = wD \end{cases}$$

for $\phi \in \mathcal{W}$.

It turns out that if ϕ is consistent with ZFC, then ϕ^* is consistent with ZFC. Moreover, in most cases, one can find a notion of forcing P such that ω_2-iteration of P over of model for CH gives a model for ϕ, while ω_1-iteration of P over a model for $MA + \neg CH$ gives a model for ϕ^*.

To give an example of our method of work, we shall describe step-by-step how we got a model for $\neg B(c)$ & $\neg U(m)$ & $\neg B(m)$ & $U(c)$ & wD & $\neg D$.

The first step is to find the appropriate support for the iteration. Because we want $\neg B(c)$, we are obliged to avoid adding Cohen reals, therefore we must use a countably supported iteration.

We thus get two restrictions: namely, we must start from a model of CH and we can not get models for the continuum being bigger than $\aleph_2$. We do not have a preservation theorem for the sentence "not adding Cohen reals." But if the forcing notion satisfies a little more than axiom A, then we are able to show that the Cohen reals are not added at limit stages.

The second restriction is to get $\neg U(m)$ in the final model. We take care of this by a preservation theorem for the sentence "the outer measure of A is one," as mentioned in the section on the Kunen-Miller chart.

The third restriction is to get a model for $\neg B(m)$, which means not adding random reals. We prove a preservation theorem for "not adding random reals."

The fourth restriction is $\neg D$. We use here a preservation theorem for the sentence "not adding dominating reals."

Now we go to the second stage. We should find forcing notions for getting $U(c)$, that is, for making the old reals a meager set. This forcing notion must also satisfy the fourth previous condition, i. e., it must not add dominating reals. (This was one of the hardest problems.)

Finally, we want to get wD. For this, we must add an unbounded real without violating the four above-mentioned restrictions. Miller rational perfect forcing is used for this.

Recently, J. Brendle [**10**] studied the cardinals associated to the Cichoń diagram, but he considered cardinals bigger than $\aleph_2$. This is a very interesting direction and the main goal is to produce a model where the cardinals associated to the Cichoń diagram are all different. It is clear that new ideas about iterated forcing are intrinsically needed for this problem.

From the sketched solution presented above, we see that in the completion of the Cichoń diagram, numerous preservation theorems were used. Each of these theorems has its own distinct proof. Today, we are working on a general iteration theory from which we can obtain the above results as a particular case. It is important to remark that all the forcing notions used in the completion of the Cichoń diagram have a simple definition, and this gives the opportunity to work with them in an abstract way, like in "Souslin Forcing" [**15**].

4. Cofinalities

It is an interesting problem to study the cofinalities of the cardinalities associated with the Kunen-Miller chart. In general, these cardinals are defined as follows.

Let $\mathcal{T}$ be a σ-ideal of Borel sets of $\mathbf{R}$, then we define

$$\begin{aligned}
\kappa_A(\mathcal{T}) &= \text{the least } \kappa \text{ such that } (\exists C \in [\mathcal{T}]^\kappa)(\bigcup C \notin \mathcal{T}),\\
\kappa_B(\mathcal{T}) &= \text{the least } \kappa \text{ such that } (\exists C \in [\mathcal{T}]^\kappa)(\bigcup C = \mathbf{R}),\\
\kappa_U(\mathcal{T}) &= \text{the least } \kappa \text{ such that } [\mathbf{R}]^\kappa \backslash \mathcal{T} \neq 0,\\
\kappa_C(\mathcal{T}) &= \text{the least } \kappa \text{ such that } (\exists \mathcal{F} \in [\mathcal{T}]^\kappa)(\forall A \in \mathcal{T})(\exists B \in \mathcal{F})(A \subseteq B).
\end{aligned}$$

Usually we drop the letter $\mathcal{T}$ if it does not lead to any confusion.

The following is part of the folklore.

(a) $\kappa_A \leq \kappa_B \cap \kappa_U \leq \kappa_B \cup \kappa_U \leq \kappa_C$,
(b) κ_A is regular,
(c) $\mathrm{cof}(\kappa_U) \cap \mathrm{cof}(\kappa_C) > \omega$.

Fremlin has proved that

(d) $\mathrm{cof}(\kappa_C) \geq \kappa_A$,
(e) $\mathrm{cof}(\kappa_U) \geq \kappa_A$.

Let $\mathcal{L}$ be the ideal of measure zero sets and $\mathcal{M}$ the ideal of meager sets. Then Miller [**29**] proved $\mathrm{cof}(\kappa_B(\mathcal{M})) > \omega$. This result was improved by Bartoszyński-Judah [**5**] where we get $\mathrm{cof}(\kappa_B(\mathcal{M})) > \kappa_A(\mathcal{L})$.

We have the following.

Conjecture. $\operatorname{cof}(\kappa_B(\mathcal{M})) \geq \kappa_A(\mathcal{M})$.

There is a large number of questions in this topic. It seems that the most intensely studied one is the following question of Fremlin:

$$\text{Is } \operatorname{cof}(\kappa_B(\mathcal{L})) > \omega?$$

A lot of effort has been devoted to solve this problem. (See [**16**], [**17**].) The best partial result was obtained by Bartoszyński [**4**]:

$$\text{If} \quad \mathfrak{b} \geq \kappa_B(\mathcal{L}), \quad \text{then} \quad \operatorname{cof}(\kappa_B(\mathcal{L})) > \omega,$$

where $\mathfrak{b}$ is the minimal cardinal of an unbounded family in ω^ω. It seems plausible that the Bartoszyński result is the best possible. We think that a solution of this problem is connected with the existence of perfect sets of random reals. The following two questions are closely related to the construction of a model for $\operatorname{cof}(\kappa_B(\mathcal{L})) = \omega$:

(1) Does the existence of a perfect set of reals random over a model M for a sufficiently large fragment of ZFC imply the existence of a dominating real over M?
(2) Assume that for each n and each $\aleph_n$-sized family of measure zero sets there is a perfect set disjoint from the union of the family. Does this imply $\kappa_B > \aleph_\omega$?

All possibilities for the cofinalities of $\kappa_C(\mathcal{L})$ and $\kappa_C(\mathcal{M})$ were completely described in Bartoszyński-Judah-Shelah [**9**], but hard questions remain concerning $\operatorname{cof}(\kappa_U)$.

Recently J. Brendle has built a model where $\operatorname{cof}(\kappa_U(\mathcal{L})) < \kappa_A(\mathcal{M})$. This result surprised me, because I was sure that a new idea on iteration was necessary to get a model of this inequality. J. Brendle's idea was to start with a ground model satisfying $\kappa_U = \mathfrak{c} = \aleph_{\omega_1}$, by adding $\aleph_{\omega_1}$ Cohen reals. Then he added ω_2 Hechler reals. In the final model, $\kappa_A(\mathcal{M}) = \omega_2$. Then he uses ideas of Bartoszyński-Judah [**6**] to show that if $P \models$ *σ-centered*, then

$$V^P \models \text{“}\kappa_U(\mathcal{L}) \geq \kappa_U(\mathcal{L})^V\text{”}.$$

Clearly this is enough to show that in his model, $\kappa_U(\mathcal{L}) = \aleph_{\omega_1}$.

I think that the study of the cofinalities associated to the Kunen-Miller chart will be an area of very interesting development in the near future.

5. Special Sets of Reals

Pathological sets of reals are always capturing the attention of mathematicians. As mentioned in §1, Bernstein's set was built in the last century.

At the beginning of this century Łuzin defined his set: an uncountable set of reals that has a countable intersection with every meager set. A set of reals X is called a *strong measure zero set* if for every sequence $\langle \epsilon_i : i < \omega \rangle \in (\mathbf{R}^+)^\omega$ there exists a sequence $\langle x_i : i < \omega \rangle \in \mathbf{R}^\omega$ such that

$$X \subseteq \bigcup_{i<\omega} (x_i - \epsilon_i, x_i + \epsilon_i).$$

It turned out that Łuzin sets have strong measure zero.

E. Borel conjectured that the strong measure zero sets are exactly the countable sets of reals. This conjecture is known as the Borel conjecture. Łuzin built up a Łuzin set from CH, and therefore the Borel conjecture fails if $2^{\aleph_0} = \aleph_1$.

Sharp results concerning the strong measure zero sets were given by Rothberger. He proved the following. Let $\mathcal{S}$ be the σ-ideal of the strong measure zero sets.

Theorem. *(a) $\mathfrak{d} = \aleph_1$ implies the Borel Conjecture fails.*
(b) $\mathfrak{b} = \kappa_U(\mathcal{S}) = 2^{\aleph_0}$ iff $\kappa_A(\mathcal{M}) = 2^{\aleph_0}$.

It is impressive how Rothberger's work done in the 40's and 50's has a strong flavor of our work in the 80's.

In his celebrated work, "On the consistency of the Borel Conjecture," R. Laver [**26**] built a model where $2^{\aleph_0} = \aleph_2$ and every strong measure zero set is countable. In this paper countably supported iterations of forcing were introduced. A complete solution of the Borel conjecture with large continuum was given independently by W. H. Woodin and Judah-Shelah (see [**18**]). In [**18**] it is proved that adding ω_2-Laver reals followed by any number of random reals gives models for the Borel conjecture. It is an open problem if we can destroy the Borel conjecture by a adding a random real.

A. Miller asked if the existence of a Ramsey filter on ω implies the negation of the Borel conjecture. This is a natural question when you know that the existence of Ramsey filters has a close relation with models having a lot of Cohen reals. The Cohen reals are the main ingredient to build Łuzin sets. In [**19**] a model for both the Borel conjecture and the existence of Ramsey filters was constructed. It can be noticed that all the constructions of strong measure zero sets of size $\aleph_2$ have used the existence of Cohen reals over L. T. Weiss and I, working independently, were looking for models where $\mathcal{S} \backslash \mathbf{R}^{<\aleph_2} \neq 0$ and no real is Cohen over L. This problem was solved recently by Goldstern-Judah-Shelah [**12**], where we produce such models.

Using Bartoszyński's [**3**] characterization of $\kappa_A(\mathcal{L})$ it is not hard to show that

$$\kappa_A(\mathcal{L}) \leq \kappa_A(\mathcal{S}).$$

Galvin asked if

$$\kappa_A(\mathcal{M}) \leq \kappa_A(\mathcal{S}).$$

In Judah-Shelah [**20**] a negative answer to Galvin's question was given by a model with $MA(\sigma\text{-}centered) + \kappa_A(\mathcal{S}) = \aleph_1 + 2^{\aleph_0} = \kappa_A(\mathcal{M}) = \aleph_2$.

J. Pawlikowski [**32**] improved our result by showing that for any model M obtained by finitely supported iteration of forcing notions having precalibre $\aleph_1$, of length $\geq \omega_1$, we have $M \models \kappa_B(\mathcal{S}) = \aleph_1$. This result uses the elegant characterization of a strong measure zero set due to Galvin-Mycielsky-Solovay (see [**30**]): $X \subseteq \mathbf{R}$ has strong measure zero iff for every meager set M there is a $y \in \mathbf{R}$ such that $(X+y) \cap M = 0$.

This characterization of strong measure zero sets suggests the definition of a strongly meager set: A set $X \subseteq \mathbf{R}$ is a strongly meager set iff for every null set M there is a $y \in \mathbf{R}$ such that $(X+y) \cap M = 0$.

Clearly the countable sets are strongly meager. It is an open question whether the strongly meager sets form an ideal!

A Sierpiński set is an uncountable set that has countable intersection with every measure zero set. Galvin asked whether Sierpiński sets are strongly meager. This too is an open problem. In this direction, Bartoszyński-Judah [**7**] proved

(a) Consistency of "every Sierpiński set is strongly meager" (by adding $\aleph_1$-random reals to a model for MA).

(b) Every Sierpiński set is the union of two strongly meager sets

In the model for (a) there are uncountable strongly meager sets (i. e., the Sierpiński sets). Also, it is possible to ask the question dual to Borel's conjecture: is it consistent that every strongly meager set is countable? A model for this dual Borel conjecture was constructed by T. Carlson [**11**], by adding $\aleph_2$ Cohen reals to any model. Carlson's result was improved in Judah-Shelah [**20**] where we gave a model for $MA(\sigma\text{-}centered) + 2^{\aleph_0} > \aleph_1 +$ Dual Borel Conjecture. Pawlikowski [**32**] also improved our result by getting the same with $MA(precalibre\ \aleph_1)$.

It would be interesting to get a model where the Borel Conjecture and Dual Borel Conjecture hold simultaneously. I think that the Laver model is a good candidate for this.

There are other interesting pathological sets besides the ones mentioned here; for an introduction you should see Miller [**30**].

During this workshop on the continuum, Shelah built a model where there is an open set of $[0,1]^2$ of measure one which does not contain a rectangle of outer measure one. This is one the nicest results obtained during the logic year. H. Friedman got a weak form of this in the Cohen model.

6. Descriptive Set Theory

Let us introduce some notation. We shall write $\Sigma^1_n(\mathcal{L})$ $(\Delta^1_n(\mathcal{L}), \Pi^1_n(\mathcal{L}))$ if every $\mathbf{\Sigma}^1_n$ $(\mathbf{\Delta}^1_n, \mathbf{\Pi}^1_n)$ set is Lebesgue measurable. $\Sigma^1_n(\mathcal{M})$ $(\Delta^1_n(\mathcal{M}), \Pi^1_n(\mathcal{M}))$ if every $\mathbf{\Sigma}^1_n$ $(\mathbf{\Delta}^1_n, \mathbf{\Pi}^1_n)$ set has the property of Baire. (We only refer to boldface sets.) MA is understood to imply $\neg CH$.

Łuzin and Sierpiński proved that

$$ZFC \vdash \Sigma^1_1(\mathcal{L}) \mathbin{\&} \Sigma^1_1(\mathcal{M}).$$

As a corollary of Gödel's work on the constructible universe we have

$$V = L \vdash \neg\Delta^1_2(\mathcal{L}) \mathbin{\&} \neg\Delta^1_2(\mathcal{M}).$$

Actually, there are $\mathbf{\Delta}^1_2$ Bernstein sets in L.

Measurability and categoricity of the $\mathbf{\Sigma}^1_2$-sets of reals were studied by R. M. Solovay in the 60's. The following characterizations were discovered:

$$\begin{aligned}\Sigma^1_2(\mathcal{L}) &\text{ iff } (\forall r \in \mathbf{R})(\{s : s \text{ is random over } L[r]\} \text{ has measure } 1)\\ \Sigma^1_2(\mathcal{M}) &\text{ iff } (\forall r \in \mathbf{R})(\{c : c \text{ is Cohen over } L[r]\} \text{ is comeager})\end{aligned}$$

Using these characterizations, Martin-Solovay [**27**] proved

$$MA \vdash \Sigma^1_2(\mathcal{L}) \mathbin{\&} \Sigma^1_2(\mathcal{M}).$$

The $\mathbf{\Delta}^1_2$-sets of reals were studied in Judah-Shelah ([**21**]). We found:

$$\begin{aligned}\Delta^1_2(\mathcal{L}) &\text{ iff } (\forall r \in \mathbf{R}\, \exists s)(s \text{ is random over } L[r])\\ \Delta^1_2(\mathcal{M}) &\text{ iff } (\forall r \in \mathbf{R}\, \exists c)(c \text{ is Cohen over } L[r])\end{aligned}$$

(In [**23**], it was proved that $MA \nvdash \Delta^1_3(\mathcal{L}), \Delta^1_3(\mathcal{M})$.)

In the mid-80's, Bartoszyński, and independently Raisonnier-Stern [**34**], discovered that $\Sigma^1_2(\mathcal{L}) \Rightarrow \Sigma^1_2(\mathcal{M})$. It is part of the folklore in set theory that this implication can not be reversed. Moreover, $\Sigma^1_2(\mathcal{M})$ does not imply $\Delta^1_2(\mathcal{L})$. By adding $\aleph_1$ random reals to L we can also see that $\Delta^1_2(\mathcal{L})$ does not imply $\Delta^1_2(\mathcal{M})$. Shelah [**36**] showed, in ZFC, that $\Delta^1_3(\mathcal{L})$ is consistent. Further study of this model proved that $\Delta^1_3(\mathcal{M})$ holds in this extension.

Using the ideas of [**36**], it was not hard to get a model for $\Delta^1_3(\mathcal{M})$ by a σ-centered forcing extension, therefore, if we start from L, we can get a model for $\Delta^1_3(\mathcal{M}) + \neg\Delta^1_2(\mathcal{L})$.

For a long time, the main problem concerning $\mathbf{\Delta}^1_3$-sets was to show that

$$\Delta^1_3(\mathcal{L}) \text{ does not imply } \Delta^1_3(\mathcal{M}).$$

In [**23**], we built a model of

$$\Delta^1_3(\mathcal{L}) + \neg\Delta^1_3(\mathcal{M}),$$

but we used the consistency of a measurable cardinal. This result did not yet make me happy. Fortunately, during this logic year, we built a model for $\Delta^1_3(\mathcal{L}) + \neg\Delta^1_3(\mathcal{M})$ using only the consistency of ZFC [-]. This construction owes a lot to technology introduced by Galvin, Laver, Shelah, Todorcevič, etc.

We think that a forcing characterization of $\Delta^1_3(\mathcal{L})$ ($\Delta^1_3(\mathcal{M})$) would give us a deeper understanding of these statements.

Concerning the $\mathbf{\Delta}^1_3$-sets, Harrington-Shelah [**13**] proved that

$$MA + \Delta^1_3(\mathcal{M}) \vdash \omega_1 \text{ is weakly compact in } L.$$

In Judah-Shelah [**22**], we showed

$$MA + \Delta^1_3(\mathcal{L}) \vdash \omega_1 \text{ is weakly compact in } L.$$

Indeed, by an unpublished result of Kunen-Solovay we have that $MA + \Delta^1_3(\mathcal{L})$ ($\Delta^1_3(\mathcal{M})$) is equiconsistent with

$$ZFC + \exists \text{ a weakly compact cardinal.}$$

As a corollary of this, we have

$$MA \text{ does not imply } \Delta^1_3(\mathcal{L}) \text{ nor } \Delta^1_3(\mathcal{M}).$$

In Judah-Shelah [**22**], we also built a model for $\Delta^1_3(\mathcal{L}) + \Delta^1_3(\mathcal{M}) + MA(I)$, starting from L. (I is the class of c.c.c. posets that satisfy c.c.c. in any c.c.c. extension.) We don't know if it is possible to improve this result by enlarging I.

The most interesting open problem concerning $\mathbf{\Delta}^1_3$-sets involves MA, mainly

$$\text{Does } MA + (\forall r \in \mathbf{R})({\omega_1}^{L[r]} < \omega_1) \text{ imply } \Delta^1_3(\mathcal{L})?$$

The present state of knowledge does not allow us to differentiate, in ZFC, between "$\Sigma^1_3(\mathcal{L})$" and "$\forall n\,\Sigma^1_n(\mathcal{L})$." Immediately after Cohen's breakthrough, R. M. Solovay built his famous model for "$\forall n\,\Sigma^1_n(\mathcal{L})$" starting from the existence of an inaccessible cardinal. Later, in the 70's, S. Shelah [**36**] proved that

$$\Sigma^1_3(\mathcal{L}) \text{ implies that } \aleph_1 \text{ is inaccessible in } L.$$

We don't know how to build models for $\Sigma^1_3(\mathcal{L}) + \neg\Sigma^1_4(\mathcal{L})$ starting from large cardinals which are possibly consistent with $V = L$. I think this must be one of the most interesting problems in the near future of set theory.

One of the first asymmetries of measure and category was found by S. Shelah [**36**] when he started from L and built a model for $\forall n\,\Sigma^1_n(\mathcal{M})$ (without using an inaccessible cardinal).

Surprisingly, this asymmetry disappears when one adds a weak assumption to ZFC, as shown by the following theorem of Raisonnier[**33**]:

$$ZFC + \Sigma^1_2(\mathcal{L}) \vdash \text{“}\Sigma^1_3(\mathcal{M}) \text{ implies } \aleph_1 \text{ is inaccessible in } L\text{”}.$$

We generalized this result in the presence of different forms of MA. Also, we proved in [**22**] the following results:

(1) The following theories are equiconsistent:
  (a) $ZFC + \exists$ weakly compact cardinal,
  (b) $MA(\textit{precalibre}\ \aleph_1) + \Sigma^1_3(\mathcal{L}) \quad (\Sigma^1_3(\mathcal{M}))$.
(2) The following theories are equiconsistent:
  (a) $ZFC + \exists$ Mahlo cardinal,
  (b) $MA(\sigma\textit{-centered}) + \Sigma^1_3(\mathcal{L}) \quad (\Sigma^1_3(\mathcal{M}))$.
(3) The following theories are equiconsistent:
  (a) $ZFC + \exists$ inaccessible cardinal,
  (b) $MA(\textit{Souslin}) + \Sigma^1_3(\mathcal{L}) \quad (\Sigma^1_3(\mathcal{M}))$.

The class of "Souslin forcing notions" is defined by the class of forcing notions which are c.c.c. and have a $\mathbf{\Sigma}^1_1$-definition. The study of this class was started in [**15**] and continued by Bagaria in his Ph.D. thesis. In [**5**], the concept of "Souslin absoluteness" was introduced: we say that a model V is Souslin absolute if for every Souslin forcing $P \in V$, we have $\mathbf{R}^V \prec \mathbf{R}^{V^P}$.

During this logic year, we proved the following theorem:

$V \models$ "Souslin absoluteness" implies
  (a) $V \models \aleph_1$ is inaccessible in L,
  (b) $V \models \Delta^1_3(\mathcal{L}) + \Delta^1_3(\mathcal{M})$.

We don't know yet if Souslin absoluteness implies projective measurability. However, we hope that this direction of research will give a forcing characterization of the statement $\forall n\, \Sigma^1_n(\mathcal{L})$.

Closely related with these results are the following open problems: Let r be a random real.

(a) Does $V \models$ "Souslin absoluteness" imply $V[r] \models$ "Souslin absoluteness"?

(b) Does $V \models \forall n\, \Sigma^1_n(\mathcal{L})$ imply $V[r] \models \forall n\, \Sigma^1_n(\mathcal{L})$?

At the same time that I was writing this note, M. Goldstern and I built a model for the Borel Conjecture where Projective measurability holds. We got this by starting from an inaccessible cardinal.

Also, we are dealing with measurability and categoricity of filters on ω. The reader can find a chapter on the subject in this proceedings. The most remarkable result is a combinatorial characterization of measurable filters.

Acknowledgments

The author would like to thank J. W. Addison and D. Martin for helping me to be at M.S.R.I. during the logic year, Richard Shore for supporting my last month under a grant for Latin America, W. Just for pressing me to write this evaluation, and M. Wiener for improving the presentation to this final form.

References

1. J. Bagaria, H. Judah, *Amoeba Forcing, Souslin Absoluteness and Additivity of Measure*, this volume.
2. H. Judah, *Δ^1_3(measurability) does not imply Δ^1_3(categoricity)*, in preparation.
3. T. Bartoszyński, *Additivity of measure implies additivity of category*, Transactions of the American Mathematical Society **281** (1984), 209–213.
4. T. Bartoszyński, *On covering of the real line by null sets*, Pacific Journal of Mathematics **131** (1988), 1–12.
5. T. Bartoszyński, H. Judah, *On the cofinality of the smallest covering of the real line by meager sets*, Journal of Symbolic logic **54** (1989), 828–832.
6. T. Bartoszyński, H. Judah, *Jumping with Random reals*, Annals of pure and applied logic (to appear).
7. T. Bartoszyński, H. Judah, *On Sierpinski Sets*, Proceedings of the American Mathematical Society **108** (1990), 507–512..
8. T. Bartoszyński, H. Judah, S. Shelah, *The Cichoń diagram*, submitted.
9. T. Bartoszyński, H. Judah, S. Shelah, *The cofinality of cardinal invariants related to measure and category*, Journal of Symbolic logic **54** (1989), 719–726.
10. J. Brendle, *Large Cardinals in Cichoń's diagram*, accepted by Journal of Symbolic logic.
11. T. Carlson, unpublished notes.
12. M. Goldstern, H. Judah, S. Shelah, *Strong measure zero sets and avoiding Cohen reals*, in preparation.

13. L. Harrington and S. Shelah, *Some exact equiconsistency results in set theory*, Notre Dame Journal of Formal Logic **26** (1985), 178–188.
14. H. Judah, S. Shelah, *The Kunen-Miller Chart*, JSL (to appear).
15. H. Judah and S. Shelah, *Souslin forcing*, Journal of Symbolic logic **53** (1988), 1188–1207.
16. H. Judah, S. Shelah, *Around random algebra*, Archive for Mathematical Logic (to appear).
17. H. Judah, S. Shelah, *Adding dominating reals with measure algebras*, submitted.
18. H. Judah, S. Shelah, H. Woodin, *The Borel Conjecture*, AnPAL (to appear).
19. H. Judah, *Strong measure zero sets and rapid filters*, Journal of Symbolic logic **53** (1988), 393–402 .
20. H. Judah, S. Shelah, *$MA(\sigma$-centered), Cohen reals, strong measure zero sets, and strongly meager sets*, Israel Journal of Mathematics **68** (1989), 1–17.
21. H. Judah and S. Shelah, *Δ^1_2-sets of reals*, Annals of pure and applied logic **42** (1989), 207–233.
22. H. Judah, S. Shelah, *Martin's axioms, measurability and equiconsistency results*, Journal of Symbolic logic **54** (1989), 78–94.
23. H. Judah, S. Shelah, *Δ^1_3-sets of reals*, submitted.
24. A. Kamburelis, *Iteration of Boolean Algebras with Measure*, Archives of Mathematical logic **29** (1989), 21–28.
25. K. Kunen, *Random and Cohen Reals*, Handbook of Set Theoretical Topology, North-Holland, 1984.
26. R. Laver, *On the consistency of Borel's Conjecture*, Acta Mathematica **137** (1976), 151–169.
27. D. Martin and R. Solovay, *Internal Cohen extensions*, Annals of Mathematical Logic **2** (1970), 143–178.
28. A. Miller, *Some properties of measure and category*, Transactions of the American Mathematical Society **266,1** (1981), 93–114.
29. A. Miller, *The Baire category theorem and cardinals of countable cofinality*, Journal of Symbolic logic **47** (1982), 275–288 .
30. A. Miller, *Special sets of reals*, Handbook of Set Theoretical Topology, North-Holland, 1984.
31. Oxtoby, *Measure and Category*, Springer-Verlag, 1971.
32. J. Pawlikowsky, *Finite support iteration and strong measure zero sets*, Journal of Symbolic logic (to appear).
33. J. Raisonnier, *A mathematical proof of S. Shelah's theorem on the measure problem and related results*, Israel Journal of Mathematics **48** (1984), 48–56.
34. J. Raisonnier, J. Stern, *The strength of measurability hypotheses*, Israel Journal of Mathematics **50** (1985), 337–349.
35. Rothberger, *Eine Äquivalenz zwischen der Kontinuumshypothese unter der Existenz der Łuzinschen und Sierpińskischen Mengen*, Fundamenta Mathematicae **30** (1938), 215–217.
36. S. Shelah, *Can you take Solovay's inaccessible away?*, Israel Journal of Mathematics **48** (1984), 1–47.

DEPARTMENT OF MATHEMATICS, BAR-ILAN UNIVERSITY, RAMAT-GAN, ISRAEL

THE STRUCTURE OF BOREL EQUIVALENCE RELATIONS IN POLISH SPACES

ALEXANDER S. KECHRIS

ABSTRACT. An exposition of recent work on Borel equivalence relations in Polish spaces is presented. This includes a general Glimm-Effros dichotomy for Borel equivalence relations and a study of countable Borel equivalence relations and their classification into subclasses such as smooth, hyperfinite, amenable, treeable etc.

1. INTRODUCTION

This article is a survey of some recent work on Borel equivalence relations in Polish spaces. The subject has interesting connections with ergodic theory and operator algebras and in fact a lot of the work reported here has been motivated by results and concepts originating in these areas.

Before getting down to specific results, it would be helpful, in order to put things in perspective, to discuss informally some aspects of the subject of "definable" equivalence relations in Polish spaces to which these results belong. One can look at this from two different but related points of view. The first we dub the "set theoretic point of view", the second one "the classification point of view". Here is what we have in mind.

1.1. The set theoretic point of view

Consider sets of "definable cardinality at most that of the continuum", i.e., sets I for which there is a "definable" surjection $f : \mathbb{R} \twoheadrightarrow I$ from the reals onto I.

We would like to study "definable cardinality theory" for such sets. The basic concepts here are

$$I \leq^D J \iff \exists \text{ "definable injection" } f : I \rightarrowtail J$$

$$I \sim^D J \iff I \leq^D J \quad \& \quad J \leq^D I$$

$$(\iff \exists \text{ "definable" bijection } f : I \rightarrowtail\!\!\!\!\twoheadrightarrow J) .$$

The appropriate context for carrying out "definable cardinality theory" is to work in an inner model of the Axiom of Determinacy (AD). In fact such a theory would be even smoother if one works in an inner model (containing $\mathbb{R}$) of the Axiom of Determinacy for reals ($AD_{\mathbb{R}}$), see for example [30, §3]. This is because $ZF + DC + AD_{\mathbb{R}}$ implies that every subset of $\mathbb{R}^2$ can be uniformized and, even more, that every subset of $\mathbb{R}$ admits a scale (Woodin). Working in $ZF+DC+AD_{\mathbb{R}}$, one is studying now *arbitrary* sets I which are surjective images of $\mathbb{R}$ and the usual notions of Cantor's cardinality theory, i.e. embedding (injection) $I \leq J$ and equivalence (bijection) $I \sim J$ of sets. However, since AC fails, cardinality theory looks quite different here. The cardinality theory of such I which are ordinals (i.e. the ordinals $< \Theta$) has been extensively studied over the last 20 years. But the theory for arbitrary I, even of the form power(α), $\alpha < \Theta$, is still very little understood. For instance, the question whether there are infinite α with $\alpha^+ \leq$ power(α) is still open.

This "definable cardinality theory" can be also viewed as a study of "definable" equivalence relations: Given a "definable" surjection $f : \mathbb{R} \twoheadrightarrow I$, let E be the corresponding equivalence relation

$$xEy \iff f(x) = f(y) .$$

Then there is a canonical bijection between I and $\mathbb{R}/E$. The embeddability relation $I \leq^D J$ corresponds then to the concept of "definable" reducibility between "definable" equivalence relations

$$E \leq^D F \iff \exists \text{ "definable" } f : \mathbb{R} \to \mathbb{R} \, \forall x, y$$

$$\big(xEy \iff f(x)Ff(y)\big) .$$

(The notions coincide if "definable" relations on $\mathbb{R}$ admit "definable" uniformizations, as for instance is the case when we work in an inner model of $ZF + DC + AD_{\mathbb{R}}$).

1.2. The classification point of view

Suppose now X is an arbitrary Polish space and E a "definable" equivalence relation on X. One is frequently interested in the problem of classifying elements of X up to E-equivalence by appropriate "invariants".

It would be best if one could find reasonably "concrete invariants", which in general could be viewed as elements of some Polish space Y. That is, one is looking for a "definable" map $f : X \to Y$, where Y is some Polish space, such that

$$xEx' \iff f(x) = f(x') \, .$$

In that case we have $E \leq^D \Delta(Y)$ (= the equality on Y). An example of this situation is the classification of $n \times n$ complex matrices under similarity by their Jordan canonical forms. (Here $X = M_n(\mathbb{C}) = Y$, $E =$ similarity and $f(A) =$ the Jordan canonical form of A).

However, quite often one has to settle for somewhat "less concrete invariants". For example, if we seek to classify up to unitary equivalence normal operators on (separable) Hilbert space, which (for simplicity) have a given spectrum Ω and are multiplicity-free, then the invariants are measure classes on Ω, i.e., equivalence classes of measures on Ω under the equivalence relation of mutual absolute continuity: $\mu \sim \nu \iff \mu \prec\prec \nu \;\&\; \nu \prec\prec \mu$.

In this and other similar situations one has a "definable" map $f : X \to Y$, where Y is some Polish space, and a "definable" equivalence relation E' on Y such that

$$\begin{aligned} xEy &\iff f(x)\, E' f(y) \\ &\iff [f(x)]_{E'} = [f(y)]_{E'} \end{aligned}$$

so that the "invariants" are now E'-equivalence classes. In that case we have of course $E \leq^D E'$.

We will concentrate in the sequel on Borel equivalence relations. Although many of the subsequent results extend appropriately under determinacy hypotheses, we will not discuss these extensions here except for some occasional remarks.

2. A Glimm-Effros Dichotomy for Borel Equivalence Relations

Let X, X' be Borel sets in Polish spaces, E, E' Borel equivalence relations on X, X' resp.

Definition 1. We say that E is **reducible** to E', in symbols

$$E \leq E'$$

if there is a Borel function $f : X \to X'$ such that

$$xEy \iff f(x)E'f(y) \, .$$

We say that E is **embeddable** in E', in symbols $E \sqsubseteq E'$, if there is a 1–1 Borel function f satisfying the above.

Definition 2. A **(countable) separating family** for E is a sequence $\{A_n\}$ of subsets of X such that

$$xEy \iff \forall n[x \in A_n \iff y \in A_n] \ .$$

Notice that E has a Borel separating family iff $E \leq \triangle(2^\omega)$, where $\triangle(S)=$ equality on S.

Definition 3. The Borel equivalence relation E is called **smooth** if it has a Borel separating family.

This means that E is smooth iff it is "concretely classifiable".

A standard non-smooth equivalence relation is the following:

$$X = 2^\omega \ ,$$

$$xE_0y \iff \exists n \, \forall m \geq n\big(x(m) = y(m)\big) \ .$$

The quotient space $2^\omega/E_0$ is canonically isomorphic to $P(\omega)/\text{fin}$. We can easily see that E_0 is not smooth by noticing that the standard probability measure on 2^ω is E_0-ergodic and E_0-non-atomic. These concepts are defined as follows.

Definition 4. A (Borel probability) measure μ on X is (E-)**non-atomic** if $\mu([x]_E) = 0$ for each equivalence class $[x]_E$.

A measure μ on X is (E-)**ergodic** if $\mu(A) = 0$ or $\mu(A) = 1$ for each μ-measurable E-invariant set $A \subseteq X$.

We have now the following

Theorem 5. (Harrington-Kechris-Louveau[16]). *For each Borel equivalence relation E on a Borel set X in a Polish space exactly one of the following holds:*

(I) *E is smooth;*
(II) *$E_0 \sqsubseteq E$.*

Remarks. 1) (I) is equivalent to the existence of a universally measurable separating family or to the existence of a C-measurable selector ($C =$ the smallest σ-algebra containing the Borel sets and closed under the Souslin operation $\mathcal{A}$; a (E-)**selector** is a map $s : X \to X$ with $xEy \Rightarrow s(x) = s(y)$ and $s(x)Ex$). In general one cannot find Borel selectors for smooth E (even closed ones), except in certain special situations, e.g., if every equivalence class $[x]_E$ is K_σ (a countable union of compact sets) or if E is induced by a Polish group acting by Borel automorphisms on X (see Burgess [4]).

Further equivalences can be proved under further assumptions on E (see [9], [15]).

2) (II) is equivalent to the existence of a continuous (from 2^ω into the Polish space in which X lives) or universally measurable embedding of E_0 into E and also to the existence of a (E-) non-atomic, ergodic measure. (This last equivalence is useful in analytic applications).

The preceding result is an outgrowth of two lines of work, one originating in analysis and the other in set theory. From the analysis side, the first such dichotomy was established by Glimm [14] for the case of equivalence relations induced by (continuous) locally compact transformation groups and then extended by Effros [9], [10] for the case of F_σ equivalence relations induced by Polish transformation groups. The Glimm-Effros work is related to the proof of the "Type I iff smooth dual" conjecture of Mackey in the representation theory of C^*-algebras and groups. Special cases of the Glimm-Effros dichotomy have been rediscovered and applied in ergodic theory, see e.g. [22], [20], [27] and [34]. Finally, in [7] a dichotomy result has been established for arbitrary F_σ equivalence relations.

From the set theory side, Silver [28] proved (in particular) that for Borel E either $E \leq \Delta(\omega)$ or $\Delta(2^\omega) \sqsubseteq E$ (via a continuous function). (This of course also easily follows from Theorem 5). Harrington (unpublished) later found a much simpler proof of Silver's Theorem using effective descriptive set theory and making use of the topology generated by the $\sum_1^1$ sets– the so-called Gandy-Harrington topology. Further development of these techniques appeared in work of Harrington-Marker-Shelah [17] as well as Louveau [23], Louveau-Saint Raymond [24] and Kada [19] on Borel partial orders. The proof of Theorem 5 uses techniques of effective descriptive set theory associated with the Gandy-Harrington topology and provides an effective version of Theorem 5. More precisely, we have:

Theorem 6. (Harrington-Kechris-Louveau [16]). *For each Δ_1^1 equivalence relation E on $\mathcal{N} = \omega^\omega$ exactly one of the following holds:*

(I) *There is a Δ_1^1 set $A \subseteq \omega \times \omega^\omega$ such that if $A_n = \{x : (n, x) \in A\}$, then $\{A_n\}$ is a separating family for E.*

(II) *$E_0 \sqsubseteq E$.*

Concerning the partial (pre)ordering $\leq$ on the Borel equivalence relations, Theorem 5 and Silver's Theorem show that $\Delta(\omega)$, $\Delta(2^\omega)$, E_0 are in increasing order the first three ones, among those that have infinitely many equivalence classes. What is happening above E_0 is unclear. It is known that there are incomparable elements (some nice examples are due to S. Jackson, W. Just and A. Louveau) and it is not hard to see that there is a cofinal $\aleph_1$ sequence $\{E_\xi\}$ of Borel equivalence relations (Harrington). However it is open to find a canonical such cofinal sequence. It is also not known if this partial (pre)ordering is a well-quasiordering. There is one interesting further result due to Friedman-Stanley [13]: For any Borel equivalence relation E there is a Borel equivalence relation E' strictly bigger than E (i.e., $E \leq E'$ but $E' \not\leq E$).

Remark. In the context of $ZF+DC+AD_{\mathbb{R}}$ the following general dichotomy seems to be true: For any set I which is a surjective image of $\mathbb{R}$ either I embeds into 2^α for some ordinal $\alpha < \Theta$ or else $P(\omega)/\text{fin}$ embeds in I. (A proof of this should combine the proof of Theorem 5 with the techniques of [12]). This and earlier results provide the following partial cardinality picture for such sets I: Either I embeds in some $\alpha < \Theta$ or else 2^ω embeds in I (this was proved in Harrington-Sami [18]). If 2^ω embeds into I either I embeds into 2^α for some $\alpha < \Theta$ or else $P(\omega)/\text{fin}$ embeds into I. Beyond that we do not understand what is happening.

3. Countable Borel Equivalence Relations

In the rest of this paper we will concentrate on the structure of countable Borel equivalence relations, where we have the following

Definition 1. Let E be a Borel equivalence relation on a Borel set X in a Polish space. We call E **countable** if every equivalence class $[x]_E$ is countable.

Examples of such E are $\equiv_T$ (Turing equivalence), $\equiv_A$ (arithmetic equivalence), E_0, the tail equivalence E_{tail} on 2^ω (where $xE_{\text{tail}}\,y \iff \exists n\,\exists m\,\forall k(x(n+k) = y(m+k))$, etc. Also, if G is a countable group, α an action of G by Borel automorphisms on X (briefly: a Borel action) and we denote by $(g, x) \mapsto x \cdot_\alpha g$ the action, the induced equivalence relation

$$xE_\alpha y \iff \exists g \in G(x = y \cdot_\alpha g)$$

is a countable Borel equivalence relation. We denote E_α by E_G when there is no danger of confusion. In particular, if we consider the canonical action of G on X^G (X a Polish space) given by

$$x \cdot g(h) = x(gh)$$

we denote by $E(X^G)$ the induced equivalence relation.

The following result shows that all countable Borel equivalence relations come from group actions.

Theorem 2. (Feldman-Moore [11]). *Let E be a countable Borel equivalence relation on a Borel set X in a Polish space. There is a countable group G and a Borel action α of G on X such that*

$$E = E_\alpha .$$

This result has the following application (see [8])

Proposition 3. *The equivalence relation $E(2^{F_2})$ is* **universal** *among countable Borel equivalence relations, i.e. for every such E, $E \sqsubseteq E(2^{F_2})$. (Here F_2 is the free group on 2 generators).*

Thus the countable Borel equivalence relations on uncountable Borel sets are exactly those in the interval

$$\Delta(2^\omega) \sqsubseteq E \sqsubseteq E(2^{F_2}) .$$

Apart from the group actions, another important ingredient in the study of countable Borel E is the type of "structures" that can be "uniformly" attached to each E-equivalence class, as it will be gradually explained below.

In terms of these ingredients one can ramify countable Borel equivalence relations in different levels of complexity.

3.1. Finite Borel equivalence relations

These are by definition the ones with finite equivalence classes, and there is not much to say about them.

3.2. Smooth (countable) Borel equivalence relations

Again these are fairly easy to understand. We only want to make here a couple of remarks: Because of the countability assumption, smoothness can be characterized by the existence of a Borel selector. Also because of Theorem 2.5 and the remarks following it, non-smoothness is characterized by the existence of a non-atomic, ergodic and **quasi-invariant** probability measure. (A measure μ is E-**quasi-invariant** if for every Borel set A, $\mu(A) = 0$ implies $\mu([A]_E) = 0$, where $[A]_E = \{x : \exists y \in A(xEy)\}$).

Before we go to the next level, recall the Feldman-Moore Theorem. One can ask various questions about a countable group generating a given equivalence relation. For example, can it always be taken to have 2 generators? This does not seem to be known. However one has the following fact proved in [8].

Proposition 4. *Let E be a countable Borel equivalence relation on a Borel set X in a Polish space. If there is a Borel equivalence relation $F \subseteq E$ which is smooth and has infinite equivalence classes, then there is a countable group G with 2 generators and a Borel action α of G on X with $E = E_\alpha$.*

This applies easily to show for example that $\equiv_T$ or $\equiv_A$ are induced by groups with 2 generators.

How about 1-generated groups, i.e. equivalences induced by $\mathbb{Z}$-actions? For each Borel automorphism $T : X \to X$, where X is a Borel set in a Polish space, we denote by E_T the equivalence relation induced by T i.e.

$$xE_T\, y \iff \exists n \in \mathbb{Z}\ (x = T^n y)$$

Definition 5. A countable Borel equivalence E on X is called **hyperfinite** if it is of the form E_T for some Borel automorphism T of X.

This is our next level of complexity.

3.3. Hyperfinite Borel equivalence relations

The term hyperfinite is justified by the following

Theorem 6. (Weiss [34], Slaman-Steel [29]). *The following are equivalent for a countable Borel equivalence relation E:*

(i) *E is hyperfinite;*
(ii) *$E = \cup_n E_n$, where $E_0 \subseteq E_1 \subseteq E_2 \subseteq \cdots$ are finite Borel equivalence relations;*
(iii) *There is a Borel map assigning to each $[x]_E = C$ a linear order $<_C$ of C of order type finite or $\mathbb{Z}$. (More precisely, to say $C \mapsto <_C$ is Borel means that the relation $x <_{[y]_E} z$ is Borel).*

Examples of hyperfinite E include E_0, $E(2^{\mathbb{Z}})$, E_{tail} (see [8]). On the other hand, $E(2^{F_2})$ is not hyperfinite. Hyperfinite Borel equivalence relations have the following closure properties

1) If $E \subseteq F$ or $E \leq F$ or $E = F \restriction A$ (A Borel) and F is hyperfinite, then so is E.
2) If the Borel set A is *full* for a countable Borel equivalence relation E and $E \restriction A$ is hyperfinite, so is E. (A set A is *full* if it meets every equivalence class).
3) [8] If $E_0 \subseteq E_1 \subseteq E_2 \subseteq \cdots$ are smooth, $\cup_n E_n$ is hyperfinite. (However it is not known if hyperfinite Borel equivalence relations are closed under increasing unions).

We have now the following basic fact for hyperfinite equivalence relations. Put

$$E \approx F \iff E \sqsubseteq F \quad \& \quad F \sqsubseteq E$$

Theorem 7. (Dougherty-Jackson-Kechris [8]). *Let E, F be hyperfinite, non-smooth Borel equivalence relations. Then $E \approx F$.*

In particular, it follows that given any two such E, F there are full sets A, B such that $E \upharpoonright A \cong F \upharpoonright B$, i.e., $E \upharpoonright A$, $F \upharpoonright B$ are Borel isomorphic. (On the other hand there are hyperfinite, non-smooth E, F with $E \not\cong F$). Thus any two hyperfinite, non-smooth Borel equivalence relations look very much alike (for example, there is a "canonical" 1–1 correspondence between their non-atomic, ergodic, quasi-invariant measure classes).

Remark. The proof of the preceding result shows also that E_{tail} is hyperfinite, and so $E_{\text{tail}} \approx E_0 \approx E(2^{\mathbb{Z}})$. (The fact that $E_0 \approx E(2^{\mathbb{Z}})$ answers a question of Mycielski, see [25], I.6, who showed that $E_0 \approx E$, where E is the equivalence relation on $\mathbb{R}$ given by $xEy \iff \exists q \in \mathbb{Q}(x + q = y)$.) In fact, more generally, if $T : X \to X$ is a Borel map and $xEy \iff \exists n \exists m \quad T^n x = T^m y$, then E is the increasing union of a sequence of smooth Borel equivalence relations (this extends a result of Vershik [32], who proved this in the measurable context).

We have thus seen that the partial (pre)order $\leq$ of hyperfinite Borel equivalence relations (on uncountable sets) has only two elements: $\Delta(2^\omega)$ and E_0.

There is one important question that is open about hyperfiniteness, namely whether the notion is effective. More precisely we have the following

Problem 8. Let E be a Δ^1_1 equivalence relation on $\mathcal{N} = \omega^\omega$. Assume E is hyperfinite. Is there a Δ^1_1 bijection $T : \mathcal{N} \to \mathcal{N}$ such that $E = E_T$?

Notice that smoothness is effective by Theorem 2.6.

We proceed now to the next level.

3.4. Amenable (countable) Borel equivalence relations

This notion was introduced in [21], by carrying over a measure theoretic notion of Zimmer [35]. We will briefly review below some facts and open problems about this notion. For more information, see [21].

Definition 9. A countable Borel equivalence relation E on X is **amenable** if there is a map assigning to each equivalence class C of E a finitely additive probability measure φ_C defined on all subsets of C such that $C \mapsto \varphi_C$ is universally measurable, i.e., for each Borel bounded $F : X^2 \to \mathbb{R}$, the function $f : X \to \mathbb{R}$ given by

$$f(x) = \int_{[x]_E} F(x,y) \mathrm{d}\varphi_{[x]_E}(y)$$

is universally measurable.

Recall that a countable group G is **amenable** if there is a G-invariant finitely additive probability measure φ defined on all subsets of G. By a result of Mokobodzki (see [6]), for each probability measure μ on 2^G one can find such a measure φ which is μ-measurable (viewing φ as a map of 2^G into [0,1]) and if the Continuum Hypothesis (CH) holds, actually φ can be taken to be universally measurable. Using this, it is easy to see from CH that every E_G, where G is amenable, is amenable. This includes the case of abelian, solvable, etc. G. In particular (from CH): hyperfinite $\Rightarrow$ amenable. This can be extended as follows.

Theorem 10. ([21]). (CH) *Let E be a countable Borel equivalence relation. If there is a Borel map assigning to each equivalence class C of E a linear order $<_C$ of C which is scattered (i.e., contains no copy of $\mathbb{Q}$), then E is amenable.*

Examples of non-amenable E include $E(2^{F_2})$ and $\equiv_T$, $\equiv_A$.

This notion and the above result were originally used in [21] to solve a problem of Slaman-Steel [29] about orderings on Turing degrees. Orderings on equivalence classes and an operator-theoretic version of the preceding theorem also came up independently in work in operator algebras of Muhly, Saito and Solel [26].

Amenability in the context of Borel equivalence relations is not yet well understood and there is even a question whether the above is the "right" definition of amenability. Some basic problems are the following:

Problem 11. Is amenability the same as hyperfiniteness?

The answer is positive in the measure-theoretic category (see Connes-Feldman-Weiss [5]). From this it follows, assuming CH, that any amenable Borel equivalence relation is induced by a universally measurable automorphism i.e., is "universally measurably" hyperfinite. As far as we know, Problem 11 is open even in the case of E_G, for G amenable (see Weiss [34]). Notice also that Sullivan-Weiss-Wright [31] (with an additional argument by Woodin) prove that if E on a perfect Polish space X has the property that every invariant Borel set is either meager or comeager, then $E \restriction A$ is hyperfinite for an invariant comeager Borel set A. In particular, $\equiv_T$ is hyperfinite on an invariant comeager Borel set.

Problem 12. Is there a Glimm-Effros type dichotomy for amenable (or perhaps hyperfinite) equivalence relations?

A strong possible formulation (that settles Problem 11 as well) is the following: Is there a non-amenable equivalence relation E_1, perhaps induced by some appropriate action of F_2, which embeds in any given non-hyperfinite E? (If such a result holds effectively this would also imply that hyperfiniteness is effective). Notice that this can be viewed as an analog of the following classical problem for groups: Does every non-amenable countable group contain F_2? (see [33]). The answer in this case is of course known to be negative (see again [33]).

Up until now we have not yet seen equivalence relations strictly between E_0 and $E(2^{F_2})$ (in $\leq$). Such examples have been pointed out to us by Zimmer and also Adams. We describe here the Borel version of Adams' notion of a treeable equivalence relation (see Adams [1]).

3.5. Treeable (countable) Borel equivalence relations

Definition 13. Let E be a countable Borel equivalence relation. We say that E is **treeable** if there is a Borel map which assigns to each equivalence class C of E a tree on C, i.e., an acyclic, connected graph on C.

Examples of treeable E include any E_α, where α is a *free* action of the free group F_n with n generators. (An action $(x, g) \mapsto x \cdot g$ is *free* if $x \neq x \cdot g$ for all $g \neq 1$ and all $x \in X$.) It immediately also follows that: hyperfinite $\Rightarrow$ treeable. We have now

Theorem 14. (Adams [2], Adams-Spatzier [3]). *There are countable Borel equivalence relations which are not treeable.*

Now one can verify that if $E \leq F$ or $E \sqsubseteq F$ and F is treeable, so is E. Thus we have

Corollary 15. *$E(2^{F_2})$, $\equiv_T$ are not treeable.*

Corollary 16. *If $E = E_\alpha$, where α is a free action of F_2 which has an invariant probability measure, then $E_0 < E < E(2^{F_2})$, ($E < F$ means $E \leq F$ & $F \not\leq E$).*

We conclude with two open problems

Problem 17. Is $\equiv_T$ universal? In other words if E is a countable Borel equivalence relation, is it true that $E \sqsubseteq \equiv_T$?

Problem 18. Find countable Borel E, F (on uncountable sets) such that $E \not\leq F$ and $F \not\leq E$.

In other words we do not know yet if $\leq$ restricted to countable Borel equivalence relations (on uncountable sets) is a (pre)-linear ordering or not.

4. Addendum

We have recently established the following Borel version of the result in J. Feldman, P. Hahn and C.C. Moore, Orbit structure and countable sections for actions of continuous groups, Adv. in Math. 28, (1978), 186-230.

Theorem 19. *Let G be a second countable locally compact group and $\alpha : G \times X \to X$ a Borel action of G on a Borel set X in a Polish space. If $E = E_\alpha$ is the induced Borel equivalence relation, then there is a Borel set $B \subseteq X$ and a nbhd U of the identity in G such that*

(i) *B is full (i.e. meets every equivalence class) and*
(ii) $\forall x \in B \quad (x \cdot U \cap B = \{x\})$.

In particular, B meets every equivalence class in an at most countable set.

It follows that for any such E there is a countable Borel equivalence relation F (namely $E \restriction B$) such that

$$E \approx^* F \Longleftrightarrow E \leq F \quad \wedge \quad F \leq E\,.$$

Thus up to $\approx^*$–equivalence countable Borel equivalence relations are the same as those induced by Borel actions of second countable locally compact groups. (This may be also useful for Problem 18).

The conjecture in the Remark at the end of §2 has now been proved by A. Ditzen and (independently) M. Foreman-M. Magidor.

References

1. S. Adams, *Trees and amenable equivalence relations*, Erg. Theory and Dyn. Systems **10** (1990), 1–14.
2. S. Adams, *Indecomposability of treed equivalence relations*, Israel J. Math **64(3)** (1988), 362–380.
3. S. Adams and R. Spatzier, *Kazhdan groups, cocycles and trees*, Amer. J. Math **112** (1990), 271–287.
4. J. Burgess, *A selection theorem for group actions*, Pac. J. Math. **80(2)** (1979), 333–336.
5. A. Connes, J. Feldman and B. Weiss, *An amenable equivalence relation is generated by a single transformation*, Erg. Theory and Dyn. Systems **1** (1981), 431–450.
6. C. Dellacherie and P-A. Meyer, *Theórie discrète du potentiel*, Hermann, 1983.
7. R. Dougherty, S. Jackson and A. Kechris, *The structure of equivalence relations on Polish spaces,* I*: An extension of the Glimm-Effros dichotomy*, circulated notes, March 1989.
8. R. Dougherty, S. Jackson and A. Kechris, *The structure of equivalence relations on Polish spaces,* II*: Countable equivalence relations or descriptive dynamics*, circulated notes, March 1989.
9. E. Effros, *Transformation groups and C^*-algebras*, Ann. of Math. **81(1)** (1965), 38–55.
10. E. Effros, *Polish transformation groups and classification problems*, General Topology and Modern Analysis, Rao and McAuley, eds. (1980), Academic Press, 217–227.
11. J. Feldman and C. C. Moore, *Ergodic equivalence relations, cohomology and von Neumann algebras,* I, Trans. Amer. Math. Soc. **234(2)** (1977), 289–324.
12. M. Foreman, *A Dilworth decomposition theorem for λ-Suslin quasi-orderings of* $\mathbb{R}$, in "Logic, Methodology and Philosophy of Science VIII", North Holland, 1989, 223–244, 223–244.
13. H. Friedman and L. Stanley, *A Borel reducibility theory for classes of countable structures*, J. Symb. Logic **54(3)** (1989), 894–914.
14. J. Glimm, *Locally compact transformation groups*, Trans. Amer. Math. Soc. **101** (1961), 124-138.
15. A. Godefroy, *Some remarks on Suslin sections*, Fund. Math. **LXXXIV** (1986), 159–167.
16. L. Harrington, A. Kechris and A. Louveau, *A Glimm-Effros dichotomy for Borel equivalence relations*, J. Amer. Math. Soc. **3**(4) (1990), 903–928.
17. L. Harrington, D. Marker and S. Shelah, *Borel orderings*, Trans. Amer. Math. Soc. **310(1)** (1988), 293-302.
18. L. Harrington and R. Sami, *Equivalence relations, projective and beyond*, Logic Colloq. **78** (North Holland, 1979), 247–264.
19. K. Kada, *A Borel version of Dilworth's theorem*, (to appear).
20. Y. Katznelson and B. Weiss, *The construction of quasi-invariant measures*, Israel J. Math. **12** (1972), 1–4.
21. A. Kechris, *Amenable equivalence relations and Turing degrees*, J. Symb. Logic (to appear).
22. W. Krieger, *On Borel automorphisms and their quasi-invariant measures*, Math. Z. **151** (1976), 19–24.
23. A. Louveau, *Two results on Borel orders*, J. Symb. Logic **34(3)** (1989), 865–874.
24. A. Louveau and J. Saint Raymond, *On the quasi-ordering of Borel linear orders under embeddability*, J. Symb. Logic **55**(2) (1990), 537–560.
25. D. Mauldin and S. Ulam, *Mathematical problems and games*, Adv. in Appl. Math **8** (1987), 281–344.

26. P. Muhly, K. Saito and B. Solel, *Coordinates for triangular operator algebras*, Ann. of Math. **127** (1988), 245–278.
27. S. Shelah and B. Weiss, *Measurable recurrence and quasi-invariant measures*, Israel J. Math. **43** (1982), 154–160.
28. J. Silver, *Counting the number of equivalence classes of Borel and coanalytic equivalence relations*, Ann. Math. Logic **18** (1980), 1–28.
29. T. Slaman and J. Steel, *Definable functions on degrees*, in "Cabal Seminar 81–85", Lecture Notes in Math. 1333, Springer-Verlag, 1988, 37–55.
30. J. Steel, *Long games*, in "Cabal Seminar 81–85", Lecture notes in Math. 1333, Springer-Verlag, 1988, 56–97.
31. D. Sullivan, B. Weiss and J.D.M. Wright, *Generic dynamics and monotone complete C^*-algebras*, Trans. Amer. Math. Soc. **295(2)** (1986), 795–809.
32. A. Vershik, *The action of $PSL(2,\mathbb{Z})$ on $\mathbb{R}^1$ is approximable*, Uspekhi Mat. Nauk. **33(1)** (1978), 209–210 (in Russian).
33. S. Wagon, *The Banach-Tarski Paradox*, Cambridge Univ. Press, 1985.
34. B. Weiss, *Measurable dynamics*, Cont. Math. **26** (1984), 395–421.
35. R. Zimmer, *Hyperfinite factors and amenable ergodic actions*, Inv. Math **41** (1977), 23–31.

DEPARTMENT OF MATHEMATICS, CALIFORNIA INSTITUTE OF TECHNOLOGY, PASADENA CA 91125

RESEARCH SUPPORTED IN PART BY NSF GRANT DMS-8718847

CLASSIFYING BOREL STRUCTURES

ALAIN LOUVEAU

1. INTRODUCTION

In what follows, a Borel structure is a first-order structure $\mathcal{A}$ (in some countable language) such that both the domain of $\mathcal{A}$, and its relations and functions, are Borel (sets or functions) in some Polish space.

In Analysis, these structures occur quite naturally, but have been much less studied than their topological counterparts. Reasons for that may be that for most practical uses the topological frame is sufficient, and also the lack, in the Borel case, of the powerful duality methods. Still there has been some investigations, for particular Borel structures, like e.g. the work of J. P. R. Christensen on Borel groups [C] or the study of Borel transformations in Ergodic theory. Moreover, there seems to be a renewal of interest in Borel structures in various parts of Analysis, e.g. in specific Borel subgroups of the circle in Harmonic Analysis (Host-Méla-Parreau [H-M-P]), or in Borel equivalence relations in Ergodic theory and in C^*-algebra theory (see the paper by Kechris [Ke], in this volume).

In the mid-seventies H. Friedman proposed a systematic model-theoretic study of the Borel structures, as an important intermediate level between the countable structures and the general abstract structures of standard model theory. He proved some general model theoretic results for Borel structures, like a completeness theorem which insures *the existence, for first order theories with infinite models, of an uncountable Borel model in which every definable relation is Borel* (see H. Friedman [F] and Steinhorn [Stn]). He also proved specific structural results, in particular on Borel linear orders, that we will discuss later.

Since then, a lot of results, concerning Borel partial orders, Borel linear orders and Borel equivalence relations have been established. Although there is no general theory relating these results, they all share the same flavour, and are proved using very similar techniques, those of Descriptive Set Theory. The aim of this paper is to give an account of what has been obtained in these last 15 years, and to organize the exposition of the results so that to stress these similarities.

When one wants to classify a family of structures, one usually defines an equivalence relation between structures, and then tries to attach "invariants" to each equivalence class. In our context, the natural equivalence relation which comes to mind is Borel isomorphism between structures. However there are very few known results for it, and we will instead use a weaker equivalence relation, Borel bi-reducibility, which is associated to the following partial order: Let $\mathcal{A}$, $\mathcal{B}$ be two Borel structures in the same language, with domains $|\mathcal{A}|$ and $|\mathcal{B}|$. A function $f : |\mathcal{A}| \to |\mathcal{B}|$ is a *reduction*, or *reduces* $\mathcal{A}$ to $\mathcal{B}$ if for all predicate symbols R and function symbols $\varphi(x_1, \dots, x_k) \in R^{\mathcal{A}} \leftrightarrow (f(x_1), \dots, f(x_k)) \in R^{\mathcal{B}}$ and $f(\varphi^{\mathcal{A}}(x_1, \dots, x_k)) = \varphi^{\mathcal{B}}(f(x_1), \dots, f(x_k))$. If moreover f is one-to-one, we say that f is an *embedding* from $\mathcal{A}$ into $\mathcal{B}$.

Let us say that $\mathcal{A}$ is Borel reducible to $\mathcal{B}$ if there is a Borel reduction $f : |\mathcal{A}| \to |\mathcal{B}|$, in notations $\mathcal{A} \leq \mathcal{B}$, and that $\mathcal{A}$ and $\mathcal{B}$ are Borel bi-reducible, $\mathcal{A} \approx \mathcal{B}$, if $\mathcal{A} \leq \mathcal{B}$ and $\mathcal{B} \leq \mathcal{A}$. Similarly we write $\mathcal{A} \leq^* \mathcal{B}$ if $\mathcal{A}$ is Borel embeddable in $\mathcal{B}$, i.e. there is a Borel one-to-one reduction $f : |\mathcal{A}| \to |\mathcal{B}|$, and $\mathcal{A} \approx^* \mathcal{B}$ if $\mathcal{A} \leq^* \mathcal{B}$ and $\mathcal{B} \leq^* \mathcal{A}$.

The terminology of "reduction" comes from the analogous terminology used in the theory of Wadge classes (where the reductions are continuous). The usefulness of this notion emerged mainly from works of Louveau and Saint-Raymond on Borel orders (where the analogy with the Wadge hierarchy is exploited) and from works on equivalence relations by Harrington, Kechris, and Louveau [H-K-L], and by H. Friedman and L. Stanley [F-S]. This notion has both the advantage of structuring the results on Borel structures, but it also relates them to older results, and to apparently barely related questions—like the Wadge ordering. In some cases, and especially for equivalence relations, it also seems to be the most natural notion to consider, or at least to lead to very natural questions in the applications.

One can of course also consider various other notions of definable reducibility, like continuous reducibility and embeddability (that we will denote $\leq_c$ and $\leq_c^*$), or projective reducibility, etc... . We will occasionally say a few words about these notions, as well as about the abstract reducibility (i.e. using arbitrary reductions), that we will denote by $\leq_a$ and $\leq_a^*$.

Our main task, given a class Γ of Borel structures (in some given language), will be to get information about the partial orderings $(\Gamma/\approx, \leq)$ and $(\Gamma/\approx^*, \leq^*)$.

The kind of results we will look for are

(a) Cofinality results: To try to find simple—and easily describable—subsets of Γ which are cofinal in it. In the sequel, these subsets will be well ordered chains, and thus will give a "natural" ranking on Γ.

(b) Dichotomy results: Typically, a dichotomy result asserts, given two structures $\mathcal{A}_0, \mathcal{A}_1$ in Γ, that any structure $\mathcal{A}$ in Γ either is Borel reducible to $\mathcal{A}_0$, or Borel reduces $\mathcal{A}_1$, i.e. $\mathcal{A} \leq \mathcal{A}_0$ or $\mathcal{A}_1 \leq \mathcal{A}$.

If $\mathcal{A}_0 < \mathcal{A}_1$, the dichotomy results not only says that $\mathcal{A}_1$ is the successor of $\mathcal{A}_0$ in Γ, but also that both $\mathcal{A}_0$ and $\mathcal{A}_1$ are *nodes* in $(\Gamma, \leq)$, i.e. are comparable to all other structures.

If $\mathcal{A}_0$ and $\mathcal{A}_1$ are incomparable, the dichotomy says that $\{\mathcal{A}_0, \mathcal{A}_1\}$ is a maximal antichain in $(\Gamma, \leq)$.

In all dichotomy results we will discuss below, the dichotomy can be strengthened in the following way: In case $\mathcal{A} \leq \mathcal{A}_0$, the Borel reduction can be found Δ^1_1 in codes for the structures $\mathcal{A}$ and $\mathcal{A}_0$; and in case $\mathcal{A}_1 \leq \mathcal{A}$, the reduction is in fact continuous and one to one, i.e. $\mathcal{A}_1 \leq^*_c \mathcal{A}$. Note that usually, these refinements *are* instrumental in proving the dichotomy results: They allow to bring in the techniques from Effective Descriptive Set Theory, especially the use of the Gandy-Harrington topology, or the equivalent notion of forcing, see [H-M-S], [Lo3].

Weak versions of dichotomy results correspond to isolating finite maximal antichains in $(\Gamma, \leq)$. This leads to the third type of results.

(c) Wqo or Bqo results: Recall that a quasi-order $(Z, \leq_Z)$ is a well-quasi-ordering, or wqo, if every $<_Z$-decreasing sequence and every antichain in Z are finite. If $(Z, \leq_Z)$ is wqo, one can in a standard way attach to each $z \in Z$ an ordinal—its height in $\leq_Z$, which is almost an invariant for z, as it is shared only by finitely many other elements. So wqo results for $(\Gamma, \leq)$ almost correspond to a complete classification of the structures in Γ, up to Borel bi-reducibility.

As usual for studying the wqo property, one in fact uses the more tractable better quasi-order (bqo) theory of Nash-Williams and Laver ([N], [La]). We won't get into this here, and refer the interested reader to the above papers, as well as [L-S4] and [vE-M-S]. One should note that these bqo results bring in another fundamental tool from descriptive set theory, the use of determinacy results for infinite games, which allow in some cases to build reductions between Borel structures from winning strategies in ad hoc games.

2. Structures with Unary Predicates

The first interesting case concerns Borel structures with only equality, i.e. the study of cardinality for Borel sets. As is well-known, any uncountable Borel set has cardinality c, and this can be made more precise by the

Perfect Set Theorem (Suslin; Harrison. See [Mo]). *For every Borel set X, either $X \leq^* \omega$ or $2^\omega \leq^*_c X$. Moreover in the first case the reduction can be found Δ^1_1 in a code for X.*

This result is a paradigm for all dichotomy results. And using the Cantor-Bernstein technique, it easily follows that *any two Borel sets of the same cardinality are Borel isomorphic*, so that for this class of structures, the notions of isomorphism, Borel isomorphism and Borel bi-reducibility coincide.

The case of finitely many unary predicates is very similar: Again isomorphism, Borel isomorphism and Borel bi-reducibility coincide, and the equivalence class of a structure $(X, A_0, \ldots, A_{n-1})$ is determined by the cardinality of each atom $A_S = \bigcap_{s(i)=1} A_i \cap \bigcap_{s(i)=0}(X \backslash A_i)$, for $s \in 2^n$.

For structures with countably many unary predicates, or with unary functions, the situation is not really known, and probably quite interesting and complicated. Note that the latter case contains the case of Borel transformations, which are studied (usually in a measure-theoretic, not descriptive set theoretic context) in ergodic theory and dynamical systems.

The situation for structures with a unary predicate is much less trivial if instead of considering Borel reducibility, one considers the partial ordering of continuous reducibility. To simplify the statements, let us consider only the case where the domain is (a closed subset of) ω^ω. One then gets the *Wadge ordering*, usually denoted by $\leq_W$, on Borel sets: $A \leq_W B$ if for some continuous $f : \omega^\omega \to \omega^\omega A = f^{-1}(B)$.

Wadge's Main Lemma [W], which uses Borel determinacy, asserts that *a Borel set A and its complement $A^c = \omega^\omega \backslash A$ always form a maximal antichain in $\leq_W$.* If $A \approx_W A^c$, A is said to be selfdual, and non self dual otherwise. Self-dual sets can easily be described in terms of non self dual ones. And for non self dual sets, Wadge's lemma can be strengthened in the following dichotomy result:

Theorem. *Let $A \subseteq \omega^\omega$ be Borel, and non self dual. Then one can find a set $A_0 \subseteq \omega^\omega$, $A_0 \approx_W A$, and a structure $(K_1, A_1) \approx_W (\omega^\omega, \check{A})$ such that for any Borel set $B \subseteq \omega^\omega$*

(i) either $B \leq_W A_0$, and in this case the continuous reduction can be found Δ^1_1 in codes for A_o and B or

(ii) $(K_1, A_1) \leq_W (\omega^\omega, B)$, and in this case the continuous reduction can be found one-to-one.

[In fact if A is $\underset{\sim}{\Delta}^0_2$ the set K_1 is a countable compact set, and if A is not $\underset{\sim}{\Delta}^0_2$, one can take $K_1 = 2^\omega$, so that in both cases, the reduction is a homeomorphism on its image.]

This theorem is the result of many investigations. For the equivalence between $B \not\leq_W A$ with (ii), the archetypical result is Hurewicz's result [Hu] *characterizing $\mathcal{G}_\delta$ sets among $\underset{\sim}{\Pi}^1_1$ sets as those for which no relatively closed subset is homeomorphic to* $\mathbb{Q}$. The general case is proved for most classes in van Engelen-Miller-Steel [vE-M-S] and for all of them in Louveau-Saint Raymond [L-S1], [L-S2]. For the equivalence between $B \leq_W A$ and (i), the archetypical result is the effective theorem of Louveau [Lo1] about the Borel hierarchy, and the general result is given in Louveau [Lo2], in rather different terms.

The other main feature of the Wadge ordering is:

Theorem (Martin). *The order $\leq_W$ is wellfounded, hence* wqo, *on the Borel sets.*

The original proof of Martin (see [vE-M-S]) uses Borel determinacy (although the result, as well as the preceding dichotomy result, can be proved in second order arithmetic, see Louveau-Saint Raymond, [L-S1] and [L-S2]). The result is extended to the case of finitely many Borel sets, (and more general situations), and strengthened to a bqo result in van Engelen-Miller-Steel [vE-M-S].

Many other results are known for the ordering $\leq_W$. One knows its ordinal length (Wadge [W]), and various descriptions of all classes (Wadge [W], Louveau [Lo2]). A structural result of Steel [Stl] allows to distinguish between the twin dual classes, and most standard structural descriptive set theoretic properties are exactly localized in the hierarchy (Louveau-Saint Raymond [L-S3]).

Although it may seem that the preceding discussion is a digression from our main concern, Borel reducibility, this is not really so. The reason is the existence of "automatic continuity" phenomena: For some important Borel structures, Borel reductions are necessarily continuous, or close to continuous. For example, a Borel homomorphism between Polish groups is necessarily continuous. Also, an increasing function from $\mathbb{R}$ into $\mathbb{R}$ is continuous except on a countable set. This last remark, together with the results above on the Wadge ordering, is the basis for the investigations about Borel orders in Louveau-Saint Raymond [L-S4].

3. Borel Equivalence Relations

In 1970, Silver [Si] proved the following "cardinality" result about Borel— and even $\underset{\sim}{\Pi}^1_1$—equivalence relations: *Each $\underset{\sim}{\Pi}^1_1$ equivalence relation either has countably many classes, or admits a perfect set of pairwise inequivalent elements.*

This result was the starting point for many investigations, especially about possible extensions to more complicated definable equivalence relations (see [Sh], [H-S]).

A later much simpler proof by Harrington of Silver's result leads to the following dichotomy result for the ordering $\leq$ on Borel equivalence relations.

Theorem (Harrington [Ha]). *Let (X, E) be a Borel equivalence relation. Then*

–either $(X, E) \leq (\omega, =)$, and in this case the reduction can be found Δ^1_1 in (a code for) (X, E)

–or $(2^\omega, =) \leq (X, E)$, and in this case the reduction can be found continuous and one-to-one.

Harrington's proof of this result (and of the natural extension to $\underset{\sim}{\Pi}^1_1$ equivalence relations) is historically very important, for it is the first place where the Gandy-Harrington forcing is used to get dichotomy results.

It follows from this result that the first $\omega + 2 \approx$-classes of Borel equivalence relations are those of $(n, =)$ for $n < \omega$, $(\omega, =)$ and $(2^\omega, =)$.

Very recently, another dichotomy result has been proved by Harrington, Kechris and Louveau.

Let E_0 be the following equivalence relation on $2^\omega : \alpha E_0 \beta \leftrightarrow \alpha$ and β are eventually equal $\leftrightarrow \exists k \forall n \geq k\, \alpha(n) = \beta(n)$.

Theorem (Harrington-Kechris-Louveau [H-K-L]). *Let (X, E) be a Borel equivalence relation. Then*

–either $(X, E) \leq (2^\omega, =)$, and in this case the reduction can be found Δ^1_1 in a code for (X, E)

–or $(2^\omega, E_0) \leq (X, E)$, and in this case the reduction can be found continuous and one-to-one.

We won't discuss here the origins of this dichotomy result, nor its relevance in Analysis—in particular for building ergodic measures. We refer the reader to Kechris' paper [Ke] in this volume.

So by this result, one gets that the $\approx$-class of $(2^\omega, E_0)$ is the $(\omega + 3)$rd class in the ordering $\leq$ on Borel equivalence relations.

Rather few other results are known for this ordering: *It is not linear, and has no maximal element*, by a result of Friedman-Stanley [F-S], which uses the Borel diagonalization results of H. Friedman, see [Sta]. It follows easily that there are chains of length ω_1 in it. And Harrington has noticed that *there is a chain* of ω_1 *Borel equivalence relations which is cofinal in* $\leq$. However, there is no known "natural" example of such a chain.

It is also not known if there are any dichotomy results above $(2^\omega, E_0)$, and the wqo problem is open.

A lot of attention has been paid to a subclass of the class of Borel equivalence relations, those with countable classes, which is of particular importance in the applications. Although there are now many results for this subclass, the situation is still rather unclear—and seems to indicate that the problem of classifying Borel equivalence classes is quite difficult. We again refer the reader to Kechris' paper [Ke] for a discussion of these results, as well as bibliographical references.

4. Borel Orderings

Let us consider first Borel partial (pre-)orders. The main result is the following dichotomy-type result, proved by Harrington and Shelah (see [H-M-S]), and which is an extension of the Silver-Harrington result on Borel equivalence relations

Theorem (Harrington-Shelah). *Let (X, R) be a Borel partial preorder. Then*

–either there is a decomposition $(X_n)_{n\in\omega}$, of X into Borel sets which are R-chains (i.e. R restricted to X_n is a linear preorder) and in this case the partition (X_n) can be found Δ^1_1 in a code for (X, R)

–or there is a perfect subset of X of pairwise R-incomparable elements.

This result refines an earlier result of Shelah [S] stating that *a Borel partial order admitting an uncountable antichain must admit a perfect antichain.* It can also be viewed as an infinite Borel analog of the classical theorem of Dilworth [D] which states that *a partial preorder for which all antichains are of cardinality bounded by $k \in \omega$ is the union of k chains.*

Recently, K. Kada has proved the following finite Borel version of Dilworth's theorem:

Theorem (Kada [Ka]). *If (X, R) is a Borel partial preorder, and all antichains in it are of cardinality bounded by $k < \omega$, then $X = \bigcup_{i=1}^{k} X_i$, where X_i are Borel R-chains. Moreover the X_i's can be found Δ^1_1 in a code for (X, R).*

For both previous theorems, the effective refinements are instrumental for the proofs.

Let us consider now the subclass BOR of Borel linear orders. In this case Borel reductions are just Borel strictly increasing functions, and $\leq$ and $\leq^*$ coincide.

For each ordinal $\xi < \omega_1$, consider the structures $(2^\xi, \text{lex})$ (resp. $2^{<\xi}, \text{lex})$ of sequences of 0's and 1's of length ξ (resp. $< \xi$), with the lexicographical ordering. These are clearly Borel (in fact $\underset{\sim}{\Delta}^0_2$) linear orders.

The first result for $(\text{BOR}, \leq)$ is a cofinality result.

Theorem (Harrington-Shelah, see [H-M-S]). *For every (X,R) in* BOR, *there is a $\xi < \omega_1$ such that $(X,R) \le (2^\xi, \text{lex})$. Moreover ξ and the Borel reduction can be found Δ^1_1 in a code for (X,R).*

This result easily implies (Harrington-Shelah [H-S]) that *in any Borel linear order there are no ω_1-chains.* Clearly the set DEN of countable linear orders forms an initial segment of $(\text{BOR}, \le)$, with maximal element $(2^{<\omega}, \text{lex})$ ($\approx \mathbb{Q}$, with its usual order). And using the perfect set theorem, one easily shows that $(2^\omega, \text{lex})$ ($\approx \mathbb{R}$) is a successor of $(2^{<\omega}, \text{lex})$ in $\le$.

The next dichotomy result is due to Marker (see [H-M-S]).

Theorem (Marker). *For every (X,R) in* BOR

–*either $(X,R) \le (2^\omega, \text{lex})$, and in this case the Borel reduction can be found Δ^1_1 in a code for (X,R)*

–*or there is a perfect set of pairwise disjoint non empty closed intervals in (X,R), and hence $(2^{\omega+1}, \text{lex}) \le (X,R)$ (in fact continuously).*

The non-effective version of this result is due to Friedman [F], and implies that *there is no Borel Souslin line* (Friedman-Shelah [F], [Stn]).

A similar situation holds at all limit countable ordinals:

Theorem (Louveau [Lo3]). *Let $\xi < \omega_1$, and $(X,R) \in$ BOR. Then*

(i) *Either $(X,R) \le (2^{<\omega\cdot\xi}, \text{lex})$, in which case the Borel reduction can be found Δ^1_1 in codes for (X,R) and ξ or $(2^{\omega\cdot\xi}, \text{lex}) \le (X,R)$, in which case the reduction can be found continuous.*

(ii) *Either $(X,R) \le (2^{\omega\cdot\xi}, \text{lex})$, in which case the Borel reduction can be found Δ^1_1 in codes for (X,R) and ξ or $(2^{\omega\cdot\xi+1}, \text{lex}) \le (X,R)$, in which case the reduction can be found continuous.*

This result says that for all ξ $(2^{<\omega\cdot\xi}, \text{lex})$, $(2^{\omega\cdot\xi}, \text{lex})$ and $(2^{\omega\xi+1}, \text{lex})$ are three consecutive nodes in the ordering $(\text{BOR}, \le)$.

The last type of results deals with the bqo property. Note that the restriction of $\le$ to the class DEN of countable linear orders is the relation called by Fraïssé "abritement", and that by Laver's celebrated result [La], solving Fraïssé's conjecture, $(\text{DEN}, \le)$ *is a better-quasi-ordering.*

This result has been extended by Louveau and Saint Raymond [L-S4]. For each $\xi < \omega_1$, set $\text{BOR}_\xi = \{(X,R) \in \text{BOR}\}$ $(X,R) \le (2^{\omega\cdot\xi}, \text{lex})$. It immediately follows from Laver's theorem that $(\text{BOR}_1, \le)$ is bqo.

Theorem (Louveau-Saint Raymond). $(\text{BOR}_2, \le)$ *is a better-quasi-ordering. Moreover on $\text{BOR}_2, \le$ coincides with $\le_a$ (the order given by arbitrary reductions).*

The case of $(\mathrm{BOR}_\xi, \leq)$ for $\xi > 2$ is entirely open. However if one accepts strong set theoretical axioms, there are some partial results for the order $\leq_a$ (and in fact for various intermediate notions of definable reducibility):

Theorem (Louveau-Saint Raymond [L-S4]).

(i) *Assume projective determinacy. Then* $(\bigcup_{n\in\omega} \mathrm{BOR}_n, \leq_a)$ *is* bqo. *In fact the class of all projective linear orders which are projectively reducible to some* $(2^{\omega\cdot n}, \mathrm{lex})$, $n \in \omega$, *is* bqo *under projective reducibility.*

(ii) *Assume hyperprojective determinacy. Then* $(\mathrm{BOR}_\omega, \leq_a)$ *is* bqo *(and again* $\leq_a$ *can be replaced by some form of definable reducibility).*

A natural conjecture is that $(\mathrm{BOR}, \leq)$ should be bqo—and that this should be provable in ZFC, maybe even in second order arithmetic. (The proof in [L-S4] for BOR_2 heavily uses Borel determinacy.)

References

[C] J. P. R. Christensen, *Topology and Borel structure*, Mathematics Studies **10**, North Holland, 1974.

[D] R. P. Dilworth, *A decomposition theorem for partially ordered sets*, Ann. Math. **51** (1950), 161–166.

[vE-M-S] F. van Engelen, A. Miller and J. Steel, *Rigid Borel sets and better quasi order theory*, Contemporary Math. **65** (1987), 199–222.

[F] H. Friedman, *Borel structures in mathematics*, Manuscript (1979), Ohio State University.

[F-S] H. Friedman and L. Stanley, *A Borel reducibility theory for classes of countable structures*, J. of Symb. Logic **54** (1989), 894–914.

[Ha] L. Harrington, *A powerless proof of a theorem of Silver*, Handwritten Notes (1976), University of California, Berkeley.

[H-K-L] L. Harrington, A. S. Kechris and A. Louveau, *A Glimm-Effros dichotomy for Borel equivalence relations*, (to appear).

[H-M-S] L. Harrington, D. Marker and S. Shelah, *Borel orderings*, Trans. Amer. Math. Soc. **310** (1988), 293–302.

[H-S] L. Harrington and S. Shelah, *Counting equivalence classes for co-Suslin relations*, Logic Colloquium '80, ed. D. Van Dalen, D. Lascar and T. J. Smiley (1982), North Holland.

[H-M-P] B. Host, J. F. Méla and F. Parreau, *Saturated subgroups, groups of quasi-invariance and spectral analysis of measures*, Preprint.

[Hu] W. Hurewicz, *Relativ perfekte Teile von Punktmengen und Mengen (A)*, Fund. Math. **12** (1928), 78–109.

[Ka] K. Kada, *Une version boiélienne d'un théorème de Dilworth*, Trans. A.M.S (to appear).

[Ke] A. S. Kechris, *The structure of Borel equivalence relations in Polish spaces*, this volume.

[La] R. Laver, *On Fraïssé's order type conjecture*, Ann. of Math. **93** (1971), 89–111.

[Lo1] A. Louveau, *A separation theorem for* Σ^1_1 *sets*, Trans. A.M.S. **260** (1980), 363–378.

[Lo2] ———, *Some results in the Wadge Hierarchy of Borel sets*, CABAL Seminar 79-81, Lecture Notes in Math. **1019** (1983), Springer-Verlag, 28–55.

[Lo3] ———, *Two results on Borel orders*, J. of Symb. Logic **54** (1989), 865–873.

[L-S1] A. Louveau and J. Saint Raymond, *Borel classes and closed games, Wadge type and Hurewicz type results*, Trans. A.M.S. **304** (1987), 431–467.

[L-S2] ———, *The strength of Borel Wadge determinacy*, CABAL Seminar 81–85, A. S. Kechris, D. A. Martin, J. R. Steel (Eds.), Lecture Notes in Math. **1333** (1988), 1–30.

[L-S3] ———, *Les propriétés de réduction et de norme pour les classes de Boréliens*, Fund. Math. **131** (1988), 223–243.

[L-S4] ———, *On the quasi-ordering of Borel linear orders under embeddability*, J. of Symbolic Logic (to appear).

[Mo] Y. N. Moschovakis, *Descriptive Set Theory*, North Holland (1980).

[N] C. St. J. A. Nash-Williams, *On better quasi-ordering transfinite sequences*, Proc. Camb. Phil. Soc. **64** (1968), 273–290.

[Sh] S. Shelah, *On co-Suslin relations*, Israel J. of Math. **47** (1984).

[Si] J. H. Silver, *Counting the number of equivalence classes of Borel and coanalytic equivalence relations*, Annals of Math. Logic **18** (1980), 1–28.

[Sta] L. Stanley, *Borel diagonalization and abstract set theory: Recent results of Harvey Friedman*, in Harvey Friedman's research on the foundations of mathematics, L. A. Harrington et al. (Eds.) (1985), North Holland.

[Stl] J. R. Steel, *Determinateness and the separation property*, J. Symbolic Logic **46** (1981), 41–44.

[Stn] C. Steinhorn, *Borel structures and measure and category logics*, Model Theoretic Logics, Ed. Barwise and Feferman, Springer-Verlag, 1985.

[W] W. W. Wadge, *Thesis*, University of California, Berkeley (1984).

C.N.R.S., Equipe d'Analyse, T 46/4e, Université Paris VI,
4 Place Jussieu, 75252 Paris Cedex 05, FRANCE

WHAT IS MAC LANE MISSING?

Adrian R.D. Mathias

A sociologist observing the 1989/90 Logic Year at the Berkeley Mathematical Sciences Research Institute would have judged it to be a typical gathering of mathematicians, exchanging ideas, running seminars to chip away at current problems, and writing papers and books. But there was one speaker who from time to time would tell the others that they were working on the wrong problems in the wrong subject. This was not the result of a momentary aversion: Professor Mac Lane has for at least twenty years been saying that "set theory is obsolete," that "measurable cardinals are bizarre," and so on, and he has written one large book ([2]) and many articles in order to present his view of mathematics.

It is the purpose of this essay to examine his stance, and to suggest that insofar as Mac Lane urges the unity of mathematics, he is to be supported, but insofar as he secretly desires the uniformity of mathematics, he is to be opposed.

Perhaps one should begin with a few reflections on the psychology of mathematics. One of the remarkable things about mathematics is that I can formulate a problem, be unable to solve it, pass it to you; you solve it; and then I can make use of your solution. There is a unity here: we benefit from each other's efforts. In this regard mathematicians interact much more than do (say) historians or composers.

But if I pause to ask *why* you have succeeded where I have failed to solve a problem, I find myself faced with the baffling fact that you have thought of the problem in a very different way from me: and if I look around the whole spectrum of mathematical activity the huge variety of styles of thought becomes even more evident.

Is it desirable to press mathematicians all to think in the same way? I say not: if you take someone who wishes to become a set theorist and force him to do (say) algebraic topology, what you get is not a topologist but a neurotic. Uniformity is not desirable, and an attempt to attain it, by (say) manipulating the funding agencies, will have unhealthy consequences.

The purpose of foundational work in mathematics is to promote the unity of mathematics; the larger hope is to establish an ontology within which all can work in their different ways.

What, then, is Mac Lane's ontology? This seems to admit a clear answer. In his book *Mathematics: Form and Function* he urges the claims of a system he calls ZBQC, which initials stand for Zermelo with Bounded Quantification and Choice, to supply all that he needs to do the mathematics he wants to do.

The axioms of this system are Extensionality, Null Set, Pairing, Power Set, Union, Infinity, Comprehension for Δ_0 formulæ, Regularity (*i.e.* Foundation) and Choice.

This system provides for the existence of the real numbers, and for ω types over them, thus yielding the complex numbers, functions from reals to reals, functionals and so on.

That this system represents a natural portion of mathematics may be seen from the way in which it keeps reappearing, first as the simple theory of types, and more recently as topos theory, with each of which it is equiconsistent. A natural model for it is $V_{\omega+\omega}$.

It is plain from Mac Lane's book that this system indeed supports a large amount of mathematics, more than I shall ever learn. Why then need we go outside it?

I suggest that an area ill supported by Mac Lane's system ZBQC is that of iterative constructions. We know from the work of Cantor onwards that there are processes which need more than ω steps to terminate; of which examples may be found even within traditional areas of mathematics. For example, within the space of continuous functions on $[0, 1]$, the class of differentiable functions forms a set which is not a Borel set but is naturally expressible as the union of $\aleph_1$ Borel sets; and this has implications for the problem of building the anti-derivative of a given function.

So therefore let us look for a moment at abstract recursion theory and ask how easily it sits within Mac Lane's system.

A well-established axiomatic framework for abstract recursion theory is the system of Kripke-Platek.

Theorem 1. *If Consis(ZBQC) then Consis(ZBQC + KP).*

The intuition behind the proof of theorem 1 is this: just as $V_{\omega+\omega}$ is a natural model for ZBQC, so $H_{\beth_\omega}$, the collection of sets which are members of transitive sets of cardinality less than $\beth_\omega$ is a natural model of ZBQC + KP; moreover each transitive set in the second model is isomorphic to some well-founded extensional relation which is a member of $V_{\omega+\omega}$. Hence the second model can be regarded as coded within the first, the building bricks being well-founded extensional relations with designated elements. To get a relative consistency proof one has to convert this semantic argument into a syntactic manipulation.

With slightly more trouble one may establish

Theorem 2. *If Consis(ZBQ) then Consis(ZBQC + KP + V=L).*

Here ZBQ is ZBQC with the axiom of choice omitted. The proof is similar to that of theorem 1, but here the building bricks are fragments of the constructible hierarchy defined along well-orderings.

Thus ZBQC has *via* suitable coding a reasonable capacity for recursive constructions; and this would support Mac Lane's thesis that it is a reasonable basis for much of mathematics. However it will, as is clear from the work of Harvey Friedman, fail to support many constructions: it will not be able to prove Borel determinacy, which requires the iteration of the power set operation through all countable ordinals; similarly it will not be able to prove Borel diagonalization.

Set theory is so rich a theory that it has been claimed for much of this century to be the foundation of mathematics. In ontological terms this claim is not unreasonable; but Mac Lane resists. I would guess that his reason is not so much that he objects to the ontology of set theory but that he finds the set-theoretic cast of mind oppressive and feels that other modes of thought are more appropriate to the mathematics he wishes to do.

One must acknowledge that ideas from category theory provide a smooth way to handle a large amount of material. However to reject a claim that set theory supplies a universal mode of mathematical thought and of mathematical existence need not compel one to declare set theory entirely valueless.

Let us therefore set aside set theory's claim to be a foundation of the whole of mathematics, it being misguided to define the worth of a subject solely in terms of its serviceability to other areas of mathematics. Instead let us define set theory to be the study of well-foundedness. As such, it is a worthy object of study; and it can scarcely be said that this is a subject of little content !

From this point of view, Mac Lane's view that "measurable cardinals are bizarre" becomes hard to defend. May we suppose him to mean that he sees no need to think about them and therefore resents a suggestion that he should think about them?

In terms of the study of well-foundedness, measurable cardinals are natural objects: just as ZBQC has resurfaced in many forms, so do measurable cardinals keep bobbing up in unexpected contexts. The hypothesis that they exist, or the hypothesis that in some inner model there are measurable cardinals may be construed as saying that in certain circumstances the direct limit of well-founded structures is well founded. Other large cardinal axioms may also be interpreted as assertions of this general kind. These hypotheses seem worthy of study: well-foundedness is important, being central to the general enterprise of constructing objects by recursion, and it is natural to ask when well-foundedness is preserved under direct limits. These questions are interesting in their own right.

This might be a good moment to challenge one of Mac Lane's opinions, which I believe to rest on a misconception. On page 359 of his book he writes, after reflecting on the plethora of independence results, that "for these reasons 'set' turns out to have many meanings, so that the purported foundation of all of Mathematics upon set theory totters." Elsewhere, on page 385, he remarks that "the Platonic notion that there is somewhere *the* ideal realm of sets, not yet fully described, is a glorious illusion."

I would suggest a contrary view: independence results within set theory are generally achieved either by examining an inner model of the universe (an inner model being a transitive class containing all ordinals) or by utilizing forcing to build a larger universe of which the original one is an inner model. The conception that begins to seem more and more reasonable with the advance of the inner model program on the one hand and a deeper understanding of iterated forcing on the other is that within one enormous universe there are many inner models, and the various "independence arguments" may be reworked to give positive information about the way the various inner models relate to each other. Far from undermining the unity of the set-theoretic view, the various techniques available for building models actually promote that unity.

In a more diplomatic mood, Mac Lane writes on page 407:

> Neither organization is wholly successful. Categories and functors are everywhere in topology and in parts of algebra, but they do not as yet relate very well to most of analysis. Set theory is a handy vehicle, but its constructions are artificial. ... We conclude that there is as yet no simple and adequate way of conceptually organizing all of Mathematics.

Let me now consider briefly whether there can be a single foundation for Mathematics. In probing this question I have found myself coming to a view that can be traced back certainly to Plato, namely that there are *two* primitive mathematical intuitions; which might be called the geometrical and the arithmetical; or, alternatively, the spatial and the temporal.

Plato did not have the advantage of modern research into the functions of the left and right half of the brain; this work suggests that the temporal mode (which would include recursive constructions) is handled in the left brain, whereas the spatial mode is handled in the right.

What can each mode of thought contribute to the understanding of the other? I believe, a lot.

Can either be reduced to the other? I should say not; certain formal translations exist, but the underlying intuitions do not translate; and these obstructions show themselves as paradoxes such as that of Banach-Tarski.

Let me refer to my contention that there are these two modes, neither reducible to the other, as positing an essential *bimodality* of mathematical thought.

In earlier pieces I have remarked how Mac Lane's choice of axioms agrees with that made by Bourbaki, at least initially; Liliane Beaulieu has recently remarked that Bourbaki's initial choice of topics was influenced by consideration of the needs of physicists (see [1]); this in turn suggests that Bourbaki attaches greater importance to the descriptive powers of mathematics than to the constructive, and prompts a speculative question: what need is there for a theory of recursion in physics?

There is certainly a need for a theory of recursion in mathematics. The recursion theorem itself is the heart of logic; it is the watershed where processes become objects. In descriptive set theory it takes the shape of the Coding Theorem of Moschovakis, and is thus the source of the strength of the axiom of determinacy.

My sense of the bimodality of mathematics is such that to suppress the ordinals or other frameworks on which to carry out recursions is to suppress half one's mathematical consciousness. I wonder therefore what physicists might be missing by using only the Bourbaki-Mac Lane portion of mathematics in their modeling. Might it be that physical time might fruitfully be modeled by an ordering other than the reals, for example by $\mathbf{R} \times \omega_2$, so that a leap ahead by ω_1 corresponds to some discontinuous event?

Such speculation prompts a further question: is it necessary for all the mathematical concepts invoked in physical explanation to have a direct physical meaning? Or might it be desirable to have abstract concepts which have the merit of making the physics easier to understand without having a perceptible physical interpretation?

But physics aside, the unity of mathematics is a desirable aim; and I would suggest as a modest first step that working in ZBQC + KP rather than ZBQC would encourage awareness of the temporal side of mathematics as well as the spatial side.

Mac Lane's set theory is weak in constructive power, but strong in manipulating the objects naturally arising in geometry. The reverse, as I expect Mac Lane would agree, is true of set theory. I suggest that category theory is as natural a framework for spatial mathematics as set theory is for temporal. I suggest therefore that we should seek an organization of mathematics that will allow the two fundamental intuitions room to develop and to interact; in doing so, we should move away from the regrettable situation so pithily described by Augustus de Morgan over a century ago and still, sadly, to be found today:

> We know that mathematicians care no more for logic than logicians for mathematics. The two eyes of exact science are mathematics and logic: the mathematical sect puts out the logical eye, the logical sect puts out the mathematical eye; each believing that it sees better with one eye than with two.

References

Beaulieu, Liliane, *Bourbaki for physicists? A glance at some unrealized projects (1934 - 1954)*, Abstracts AMS 73 **12**(1) (Jan. 1991), # 863-01-79.

Mac Lane, Saunders, *Mathematics: Form and Function*, Springer-Verlag, 1986.

Institute of Mathematics, University of Warsaw, ul. Banacha 2
02-097 Warszawa, POLAND

IS MATHIAS AN ONTOLOGIST?

SAUNDERS MAC LANE

I am glad to see that Adrian Mathias has taken me to task. Yes, I once gave a lecture with the flamboyant title, "Set theory is obsolete." In this and in a few other contentious articles, I have violated one of the cardinal principles of mathematical activity: Mathematicians do not make pronouncements; they prove theorems. My apologies.

Mathias also argues correctly that there are at least two modes of mathematical thought: the geometrical and the arithmetical. I doubt that this has much to do with the two halves of the brain because I would include at least two more modes: the algebraical and the analytical.

My "one large book" (*Mathematics, Form and Function*, Springer, 1986), is said just to present "my" view of mathematics. I had a wider aim. The first ten chapters try to summarize many of the basic constructions of mathematics up through manifolds, mechanics, complex analysis and topology, in a form that might be of use to beginning mathematicians, including those with no interest in foundations, ontology, or philosophy.

That shaky subject of foundations does then appear in Chapter XI of the book, where I discuss ZBQC (Zermelo set theory with bounded quantifiers). I claim that this does better fit what most mathematicians do because their quantifiers are almost always bounded. As Mathias notes, this system ZBQC is not adequate for Borel determinacy or even for a good theory of ordinals. For that there are other foundations. But I see no need for a single foundation—on any one day it is a good assurance to know what the foundation of the day may be—with intuitionism, linear logic or whatever left for the morrow.

Yesterday, when I wrote that chapter, I suspected that the Kripke-Platek approach might be somehow used. I am delighted to see Mathias propose this, and I hope that he will publish his relative consistency results. The only sources I found yesterday on KP were so buried in technicalities that I failed to see this possibility.

Incidentally, that was one of my earlier flamboyant criticisms: logicians have isolated themselves too much from the rest of mathematics and often present the technique and not the meaning of their theorems. I am now inclined to apologize to my friends the logicians—other branches of mathematics, including some categorists, are even more isolated, and the algebraic geometers are accomplished experts at obscuring their ideas behind mountains of technique.

Mathias seems to claim that having just one foundation promotes the unity of mathematics. I disagree; it is still the case that most mathematicians don't think much about foundations. Real unity is fine, and unity is promoted more by cross connections, especially the unexpected ones. For example, categorical coherence theorems for tensored categories cropped up in Tanaka duality for groups and then in conformal field theory. Again, set theoretical forcing turned out to be related to Kripke semantics for intuitionistic logic, then to Kripke-Joyal semantics for topoi and then to sheafification for Grothendieck topologies. This latter connection seems to me illuminating, but is one as yet little noted by logicians.

In this case, the neglect of this remarkable connection may arise because the available categorical presentations are obscure. A forthcoming book by Mac Lane and Moerdikj on topos theory will, I hope, serve to rectify this situation.

A final word about foundations: my flashy title "Set theory is obsolete" was intended to draw attention to that remarkable observation by F.W. Lawvere: axiomatics for sets is no longer the only effective way to a foundation—one may instead start with axioms on functions—that is on the category of sets.

The last chapter of my "big book" deals with the philosophy of mathematics, with the hope of perhaps reviving this moribund field. My first claim was that too many philosophers of mathematics pay too little heed to what there really is in mathematics. This applies in particular to Wittgenstein and Lakatos, but for now I take on the biggest living target. My learned and articulate friend Van Quine has claimed that ontology is served by observing that "to be" is to be existentially quantified. I disagree, and I also doubt if Van realizes that the existential quantified is a left adjoint—an important observation, again due to Lawvere.

My last chapter attempted to use the earlier survey of the content of the mainstream of mathematics to draw some philosophical conclusions. Today, I would put my view as follows: Mathematics is that branch of science in which the concepts are protean: each concept applies not to one aspect of reality, but to many. The real numbers are both analytical and geometrical, natural numbers are both cardinal and ordinal, and so on in many, many cases. Mathematical form fits varied substance.

This view, if correct, has consequences. For example, the familiar set theoretic explanation of the ordered pair is a convenience and not an ontology. The same idea is formulated differently by observing that a product $A \times B$ is something with projections to the objects A and B which are "universal;" in this case the ordered pair has been swallowed by the syntactic order. Again, a real number is not a Dedekind cut; that cut is just one possible model of a protean idea of the reals.

Long ago, mathematicians recognized that "Space" was not unique. There was the Euclidean plane and the hyperbolic one, as well as elliptic planes. Now there are many types of space—Hausdorff, metric, uniform and so on, each with various contacts with different realities. Much the same now applies to sets. The notions arise variously from finite sets, infinite sets, combinatorial properties of sets, sets as extensional representations of properties, and so on. ZFC had different models. Mathias observes that one model of sets is often inner with respect to another. I am not persuaded that this circumstance argues for the existence of "One enormous universe." Evidently, what one has is different universes, perhaps with different axioms, and connected with each other. These differences match the different purposes of set theory. Moreover, the connections by the inner model relation can be described with sheaf theory more clearly by observing that the new model may consist of sheaves for a suitable "site" of the given model and that then there often is a geometric morphism form one model to the other (For definition, MacLane-Moerdijk, *loc. cit.*). This view of the matter does give a better understanding because it ties the relations between different models of set theory to the continuous functions between different models of space. This promotes the unity of mathematics.

Mathias asks "What, then, is Mac Lane's ontology?" Since mathematics is protean, I can answer easily: Ontology has to do with the nature of the reality at issue. Each mathematical notion is protean, thus deals with different realities, so does not have an ontology.

In closing, may I count my advantages. About 1940, when Bertrand Russell lectured at the mathematical colloquium at Harvard, I was in a position to berate him for his ignorance of the progress in foundational studies. In the 1970's when I was a member of the National Science Board, I was able to tell my colleagues that Kurt Gödel was the greatest logician since Aristotle; soon thereafter, Gödel was awarded the National Medal of Science...I admire Gödel's accomplishments, but I suspect that it is futile to wonder now what he imagined to be the "real" cardinal of the continuum. Those earnest specialists who still search for that cardinal may call to mind that infamous image of the philosopher—a blind man in a dark cellar looking for a black cat that is not there.

Set theory, like the rest of mathematics, is protean, shifting and working in different ways for different uses. It is subordinate to mathematics and not its foundation. The unity of mathematics is real and depends on wonderful new connections which arise all around us. I urge my friends in logic to look around.

Department of Mathematics, University of Chicago, Chicago IL

DEGREES OF CONSTRUCTIBILITY

Richard A. Shore

This paper is a written version of a talk given at the MSRI Workshop on Set Theory and The Continuum. Other than some introductory material, it is an exposition of the work in Groszek and Shore [12]. Its subject matter is certainly based in set theory and deals with the continuum but with a decidedly recursion theoretic bent. We are concerned with the ordering of reals under relative constructibility. For reals f and g, *i. e.* functions from ω into ω, we say that f **is constructible from** g, $f \leq_c g$, iff $f \in L[g]$. This defines a partial ordering and, in the usual way, we form equivalence classes which are called degrees of constructibility and are ordered by the induced ordering to produce a structure $\mathcal{D}_c$. (We use boldface symbols to stand for the degree of a function named by a lightface symbol as in $d \in \mathbf{d}$.) This structure is obviously highly non-absolute. If $V = L$, it consists simply of the singleton containing the constructible (and so all) reals. Other set theoretic assumptions, however, tend to make the structure very rich. One can take the view that investigations into the possible nature of $\mathcal{D}_c$ are simply consistency results. We prefer the attitude that the universe is rich and we are analyzing the structure of the reals under relative constructibility.

Early on Solovay, as reported for example in Sacks [17], suggested that a sufficiently strong assumption such as the existence of a measurable cardinal might determine the structure of the degrees of constructibility or at least their theory. This conjecture turned out to be technically far from correct: the theory of $\mathcal{D}_c$, under only mild set-theoretic assumptions, interprets second order arithmetic and so is as non-absolute as possible. On the other hand it was morally true in that the theorems describing the structure are all proven from quite weaker assumptions. We will typically assume that $\aleph_1^{L[f]}$ is countable for every real f although often less suffices for our constructions and much more is probably true.

The study of the structure of most reducibilities from 1-1 on up with the Turing degrees, $\mathcal{D}_T$, being the prime example, followed a path of extensive exploration of local properties of the ordering such as embeddings, extension of embeddings, initial segments and the like. These early investigations then played crucial roles in the analysis of the global structure of the orderings in considering such questions as automorphisms, homogeneity and definability. (Although as reported in the talk at this Workshop by Slaman, he and Woodin have now developed a new approach to such global questions which eliminates the dependence on much of the earlier work.) The development of the analysis of $\mathcal{D}_c$ has been similar to that of the Turing degrees, $\mathcal{D}_T$, with some noticeable differences. The primary source or these differences is the fact that $\leq_c$ is a constructible relation while Turing reducibility is not recursive. This makes coding arguments much simpler for $\mathcal{D}_c$ than for $\mathcal{D}_T$ and leads to a much easier approach to global results about its structure.

Our major concern in this paper will be with an unfinished chapter in the analysis of the local structure of $\mathcal{D}_c$: initial segments. Before delving into this problem, however, we would like to mention some of the early local results and describe the current status of the global analysis of $\mathcal{D}_c$. We will also very briefly indicate the nature of the proofs.

To begin, note that, like the Turing degrees, the constructibility degrees form an upper semilattice of size the continuum with least element and the countable predecessor property, *i. e.* every degree has at most countably many predecessors. (Remember we are assuming that $\aleph_1$ is inaccessible from reals and so as there are $\aleph_1^{L[f]}$ many reals constructible from f, there are at most countably many c-degrees below that of f.)

Theorem 1. *(Cohen [6]): Every countable partial ordering is embeddable in $\mathcal{D}_c$.*

Proof. Any infinite set of mutually generic Cohen reals generates an independent set of c-degrees (*i. e.* none of them are constructible from any finite join of the others) and so generates a universal countable partial ordering. This argument is essentially like that of Kleene and Post [13] for the Turing degrees.

Theorem 2. *(Sacks [17]): There is a minimal c-degree.*

Proof. Use Sacks forcing, *i.e.* forcing with perfect trees in the style of Spector's [20] construction of a minimal T-degree.

Theorem 3. *(Balcar and Hajek [5], Truss [22]): $\mathcal{D}_c$ is not a lattice.*

Proof. Use Cohen style forcing to build an ascending sequence $\langle \mathbf{c}_i \rangle$ of degrees with an exact pair $\mathbf{a}, \mathbf{b}$, *i. e.* any $\mathbf{d}$ below both $\mathbf{a}$ and $\mathbf{b}$ is below some $\mathbf{c}_i$. As in the construction of Kleene and Post [13] for $\mathcal{D}_T$, no such pair can have a greatest lower bound.

Theorem 4. *(Adamowicz [3]): All finite lattices are isomorphic to initial segments of $\mathcal{D}_c$.*

Proof. This proof is quite complicated. It uses trees in the style of Lerman [14] to force the desired results.

Theorem 5. *The $\forall\exists$-theory of $\mathcal{D}_c$ is decidable but not the $\forall\exists\forall$-theory.*

Proof. All the ingredients of the proofs of Shore and Lerman of decidability of the $\forall\exists$-theory and of Schmerl for the undecidability result (both of which are presented in Lerman [14, VII.4]) are supplied for the c-degrees by Theorem 4 and a suitable relativization of the arguments for Theorem 1.

Theorem 6. *(Farrington [8]): The first order theory of $\mathcal{D}_c$ is recursively isomorphic to the second order theory of arithmetic.*

Proof. The coding scheme is like that used by Simpson [19] for $\mathcal{D}_T$ but a few additional complications arise.

Theorem 7. *(Farrington [7], Groszek [9], Abraham and Shore [1]): There are no non-trivial automorphisms of $\mathcal{D}_c$. Indeed no two distinct cones of c-degrees, $\mathcal{D}_c(\geq \mathbf{a})$ and $\mathcal{D}_c(\geq \mathbf{b})$, for $\mathbf{a} \neq \mathbf{b}$, are isomorphic.*

Proof. One can code any Cohen real $d \in \mathbf{d}$ by Cohen reals in the c-degrees below $\mathbf{d}$. As the Cohen reals generate all the c-degrees (for every $\mathbf{d}$ there are (degrees of) Cohen reals $\mathbf{c}_1, \mathbf{c}_2, \mathbf{c}_3$, and $\mathbf{c}_4$ such that $\mathbf{d} = (\mathbf{c}_1 \vee \mathbf{c}_2) \wedge (\mathbf{c}_3 \vee \mathbf{c}_4)$), the structure is rigid. The result on cones follows by relativization.

Theorem 8. *Every projective relation on $\mathcal{D}_c$ is definable in $\mathcal{D}_c$ (from just the ordering and without parameters).*

Proof. Following the style of the definability results for $\mathcal{D}_T$ in Simpson [19] and Shore [18], it suffices to be able to define the relation $R(\bar{\mathbf{x}}, \bar{\mathbf{y}}, \bar{\mathbf{a}})$ which says that the degrees $\bar{\mathbf{y}}$ code sets of c-degree $\bar{\mathbf{x}}$ in the model of arithmetic coded by the parameters $\bar{\mathbf{a}}$. Once we have this relation, anything that we might wish to say about the degrees $\bar{\mathbf{x}}$ can simply be translated into sentences of second order arithmetic about the sets coded by the $\bar{\mathbf{y}}$ in the model given by $\bar{\mathbf{a}}$. As the c-degrees of the Cohen reals generate all the c-degrees, it suffices to define this relation for Cohen reals $\bar{\mathbf{x}}$ as long as we can also define the property of being the c-degree of a Cohen real.

Lemma 9. *The property of containing a Cohen real is definable in $\mathcal{D}_c$.*

Proof. We claim that a c-degree $\mathbf{c}$ contains a Cohen real iff there is a model of arithmetic coded below $\mathbf{c}$ and there is a (code of) a set C (not necessarily below $\mathbf{c}$) which is the supremum of all c-degrees coded in this model by degrees below $\mathbf{c}$ and C is a Cohen real. (Remember that we have, relative to this standard model, all of second order arithmetic at our disposal in $\mathcal{D}_c$.) The codings of arithmetic used here are those of Slaman and Woodin [21] for $\mathcal{D}_T$ and are carried out for $\mathcal{D}_c$ in Abraham and Shore [1]. The latter paper also contains the analysis needed to see that the degree of a Cohen real satisfies the above property. For the converse, the analysis there shows that the given property first implies the existence of a Cohen real a below c. The join theorem of Farrington [7] then says that there is another Cohen real $\mathbf{b}$ such that $\mathbf{c} = \mathbf{a} \vee \mathbf{b}$. As both $\mathbf{a}$ and $\mathbf{b}$ are code below $\mathbf{c}$ and nothing more complicated than $\mathbf{c}$ can be so coded, the property says precisely that $\mathbf{c}$ contains a Cohen real.

In contrast to the Turing degrees, the last few results mentioned show that the global structure of $\mathcal{D}_c$ is well understood. On the other hand the local analysis is not as well developed for $\mathcal{D}_c$. In particular, compared to our knowledge of $\mathcal{D}_T$, we are far from a complete characterization of the possible initial segments (or equivalently, ideals) of $\mathcal{D}_c$. Of course any ideal in either structure is an upper semilattice (usl) of size at most the continuum with a least element and the countable predecessor property. For the Turing degrees, Abraham and Shore [2] show that every such usl of size at most $\aleph_1$ is in fact isomorphic to an ideal of $\mathcal{D}_T$. On the other hand, Groszek and Slaman [13] show that no more is provable: It is consistent (with ZFC) that the continuum be large but that there are usl's of size the continuum which are not isomorphic to ideals of $\mathcal{D}_T$. Our knowledge about $\mathcal{D}_c$ is much less complete. What we do know, however, indicates that the story here is much more complicated.

The first reasonably comprehensive positive results (following the path broken by the construction of a minimal c-degree in Sacks [17]) are due to Adamowicz:

Theorem 10. *(Adamowicz [4]): Every countable constructible well-founded usl is isomorphic to an initial segment of $\mathcal{D}_c$.*

The restriction to countable usl's is natural at least as a starting point (and indeed for the rest of the paper we will, in addition, restrict ourselves to lattices rather than usl's simply as a reflection of the state of our knowledge, or better, lack thereof); but what of the other restrictions required in this result? It is fairly easy to see that some assumption of constructibility is necessary as indicated by the coding argument used in the following result which, contrary to our standing conventions, is proved in ZFC alone.

Theorem 11. *(Abraham and Shore [1]): (ZFC) Not every countable well-founded distributive lattice is isomorphic to an initial segment of $\mathcal{D}_c$.*

Proof. The proof is non-uniform and like other coding results exploits the constructibility of the ordering relation $\leq_c$. Either the diamond is not an initial segment of $\mathcal{D}_c$ or it is with top $\mathbf{d}$. In the former case we are done. In the latter no lattice coding a set $D \in \mathbf{d}$ which does not begin with a diamond can be an initial segment of $\mathcal{D}_c$.

To avoid such coding problems we will restrict our attention in this paper to constructible lattices. On the other hand, there is more leeway in relaxing the restriction of well-foundedness. The first constructions of nonwell-founded initial segments of $\mathcal{D}_c$ can be found in Groszek [10] where all orderings $\alpha*$ for $\alpha \leq \omega_1$ are embedded as initial segments of $\mathcal{D}_c$. Some restriction along these lines, however, is necessary. The first serious demonstration of such restrictions on possible initial segments of $\mathcal{D}_c$ are due to Lubarsky:

Theorem 12. *(Lubarsky [15]): Every countable lattice isomorphic to an initial segment of $\mathcal{D}_c$ is complete.*

Proof. We illustrate the starting idea for the result by considering the lattice $\omega + \omega*$. Let $\langle \mathbf{a}_i \rangle$ be the ascending chain and $\langle \mathbf{b}_i \rangle$ the descending one in a purported realization of $\omega + \omega*$ as an initial segment of $\mathcal{D}_c$. We will build a degree c strictly in between the chains for a contradiction. The point is that we can define representatives A_i from $\mathbf{a}_i$ in a canonical way that can be recovered from each $\mathbf{b}_i$: We start with $A_0 \in \mathbf{a}_0$ and $B_0 \in \mathbf{b}_0$. Suppose we have defined $A_i \in \mathbf{a}_i$ and $B_i \in \mathbf{b}_i$. We then choose representatives $A_{i+1} \in \mathbf{a}_{i+1}$ and $B_{i+1} \in \mathbf{b}_{i+1}$ which are least in the canonical ordering of $L[B_i]$. It is clear that the sequence $\langle A_i : i \geq j \rangle$ is uniformly constructible in B_j. Thus the entire sequence $\langle A_i : i \in \omega \rangle$ is constructible in each B_i but strictly above, in c-degree, each A_i.

The question now is how far can we go towards embedding every countable complete constructible lattice $\mathcal{L}$ (with ordering $\preceq$) as an initial segment of $\mathcal{D}_c$. Of course any such lattice has a least element, 0. As we consider only countable lattices, we may also assume without loss of generality that $\mathcal{L}$ has a greatest element, 1, as well. The results that we report on here are joint work with Marcia Groszek and appear in full in Groszek and Shore [12]. Our work shows that a much larger subclass of the complete countable lattices than the well founded or reverse well founded ones can be embedded as initial segments of $\mathcal{D}_c$. We also show, however, that there are inherent limitations on the available technology that imply that any significant further positive results will require a new approach.

We begin with a general description of the techniques employed. Our constructions are, like all known initial segment procedures for any reducibilities, basically forcing arguments with trees. The trees, as usual, consist of finite sequences of elements from some countable set Θ. The construction produces a generic object, g, which is a path through all the trees in the generic filter and so defines a map from ω into Θ. The idea is to define a structure on Θ that reflects that of the given lattice $\mathcal{L}$ in such a way as to facilitate the proof that $(\mathcal{D}_c)^{L[g]} \cong \mathcal{L}$. We want g to code in some simple way representatives for the degrees corresponding to the elements i of $\mathcal{L}$. Moreover we would like this coding to guarantee on its own at least some of the properties required of the isomorphism.

We present the notions needed for our lattice representations in §1, the forcing notions and the outline of the argument in §2 and a discussion of limitations on the methods and open problems in §3.

1. Lattice Representations

We follow the style for representations, Θ, of lattices introduced by Lerman for embeddings in the Turing degrees as presented for example in Lerman [14]. The elements of Θ will be maps $\alpha : \mathcal{L} \to \omega$. We will define maps $h_i : \omega \to \omega$ for each $i \in \mathcal{L}$ with the intention that the map sending i to the c-degree of h_i will be the desired isomorphism between $\mathcal{L}$ and $\mathcal{D}_c$ in $L[g]$. The coding of the h_i in our generic map g from ω into Θ is straightforward:

$$h_i(n) = (g(n))(i).$$

We now wish to impose some structure on Θ so as to make it into a standard lattice representation of $\mathcal{L}$. We also need some additional properties that are essentially dictated by the needs of the construction and verification that the intended map is in fact an isomorphism. We use $\alpha \equiv_i \beta$ to mean $\alpha(i) = \beta(i)$ and consider the following conditions on Θ for every $\alpha, \beta \in \Theta$ and $i, j, k \in \mathcal{L}$:

1.0) *Zero*: $\alpha \equiv_0 \beta$. Here 0 is the zero of the lattice. This guarantees that h_0 is a constant and so in L as required.

1.1) *Ordering*: $i \preceq j \& \alpha \equiv_j \beta \Rightarrow \alpha \equiv_i \beta$. This guarantees that if $i \preceq j$ (in $\mathcal{L}$) then $h_i \leq_c h_j$. To calculate $h_i(n)$ it suffices to know $h_j(n)$ and Θ as this requirement says that $h_i(n) = \beta(i)$ for any β such that $\beta(j) = h_j(n)$. As Θ will be constructible, we will have $h_i \leq_c h_j$.

1.2) *Non-ordering*: $i \not\preceq j \Rightarrow \exists \alpha, \beta \in \Theta(\alpha \equiv_j \beta \& \alpha \not\equiv_i \beta)$. This property allows us to find alternate extensions of g which keep h_j the same but give different values to h_i. This possibility will allow us to generically guarantee that if $i \not\leq j$ then $h_i \not\leq_c h_j$.

1.3) *Join*: $(i \vee j = k) \& \alpha \equiv_i \beta \& \alpha \equiv_j \beta \Rightarrow \alpha \equiv_k \beta$. With this property we guarantee that, if $i \vee j = k$, then $h_k \leq_c h_i \vee h_j$ (and so by (1.1) that $h_k \equiv_c h_i \vee h_j$). As in (1.1), if one knows $h_i(n)$, $h_j(n)$ and Θ one can calculate $h_k(n)$ by finding any β with $\beta(i) = h_i(n)$ and $\beta(j) = h_j(n)$. We then have $h_k(n) = \beta(k)$ by this property.

The next two properties of arbitrary subsets I of $\mathcal{L}$ are ones that do not occur in the Turing degree arguments. They are introduced here to enable us to use infinite representations rather than the finite ones basic to the recursion theoretic arguments. Each converts an infinitary meet or sup into a finitary one.

1.4) *Completeness*: $i = \vee I \& \forall j \in I(\alpha \equiv_j \beta) \Rightarrow \alpha \equiv_i \beta$.

1.5) *Compactness*: $i = \wedge I \& \alpha \equiv_i \beta \Rightarrow \exists$ finite $F \subset I$ with $j = \wedge F$ such that $\alpha \equiv_j \beta$.

The next property is the standard one for lattice representations that reflects the meet structure of the lattice. It plays a crucial role in the argument that our map from $\mathcal{L}$ is onto the c-degrees of $L[g]$.

1.6) *Meet*: $(i \wedge j = k)$ & $\alpha \equiv_k \beta \Rightarrow (\exists \gamma_1, \gamma_2, \gamma_3 \in \Theta)(\alpha \equiv_i \gamma_1 \equiv_j \gamma_2 \equiv_i \gamma_3 \equiv_j \beta)$.

The final property we consider is one introduced for the Turing degree constructions to facilitate certain fusion arguments. The particular form it takes is best ignored on first (and even second) reading.

1.7) *Homogeneity*: For every finite $\Theta' \subset \Theta$ and every α_0, α_1, β_0, $\beta_3 \in \Theta$ such that $\forall i \in \mathcal{L}(\alpha_0 \equiv_i \alpha_1 \rightarrow \beta_0 \equiv_i \beta_3)$, there are β_1 and β_2 in Θ and f_0, f_1, $f_2 : \Theta' \rightarrow \Theta$ such that, for $m = 0, 1, 2$, $f_m(\alpha_0) = \beta_m$, $f_m(\alpha_1) = \beta_{m+1}$ and $\forall \alpha, \beta \in \Theta' \forall i \in \mathcal{L}(\alpha \equiv_i \beta \rightarrow f_m(\alpha) \equiv_i f_m(\beta))$.

In fact we need a bit more than is expressed even by all the conditions (1.0) — (1.7). We have to use approximations to the given lattice $\mathcal{L}$ and the desired representation Θ. To be precise, we will express $\mathcal{L}$ as an increasing union of finite subusl's $\mathcal{L}_n$ and Θ as an increasing union of representations Θ_n. We require not only that Θ satisfies (1.0)-(1.7) but also that, for each n, Θ_n contains the witnesses required in (1.2) for any $i, j \in \mathcal{L}_n$ and that Θ_{n+1} contains those required in (1.6) and (1.7) for elements (and finite subsets) of Θ_n and $\mathcal{L}_n$. It should be clear that if Θ satisfies (1.0)-(1.7) then we can find a decomposition of this sort. We call such a Θ with decomposition $\cup \Theta_n = \Theta$ a **sequential algebraic representation** of $\mathcal{L}$.

These properties were actually designed to make the forcing argument that we will describe in the next section work. They turn out to correspond to a well known class of complete lattices.

Definition 1.8. An element i of $\mathcal{L}$ is **compact** if for every $I \subseteq \mathcal{L}$ with $\wedge I \preceq i$, there is a finite $F \subset I$ such that $\wedge F \preceq i$.

Definition 1.9. An **algebraic lattice** is a countable complete lattice which is **compactly generated**, that is, every $i \in \mathcal{L}$ is the infimum of the compact elements above it.

Theorem 1.10. *(Groszek and Shore [12]): i) Every complete countable lattice with a representation Θ satisfying (1.0)-(1.6) is algebraic.*

ii) Every algebraic lattice has a representation satisfying (1.0)-(1.7) and so a sequential algebraic representation.

2. The Forcing Notions

We fix a decomposition of our given countable lattice $\mathcal{L}$ into a sequence of finite subusls $\mathcal{L}_n$. Although we will need all of the properties (1.0) — (1.7) of Θ to carry out our proof, we define notions of forcing $\mathcal{P}(\Theta)$ for any $\Theta \subseteq \omega^{\mathcal{L}}$ with any decomposition into an increasing nested sequence of finite subsets Θ_n. This will enable us to see what the limitations are on such forcing constructions in terms of the possible lattices of c-degrees that can be produced.

We begin by describing the trees that will be the elements of our forcing relations. Recall that $\langle \Theta_n : n \in \omega \rangle$ is a nested increasing sequence of finite sets with union Θ an arbitrary subset of $\omega^{\mathcal{L}}$.

Definition 2.1. A **Θ-tree** is a downward closed subset T of $\Theta^{<\omega}$ (the finite sequences from Θ) ordered by extension such that every element of T has incomparable extensions in T. The elements of T are called its **nodes**.

Definition 2.2. A node $\sigma \in T$ **splits** in T iff σ has at least two immediate successors, $\sigma\hat{\ }\alpha$ and $\sigma\hat{\ }\beta$, in T.

Definition 2.3. $L_n(T)$, the nth **splitting level** of T, is

$$\{\sigma \in T | \sigma \text{ splits in } T \text{ and } |\{\tau | \tau \subset \sigma \text{ and } \tau \text{ splits in } T\}| = n\}.$$

Definition 2.4. Suppose $\sigma \in L_n(T)$ and $\sigma\hat{\ }\alpha \in T$. We define $\sigma^-\alpha$ to be the unique extension of $\sigma\hat{\ }\alpha$ in $L_{n+1}(T)$.

Definition 2.5. The forcing partial order $\mathcal{P}(\Theta)$ is defined by imposing the usual ordering for trees ($S \leq T$ iff $T \supseteq S$) on the set of Θ-trees satisfying the following properties:

(1) *Splitting*: If σ is in $L_n(T)$ then the immediate successors of σ in T are $\{\sigma\hat{\ }\alpha | \alpha \in \Theta_n\}$.
(2) *Congruence*: Suppose σ is in $L_n(T)$, and α and β are in Θ_n. Then for all i in $\mathcal{L}_n$, if $\alpha \equiv_i \beta$ then $\sigma^-\alpha \equiv_i \sigma^-\beta$.

(This condition says that, for σ in $L_n(T)$ and i in $\mathcal{L}_n$, if two immediate successors α and β of σ are congruent mod i, then their extensions up to $L_{n+1}(T)$ must respect this congruence. The effect of this requirement is that, if g and g' are two paths through T, $g \upharpoonright m = \sigma^- \alpha$, and $g' \upharpoonright m = \sigma^- \beta$, then $h_i \upharpoonright m = h_i' \upharpoonright m$. Thus h_i carries less information than g.)

(3) *Uniformity 1*: For all n, all nodes on $L_n(T)$ have the same length.

(4) *Uniformity 2*: If σ and τ are both in $L_n(T)$ then for all ρ, $\sigma\hat{\ }\rho \in T$ iff $\tau\hat{\ }\rho \in T$.

(These uniformity conditions guarantee that, in the situation described in (2), h_i truly carries less information than g. Since T above $\sigma^-\alpha$ and T above $\sigma^-\beta$ are identical, it may well be that g above $\sigma^-\alpha$ and g' above $\sigma^-\beta$ are identical, in which case $h_i = h_i'$ but $g \neq g'$.)

We can now describe the plan of the proof of the main theorem on initial segments of $\mathcal{D}_c$.

Theorem 2.6. *(Groszek and Shore [12]): Every countable constructible algebraic lattice is isomorphic to an initial segment of $\mathcal{D}_c$.*

Note that as being compact is a Π^1_1 property and being "the sup of the compact elements below" is Σ^1_1, the property of a countable lattice being algebraic is absolute. As $\mathcal{L}$ is constructible Theorem 1.10 tells us that it has a sequential algebraic representation $\langle \Theta_n \rangle$ in L. (An absoluteness argument for properties (1.4) and (1.5) would also then tells us that this constructible representation is an algebraic one in V as well. As we do our forcing construction over L, this observation is not, however, needed at this point.)

We force over L with the notion of forcing $\mathcal{P}(\Theta)$ given by this representation. Our underlying assumption that $\aleph_1^{L[f]}$ is countable for every real f (actually one only needs $\aleph_2^L < \omega_1$) allows us to produce a g generic for this forcing. We claim that the map defined in §1 sending $i \in \mathcal{L}$ to the degree of h_i (where $h_i(n) = (g(n))(i)$) defines the desired isomorphism. The explanations given in §1 when the appropriate properties of a representation were presented show that this map automatically preserves order and join. Once one shows that the values of terms for reals constructible from h_i can be made, for each n, to depend on only finitely much of h_i and so of g (essentially a local Cohen forcing argument), the non-ordering property allows us to show that if $i \not\leq j$ then $h_i \not\leq_c h_j$. Thus our map is one-one. The difficult part of the argument is to show that the map is in fact onto the c-degrees below g.

The proof has two main parts. First we show that

a) For every real $t \leq_c g$, there is a least $j \in \mathcal{L}$ such that $t \leq_c h_j$.

We then prove that

b) For this j, $h_j \leq_c t$.

The argument for (a) proceeds in three steps.

i) We use the meet and homogeneity properties of the representation to prove that the set $X_t = \{i \in \mathcal{L} \mid t \leq_c h_i\}$ is closed under $\wedge$.

ii) We then use a fusion argument and the compactness property to show that X_t is closed under arbitrary infima in L, *i. e.* even though X_t may not be in L the infimum of a constructible subset of X_t is itself in X_t.

iii) Finally we get the existence of j as the infimum of the compact elements above $\wedge X_t$. In addition to the fact fact that $\mathcal{L}$ is algebraic, we need to use the absoluteness of the completeness of $\mathcal{L}$ and of the compactness of individual elements of $\mathcal{L}$.

To complete the proof, we establish (b) by a complex fusion argument that relies on the completeness and homogeneity properties of the representation for its combinatorial details.

We should point out that, together with Lubarsky's Theorem 12, this result precisely characterizes the countable constructible linearly ordered initial segments of $\mathcal{D}_c$.

Theorem 2.7. *A countable constructible linear ordering $\mathcal{L}$ is isomorphic to an initial segment of $\mathcal{D}_c$ if and only if it is complete.*

Proof. The necessity of completeness is Theorem 12. For its sufficiency note that any complete countable linear order is algebraic as a lattice: If not there is an $a \in \mathcal{L}$ which is not the infimum of the compact elements above it. It must then be the infimum of a strictly descending chain a_i in $\mathcal{L}$. Moreover, by going to a subsequence if necessary, we may assume that each a_i is also not the inf of the compact elements above it. Thus we may, for each a_i, choose a strictly descending sequence of elements $a_{i,j}$ each of which is again not the inf of the compact elements above it. Continuing in this way to form the sequences $a_{i_1,\ldots,i_n}$ for each n, we build a countable set A of elements of $\mathcal{L}$ every one of which is also a limit point of A. The closure of A is then a perfect subset of $\mathcal{L}$ in the topological sense. As every perfect subset of a linear ordering has size at least the continuum, we have our desired contradiction.

3. Limitations and Open Questions

The obvious question left open by our results and those of Lubarsky [15] is whether complete but not algebraic countable constructible lattices are isomorphic to initial segments of $\mathcal{D}_c$. While we have no further theorems one way or the other, we do know that the technology described here has severe limitations.

Theorem 3.1. *(Groszek and Shore [12]): Let $\mathcal{L}$ be any countable constructible lattice and Θ any constructible subset of $\omega^{\mathcal{L}}$. If g is generic for the notion of forcing $\mathcal{P}(\Theta)$ defined in §3 and $\mathcal{L}$ is isomorphic to $\mathcal{D}_c^{L[g]}$, then $\mathcal{L}$ is algebraic.*

Thus the technology used here can be pushed no further at least not in precisely its current form. A natural question to ask here is how are these limitations overcome for the Turing degrees. The answer is that for initial segments of $\mathcal{D}_T$ one uses a sequence of finite approximations to the representation Θ. That is, each forcing condition consists of a pair $\mathcal{L}_n$ and Θ_n where $\langle\Theta_n\rangle$ is a sequential representation for $\mathcal{L}$ but each Θ_n is a finite representation for the finite usl $\mathcal{L}_n$. Of course any attempt to mimic this in the set theoretic case will give a notion of forcing with finite conditions. As any such forcing will add Cohen reals, it cannot produce the desired initial segments.

Indeed even much weaker assumptions than the ones of Theorem 3.1 on the forcing notions that might construct an initial segment of $\mathcal{D}_c$ seem to impose restrictions on the types of lattices that could be produced. Consider the following lattice $\mathcal{R}$ as a test case for further progress and as an illustration of the limitations of existing methods:

Test Problem: Let $\mathcal{R}$ be the lattice with least element 0, greatest element 1, a descending sequence of elements $a_{i+1} < a_i$ for $i > 1$ and another element b incomparable with all the a_i. The join and meet relations of $\mathcal{R}$ are determined by requiring that $b \vee a_i = 1$ and $b \wedge a_i = 0$ for every $i > 1$.

Suppose that $\mathcal{R}$ is embedded (as a lattice) into the c-degrees below a g that is generic for some notion of forcing by a map sending the elements of $\mathcal{R}$ to the c-degrees of functions f and h_i for $i \in \omega$. (The map sends b to $\deg_c(f)$, 0 to $\deg_c(h_0)$, 1 to $\deg_c(h_1)$ and a_i to $\deg_c(h_i)$ for $i > 1$.) If the coding procedures inherent in the forcing construction satisfy the ordering and join properties, then the embedding cannot be onto an initial segment of $\mathcal{D}_c$. To be a bit more precise, if there is a "constructible procedure" (think of it as a form of generalized truth table reduction applied to the graphs of the h_j) which determines, for each n, $h_i(n)$ from $h_j(n)$ if $i \preceq j$, and $h_1(n)$ from $h_j(n)$ and $f(n)$ for $j > 1$, then there is a non-zero c-degree h below all those of all the h_i. The proof is like that of Theorem 12: Define $h(n) = h_n(n)$. It is clear from our coding assumptions that for each $i > 1$, $h_i \leq_c h$. On the other hand, it is also clear that $h_1 \leq_c h \vee f$. Thus h is the desired witness that the embedding is not onto an initial segment of $\mathcal{D}_c$.

Thus some new approach is necessary to realize lattices such as $\mathcal{R}$ as initial segments of $\mathcal{D}_c$. Perhaps all that is needed is a way to build the coding machinery (*i. e.* the lattice representation) generically along with the construction of the top of the initial segment. On the other hand it would be even more interesting to find some way other than forcing with trees to produce the missing initial segments (or indeed to produce any initial segments at all). Of course, the other possibility is to try to turn arguments like the one above for $\mathcal{R}$ into ones like those of Theorem 12 to show that non-algebraic lattices cannot be initial segments of $\mathcal{D}_c$, or at least that they cannot be realized by any forcing extensions.

In closing, we would like to raise one related embedding type question about the "local" structure of $\mathcal{D}_c$. Under our assumptions, $\mathcal{D}_c$, like $\mathcal{D}_T$, is a partial order of size the continuum with the countable predecessor property. Is it a universal such partial order, *i. e.* is every partial order of size at most the continuum with the countable predecessor property embeddable (as a partial ordering) in $\mathcal{D}_c$? The corresponding question for $\mathcal{D}_T$ was first raised by Sacks [16] and remains open.

REFERENCES

1. U. Abraham and R. Shore, *The degrees of constructibility below a Cohen real*, J. Lon. Math. Soc. (3) **53** (1986), 193-208.
2. U. Abraham and R. Shore, *Initial segments of the Turing degrees of size* $\aleph_1$, Is. J. Math. **55** (1986), 1-51.
3. Z. Adamowicz, *On finite lattices of degrees of constructibility of reals*, J. of Symb. Logic **41** (1976), 313-322.
4. Z. Adamowicz, *Constructible semi-lattices of degrees of constructibility*, Set Theory and Hierarchy Theory V, Lachlan, Srebrny and Zarach eds., Lecture Notes in Mathematics 619, Springer-Verlag, Berlin, 1977, pp. 1-44.
5. B. Balcar and P. Hajek, *On sequences of degrees of constructibility*, Z. f. Math. Logik **24** (1978), 291-296.
6. P. J. Cohen, *Set Theory and the Continuum Hypothesis*, Benjamin, New York, 1966.
7. P. Farrington, *Hinges and automorphisms of the degrees of non-constructibility*, J. Lon. Math. Soc. (2) **28** (1983), 193-202.
8. P. Farrington, *First order theory of the c-degrees*, Z. f. Math. Logik **30** (1984), 437-446.
9. M. Groszek, *Applications of iterated perfect set forcing*, Ann. Pure and App. Logic **39** (1988), 19-53.
10. M. Groszek, ω_1^* *as an initial segment of the degrees of constructibility*, J. Symb. Logic (to appear).
11. M. Groszek and T. Slaman, *Independence results on the global structure of the Turing degrees*, Trans. Am. Math. Soc. **277** (1983), 579-588.
12. M. Groszek and R. A. Shore, *Initial segments of the degrees of constructibility*, Is. J. Math. **63** (1988), 149-177.
13. S. C. Kleene and E. L. Post, *The upper semi-lattice of degrees of unsolvability*, Ann. Math. **59** (1954), 379-407.
14. M. Lerman, *Degrees of Unsolvability*, Springer-Verlag, Berlin, 1983.
15. R. Lubarsky, *Lattices of c-degrees*, Ann. Pure and App. Logic **36** (1987), 115-118.
16. G. Sacks, *Degrees of Unsolvability*, Annals of Math. Studies **55**, Princeton University Press, Princeton NJ, 1963.
17. G. Sacks, *Forcing with perfect closed sets*, Axiomatic Set Theory, D. Scott, ed., Proc. Symp. Pure Math. **13**(1), Am. Math. Soc., 1971, pp. 331-355.
18. R. A. Shore, *On homogeneity and definability in the first order theory of the Turing degrees*, J. Symb. Logic **47** (1982), 8-16.
19. S. G. Simpson, *First order theory of the degrees of recursive unsolvability*, Ann. Math. **105** (1977), 121-139.
20. C. Spector, *On degrees of recursive unsolvability*, Ann. Math. **64** (1956), 581-592.
21. T. Slaman and H. Woodin, *Countable relations in the Turing degrees*, Ill. J. Math. **30** (1986), 320-334.
22. J. Truss, *A note on increasing sequences of constructibility degrees*, Set Theory and Hierarchy Theory V, Lachlan, Srebrny and Zarach eds., Lecture Notes in Mathematics 619, Springer-Verlag, Berlin, 1977, pp. 473-476.

DEPARTMENT OF MATHEMATICS, CORNELL UNIVERSITY, ITHACA NY 14853

APPLICATIONS OF THE OPEN COLORING AXIOM

Boban Velickovic

1. Introduction

Standard forcing axioms are usually stated in the form which asserts the existence of sufficiently generic filters in every partial order $\mathcal{P}$ which belongs to a given class $\mathcal{K}$ of forcing notions. This approach, which is derived by "internalizing" generic extensions, has been very successful in providing strong forcing axioms and proving their consistency; in [FMS] a maximal axiom of this sort is proved consistent for the case when one wishes to consider only generic filters for families of at most $\aleph_1$ dense sets. However, when applying these axioms we need to know when there is a partial order in the class $\mathcal{K}$ which introduces the object we wish to find. Of course, there is no easy general answer to this question and even some of the most basic instances are still open.

Following the realization that many applications of forcing axioms involve finding homogeneous sets in certain kinds of partitions, in [TV] a study of the so-called Ramsey forcing axioms was initiated. The idea is that these statements would provide a combinatorial intermediary between the abstract forcing axioms and their applications. It turned out that in some cases they are equivalent to the axioms from which they are derived.

To be more specific suppose we are given an uncountable set S and a partition of the form:

$$[S]^n = K_0 \cup K_1 \tag{1}$$

or

$$[S]^{<\omega} = K_0 \cup K_1 \tag{2}$$

together with a class $\mathcal{K}$ of partial orders. Let us say that this partition is $\mathcal{K}$-*destructible* if there is a poset $\mathcal{P}$ in $\mathcal{K}$ which forces an uncountable subset H of S which is 0-*homogeneous* (i.e. $[H]^n \subseteq K_0$ for (1) and $[H]^{<\omega} \subseteq K_0$ for (2)) and in addition every $s \in S$ is forced by some condition in $\mathcal{P}$ to be in H. Let $\mathrm{RFA}^n(\mathcal{K})$ ($\mathrm{RFA}^{<\omega}(\mathcal{K})$) be the statement that every $\mathcal{K}$-destructible partition of form (1) ((2)) has an uncountable 0-homogeneous set. The following results are proved in [TV].

Theorem 1.1. *If κ is an uncountable cardinal* MA_κ *is equivalent to the statement that for every ccc destructible partition of the form* (2) *with* card$(S) \geq \kappa$ *there are countably many* 0*-homogeneous sets whose union covers* S.

Theorem 1.2. $\mathrm{MA}_{\aleph_1}$ *is equivalent to* $\mathrm{RFA}^{<\omega}(\mathrm{CCC})$.

These results raise the following questions.

Question 1.1. Can the assertion in Theorem 1.1 be weakened to say that if κ is regular then for every ccc destructible partition of form *(2)* such that card$(S) \leq \kappa$ there is an 0-homogeneous subset of S of size κ?

Question 1.2. Is there $n < \omega$ such that $RFA^n(CCC)$ is equivalent to $MA_{\aleph_1}$?

It is possible that these statements provide a natural hierarchy of axioms whose limit is $\mathrm{MA}_{\aleph_1}$. These questions were further studied in [To2].

Now, turning to stronger axioms much less is known. When is there a proper poset forcing an uncountable homogeneous set for a partition of the form (1) or (2)? For our purposes we only need to know that iterations of σ-closed and ccc posets are proper. While for a given ccc destructible partition there always exists a ccc poset of size at most $\aleph_1$ which adds an uncountable 0-homogeneous set and, in fact, there is a poset of finite 0-homogeneous sets which does this, there is no known such bound in the case of proper posets. Thus, for example, the following is open.

Question 1.3. If there is a proper poset forcing an uncountable 0-homogeneous set to a partition of form (1) or (2), is there such a poset of size $< \beth_\omega(S)$?

It is known though that $\mathrm{RFA}^{<\omega}$(proper) has roughly the same consistency strength as PFA. Given these limitations of our knowledge we adopt a more modest approach by trying to find sufficient conditions for the existence of proper posets adjoining an uncountable 0-homogeneous set. This approach was taken by Todorcevic in [To1] where it was pursued in connection with the well-known (S) and (L) problems from general topology. The thesis is that this line of work would provide partition-type statements which lie at the core of many diverse problems and are thus more suitable for applications than the abstract forcing axioms. In this paper we offer further evidence for this point of view by focusing on one particular axiom of this kind which has been very successful in resolving questions about sets of reals. We present a survey of applications of this statement, study possible extensions and indicate directions for further research.

Thus, let us consider partitions of form (1) for $n = 2$. The idea is to put a topology on S and require the color classes to be open and closed respectively. It was first formulated explicitly by Abraham, Rubin, and Shelah ([ARS]) who were working on the extension of Baumgartner's consistency result that every two $\aleph_1$-dense sets without endpoints are isomorphic. Their work was further extended and refined by Todorcevic ([To1]) who formulated and proved the relative consistency with ZFC + $\mathrm{MA}_{\aleph_1}$ of the following version of the Open Coloring Axiom (OCA):

> *If S is a set of reals and*
>
> $$[S]^2 = K_0 \cup K_1$$
>
> *is a partition with K_0 open in the product topology then either there exists an uncountable 0-homogeneous subset of S, or else S can be covered by countably many 1-homogeneous sets.*

The statement of the original ARS-axiom was symmetric and required only the existence of an uncountable homogeneous set in one of the colors. As it turns out this amplification yields a much more useful axiom which has a particularly strong influence on $\mathcal{P}(\omega)/\mathrm{fin}$ and related structures. Its additional advantage is that applying it does not require any knowledge of the niceties of forcing and is thus suitable for use by topologists, analysts, and other non-specialists in set theory working on subjects related to $\beta\omega$.

The paper is organized as follows. In section 2 we present a proof of the consistency of OCA, in fact, we derive it from PFA. Sections 3, 4 and 5 consist of applications of OCA. In section 3 we present some combinatorial consequences and show, for example, that OCA has strong influence on the partial order of all functions from ω to ω, ordered under eventual dominance. In particular, it implies that the least size of an unbounded subset of $\omega^\omega, <_*$ is $\aleph_2$. This gives evidence for the conjecture that OCA implies that the continuum is $\aleph_2$. In section 4 we turn to the study of automorphisms of $\mathcal{P}(\omega)/\mathrm{fin}$. We show that OCA can be used to prove that every automorphism of $\mathcal{P}(\omega)/\mathrm{fin}$ is trivial, i.e. is induced by an almost permutation of ω. In section 5 OCA is used to prove that a particular kind of topological space designated by $\gamma\omega$ cannot be completely normal. This implies that under PFA a version of Tychonoff's product theorem holds for countably compact spaces. Finally, in section 6, we consider possible extensions of OCA, and show, for example, that it cannot be generalized to dimensions bigger than 2. Then we raise some open problems and indicate areas for further research.

We believe that our notation is mostly standard, as for example in [Ku], or self explanatory.

2. Consistency of OCA

In this section we present the proof of the consistency of OCA ([To1, Theorems 4.4 and 8.0]) . We start with a ZFC result which is a natural generalization of the classical diagonalization argument of Sierpinski-Zygmund ([SZ]).

Theorem 2.1. *Let S be a set of reals and suppose*

$$[S]^2 = K_0 \cup K_1$$

is a given coloring where K_0 is open in the product topology. Assume that S is not the union of $< 2^{\aleph_0}$ 1-homogeneous sets. Then there is $Y \subseteq S$ of size $2^{\aleph_0}$ such that the poset of finite 0-homogeneous subsets of Y ordered by reverse inclusion has the $2^{\aleph_0}$-chain condition.

Proof. For $p \in S^n$ and open $U \subseteq S^n$ such that $p \in U$ let:

$$U_p = \{q \in U : q_i \neq p_i \text{ and } \{p_i, q_i\} \in K_0, \text{ for all } i < n\}.$$

If f is a function from $A \subseteq S^n$ into S and $p \in S^n$ let:

$$\omega_f(p) = \bigcap \{\mathrm{cl}(f(U_p \cap A)) : U \subseteq S^n \text{ open and } p \in U\}.$$

Let $\{f_\xi : \xi < 2^{\aleph_0}\}$ enumerate all countable functions from a finite power of S into S, and let $\{T_\xi : \xi < 2^{\aleph_0}\}$ enumerate all closed 1-homogeneous subsets of S. Build Y as the set $\{x_\xi : \xi < 2^{\aleph_0}\}$ such that:

(a) $x_\alpha \in S \setminus \{x_\xi : \xi < \alpha\}$,
(b) $x_\alpha \notin T_\xi$, for $\xi < \alpha$,
(c) x_α does not belong to any 1-homogeneous set which has the form $\omega_{f_\xi}(p) \cap S$, where $\xi < \alpha$ and p is a finite sequence from $\{x_\xi : \xi < \alpha\}$.

To prove Y works, assume that $\mathcal{F}$ is a disjoint family of $2^{\aleph_0}$ many finite 0-homogeneous subsets of Y. Without loss of generality we may assume that all elements of $\mathcal{F}$ have the same size $n \geq 1$. We prove, by induction on n, there there are two members of $\mathcal{F}$ whose union is 0-homogeneous. Case $n = 1$ is handled by (b). Suppose $n > 1$. For $s \in \mathcal{F}$ let $s = \{x_{s(0)}, \ldots, x_{s(n-1)}\}_<$ be the enumeration of s in the increasing order of indices, i.e. $s(0) < \ldots s(n-1) < 2^{\aleph_0}$. Identifying each s with an element of S^n we may assume that some fixed basic open set U in S^n separates all elements of $\mathcal{F}$. Thinking of $\mathcal{F}$ as a graph of an $(n-1)$-ary function g, where $g(s \restriction (n-1)) = x_{s(n-1)}$, for all $s \in \mathcal{F}$, let:

$$\mathcal{F}_0 = \{s \in \mathcal{F} : x_{s(n-1)} \in \omega_g(s \restriction (n-1)\}.$$

Claim 1. *$\mathcal{F} \setminus \mathcal{F}_0$ has size $< 2^{\aleph_0}$.*

Proof. Assume otherwise and for each $s \in \mathcal{F} \setminus \mathcal{F}_0$ pick a rational open interval I^s which contains $x_{s(n-1)}$ and is disjoint from $\omega_g(s \restriction (n-1))$. Fix also a basic open subset U^s of S^{n-1} containing $s \restriction (n-1)$ such that if $q \in U^s_{s\restriction(n-1)}$ then $g(q) \notin I^s$. Then there is a subset Z of $\mathcal{F} \setminus \mathcal{F}_0$ of size $2^{\aleph_0}$ such that the I^s for $s \in Z$ are all equal to some I and the U^s for $s \in Z$ are all equal to some U. By the inductive assumption there are $s, t \in Z$ such that $s \cup t$ is 0-homogeneous. But then $t \restriction (n-1) \in U_{s\restriction(n-1)}$ and $g(t \restriction (n-1)) \in I$, a contradiction. $\square$

Let now g_0 be a countable dense subfunction of g. Then $g_0 = f_\xi$ for some ξ. Pick $s \in \mathcal{F}_0$ with all indices above ξ and above all the indices of elements of g_0. Then,

$$x_{s(n-1)} \in \omega_{f_\xi}(s \restriction (n-1))$$

and hence, by (c), $\omega_{f_\xi}(s \restriction (n-1))$ is not 1-homogeneous. We can now pick $u, v \in \omega_{f_\xi}(s \restriction (n-1))$ such that $\{u, v\} \in K_0$ and find open intervals I and J such that $u \in I$, $v \in J$, and $I \times J \subseteq K_0$. By the definition of $\omega_{g_0}(s \restriction (n-1))$, there is $p \in \mathrm{dom}(g_0)$ such that $p \cup s \restriction (n-1)$ is 0-homogeneous and $g_0(p) \in I$. Pick $U \subseteq S^{n-1}$ such that $s \restriction (n-1) \in U$ and for every $q \in U$ $p \cup q$ is 0-homogeneous. Now, pick $q \in U$ such that $g_0(q) \in J$. Then $p \cup \{g_0(p)\}$ and $q \cup \{g_0(q)\}$ are two members of $\mathcal{F}$ whose union is 0-homogeneous. $\square$

Theorem 2.2. PFA *implies* OCA.

Proof. Fix a partition $[S]^2 = K_0 \cup K_1$ as in OCA and assume that S cannot be covered by countably many 1-homogeneous sets. This remains to hold in $V^{\mathcal{P}}$ where $\mathcal{P}$ is the σ-closed collapse of $2^{\aleph_0}$ to $\aleph_1$. In $V^{\mathcal{P}}$ CH holds so there is $Y \subseteq S$ such that the poset $\mathcal{Q}$ of finite 0-homogeneous subsets of Y is ccc. Some conditions in $\mathcal{Q}$ forces the generic homogeneous set to be uncountable and we may assume that the maximal condition does so. Thus, in $V^{\mathcal{P}*\mathcal{Q}}$ there is an uncountable 0-homogeneous set. By forcing internally with $\mathcal{P} * \mathcal{Q}$ we can produce such a set in V. $\square$

Let us point out that although large cardinals are needed to prove the consistency of PFA this is not the case with OCA + $\mathrm{MA}_{\aleph_1}$. Namely, we can start with a model of $V = L$ and perform a finite support iteration of ccc posets forcing $\mathrm{MA}_{\aleph_1}$. Along the way, we use $\diamondsuit(\{\alpha < \omega_2 : \mathrm{cof}(\alpha) = \omega_1\})$ to guess potential open colorings on a set of reals S and, if possible, force with the poset from Theorem 2.1 to obtain an uncountable 0-homogeneous set. The resulting model then satisfies OCA + $\mathrm{MA}_{\aleph_1}$. In [To1] OCA is shown to be equivalent to the following closed set-mapping axiom (CSM):

> *If F is a closed set-mapping on a set of reals, then either there is an uncountable F-free subset of dom(F), or else F is the union of countably many connected subfunction.*

Note that the strength of OCA comes from the fact that, although the partition is assumed to be open, S is allowed to be an arbitrary set of reals. Qi Feng ([Fe]) has studied versions of OCA obtained by restricting the complexity of the set S and has shown that the restriction of OCA to projective sets of reals follows from PD.

3. Combinatorial Applications

We start by presenting some consequences of the weak version of OCA. The following is [To1, Theorem 8.4]; but see also [Ba, Theorems 6.13 and 6.14].

Theorem 3.1. (OCA)

(a) *Every uncountable subset of $\mathcal{P}(\omega)$ contains an uncountable chain or an uncountable antichain.*

(b) *Every function from an uncountable set of reals into the reals in monotonic on an uncountable set.*

(c) *If X and Y are two uncountable sets of reals then there is a strictly increasing mapping from an uncountable subset of X into Y.*

(d) *Every uncountable Boolean algebra contains an uncountable antichain.*

(e) *Every subset of ω^ω of size $\aleph_1$ is bounded under $<_*$.*

Proof. To see (a), (b), and (c) observe that the inclusion is an closed relation on $\mathcal{P}(\omega)$, and that strictly increasing is an open relation in the plain. For (d), first show that if $\mathcal{B}$ is an uncountable Boolean algebra with no uncountable antichains then $\mathcal{B}$ can be embedded into $\mathcal{P}(\omega)$. Then use (a) and (b). For (e) let $\mathcal{F}$ be a subset of ω^ω of size $\aleph_1$. We may assume that each function in $\mathcal{F}$ is strictly increasing and that $\mathcal{F}$ is well-ordered by $<_*$ of order type ω_1. The everywhere dominance $<$ is a closed relation on ω^ω.

Since there are no uncountable linearly ordered sets under $<$, by OCA $\mathcal{F}$ has an uncountable pairwise incomparable subset A. Then by [To1, §1] A and hence $\mathcal{F}$ is bounded under $<_*$. □

The following result is implicit in [To1]. It shows that OCA has a strong influence on the partial ordering ω^ω and gives support for the conjecture that OCA implies that the continuum is $\aleph_2$.

Theorem 3.2. OCA *implies that the least size of an unbounded subset of* ω^ω *under* $<_*$ *is* $\aleph_2$.

Proof (see [To1, Theorem 3.7]). By Theorem 3.1(e) every subset of ω^ω of size $\aleph_1$ is bounded under $<_*$. To produce an unbounded subset of size $\aleph_2$ we shall need the following result which is of independent interest. Recall that a *gap* in ω^ω is a pair $\langle A, B\rangle$ of subsets of ω^ω such that:

(a) the order type of $A, <_*$ is a regular infinite cardinal,
(b) the order type of $B, <_*$ is the converse of a regular infinite cardinal,
(c) $f <_* g$ for all $f \in A$ and $g \in B$,
(d) there is no $h \in \omega^\omega$ such that $f <_* h <_* g$ for all $f \in A$ and $g \in B$.

Lemma 3.1. (OCA) *Let* $\langle A, B\rangle$ *be a gap in* ω^ω. *If* A *and* B *are uncountable then they both have size* $\aleph_1$.

Proof (see [To1, Theorem 8.6]). Suppose, for example, that the size of A is $> \aleph_1$. Given $f, g \in \omega^\omega$ such that $f <_* g$ let:

$$\Gamma(f,g) = \min\{m : f(n) < g(n) \text{ for all } n \geq m\}.$$

By shrinking A if necessary we may assume that there is a fixed n_0 and for all $f \in A$ an unbounded subset B_f of B such that $\Gamma(f,g) = n_0$, for all $g \in B_f$. Let $X = \{\langle f,g\rangle : f \in A \text{ and } g \in B_f\}$ and consider the partition

$$[X]^2 = K_0 \cup K_1$$

defined by

$$\{\langle f,g\rangle, \langle \bar{f},\bar{g}\rangle\} \in K_0 \text{ iff } \max\{\Gamma(f,\bar{g}), \Gamma(\bar{f},g)\} > n_0.$$

Then K_0 is open in the product topology. Let us show that X is not the union of countably many 1-homogeneous sets. Suppose towards contradiction that $X = \bigcup_{n<\omega} X_n$, where each X_n is 1-homogeneous. Then for some n the set $\bar{A}$ of all $f \in A$ such that the set

$$\bar{B_f} = \{g \in B_f : \langle f,g\rangle \in X_n\}$$

is unbounded in B is unbounded in A. Let f be a minimal element of $\bar{A}$. Define the function h in ω^ω by:

$$h(k) = \min\{g(k) : g \in \bar{B}_f\}.$$

Then it follows that h splits the gap $\langle A, B\rangle$, a contradiction.

Now, by OCA, there is an uncountable 0-homogeneous subset Y of X. We may assume that Y is of the form $\{\langle f_\alpha, g_\alpha\rangle : \alpha < \omega_1\}$, where the f_α are $<_*$-increasing. Note that the g_α must be distinct and thus we may assume that they are $<_*$-decreasing. Since A has cofinality $> \aleph_1$ there is $f \in A$ above all the f_α. Since $f_\alpha <_* g_\alpha$ for all $\alpha < \omega_1$ we can find an uncountable subset I of ω_1, an $n_1 < \omega$, and $p, q \in \omega^{n_1}$ such that for all $k \geq n_1$ $f_\alpha(k) < f(k) < g_\alpha(k)$, $f_\alpha \restriction n_1 = p$ and $g_\alpha \restriction n_1 = q$, for all $\alpha \in I$. It then follows that for every distinct $\alpha, \beta \in I$ $\{\langle f_\alpha, g_\alpha\rangle, \langle f_\beta, g_\beta\rangle\} \in K_1$, contradicting the fact that Y is 0-homogeneous. □

To finish the proof of Theorem 3.2, following [Ba, Theorem 4.4] fix a subset A of ω^ω such that the order type of $A, <_*$ is $\aleph_2$. Extend A to a $\subseteq$-maximal $<_*$-linearly ordered set L. Then A will determine a gap in L whose coinitiality, by Lemma 3.1, cannot be a regular uncountable cardinal. Also, it cannot be 1 since if g bounds A then so does $g - 1$. Thus the coinitiality of A in L is either 0 or ω. If it is 0 then A is already unbounded. If it is ω then by [Ro1] one can produce an unbounded subset of ω^ω of size $\aleph_2$. This is done as follows. Let $B = \{g_n : n < \omega\}$ be a subset of L such that $\langle A, B\rangle$ forms a gap. We may assume that $n \leq m$ implies $g_m(k) \leq g_n(k)$, for all k. For $f \in A$ let h_f be defined as follows

$$h_f(n) = \min\{k : f(l) < g_n(l) \text{ for all } l \geq k\}.$$

Then the family $\{h_f : f \in A\}$ is unbounded in ω^ω. □

Todorcevic and the author have shown that PFA implies that $2^{\aleph_0} = \aleph_2$ (see [Ve2] for the proof and the history involving this result). Similarly we conjecture that the answer to the following question is positive.

Question 3.1. Does *OCA* imply that $2^{\aleph_0} = \aleph_2$?

4. Automorphisms of $\mathcal{P}(\omega)/\text{fin}$

We now turn to the study of automorphisms of the Boolean algebra $\mathcal{P}(\omega)/\text{fin}$. Under the Continuum Hypothesis $\mathcal{P}(\omega)/\text{fin}$ has $2^{2^{\aleph_0}}$ automorphisms. On the other hand Shelah ([Sh]) proved the consistency that every automorphism φ of $\mathcal{P}(\omega)/\text{fin}$ is *trivial*, i.e. there exist finite sets $a, b \subseteq \omega$

and a bijection $e : \omega \setminus a \to \omega \setminus b$ such that for every $x \subseteq \omega$, $\varphi[x] = [e''(x)]$, where $[y]$ denotes the equivalence class of y modulo the ideal of finite subsets of ω. Clearly, there are only $2^{\aleph_0}$ such automorphisms. Subsequently, Shelah and Steprans ([SS]) have shown that the same conclusion follows from PFA. We now show how OCA was used in [Ve1] to derive the same result.

Theorem 4.1. (OCA + $\mathrm{MA}_{\aleph_1}$) *Every automorphism of $\mathcal{P}(\omega)/\mathrm{fin}$ is trivial.*

Proof. We indicate the main parts of the argument. To begin let us fix an automorphism φ and a function $F : \mathcal{P}(\omega) \to \mathcal{P}(\omega)$ such that $\varphi[x] = [F(x)]$, for every subset x of ω. We shall write $\varphi \restriction a$ for $\varphi \restriction \mathcal{P}(a)/\mathrm{fin}$ and say that φ is *trivial on* a provided $\varphi \restriction a$ is induced by some function $e : a \to \omega$. We shall refer ambiguously to $\mathcal{P}(a)$ and 2^a by identifying a set with its characteristic function. We shall need the following ZFC result, for the proof see [Ve1].

Theorem 4.2. *Suppose there exist Borel functions $F_n : \mathcal{P}(\omega) \to \mathcal{P}(\omega)$, for $n < \omega$ such that for every $a \subseteq \omega$ there exists $n < \omega$ such that $F(a) =_* F_n(a)$. Then φ is trivial.* □

The first step of the proof is to show that φ is somewhere trivial, i.e. there is an infinite set a such that $\varphi \restriction a$ is trivial. Let us say that a family $\mathcal{A}$ of almost disjoint infinite subsets of ω is *neat* if there is a 1-1 map $e : \omega \to 2^{<\omega}$ such that if $a \in \mathcal{A}$ and $n, m \in a$ then $e(n) \subseteq e(m)$ or $e(m) \subseteq e(n)$. Thus, $\bigcup e''(a)$ is an infinite branch through $2^{<\omega}$, for every $a \in \mathcal{A}$. The following lemma is the key application of OCA in the proof.

Lemma 4.1. *Let $\mathcal{A}$ be a neat almost disjoint family. Then φ is trivial on all but countably many $c \in \mathcal{A}$.*

Proof. Let $e : \omega \to 2^{<\omega}$ be a function witnessing that $\mathcal{A}$ is neat. Let X be the set of all pairs $\langle a, b\rangle$ of subsets of ω such that there exists $c \in \mathcal{A}$ such that $b \subseteq a \subseteq c$, and define the partition:

$$[X]^2 = K_0 \cup K_1$$

by $\{\langle a, b\rangle, \langle \bar{a}, \bar{b}\rangle\} \in K_0$ iff

(a) $\bigcup e''a \neq \bigcup e''\bar{a}$,
(b) $a \cap \bar{b} = \bar{a} \cap b$,
(c) $F(a) \cap F(\bar{b}) \neq F(\bar{a}) \cap F(b)$.

Then K_0 is open in the product of the separable metric topology τ on X obtained by identifying $\langle a, b\rangle$ with $\langle a, b, F(a), F(b)\rangle$. □

Claim: There are no uncountable 0*-homogeneous subsets of* X.

Proof. Suppose Y is an uncountable 0-homogeneous set. Let d be the union of all b such that for some a the pair $\langle a,b\rangle$ belongs to Y. Let $\langle a,b\rangle$ be such a pair. By (b) in the definition of K_0 it follows that $d\cap a=b$ and hence $F(d)\cap F(a)=_* F(b)$. We can find an uncountable $Z\subseteq Y$ and $n<\omega$ such that for every $\langle a,b\rangle\in Z$, $(F(d)\cap F(a))\Delta F(b)\subseteq n$ and $F(b)\setminus n\subseteq F(a)$. Then there are distinct $\langle a,b\rangle$ and $\langle\bar a,\bar b\rangle$ in Z such that $F(a)\cap n=F(\bar a)\cap n$ and $F(b)\cap n=F(\bar b)\cap n$. It then follows that $F(a)\cap F(\bar b)=F(\bar a)\cap F(b)$ which contradicts the fact that $\{\langle a,b\rangle,\langle\bar a,\bar b\rangle\}\in K_0$. □

Now, by OCA we can find a decomposition $X=\bigcup_{n<\omega}X_n$ where X_n is 1-homogeneous for all n. Fix for each n a countable subset D_n of X_n which is dense in X_n in the sense of τ. For each $\langle a,b\rangle\in X$ pick $\sigma(a)\in\mathcal{A}$ such that $b\subseteq a\subseteq\sigma(a)$. Let

$$\mathcal{B}=\{\sigma(a):\ \langle a,b\rangle\in D_n \text{ and } n<\omega\}.$$

We shall show that φ is trivial on every $c\in\mathcal{A}\setminus\mathcal{B}$. Thus, fix any such c and decompose it into two disjoint sets $c=c_0\cup c_1$ such that for every $i\in\{0,1\}$, $n<\omega$, and $\langle a,b\rangle\in X_n$ if $a\subseteq c_i$ then for every $m<\omega$ there exists $\langle\bar a,\bar b\rangle\in D_n$ such that:

(a) $a\cap\bar b=\bar a\cap b$,

(b) $a\cap m=\bar a\cap m$ and $b\cap m=\bar b\cap m$,

(c) $F(a)\cap m=F(\bar a)\cap m$ and $F(b)\cap m=F(\bar b)\cap m$.

This is done as follows. An increasing sequence $\langle n_i:\ i<\omega\rangle$ is constructed by induction. Let $n_0=0$. Suppose $\langle n_i:\ i\le k\rangle$ has been defined. Then n_{k+1} is chosen sufficiently large such that for every $x,y,u,v\subseteq n_k$ and every $i\le k$ if there exist $\langle a,b\rangle\in X_i$ such that $a\cap n_k=x$, $b\cap n_k=y$, $F(a)\cap n_k=u$ and $F(b)\cap n_k=v$ then there exist $\langle a,b\rangle\in D_i$ with the same property such that in addition $a\cap c\subseteq n_{k+1}$. This is possible since a is almost disjoint from c whenever there is b such that $\langle a,b\rangle\in D_n$. Finally, let

$$c_0=\bigcup\{c\cap[n_k,n_{k+1}):\ k \text{ is even}\}$$

and let $c_1=c\setminus c_0$. Define the function $F_n:\mathcal{P}(c_0)\to\mathcal{P}(\omega)$, for $n<\omega$, by:

$$F_n(b)=\bigcup\{F(c_0)\cap F(\bar b):\ \langle\bar a,\bar b\rangle\in D_n \text{ and } \bar a\cap b=c_0\cap\bar b\}.$$

Clearly, F_n is a Borel function for all n. We claim that if $\langle c_0,b\rangle\in X_n$ then $F_n(b)=_* F(b)$. This follows easily from the properties of the decomposition $c=c_0\cup c_1$. Thus, by Theorem 4.2, φ is trivial on c_0. A similar argument shows that φ is trivial on c_1, and hence it is also trivial on c. □

Now consider the following set:

$$\mathcal{I} = \{\, a \subseteq \omega : \ \varphi \text{ is trivial on } \ a\}.$$

Fix, for the rest of the proof, for each a in $\mathcal{I}$ a function $e_a : a \to \omega$ inducing $\varphi \restriction a$. Recall that an ideal on ω containing all finite sets is called *dense* provided every infinite subset of ω contains an infinite member of the ideal. Then $\mathcal{I}$ is a dense ideal on $\mathcal{P}(\omega)$. An ideal on ω is called a P-*ideal* if it is countably directed under $\subseteq_*$, and, in general, it is called a P_κ-*ideal* if it is $< \kappa^+$-directed. We shall consider two cases according to whether $\mathcal{I}$ is a P-ideal or not.

Case 1: *$\mathcal{I}$ is a dense P-ideal.*

Define the partition

$$[\mathcal{I}]^2 = K_0 \cup K_1$$

by $\{a, b\} \in K_0$ iff there exists $n \in a \cap b$ such that $e_a(n) \neq e_b(n)$. Note that K_0 is open in the topology on $\mathcal{I}$ obtained by identifying a with e_a. Now using $\mathrm{MA}_{\aleph_1}$ one can prove the following (see [Ve1, Lemma 4]).

Claim: There are no uncountable 0*-homogeneous subsets.*

By OCA, there is a decomposition $\mathcal{I} = \bigcup_{n<\omega} \mathcal{I}_n$ where for every $n < \omega$ $\mathcal{I}_n$ is 1-homogeneous. Since $\mathcal{I}$ is a P-ideal, there is $n < \omega$ such that $\mathcal{I}_n$ is cofinal in $\mathcal{I}, \subseteq_*$. Let e be the union of the e_a, for $a \in \mathcal{I}_n$. It follows that for every $a \in \mathcal{I}$ $e \restriction a =_* e_a$, and, since $\mathcal{I}$ is dense and φ is an automorphism, that e induces φ. $\square$

Case 2: *$\mathcal{I}$ is not a P-ideal.*

Find a decomposition decomposition $\omega = \bigcup_{n<\omega} a_n$ into disjoint infinite sets from $\mathcal{I}$ such that there does not exist a in $\mathcal{I}$ almost containing a_n for all n. Given $f \in \omega^\omega$ let $b_f = \bigcup\{a_n \cap f(n) : \ n < \omega\}$.

Claim: There exists $f \in \omega^\omega$ such that φ is nontrivial on b_f.

Proof. Assume otherwise and let $\mathcal{J}$ be the collection of all $b \subseteq \omega$ which are almost disjoint from the a_n. Then it follows from either Theorem 3.1(e) or by a simple application of $\mathrm{MA}_{\aleph_1}$ $\mathcal{J}$ is a $\mathrm{P}_{\aleph_1}$-subideal of $\mathcal{I}$. Then as is easily seen the partition considered in Case 1 restricted to $\mathcal{J}$ has no uncountable 0-homogeneous sets. Thus, there exists $e : \omega \to \omega$ such that $e \restriction b =_* e_b$ for every $b \in \mathcal{J}$. We claim that there exists $k < \omega$ such that e induces φ on $\omega \setminus \bigcup_{i<k} a_i$, which contradicts the nontriviality of φ. To see this, it suffices

to show that the set

$$T = \{m < \omega : \; e \restriction a_m \text{ does not induce } \; \varphi \restriction a_m\}$$

is finite. For then e induces $\varphi \restriction a$ for every a in the ideal generated by $\mathcal{J}$ and $\{a_m : m \notin T\}$. Since this ideal is dense in $\mathcal{P}(u)$, where $u = \omega \setminus \{a_m : m \in T\}$, and φ is an automorphism it follows that e induces φ on u.

Now, suppose T were infinite. For each $m \in T$ we pick an infinite subset c_m of a_m such that $e''(c_m) \cap F(c_m) =_* \emptyset$. By shrinking the c_m we can arrange that, furthermore, for every $m, k \in T$ $e''(c_m) \cap F(c_k) =_* \emptyset$. We then find d such that for every $m \in T$ $F(c_m) \subseteq_* d$ and $e''(c_m) \cap d =_* \emptyset$ and let c be such that $F(c) =_* d$. It follows that $c_m \subseteq_* c$, for each $m \in T$ and hence we can pick $i_m \in c_m \cap c$ such that $e(i_m) \notin F(c)$. Let $b = \{i_m : m \in T\}$. Then $b \in \mathcal{J}$ and hence $F(b) =_* e''(b)$. On the other hand $b \subseteq c$ and hence $F(b) \subseteq_* F(c)$. But $e''(b) \cap F(c) = \emptyset$. Contradiction. □

Note that Claim actually shows that for every $f \in \omega^\omega$ there exists $g \in \omega^\omega$ such that $b_g \setminus b_f$ is nontrivial. We can then easily construct an $<_*$-increasing sequence f_α; $\alpha < \omega_1$ in ω^ω such that φ is nontrivial on $b_{f_{\alpha+1}} \setminus b_{f_\alpha}$ for every $\alpha < \omega_1$. Let $a_\alpha = b_{f_{\alpha+1}} \setminus b_{f_\alpha}$. By another application of $\mathrm{MA}_{\aleph_1}$ (see [Ve1, Lemma 3]) we can split each a_α into two disjoint sets a^0_α and a^1_α such that $\mathcal{A}^i = \{a^i_\alpha : \alpha < \omega_1\}$ is neat, for $i = 0, 1$. By Lemma 4.1 there is $\alpha < \omega_1$ such that φ is trivial on both a^0_α and a^1_α, and hence on a_α. Contradiction. □

Some of these ideas have been used by Just ([Ju]) in the proof of the following.

Theorem 4.3. (OCA)

(a) *$(\omega^*)^{(n+1)}$ is not a continuous image of $(\omega^*)^n$, for every $n < \omega$.*

(1) *If $\mathcal{I}$ is a dense P-ideal then $\mathcal{P}(\omega)/\mathcal{I}$ is not isomorphic to $\mathcal{P}(\omega)/\mathrm{fin}$.*

(b) *If all Σ^1_{n+2} sets are measurable and $\mathcal{I}$ is a Σ^1_n ideal containing all finite sets such that $\mathcal{P}(\omega)/\mathcal{I}$ is embeddable into $\mathcal{P}(\omega)/\mathrm{fin}$ then $\mathcal{I}$ is generated over the Fréchet ideal by at most one set.*

(c) *If $\mathcal{I}$ is the ideal of sets density 0 and $\mathcal{J}$ is the ideal of sets of logarithmic density 0, then $\mathcal{P}(\omega)/\mathcal{I}$ and $\mathcal{P}(\omega)/\mathcal{J}$ are not isomorphic.*

5. Complete Normality of $\gamma\omega$

We now present an application of OCA in the study of countably compact topological spaces. Recall that a topological space X is called *completely normal* if for every two subsets A and B of X which are separated (i.e. $\mathrm{cl}A \cap B = \emptyset = A \cap \mathrm{cl}B$) there are disjoint open sets containing A and

B, respectively. Hausdorff spaces satisfying this property are designated T_5. How well-behaved can countably compact T_5 spaces be? Assuming $V = L$ they can be quite pathological, but assuming PFA it was shown in [NV] that every countably compact T_5 space is sequentially compact, in fact every countable subset has compact, Fréchet-Urysohn closure. [A space is called *Fréchet-Urysohn* if whenever a point x is in the closure of a subset A, then there is a sequence from A converging to x.] Hence, in particular, a separable subspace can have cardinality at most $2^{\aleph_0}$. A consequence of this is a version of Tychonoff's theorem for countably compact spaces: under PFA the product of any number of countably compact T_5 spaces is countably compact, although the T_5 property may be lost. The key application of OCA is to show that certain kind of spaces commonly denoted by $\gamma\omega$ cannot be completely normal. Here $\gamma\omega$ is the generic symbol for a locally compact Hausdorff space X with a countable dense set of isolated points, identified with the set ω of positive integers, such that $X \setminus \omega$ is homeomorphic to ω_1. We will also identify $X \setminus \omega$ with ω_1 using a definition of ω that makes it disjoint from ω_1.

Theorem 5.1. *Under* OCA *no version of* $\gamma\omega$ *can be completely normal.*

Proof. For each $\alpha < \omega_1$ let $a_\alpha \subset \omega$ be such that $a_\alpha \cup [0, \alpha]$ is a compact neighborhood of $[0, \alpha]$. It is easily seen that $a_\alpha \subset_* a_\beta$ and $a_\beta \setminus a_\alpha \subset_* U$, for every neighborhood U of $(\alpha, \beta]$ whenever $\alpha < \beta$. Let S be the set of all $\langle a_\xi, a_\eta, a_\mu \rangle$ such that $\xi < \eta < \mu$ and define the partition

$$[S]^2 = K_0 \cup K_1$$

by $\{\langle a, b, c\rangle\langle \bar{a}, \bar{b}, \bar{c}\rangle\} \in K_0$ iff

$$a \neq \bar{a} \text{ and } [(a \setminus b) \cap (\bar{c} \setminus \bar{b}) \neq \emptyset \text{ or } (c \setminus b) \cap (\bar{b} \setminus \bar{a}) \neq \emptyset].$$

Then K_0 is open in the product topology.

Suppose first that $\{S_n : n < \omega\}$ is a sequence of 1-homogeneous sets whose union covers S. Let T_n be the set of all ξ for which there are uncountably many η such that $\langle a_\xi, a_\eta, a_\mu\rangle \in S_n$, for some μ. Clearly some T_n must be uncountable. Fix such n and some $\xi \in T_n$. Let $\langle a_{\bar{\xi}}, a_{\bar{\eta}}, a_{\bar{\mu}}\rangle \in S_n$ be such that $\xi < \bar{\xi}$ and find $\mu > \eta > \bar{\mu}$ such that $\langle a_\xi, a_\eta, a_\mu\rangle \in S_n$. Since $\xi < \bar{\eta} < \bar{\mu} < \eta$ we have $a_{\bar{\mu}} \setminus a_{\bar{\eta}} \subset_* a_\eta \setminus a_\xi$. Thus, $\{\langle a_\xi, a_\eta, a_\mu\rangle, \langle a_{\bar{\xi}}, a_{\bar{\eta}}, a_{\bar{\mu}}\rangle\} \in K_0$, which contradicts the fact that S_n is 1-homogeneous.

Now, by OCA, there is an uncountable 0-homogeneous subset H of S. By cutting H down if necessary we may assume $\mu < \bar{\xi}$ whenever $\langle a_\xi, a_\eta, a_\mu\rangle$ and $\langle a_{\bar{\xi}}, a_{\bar{\eta}}, a_{\bar{\mu}}\rangle$ are two distinct members of H such that $\xi < \bar{\xi}$. Then

$$A = \bigcup \{(\xi, \eta] : \langle a_\xi, a_\eta, a_\mu \rangle \in H\}$$

and

$$B = \bigcup \{(\eta, \mu] : \langle a_\xi, a_\eta, a_\mu \rangle \in H\}$$

are separated in $\gamma\omega$. If there were an open subset U of $\gamma\omega$ such that $A \subset U$ and $\mathrm{cl}U \cap B = \emptyset$, we could let $c = U \cap \omega$ and have $a_\eta \setminus a_\xi$ almost contained in c and $a_\mu \setminus a_\eta$ almost disjoint from c whenever $\langle a_\xi, a_\eta, a_\mu \rangle \in H$. Now, for every ξ there are at most one η and μ such that $\langle a_\xi, a_\eta, a_\mu \rangle \in H$. If this happens choose $n(\xi) \in \omega$ such that

$$[(a_\eta \setminus a_\xi) \setminus c] \cup [(a_\mu \setminus a_\eta) \cap c] \subseteq [0, n(\xi)].$$

Then there is an uncountable subset I of H, $n \in \omega$, and $a \subseteq [0, n]$ such that whenever $\langle a_\xi, a_\eta, a_\mu \rangle \in I$ then $n(\xi) = n$ and $a_\eta \cap [0, n] = a$. But then any pair of distinct elements of I is in K_1, a contradiction. □

6. Generalizations of OCA

How can the Open Coloring Axiom be strengthened or generalized? It turns out that there are some strong limitations on the possible generalizations. We first present an example from [To3] which shows that one cannot reverse *open* and *closed* in the statement of OCA.

Theorem 6.1. *There is a coloring*

$$[\omega^\omega]^2 = K_0 \cup K_1$$

with K_0 open in the product topology such that there are no uncountable 1-homogeneous sets and ω^ω is not the union of countably many 0-homogeneous sets.

Proof. For every f in ω^ω associate a sequence $\{f_i : i < \omega\}$ converging to f as follows. Let $n_0 < n_1 < \ldots$ be the list of n such that $f(2n+1) \neq 0$. For a given i the real f_i is determined by letting $f_i \restriction n_k = f \restriction n_k$ and

$$f_i(n_k + j) = f(2^{i+1}(2n_k + 2j + 1)),$$

where $k = k(i)$ is minimal such that

$$f(2n_0 + 1) + \ldots f(2n_k + 1) > i$$

if such k exists, otherwise let $f_i = f$. Define the partition $[\omega^\omega]^2 = K_0 \cup K_1$ by:

$$\{f, g\} \in K_0 \text{ iff } f \neq g_i \text{ and } g \neq f_i, \text{ for all } i < \omega.$$

Then K_0 is open in the product topology. □

Claim 1: There are no uncountable 1*-homogeneous sets.*

Proof. Suppose Y is an uncountable subset of ω^ω. Let D be a countable dense subset of Y and let

$$\bar{D} = \{f_i : f \in D \text{ and } i < \omega\}.$$

Pick $g \in Y \setminus \bar{D}$ and find $h \in Y$ such that $h \neq g_i$, for all $i < \omega$. Then there is an open interval I containing h such that $g_i \notin I$, for all $i < \omega$. Since D is dense in Y there is $f \in D \cap I$. Then $\{f, g\} \in K_0$. □

Remark: A similar argument can be used to show that the poset of finite 0-homogeneous sets, ordered under reverse inclusion is ccc.

Claim 2: ω^ω is not the union of countably many 0*-homogeneous sets.*

Proof. Let $\{H_n : n < \omega\}$ be a sequence of 0-homogeneous subsets of ω^ω. Define the function f in ω^ω as follows. First let $f(2i+1) = 1$, for all $i < \omega$. Then define inductively $f_i \in \omega^\omega$ and $f(2i)$, for $i < \omega$. Suppose $f \restriction 2l$ has been defined as well as f_i, for all $i < l$. If $2l = 2^{i+1}(2i + 2j + 1)$ for some $i < l$ and $j < \omega$ let $f(2l) = f_i(i + j)$. Otherwise choose $f(2l)$ to be any number different from $f_i(2l)$, for all $i < l$. If there is $g \in H_l$ such that $g \restriction l = f \restriction l$ let f_l be such a g. Otherwise let f_l be any function such that $f \restriction l \subseteq f_l$. Then thus constructed f does not belong to H_n, for any $n < \omega$. □

Can OCA be generalized to dimensions bigger than two? The following example of Blass shows that it cannot. Given distinct reals x and y in ω^ω let

$$\Delta(x, y) = \min\{n : x(n) \neq y(n)\}$$

and define the partition

$$[\omega^\omega]^2 = K_0 \cup K_1$$

as follows. Given $x, y, z \in 2^\omega$ with $x < y < z$ let

$$\{x, y, z\} \in K_0 \text{ iff } \Delta(x, y) < \Delta(y, z).$$

It is easy to see that both K_0 and K_1 are open in the product topology and that there are no uncountable homogeneous sets in either color. Generalizing this example one can construct an open coloring of n-tuples of reals into $(n-1)!$ colors such that every uncountable set has n-tuples of each of the colors. Is this example in some sense optimal? Is it consistent that for every open coloring of triples of an uncountable set of reals S into finitely many colors there is an uncountable subset of S which hits at most

2 colors? This question was asked in [ARS]. We now present an example which shows that this is not possible.

Theorem 6.2. *There is an uncountable set of reals X and a continuous function $f : [X]^3 \to \omega$ such that if Y is an uncountable subset of X then $f''[Y]^3 = \omega$.*

Proof. Fix a coloring $k : \omega^{<\omega} \times \omega^{<\omega} \to \omega$ such that for every $m > 0$, for every $s \in \omega^m$, every finite $D \subseteq \omega^{<\omega}$, and every function σ which maps D to ω exists n such that for all $t \in D$

$$k(t, s \cup \{\langle \mathrm{lh}(s), n\rangle\}) = \sigma(t).$$

Such a k can be obtained, for example, as follows. Fix an enumeration $\langle \sigma_i : i < \omega\rangle$ of all finite functions from a subset of $\omega^{<\omega}$ to ω. Given $s, t \in \omega^{<\omega}$ such that $\mathrm{lh}(s) = m > 0$ let $n = s(m-1)$ and define $k(t,s)$ to be $\sigma_n(t)$ if $t \in \mathrm{dom}(\sigma_n)$, otherwise let $k(t,s) = 0$. The following lemma is a variation on the main result from [Ro2].

Lemma 6.1. *Suppose a coloring $c : [\omega_1]^2 \to \omega$ is given. Then there exists a sequence of distinct reals $\langle r_\alpha : \alpha < \omega_1\rangle$ such that for every $\alpha < \beta < \omega_1$ there exists $n < \omega$ such that $k(r_\alpha \restriction m, r_\beta \restriction m) = c(\alpha, \beta)$, for all $m \geq n$.*

Proof. The reals r_α are constructed inductively. Suppose r_ξ has been defined for all $\xi < \alpha$. To construct r_α fix a 1-1 function $e_\alpha : \alpha \to \omega$ and let

$$F_n(\alpha) = \{\xi < \alpha : e_\alpha(\xi) < n\}.$$

Define recursively $r_\alpha(m)$ as follows. Given $r_\alpha \restriction m$ let l be the largest integer $\leq m$ such that if ξ and η are distinct elements of $F_l(\alpha)$ then $r_\xi \restriction (m+1) \neq r_\eta \restriction (m+1)$.

Now, apply the property of k to $r_\alpha \restriction m$ and $\{r_\xi \restriction (m+1) : \xi \in F_l(\alpha)\}$ to find n such that for all $\xi \in F_l(\alpha)$

$$k(r_\xi \restriction (m+1), r_\alpha \restriction m \cup \{\langle m, n\rangle\}) = c(\xi, \alpha).$$

Then let $r_\alpha(m) = n$. Then thus constructed sequence $\langle r_\alpha : \alpha < \omega_1\rangle$ works. $\square$

Now, fix a coloring $c : [\omega_1]^2 \to \omega$ witnessing $\aleph_1 \not\to [\aleph_1]^2_\omega$, i.e. such that $c''[U]^2 = \omega$, for every uncountable $U \subseteq \omega_1$, (see [To4]). Let $\langle r_\alpha : \alpha < \omega_1\rangle$ be a sequence of reals as in Lemma 6.1 and let $X = \{r_\alpha : \alpha < \omega_1\}$. Let $f : [X]^3 \to \omega$ be defined as follows. Given $x, y, z \in X$ with $x < y < z$, where $<$ is the lexicographical ordering on ω^ω, let

$$f(\{x,y,z\}) = k(x \restriction \Delta(y,z), y \restriction \Delta(y,z)).$$

Clearly, f is continuous.

Now suppose Y is an uncountable subset of X. We may assume that Y is dense in itself. Given $i < \omega$ we find $x, y, z \in Y$ such that $f(\{x, y, z\}) = i$. Using the fact that c witnesses $\aleph_1 \not\twoheadrightarrow [\aleph_1]^2_\omega$ and the property of k, find $x, y \in Y$ and $n \in \omega$ such that $x < y$ and $k(x \restriction m, y \restriction m) = i$, for all $m \geq n$. Since Y is dense in itself there exists $z \in Y$ such that $\Delta(y, z) \geq n$. It follows that $f(\{x, y, z\}) = i$, as desired. $\square$

We finish by posing two open problems concerning generalizations of OCA. The first one, which was stated as a conjecture in [To1 §8], asks to weaken the topological assumptions on the space S to essentially the best possible.

Question 6.1. Is the following version of *OCA* consistent?

> If S is a regular topological space with no uncountable discrete subsets and
>
> $$[S]^2 = K_0 \cup K_1$$
>
> a partition with K_0 open in the product topology then either there is an uncountable 0-homogeneous set or else S can be covered by countably many 1-homogeneous sets.

We have not discussed generalizations of OCA to cardinals bigger than $\aleph_1$ but the following problem would certainly require new techniques.

Question 6.2. Is the following consistent with the negation of the Continuum Hypothesis?

> If S is a set of reals of size $> \aleph_1$ and
>
> $$[S]^2 = K_0 \cup K_1$$
>
> is a partition with both K_0 and K_1 open then there exists a homogeneous subset of S of size $> \aleph_1$.

References

[ARS] U. Abraham, M. Rubin, and S. Shelah, *On the consistency of some partition theorems for continuous colorings, and the structure of* $\aleph_1$*-dense real order types*, Annals of Pure and Applied Logic, **29**, (1985), pp. 123–206.

[Ba] J. Baumgartner, *Applications of the Proper Forcing Axiom*, in "Handbook of set-theoretic topology", (K. Kunen, J. Vaughan eds.), North-Holland Publ.Co., Amsterdam, 1984, pp. 913–959.

[Fe] Qi Feng, *Homogeneous sets for open partitions of pairs of reals*, to appear.

[FMS] M. Foreman, M. Magidor, and S. Shelah, *Martin's maximum, saturated ideals, and nonergular ultrafilters, part I*, Annals of Mathematics, **127(1)**, (1988), pp. 1–47.

[Ju] W. Just, *A weak form of* AT *from* OCA, these proceedings.

[Ku] K. Kunen, "Set theory", North-Holland, Amsterdam, 1980.

[vM] J. van Mill, *Introduction to* $\beta\omega$, in "Handbook of set theoretic topology", eds. K. Kunen, and J. Vaughan, North-Holland, Amsterdam, 1984. pp. 503–567.

[Mo] Y. Moschovakis, "Descriptive set theory", Studies in Logic and the Foundations of Mathematics, vol. 100, North-Holland, Amsterdam, 1980.

[NV] P. Nyikos and B. Velickovic, *Complete normality and countable compactness*, Bulletin of the AMS, to appear.

[Ro1] F. Rothberger, *Sur les famillies indénombrables des suites de nombres naturels et les problèmes concernant la propriété C*, Proc. Cambridge Math Society, **37**, (1941), pp. 621–626.

[Ro2] F. Rothberger, *On families of real functions with denumerable bases*, Annals of Math., **45 (3)**, (1944), pp. 397–406.

[Sh] S. Shelah, "Proper forcing", Lecture Notes in Math., vol. 940, Springer Verlag, Berlin, 1982.

[SS] S. Shelah, and J. Steprans, PFA *implies all automorphisms are trivial*, Proc. A.M.S., **104** (1988), pp. 1220–1225.

[SZ] W. Sierpinski and A. Zygmund, *Sur une fonction qui est discontinue sur tout ensemble de puissance du continuum*, Fund. Math., **4**, (1928), pp. 316–318.

[To1] S. Todorcevic, "Partition problems in Topology", Contemporary Mathematics vol. 84, American Math Society, Providence, R.I. 1989.

[To2] S. Todorcevic, *Martin's Axiom and Continuum Hypothesis*, to appear.

[To3] S. Todorcevic, *Two examples of Borel partial orderings with the countable chain condition*, Proceedings of the AMS, to appear.

[To4] S. Todorcevic, *Partitioning pairs of countable ordinals*, Acta Math., **159**, (1987), pp. 261–294.

[TV] S. Todorcevic and B. Velickovic, *Martin's axiom and partitions*, Compositio Math, **63**, (1987), pp. 397–408.

[Ve1] B. Velickovic, OCA *and automorphisms of* $\mathcal{P}(\omega)/fin$, Topology and its Applications, to appear.

[Ve2] B. Velickovic, *Forcing axioms and stationary sets*, Advances in Mathematics, to appear.

Department of Mathematics, University of Toronto,
Toronto, Ontario M5S 1A1, CANADA

AMOEBA FORCING, SUSLIN ABSOLUTENESS AND ADDITIVITY OF MEASURE

JOAN BAGARIA AND HAIM JUDAH

ABSTRACT. We show that Additivity of Measure does not imply MA(Suslin), thus answering an open question in [J-S 1]. We define the notion of Suslin absoluteness and we show that the existence of a Suslin absolute model of ZFC is equiconsistent with the existence of an inaccessible cardinal. Finally, we give a combinatorial characterization of MA(Amoeba) which is also equivalent to Additivity of Measure.

1. INTRODUCTION

Suslin forcing, i.e., forcing notions in which the set of conditions, the ordering, and the incompatibility relation are κ-Suslin sets of reals in the sense of descriptive set theory, was first studied by H. Judah and S. Shelah in [J-S 1]. In their paper, they show that Suslin forcing notions admit a systematic treatment and specially so for $\aleph_0$-Suslin, i.e., membership, ordering and incompatibility are $\underset{\sim}{\Sigma}^1_1$. (In what follows, "Suslin" will mean $\aleph_0$-Suslin).

Also, Suslin forcing is being considered in [J-S 2], where they study the problem of the consistency strength of regularity properties of the projective sets of reals together with some variants of MA, giving exact equiconsistency results. In particular, they show that the following are equiconsistent:

(1) ZFC + There exists an inaccessible cardinal.
(2) ZFC + MA(Suslin) + Every projective set of reals is Lebesgue measurable, has the property of Baire and is Ramsey.

We recall some definitions and basic facts about Suslin partial orderings:

1.1. Basic facts

Definition 1.1.1. A poset P is *Suslin* iff P is a $\underset{\sim}{\Sigma}^1_1$-subset of $\mathbb{R}$ and both $\leq_P$ and $\perp_P$ (the incompatibility relation) are $\underset{\sim}{\Sigma}^1_1$-subsets of $\mathbb{R}\times\mathbb{R}$. (Notice that this implies $\perp_P$ is Borel.)

Fact 1.1.2. *If P is a Suslin ccc poset, then the predicate "x codes a maximal antichain of P" is $\underset{\sim}{\Pi}^1_1$ and hence absolute for transitive models of ZF.* $\square$

Definition 1.1.3. Let P be a poset. P is *indestructible ccc* if for every poset Q satisfying the ccc,

$$\Vdash_Q \text{“}\langle P, \leq\rangle \models ccc\text{”}$$

Fact 1.1.4. *If P is Suslin ccc, then P is indestructible ccc.* $\square$

This Fact is an immediate corollary of the following theorem (see [J-S 1] 3.14):

Theorem 1.1.5. *Let $V_1 \subseteq V_2$ be models of a part of ZFC, and suppose that P is Suslin with parameters in V_1. Then,*

$$V_1 \models \text{“}P \text{ satisfies the ccc”} \text{ iff } V_2 \models \text{“}P \text{ satisfies the ccc”}. \quad \square$$

1.2. Martin's Axiom for Suslin posets

MA(Suslin) was introduced in [J-S 1] where they notice it implies, among other things, the Additivity of the Lebesgue measure (Add(L)). Since all the consequences of MA(Suslin) that appear in [J-S 1] turn out to be more or less direct consequences of Add(L), they asked whether MA(Suslin) and Add(L) are in fact equivalent.

In section 2 we answer this question in the negative by giving a model for Add(L) in which MA(Suslin) fails. In fact, it fails for a poset of very low complexity in the Borel hierarchy, thus showing that Add(L) is a fairly weak assumption compared to MA(Suslin).

In section 3 we define the notion of Suslin absoluteness and we show that the existence of a Suslin absolute model of ZFC is equiconsistent with the existence of an inaccessible cardinal. In fact, we show:

(1) Absoluteness for Amoeba (Am) and Cohen forcing implies that ω_1 is inaccessible in L.
(2) If κ is an inaccessible cardinal in V, then $V[H]$ is Suslin absolute for $H \subseteq \text{Coll}(\aleph_0, {<}\kappa)$, generic over V.

We also show that if P is the Cohen (or random) forcing, then the following are equivalent:

(1) For all $x \in \mathbb{R}$ there exists a P-generic filter over $L[x]$.
(2) $\underset{\sim}{\Sigma}^1_3$-absoluteness for P.

As a Corollary we give a partial answer to a question of H. Woodin on the preservation of Δ^1_2-determinacy under Cohen forcing extensions.

Definition 1.2.1. For P a poset, MA(P) is the following sentence:
For every family $\langle D_i : i < \kappa \rangle$, $\kappa < 2^{\aleph_0}$, of maximal antichains of P, there exists $G \subseteq P$ directed such that for every $i < \kappa$, $G \cap D_i \neq \emptyset$.

T. Bartoszyński and H. Judah gave in [B-J], under some additional assumptions, the following characterization of MA(B), where B is the random algebra:

$$\mathrm{MA}(B) \Leftrightarrow \forall P \subseteq B, \text{ if } |P| < 2^{\aleph_0}, \text{ then } P \text{ is } \sigma\text{-centered}.$$

In section **4** we show that no additional assumptions are needed in the case of Amoeba. Thus, we prove:

$$\mathrm{MA}(Am) \Leftrightarrow \forall P \subseteq Am, \text{ if } |P| < 2^{\aleph_0}, \text{ then } P \text{ is } \sigma\text{-centered}.$$

2. MA(Suslin) and Add(L)

2.1. Definitions

We recall some definitions:

Definition 2.1.1. MA(Suslin) is the following statement:
For every Suslin partial ordering P satisfying the ccc and for every family $\langle D_i : i < \kappa \rangle$, $\kappa < 2^{\aleph_0}$, of maximal antichains of P, there exists $G \subseteq P$ directed such that for every $i < \kappa$, $G \cap D_i \neq \emptyset$.

Definition 2.1.2. Add(L) is the following statement:
For every $\kappa < 2^{\aleph_0}$ and for every disjoint collection $\{X_\alpha : \alpha < \kappa\}$ of Lebesgue measurable subsets of the interval $[0,1]$, we have

$$\mu(\bigcup_{\alpha<\kappa} X_\alpha) = \sum_{\alpha<\kappa} \mu(X_\alpha) = \sup\{\sum_{\alpha\in S} \mu(X_\alpha) : S \text{ is a finite subset of } \kappa\}$$

where μ is the Lebesgue measure on $[0,1]$.

Fact 2.1.3. *The following are equivalent:*

(1) Add(L)
(2) *The union of less than $2^{\aleph_0}$ measure zero sets has measure zero.*
(3) *For every $\kappa < 2^{\aleph_0}$ and for every collection $\{X_\alpha : \alpha < \kappa\}$ of Lebesgue measurable subsets of the interval* [0,1], $\bigcup_{\alpha<\kappa} X_\alpha$ *is Lebesgue measurable.*

Proof. See [J], Lemma 27.4. □

The main result in this section is the following Theorem:

Theorem 2.1.4. Add(L) $\not\Rightarrow$ *MA(Suslin)*

Proof. We will begin by giving an example, due to S. Todorčević [T1], of a Suslin partial ordering which satisfies the ccc and is not σ-linked.

Definition 2.1.5. A partial ordering P is *σ-linked* iff there exists $f: P \to \omega$ such that for all p, q in P, $f(p) = f(q)$ implies p, q are compatible. We call the partition induced on P by f a *σ-linking* partition.

Then, we will show that if we iterate ω_2-times Am with finite support over $V \models CH$, then in the generic extension we have Add(L) but MA(Suslin) fails for Todorčević's poset.

2.2. A Suslin ccc poset that is not σ-linked

Let $\pi\mathbb{Q}$ be the power set of the rationals with ordering $X < Y$ if and only if X is an initial part of Y, under the natural ordering of $\mathbb{Q}$, and $\min(Y - X)$ exists. Let P be the set of all finite antichains of $\pi\mathbb{Q}$ ordered by inclusion.

Fact 2.2.1. *P satisfies the ccc.*

Proof. Assume the contrary. So, let $I = \langle t_\alpha : \alpha < \omega_1 \rangle$ be an antichain. By a Δ-system argument we can assume that $\forall \alpha, \beta < \omega_1$, $\alpha \neq \beta$ implies $t_\alpha \cap t_\beta = \emptyset$. Also, we can assume that $|t_\alpha| = n$ for all $\alpha < \omega_1$.

Claim. We can find $I' \subseteq I$, $I' = \langle t'_\alpha : \alpha < \omega_1 \rangle$ such that for all t'_α, t'_β, if $\alpha < \beta$, then there exists $i, j < n$ such that $t'_\alpha(i) \leq t'_\beta(j)$.

Proof of Claim. Let $t'_0 = t_0$. Given $\langle t'_\alpha : \alpha < \lambda < \omega_1 \rangle$, let t'_λ be the first t_η such that:

(i) $\exists i < n \; \exists j < n$ with $t'_\alpha(i) \leq t_\eta(j)$
(ii) For all $\alpha < \lambda$, if $t'_\alpha = t_\beta$, then $\eta > \beta$.

t'_λ exists since:

(a) $\forall \alpha, \beta < \omega_1 \; \exists i < n \; \exists j < n$ such that either $t_\alpha(i) \leq t_\beta(j)$ or $t_\alpha(i) \geq t_\beta(j)$
(b) $\forall \alpha < \omega_1$ the set $\{\beta : \exists i < n \; \exists j < n \;\; t_\beta(i) \leq t_\alpha(j)\}$ is countable.

□

Let $\mathcal{U}$ be a uniform ultrafilter on ω_1. For $\delta < \omega_1$, $i, j < n$, define:

$$A_\delta(i,j) = \{\alpha < \omega_1 : t'_\delta(i) \leq t'_\alpha(j)\}$$

Since for each $\delta < \omega_1$, $\bigcup_{i,j<n} A_\delta(i,j) \in \mathcal{U}$, it follows that for each $\delta < \omega_1$ we can find i_δ, j_δ so that $A_\delta(i_\delta, j_\delta) \in \mathcal{U}$.

Hence, there is $B \subseteq \omega_1$ uncountable and $i, j < n$ such that $i_\delta = i$ and $j_\delta = j$ for all $\delta \in B$. Now fix $\delta < \gamma$ in B and choose $\alpha \in A_\delta(i,j) \cap A_\gamma(i,j)$ such that $\alpha > \gamma$. Then, $t'_\delta(i)$, $t'_\gamma(i) \leq t'_\alpha(j)$. Hence, $t'_\delta(i)$ and $t'_\gamma(i)$ are comparable. Thus, $\{t'_\delta(i) : \delta \in B\}$ is an uncountable chain in $\pi\mathbb{Q}$. But since for all $X, Y \in \pi\mathbb{Q}$, $X < Y$ implies that $\min(Y - X)$ exists, this gives an uncountable chain in $\mathbb{Q}$. Contradiction. □

Fact 2.2.2. *For every $t \in P$, $P_t = \{t' : t' \in P \wedge t \leq t'\}$ is not σ-linked.*

Proof. Suppose $P_t = \bigcup_{n\in\omega} P_t^n$ is a σ-linking partition of P_t, some $t \in P$.

Let $q_0 \in \mathbb{Q}$. Say $q_0 = 0$. If there is $t \cup \{X\} \in P_t^0$ with $\sup X < q_0$, then pick such an X and call it X_0; otherwise, pick $q \in \mathbb{Q}$, $q < q_0$, such that $\{q\} \notin t$ and $t \cup \{\{q\}\} \in P_t$ (since t is finite, this is always possible) and let $X_0 = \{q\}$

Pick $q_1 \in \mathbb{Q}$ such that $supX_0 < q_1 < q_0$. If there is $t \cup \{X\} \in P_t^1$ such that $X_0 < X$ and $\sup X < q_1$, pick such an X and call it X_1; otherwise, pick $q \in \mathbb{Q}$ strictly between $\sup X_0$ and q_1 and such that $X_0 \cup \{q\} \notin t$. Now, let $X_1 = X_0 \cup \{q\}$. And so on.

Claim 2.2.3. For every $n \in \omega$, $t \cup \{X_n\} \in P_t$.

Proof of Claim. By induction on $n \in \omega$.

$n = 0$: clear.

Assume it true for n. If there is $t \cup \{X\} \in P_t^{n+1}$ with $X_n < X$ and $\sup X < q_{n+1}$, then X_{n+1} is such an X and we are done. Otherwise, $X_{n+1} = X_n \cup \{q\}$, where $\sup X_n < q < q_{n+1}$. Suppose there is $X \in t$ such that X, X_{n+1} are compatible. Then, either $X < X_{n+1}$ or $X_{n+1} < X$. If $X < X_{n+1}$, then either $X = X_n$, which is impossible by the construction of X_n, or $X < X_n$, which is impossible by induction hypothesis. Similarly, if $X_{n+1} < X$, then $X_n < X$, which, by induction hypothesis, is also impossible. □

Let $X_\infty = \bigcup_{n\in\omega} X_n$.

Claim 2.2.4. $t \cup \{X_\infty\} \in P_t$.

Proof of Claim. Otherwise, there is $X \in t$ such that either $X < X_\infty$ or $X_\infty < X$. If $X < X_\infty$, then there is $n \in \omega$ such that $X < X_n$, which contradicts the previous Claim. If $X_\infty < X$, then for every $n \in \omega$, $X_n < X$, which also contradicts the previous Claim. □

Claim 2.2.5. For every $n \in \omega$, $t \cup \{X_\infty\} \notin P_t^n$.

Proof of Claim. Suppose $t \cup \{X_\infty\} \in P_t^n$, some $n \in \omega$. Then it is true that there exists $t \cup \{X\} \in P_t^n$ with $X_i < X$, all $i < n$, and $\sup X < q_n$. Hence, X_n is such an X. But since $X_n < X_\infty$, $t \cup \{X_n\}$ and $t \cup \{X_\infty\}$ are incompatible. Contradiction. □

This ends the proof of the Fact. □

Fact 2.2.6. *P is Suslin.*

Proof. Each $p \in P$, being a finite set of subsets of $\mathbb{Q}$, can be coded by a real number. That P, $\leq_P$, and $\perp_P$ are Σ_1^1 clearly follows from their definition. □

2.3. Amoeba forcing

We recall the definition of Amoeba forcing (see [S]):

Definition 2.3.1. Amoeba forcing (Am) is the following partial ordering: Conditions are open subsets of the Cantor space 2^ω of measure $< 1/2$. The ordering is $\subseteq$.

So, forcing with Am adds an open subset of 2^ω of measure $1/2$.

Fact 2.3.2.

(1) *Am is Suslin.*
(2) *Am is σ-linked.* (See [J] p.564). $\square$

Fact 2.3.3. ([M-S]) *In any forcing extension by Am, the set of random reals over the ground model has measure one.* $\square$

Lemma 2.3.4. *Let V be a transitive model of ZFC+CH. Let P_{ω_2} be an iteration of length ω_2 of Amoeba forcing with finite support. Then, $V^{P_{\omega_2}} \models$ Add(L).*

Proof. It is enough to show:

$$V^{P_{\omega_2}} \models \text{"The union of } \aleph_1\text{-many null Borel sets is null"}$$

Each Borel set in $V^{P_{\omega_2}}$ appears at some stage $\alpha < \omega_2$ of the iteration. (See [J], Lemma 23.8). Hence, if $\{S_\alpha : \alpha < \aleph_1\}$ is an $\aleph_1$-collection of Borel null sets, then $\exists \beta < \omega_2$ such that $\{S_\alpha : \alpha < \aleph_1\} \subseteq V_\beta$. But, by the Fact above, forcing with Amoeba makes the union of all old Borel sets of measure zero into a measure zero set. $\square$

We will see that if P_{ω_2} is a finite support iteration of length ω_2 of Amoeba and $V \models CH$, then in $V^{P_{\omega_2}}$ MA(Suslin) fails for Todorčević's poset P.

We need the following Lemma:

Lemma 2.3.5. *Let $P_\beta = \langle P_\alpha; \dot{Q}_\alpha : \alpha < \beta \rangle$, $|\beta| \leq 2^{\aleph_0}$, be an iteration with finite support of forcing notions satisfying:*

$$\forall \alpha < \beta,\ \Vdash_{P_\alpha} \text{"}\dot{Q}_\alpha \text{ is } \sigma\text{-linked"}$$

Then, P_β is σ-linked.

Proof. By induction on $\gamma \leq \beta$:
To each $p \in P_\beta$ we associate the function $s_p : 2^{\aleph_0} \to \{0, 1\}$, where for $\alpha < 2^{\aleph_0}$, $s_p(\alpha) = 1$ iff $\alpha \in \text{supp}(p)$.

Fact 2.3.6. 2^β *with product topology contains a countable dense subset* D. (See [K], exercise II,3. Also, [E] for a complete proof.) $\square$

Now, let $D^* = \{s_p : p \in P_\beta \text{ and } \exists f \in D \text{ such that } s_p \leq_l f\}$, where $\leq_l$ is the lexicographic ordering. Notice that since supp(p) is finite, for each $f \in D$, the set $\{s_p : p \in P \text{ and } s_p \leq_l f\}$ is countable. Hence, D^* is countable.
We first prove the following Fact:

Fact 2.3.7. *If P is σ-linked and $\Vdash_p$ "$\dot{Q}$ is σ-linked", then $P*\dot{Q}$ is σ-linked.*

Proof. Let $P = \bigcup_{n\in\omega} P(n)$ be a σ-linking partition, and let

$$\Vdash_P \text{“}\dot{Q} = \bigcup_{m\in\omega} \dot{Q}(m) \text{ is a } \sigma - \text{ linking partition ”}.$$

Define

$$R_{n,m} = \{\langle p, q\rangle \in P * \dot{Q} : p \in P(n) \text{ and } p \Vdash_P \text{“}\dot{q} \in \dot{Q}(m)\text{”}\}$$

Clearly, $\bigcup_{n,m\in\omega} R_{n,m}$ is dense in $P * \dot{Q}$. Now, for all $n, m \in \omega$, let

$$R'_{n,m} = \{\langle p, \dot{q}\rangle \in P * \dot{Q} : \exists \langle p', \dot{q}'\rangle \in R_{n,m}, \langle p', \dot{q}'\rangle \geq \langle p, \dot{q}\rangle\}$$

$\bigcup_{n,m\in\omega} R'_{n,m}$ gives a σ-linking partition of $P * \dot{Q}$. □

For γ a successor ordinal, we do the same construction just given for one-step iteration.
So, let γ be a limit ordinal $\leq \beta$.
Case 1: $cf(\gamma) > \omega$.

In this case there exists $\delta < \gamma$ such that for all $s_p \in D^* \restriction \gamma$, supp($p$) $\subseteq \delta$. Then, any σ-linking partition of P_δ induces a σ-linking partition of P_γ.
Case 2: $cf(\gamma) = \omega$.

Without loss of generality, $\gamma = \omega$.

Let $p \in P_\omega$. For all $0 < k < \omega$, there exists $p'_k \in P_\omega$ and $n_k \in \omega$ such that

$$p'_k \restriction k \Vdash \text{“}p(k) \in \dot{Q}_k(n_k)\text{”}$$

We can assume that p'_k is such that for all $l < k$, $p'_l \restriction l \leq p'_k \restriction l$.

Also, since for every $k \in \omega$ and every $p \in P_k$, p is compatible with 0, we can assume that $0 \in P_0(0)$ and $\Vdash_{P_k}$ "$\dot{0} \in \dot{Q}_k(0)$", all $k \in \omega$. Hence, since supp(p) is finite, there is $l \in \omega$ such that for all $k \geq l$, $\Vdash_{P_k}$ "$p(k) \in \dot{Q}_k(0)$". Therefore, $\forall p \in P_\omega, \exists p' \in P_\omega$ such that

(i) $p'(0) = p(0)$
(ii) $\forall k \in \omega, k > 0, \exists n_k$ such that $p' \restriction k \Vdash$ "$p(k) \in \dot{Q}_k(n_k)$".

For $f \in \omega^\omega$, f eventually constant and equal to 0, define: $R_f = \{p \in P_\omega : p(0) \in P_0(f(0))$ and $\exists p' \geq p$ such that $p'(0) = p(0)$ and $\forall k \in \omega, k > 0,\ p' \restriction k \Vdash$ "$p(k) \in \dot{Q}_k(f(k))$"$\}$.
It is easy to see that $\bigcup R_f$ gives a σ-linking partition. □

Back to the Proof of the Theorem:

Let $I = \{A : A \subseteq P^V$ a maximal antichain , $A \in V\}$ and let $G \subseteq P_{\omega_2}$ be V-generic.

Claim. (i) $|I| = \aleph_1$

(ii) $\forall A \in I$, A is a maximal antichain of $P^{V[G]}$.

Proof of the Claim. (i) Clear, since $V \models CH$ and P satisfies the ccc.
(ii) See Fact 1.1.2. in the Introduction.
□

Suppose MA(Suslin) holds in $V[G]$. Then, there is a P^V-generic $g \subseteq P$ over V.

Notice that g is a collection of $\aleph_1$-many finite antichains of $\pi\mathbb{Q}$. Hence, we may assume, by coding each finite antichain of $\pi\mathbb{Q}$ into a real number, that g is a sequence of $\aleph_1$-many reals.

So, g has a simple P_{ω_2}-name $\dot{g}$. (i.e., The elements of $\dot{g}$ are of the form $\langle p, \check{n}, \check{\alpha}\rangle$ where $\check{n}$ is the standard name for $n \in \omega$, $\check{\alpha}$ is the standard name for $\alpha < \aleph_1$, and for every $\alpha < \aleph_1$ and every $n \in \omega$, $\{p : \langle p, \check{n}, \check{\alpha}\rangle \in \dot{g}\}$ is an antichain.)

Now, since P_{ω_2} is ccc, $\exists \alpha < \omega_2$ such that $\dot{g}$ is a P_α-name. So, P_α adds a P^V-generic g over V.

Let $R = \{t \in P : \exists p \in P_\alpha (p \Vdash$ "$\check{t} \in \dot{g}$"$)\}$.

Claim 2.3.8. There exists $t \in P$ such that R is dense in $P_t = \{t' : t' \in P \wedge t \leq t'\}$.

Proof of Claim. If not, then $P \setminus R$ is dense in P. Therefore, there is $p \in P_\alpha$ and $t \in P \setminus R$ such that $p \Vdash t \in \dot{g}$. Contradiction. □

Fix $P_\alpha = \bigcup_{n \in \omega} P_\alpha^n$ a σ-linking partition of P_α.
For each $n \in \omega$, let

$$R_n = \{t' \in P : \exists t \in P(t \geq t') \wedge \exists p \in P_\alpha^n (p \Vdash \text{“}\check{t} \in \dot{g}\text{”})\}$$

Claim 2.3.9. $R = \bigcup_{n \in \omega} R_n$ is a σ-linking partition of R.

Proof of Claim. Fix $n \in \omega$. Let $t'_0, t'_1 \in R_n$. We need to find $t \in P$ such that $t \geq t'_0, t'_1$. Pick $t_0 \geq t'_0$, $t_1 \geq t'_1$ and $p_0, p_1 \in P^n_\alpha$ such that $p_0 \Vdash$ "$\check{t}_0 \in \dot{g}$" and $p_1 \Vdash$ "$\check{t}_1 \in \dot{g}$".

Since P^n_α is pairwise compatible, there is $p \in P$ $p \geq p_0, p_1$ such that $p \Vdash$ "$\check{t}_0, \check{t}_1 \in \dot{g}$". Moreover, since $\dot{g}$ is forced to be a filter, there is $p' \geq p$ such that $p' \Vdash$ "$\exists t \in P(t \geq \check{t}_0, \check{t}_1)$". Thus, if $G \subseteq P_\alpha$ is generic with $p' \in G$, then

$$V[G] \models \exists t \in P(t \geq t_0, t_1)$$

But since P is a Suslin poset, the right hand side is a Σ^1_1 statement with parameters in V. Therefore, it holds in V. □

Now, fix $t \in P$ such that R is dense in P_t. For each $n \in \omega$, let $P^n_t = \{t' \in P : \exists t'' \in R_n(t'' \geq t')\}$. Then, $P_t = \bigcup_{n \in \omega} P^n_t$ is a σ-linking partition of P_t. But this contradicts Fact 2.2.3 above. □

3. Suslin Absoluteness

3.1. The consistency strength of Suslin Absoluteness

Definition 3.1.1. Let P be a forcing notion. Let V be a model of a part of ZFC and let $n \geq 1$. V *is* $\underset{\sim}{\Sigma}^1_n$*-absolute for* P if for every $\underset{\sim}{\Sigma}^1_n$-formula $\phi(x)$ with parameters in V and for every $r \in \mathbb{R}$,

$$V \models \phi(r) \text{ iff } V^P \models \phi(\check{r})$$

($\underset{\sim}{\Pi}^1_n$*-absolute for* P and $\underset{\sim}{\Delta}^1_n$*-absolute for* P are defined analogously.)

Definition 3.1.2. Let V be a model of a part of ZFC. V is *Suslin Absolute* iff for every Suslin partial ordering P satisfying the ccc,

$$Th(\mathbb{R})^V \prec Th(\mathbb{R})^{V^P}$$

i.e., for every $n \geq 1$, V is $\underset{\sim}{\Sigma}^1_n$-absolute for P.

Theorem 3.1.3. *The following are equiconsistent:*

1. *There exists a Suslin absolute model of ZFC.*
2. *There exists an inaccessible cardinal.*

Proof. 1⇒2. Let V be Suslin absolute and suppose $\aleph_1$ is not inaccessible in L. Then, for some $x \in \mathbb{R}$, $\aleph_1 = \aleph_1^{L[x]}$. Let $X = L[x] \cap \mathbb{R}$. So, $|X| = \aleph_1$.

From **Fact 2.3.3** above we have that for every $r \in \mathbb{R}$,

$$V^{Am} \models \text{“There is a Borel measure one set of random reals over } L[x][r]\text{”}$$

But this is a $\underset{\sim}{\Sigma}^1_3$ statement with parameters x and r. Hence, since Am is Suslin and V is Suslin absolute, it holds in V.

Now, suppose c is a Cohen real over V.

Claim. $V[c] \models$ "There is a Borel measure one set of random reals over $L[x][c]$".

Proof of Claim. We have just seen that $V \models$"For every $r \in \mathbb{R}$, there is a Borel measure one set of random reals over $L[x][r]$".

This is a $\underset{\sim}{\Pi}^1_4$ statement with parameter x. Hence, it holds in every Suslin extension of V. $\square$

But this contradicts the following Lemma:

Lemma 3.1.4. (H. Woodin): *Suppose X is an uncountable sequence of reals and suppose that c is Cohen over V. Then, in $V[c]$, there is no random real over $L(X, c)$.*

Proof. See [W], Lemma 4. $\square$

2$\Rightarrow$1. Let $V \models \kappa$ is an inaccessible cardinal. We will show that if $H \subseteq \mathrm{Coll}(\aleph_0, {<}\kappa)$ is generic over V, then $V[H]$ is Suslin absolute.

Let $\underset{\sim}{R}$ be a $\mathrm{Coll}(\aleph_0, {<}\kappa)$-name for a Suslin poset satisfying the ccc and let $P = \mathrm{Coll}(\aleph_0, {<}\kappa) * \underset{\sim}{R}$. Without loss of generality, $\underset{\sim}{R}$ consists of simple names. i.e., each $\underset{\sim}{r} \in \underset{\sim}{R}$ is essentially a countable sequence of antichains of $\mathrm{Coll}(\aleph_0, {<}\kappa)$.

Since $\mathrm{Coll}(\aleph_0, {<}\kappa)$ satisfies the κ-cc, and since $\Vdash_{\mathrm{Coll}(\aleph_0, <\kappa)}$ "$\underset{\sim}{R}$ satisfies the ccc", P satisfies the κ-c.c.

Definition 3.1.5. Let P, Q be posets. A *complete embedding of P into Q* is an embedding (i.e., a one-to-one order preserving function) from P into Q that preserves maximal antichains. We write $P \lessdot Q$ when $P \subseteq Q$ and the identity map is a complete embedding of P into Q.

Lemma 3.1.6. $\forall Q \in [P]^{<\kappa} = \{Q \subseteq P : |Q| < \kappa\}$, $\exists P^* \in [P]^{<\kappa}$ *such that* $Q \subseteq P^* \lessdot P$.

Proof. Fix $Q \in [P]^{<\kappa}$. There exists $\eta < \kappa$ with $\{p : \exists \underset{\sim}{r} \langle p, \underset{\sim}{r}\rangle \in Q\} \subseteq \mathrm{Coll}(\aleph_0, {<}\eta)$. Also, since $\mathrm{Coll}(\aleph_0, {<}\kappa)$ satisfies the κ-cc, there exists $\delta < \kappa$ such that every $\underset{\sim}{r} \in \{\underset{\sim}{r} : \langle p, \underset{\sim}{r}\rangle \in Q\}$ is a $\mathrm{Coll}(\aleph_0, {<}\delta)$-name. Moreover, there exists $\theta < \kappa$ such that all the parameters appearing in the definition of R have $\mathrm{Coll}(\aleph_0, {<}\theta)$-names.

Let $\gamma = (\max\{|\eta|, |\delta|, |\theta|\})^+$.

Let $\underset{\sim}{R}' = \underset{\sim}{R} \cap V^{\mathrm{Coll}(\aleph_0, <\gamma)}$ and let $P^* = \mathrm{Coll}(\aleph_0, {<}\gamma) * \underset{\sim}{R}'$.

Claim. (i) $|P^*| < \kappa$
(ii) $Q \subseteq P^* \lessdot P$

Proof of the Claim. (i) Suppose $|\,\mathrm{Coll}(\aleph_0, {<}\gamma)| = \lambda$ (so, $\lambda < \kappa$). Since κ is inaccessible, $2^\lambda < \kappa$. Hence, there are $< \kappa$ many antichains in $\mathrm{Coll}(\aleph_0, {<}\gamma)$ and, therefore, $< \kappa$ many names for reals.

(ii) Clearly, $Q \subseteq P^* \subseteq P$. It is well-known that $\mathrm{Coll}(\aleph_0, <\gamma) \lessdot \mathrm{Coll}(\aleph_0, <\kappa)$. Now, recall that for a Suslin ccc poset the property of being a maximal antichain is absolute for transitive models of ZF containing the parameters of its definition (see 1.1.2 above). Hence,

$$\Vdash_{\mathrm{Coll}(\aleph_0,<\kappa)} \text{“}\underset{\sim}{R}' \lessdot \underset{\sim}{R}\text{”}.$$

This proves the Lemma. □

Lemma 3.1.7. *Suppose that:*

(i) *κ is an inaccessible cardinal in V.*
(ii) *P satisfies the κ-c.c.*
(iii) $\Vdash_P$ “$\kappa = \aleph_1$”
(iv) $\forall Q \in [P]^{<\kappa}\ \exists P^* \in [P]^{<\kappa}$ *such that* $Q \subseteq P^* \lessdot P$.

Then, for every $G \subseteq P$ generic over V, there exists $H \subseteq \mathrm{Coll}(\aleph_0, <\kappa)$ generic over V such that

$$\mathbb{R}^{V[G]} = \mathbb{R}^{V[H]}$$

Proof. Fix $G \subseteq P$ generic over V. Let $G' \subseteq Coll(\aleph_0, 2^{|P|})$ be a generic filter over $V[G]$. In $V[G][G']$, let $\langle \underset{\sim}{r}_n : n \in \omega \rangle$ be an enumeration of all P-names for real numbers which belong to V.

Claim 3.1.8. For every $n \in \omega$, we can find P_n in V such that:

(1) $P_n \lessdot P$
(2) $|P_n| < \kappa$
(3) If $m \leq n$, then $P_m \lessdot P_n$.
(4) $\underset{\sim}{r}_n$ is a P_n-name.
(5) $\Vdash_{P_{n+1}}$ “$|P_n| = \aleph_0$”

Proof of Claim. By induction on $n \in \omega$.

For every $Q \in [P]^{<\kappa}$ and every cardinal $\lambda < \kappa$, there is $P_{Q,\lambda}$ such that $\Vdash_{P_{Q,\lambda}}$ “$\lambda = \aleph_0$”, $|P_{Q,\lambda}| < \kappa$, and $Q \subseteq P_\lambda \lessdot P$. Also, for every simple P-name $\underset{\sim}{r}$ for a real number, there exists a subalgebra $Q_{\underset{\sim}{r}} \lessdot P$ such that $|Q_{\underset{\sim}{r}}| < \kappa$ and $\underset{\sim}{r}$ is a $Q_{\underset{\sim}{r}}$-name.

Let $P_0 = Q_{\underset{\sim}{r}_0}$. We may assume $|P_0| > \aleph_0$.

Given P_n, use (iv) above to get P_{n+1} such that $|P_{n+1}| < \kappa$, $P_{n+1} \lessdot P$, and P_{n+1} contains $P_n, Q_{\underset{\sim}{r}_{n+1}}$ and $P_{P_n,|P_n|}$ as subalgebras.

To check 3. it is enough to see that $P_n \lessdot P_{n+1}$. But this is clear since $P_n \subseteq P_{n+1}$, $P_n \lessdot P$, and $P_{n+1} \lessdot P$. Also, since $Q_{\underset{\sim}{r}_{n+1}} \subseteq P_{n+1}$, $\underset{\sim}{r}_{n+1}$ is a P_{n+1}-name. Finally, since $P_{P_n,|P_n|} \subseteq P_{n+1}$, $P_{P_n,|P_n|} \lessdot P$ and $P_{n+1} \lessdot P$, we have $P_{P_n,|P_n|} \lessdot P_{n+1}$, which gives $\Vdash_{P_{n+1}}$ “$|P_n| = \aleph_0$”. □

Each P_n can be embedded into $Coll(\aleph_0, < |P_n|+1)$, all $n \in \omega$. Moreover, inductively on $n \in \omega$, we can extend the embedding f_n of P_n into $Coll(\aleph_0, < |P_n| + 1)$ to an embedding f_{n+1} of P_{n+1} into $Coll(\aleph_0, < |P_{n+1}| + 1)$ (see

[J], p.278). By identifying P_n with its image under f_n, we can assume $P_n \subseteq Coll(\aleph_0, < |P_n| + 1)$, all $n \in \omega$.

Let $G_n = G \cap P_n$. By the Claim above, for every $n \leq m < \omega$, $\underset{\sim}{r}_n[G] = \underset{\sim}{r}_n[G_n] = \underset{\sim}{r}_n[G_m]$.

Since for every $n \in \omega$, P_{n+1} collapses $|P_n|$ onto $\aleph_0$, we can find, by induction on $n \in \omega$, $H_n \subseteq Coll(\aleph_0, < |P_n| + 1)$ generic over V such that $G_n = H_n \cap P_n$ and if $n \leq m$, then $H_m \cap Coll(\aleph_0, < |P_n| + 1) = H_n$.

Then, $H = \bigcup_{n \in \omega} H_n$ is generic over V for $Coll(\aleph_0, < \kappa)$. Hence, since for every $\lambda < \kappa$, $Coll(\aleph_0, < \lambda) \lessdot Coll(\aleph_0, \kappa)$, we have $\underset{\sim}{r}_n[G] = \underset{\sim}{r}_n[G_n] = \underset{\sim}{r}_n[H_n] = \underset{\sim}{r}_n[H]$. $\square$

To prove the Theorem, fix $H \subseteq \mathrm{Coll}(\aleph_0, {<}\kappa)$ generic over V and R a ccc Suslin poset in $V[H]$.

Let $\varphi(a, x)$ be a $\underset{\sim}{\Sigma}^1_n$ formula with parameter a and let $r \in \mathbb{R}$ be such that

$$V[H] \models \varphi(a, r)$$

Fix $G \subseteq R$ generic over $V[H]$. We want to show

$$V[H][G] \models \varphi(a, r)$$

Let b be a real number which encodes all the parameters appearing in the definition of R. Thus, R is a forcing notion living in any universe containing b.

Let $V' = V[a, b, r]$.

In V', κ is still an inaccessible cardinal. So, by the Factor Lemma for the Lévy Collapse (see [J], ex. 25.11), we can find $H' \subseteq \mathrm{Coll}(\aleph_0, {<}\kappa)$ generic over V' such that

$$V'[H'] = V[H]$$

Therefore, $V'[H'] \models \varphi(a, r)$.

Now, by the above Lemmas, there exists $H'' \subseteq \mathrm{Coll}(\aleph_0, {<}\kappa)$ generic over V' such that $V'[H'']$ and $V'[H'][G]$ have the same reals.

But since $\mathrm{Coll}(\aleph_0, {<}\kappa)$ is homogeneous, $V'[H''] \models \varphi(a, r)$.

Therefore, $V'[H'][G] \models \varphi(a, r)$. $\square$

The following is an open question:

Suppose V is Suslin absolute. Does $V \models$"Every projective set of reals is Lebesgue measurable"?

3.2. Σ^1_3-absoluteness

Theorem 3.2.1. *Let P be the Cohen or the random forcing. Then, the following are equivalent:*

a. $V \models \forall x \in \mathbb{R}$ *there exists a P-generic filter over $L[x]$.*
b. V *is $\underset{\sim}{\Sigma}^1_3$-absolute for P.*

Proof. $a \Rightarrow b$. Let $\varphi(x)$ be a $\underset{\sim}{\Sigma}^1_3$-formula with parameters in V. Without loss of generality, a is the only parameter. Let $r \in \mathbb{R}$. So, $\varphi(r)$ is of form $\exists x\, \psi(y, r, a)$ where ψ is $\underset{\sim}{\Pi}^1_2$.
If $V \models \varphi(r)$, then $V^P \models \varphi(r)$.
So, suppose $V^P \models \varphi(r)$ and let G be P-generic over V. Then, $V[G] \models \exists y\, \psi(y, r, a)$.
Let b be a witness and let $\underset{\sim}{\tau}$ be a P-name for b.
$\underset{\sim}{\tau}$ can be chosen as a Borel function f such that, for $G \subseteq P$ generic over V, $V[G] \models \psi(f(G), r, a)$.

Now, let $p \in P$ be such that $p \Vdash_P \psi(\underset{\sim}{\tau}, r, a)$. Without loss of generality, $p = 0$.

Claim 3.2.2. $L[\underset{\sim}{\tau}, r, a] \models \Vdash_P \psi(\underset{\sim}{\tau}, r, a)$.

Proof of Claim. Since ψ is $\underset{\sim}{\Pi}^1_2$, so is $\Vdash_P \psi(\underset{\sim}{\tau}, r, a)$. $\square$

So, if G is P-generic over $L[\underset{\sim}{\tau}, r, a]$, then $L[\underset{\sim}{\tau}, r, a][G] \models \psi(\underset{\sim}{\tau}[G], r, a)$. But, by assumption, there is H P-generic over $L[\underset{\sim}{\tau}, r, a]$. Hence, $L[\underset{\sim}{\tau}, r, a] \models \psi(\underset{\sim}{\tau}[H], r, a)$.
Therefore, by $\underset{\sim}{\Pi}^1_2$-absoluteness, $V \models \psi(\underset{\sim}{\tau}[H], r, a)$. i.e., $V \models \varphi(r)$.

$b \Rightarrow a$. It follows immediately from the fact that, for every real x, the sentences

"There exists a Cohen real over $L[x]$"

and

"There exists a random real over $L[x]$"

are both $\Sigma^1_3(x)$. $\square$

The following Corollary gives a partial answer to a question of H. Woodin. Namely, suppose $V^{Cohen} \models \Delta^1_2$-determinacy. Does $V \models \Delta^1_2$-determinacy ?

Corollary 3.2.3.

(1) *Assume $\underset{\sim}{\Delta}^1_2(B)$ (i.e., all $\underset{\sim}{\Delta}^1_2$ sets have the Baire property). Then,*

$$V \models \Delta^1_2\text{-determinacy iff } V^{Cohen} \models \Delta^1_2\text{-determinacy}$$

(2) *Assume $\underset{\sim}{\Delta}^1_2(L)$ (i.e., all $\underset{\sim}{\Delta}^1_2$ sets are Lebesgue measurable). Then,*

$$V \models \Delta^1_2\text{-determinacy iff } V^{Random} \models \Delta^1_2\text{-determinacy}$$

Proof. H. Judah and S. Shelah gave in [J-S 3] the following characterization of $\underset{\sim}{\Delta}^1_2(B)$ and $\underset{\sim}{\Delta}^1_2(L)$:

$$\underset{\sim}{\Delta}^1_2(B) \Leftrightarrow \forall r \in \mathbb{R} \text{ there exists a Cohen real over } L[r]$$

$$\underset{\sim}{\Delta}^1_2(L) \Leftrightarrow \forall r \in \mathbb{R} \text{ there exists a random real over } L[r]$$

Now, 1 and 2 follow immediately from the theorem above since for any Δ^1_2 set A, "A is determined" is a Σ^1_3-statement. □

4. MA(AM)

4.1. A combinatorial characterization of MA(Am)

Definition 4.1.1. A poset P is *σ-centered* if there exists $h : P \longrightarrow \omega$ such that for every $p_1, p_2, .., p_n$ in P, $n < \omega$, if $h(p_1) = h(p_2) = .. = h(p_n)$, then there exists $q \in P$ such that $p_i \leq q$, all $1 \leq i \leq n$. We call the partition induced on P by h a *σ-centering* partition of P.

Theorem 4.1.2. *The following are equivalent:*

(1) Add(L)
(2) *MA(Am)*
(3) $\forall P \subseteq Am$, *if* $|P| < 2^{\aleph_0}$, *then* P *is σ-centered.*

Proof. $1 \Rightarrow 2$: Suppose λ is an uncountable cardinal $< 2^{\aleph_0}$ and suppose $\{A_\alpha : \alpha < \lambda\}$ is a collection of maximal antichains of Am. Since Am is ccc, each A_α can be coded by a real number r_α. Let $M = L(\langle r_\alpha : \alpha < \lambda \rangle)$. Note that $|\omega^\omega \cap M| = \lambda$. Hence, by *Add*($L$), the union of all Borel null sets with code in M is null.

Definition 4.1.3. Let $C = \{s \in ([\omega]^{<\omega})^\omega : \forall n |s(n)| < 2^n\}$. For $f \in \omega^\omega$ and $s \in C$, we write $f \subseteq^* s$ if $f(n) \in s(n)$ for all but finitely many $n \in \omega$.

Lemma 4.1.4. (T.Bartoszyński) *Let N denote the ideal of the null sets. There are maps $\phi : \omega^\omega \to N$ and $\phi^* : n \to C$ such that for any $f \in \omega^\omega$ and $X \in N$,*

$$f \subseteq^* \phi^*(X) \text{ whenever } \phi(f) \subseteq X$$

Proof. See [B]. □

Let $\{B_\alpha : \alpha < \lambda\}$ be a fixed enumeration of all $B \in Am$ coded in M.

Claim 4.1.5. There exists an *Amoeba*-condition A such that for every $\alpha < \lambda$ there exists C^*_α, a finite union of open intervals with rational endpoints, satisfying $A \cup B_\alpha = A \cup C^*_\alpha$.

Proof. By the Lemma above, and since the union of all Borel null sets with code in M is null, there exists $s \in C$ such that for every $f \in \omega^\omega \cap M$, $f(n) \in s(n)$ for all but finitely many $n \in \omega$. For every $n < \omega$, let $\{C^n_m : m < \omega\}$ be an enumeration of all finite unions of open intervals with rational endpoints such that $\mu(C^n_m) \leq \frac{1}{2^{2n+2}}$, all $m < \omega$. Let $A = \bigcup_{n\in\omega} \bigcup_{m\in s(n)} C^n_m$. So, $\mu(A) \leq \sum_{n\in\omega}(\sum_{m\in s(n)} \mu(C^n_m)) < \sum_{n\in\omega}(2^n \cdot \frac{1}{2^{2n+2}}) = \frac{1}{2}$. i.e., $A \in Am$. Now, for every $\alpha < \lambda$, we can find $f_\alpha \in \omega^\omega$ such that $B_\alpha = \bigcup_{n\in\omega} C^n_{f_\alpha(n)}$. So, since by the lemma above $f_\alpha \subseteq^* s$, we are done. □

Claim. There exists x a Cohen real over $M[A]$.

Proof of Claim. By [B] Add(L) $\Rightarrow$ Add(B). Hence, since $M[A] \models$ "$2^\omega = \lambda$", there are only λ Borel meager sets coded in $M[A]$. Therefore, their union is meager. □

Thus, we can apply the following Lemma:

Lemma 4.1.6. (J.Truss) *Suppose $A \in Am$ is such that for every $B \in Am$ coded in M there is a finite union C of open intervals with rational endpoints satisfying $A \cup B = A \cup C$. Then, for any Cohen $M[A]$-generic real x, $M[A][x]$ contains a M-generic ultrafilter on Amoeba.*

Proof. Let Q be the subset of Am consisting of all those p which are finite unions of open intervals with rational endpoints and $\mu(A \cup p) < 1/2$, ordered by inclusion. Q is a countable forcing notion in $M[A]$. Hence, $M[A][x]$ contains a $M[A]$-generic subset of Q. Call it g and let $\bigcup g = B$.

Claim. $\{p \in Am : p \subseteq A \cup B\}$ is an M-generic subset of Am.

Proof of Claim. Let $D \subseteq Am$ be dense, $D \in M$. Let $D' = \{q \in Q : \exists p \in D, p \subseteq A \cup q\}$. We show that D' is dense in Q.

Let $q \in Q$ be arbitrary. So, $\mu(A \cup q) < 1/2$. Hence, $A \cup q \in Am$. Let $C \subseteq D$ be a maximal antichain above q (in M). Since to be a maximal antichain of Am is an absolute notion (see **Fact 1.1.2** in the Introduction), C is also a maximal antichain above q in $(Am)^{M[A]}$. But $A \cup q \supseteq q$. So, $A \cup q$ is compatible with some $A' \in C$. Since A' is coded in M and $A' \in Am$, there is a finite union q' of open intervals with rational endpoints so that $A \cup A' = A \cup q'$.
We claim that $q \cup q' \supseteq q$ in Q and $q \cup q' \in D'$. Note that since $A \cup q$ and A' are compatible, $q \cup q' \in Q$. To see that $q \cup q' \in D'$, note that $A' \in D$ and $A' \subseteq A \cup q \cup q'$.
Hence, D' is dense in Q.
So, $\exists q \in D'$ such that $q \subseteq B$. Take $p \in D$ such that $p \subseteq A \cup q$. So, $p \subseteq A \cup B$. □

$2 \Rightarrow 3$

The following Lemma is an unpublished result of S. Shelah (implicit in [G-S]) which is included here with his permission.

Lemma 4.1.7. *Suppose P is a forcing notion and $\{f_n : n < \omega\}$ are partial automorphisms of P (i.e., partial order preserving functions) with domain dense in P, and satisfying*

$$(*)\ \forall p, q \in P \text{ there are } n, r \text{ such that } r \in dom f_n,\ r \geq p, \text{ and } f_n(r) \geq q.$$

Then, $\Vdash_P$ "$\check{P}$ *is σ-centered*".

Proof. Let $\underset{\sim}{G}$ be the name of the generic set. Define

$$\underset{\sim}{G}_n = \{p : \exists r \in \underset{\sim}{G},\ p \leq f_n(r)\}$$

For each n, $\Vdash_P$ "$\underset{\sim}{G}_n$ is a centered subset of P": If $p_1, p_2 \in \underset{\sim}{G}_n$ witnessed by $r_1, r_2 \in \underset{\sim}{G}$, then $\exists r \in \underset{\sim}{G}$ such that $r_1 \leq r$, $r_2 \leq r$ and $r \in dom f_n$. So, $p_1 \leq f_n(r)$, $p_2 \leq f_n(r)$, as f_n is order-preserving, and $f_n(r) \in \underset{\sim}{G}_n$.
Now, by (*), for every $p, q \in P$, $p \Vdash_P$ "$\exists n$ such that $q \in \underset{\sim}{G}_n$". Hence

$$\Vdash_P \text{“}\check{P} = \bigcup \underset{\sim}{G}_n \text{ is a } \sigma\text{-centering partition”}.$$

□

Lemma 4.1.8. *Amoeba satisfies the conditions of the previous Lemma.*

Proof. We can assume each $p \in Am$ is a ω-sequence $\langle \eta_i : i < \omega \rangle$ of finite sequences of zeroes and ones, each corresponding to a clopen set of 2^ω. Let $n < \omega$ and let σ be a permutation of 2^n.

Claim. σ induces a (total) automorphism F_σ of Amoeba.

Proof of Claim. Let $p \in Am$. We can assume $\forall \eta \in p,\ length(\eta) \geq n$.

If $p = \langle \eta_k : k < \omega \rangle$, let $F_\sigma(p) = \langle \sigma(\eta_k | n)^\frown \langle \eta_k(n), ..., \eta_k(length(\eta_k) - 1) \rangle : k < \omega \rangle$. i.e., We permute under σ the first n digits of η_k and leave the remaining ones (if any) the same. It is easy to see that F_σ is order-preserving. □

We show that $\{F_\sigma : \sigma \text{ a permutation of } 2^n, n \in \omega\}$ is as required. So, let $p, q \in Am$ and let $0 < \varepsilon < min(1/2 - \mu(p), 1/2 - \mu(q))$.

We can find $n \in \omega$ and $w_p \subseteq 2^n$ such that:

(i) $\forall \eta \in w_p$, $\eta \supseteq \eta'$ for some $\eta' \in p$.
(ii) $\mu(p) - \mu(w_p) < \varepsilon$.

Similarly, can find $m \in \omega$ and $w_q \subseteq 2^m$ such that:

(i) $\forall \delta \in w_q\ \delta \supseteq \delta'$ for some $\delta' \in q$.
(ii) $\mu(q) - \mu(w_q) < \varepsilon$

Without loss of generality, $n = m$. Consider the case where $|w_p| \geq |w_q|$. (The other case is symmetric.)

Let σ be a permutation of 2^n such that $\forall \delta \in w_q\ \exists \eta \in w_p$ so that $\sigma(\eta) = \delta$, and $\sigma(\eta) = \eta$ for all other $\eta \in 2^n$.

Hence, $\sigma(w_p) \geq w_q$. Define:
$p' = \{\eta' : \exists \eta \in p\, \exists k \in \omega\ \eta' \restriction k = \eta \text{ and } \eta' \restriction n \notin w_p\}$
$q' = \{\delta' : \exists \delta \in q\, \exists l \in \omega\ \delta' \restriction l = \delta \text{ and } \delta' \restriction m \notin w_q\}$
Let $r = w_p \cup p' \cup F_{\sigma^{-1}}(q')$. Since $\mu(q') < \varepsilon$, $\mu(r) < 1/2$.
Also $r \geq p$, since $\bigcup p = \bigcup w_p \cup \bigcup p'$. But $F_\sigma(r) = F_\sigma(w_p) \cup F_\sigma(p') \cup q' \geq w_q \cup p' \cup q' \geq q$. □

To show 3., let $P \subseteq Am$, $|P| < 2^{\aleph_0}$. Without loss of generality, $|P| > \omega$.

Let $M = L(P)$.

The conditions of Amoeba are open sets. So, since Am satisfies the ccc, each antichain of Am can be coded by a real number. But in M there are at most $|P|$-many reals. Hence, by MA(Am), there is a generic filter G for Am over M. Thus, by the above Lemmas,

$$M[G] \models \text{“}(Am)^M \text{ is } \sigma\text{-centered”}$$

Therefore, P is σ-centered. □

3⇒1

It is enough to show that for any $\{A_i : i < \lambda < 2^{\aleph_0}\}$ a collection of Borel measure zero sets where λ is an uncountable cardinal less than $2^{\aleph_0}$, $\bigcup_{i<\lambda} A_i$ is of measure zero.

So, let $\{A_i : i < \lambda < 2^{\aleph_0}\}$ be such a collection and let $M = L(\{A_i : i < \lambda < 2^{\aleph_0}\})$. We have $M \models 2^{\aleph_0} = \lambda$. Hence, Am^M is σ-centered.

Let $Am^M = \bigcup_{n\in\omega} P_n$ be a σ-centering partition and, for each n, let $B_n = \bigcup\{p \in P_n\}$.

$\mu(B_n) < \frac{1}{2}$. Otherwise, there would be $p_0, ..., p_m \in P_n$ with $\mu(\bigcup_{i\leq m} p_i) \geq \frac{1}{2}$, which is impossible since every finite subset of P_n has a supremum. Without loss of generality, $A_i + q = A_i$ for any rational number q. Given $i < \lambda$, there is $n \in \omega$ such that $A_i \subseteq B_n$. Hence, $A_i \subseteq \bigcap_{q\in\mathbb{Q}}(B_n + q)$. Since for every $n \in \omega$, $\mu(B_n) < \frac{1}{2}$, $\bigcap_{q\in\mathbb{Q}}(B_n + q))$ is of measure zero, all $n \in \omega$. So, since $\bigcup_{i<\lambda} A_i \subseteq \bigcup_{n\in\omega}\bigcap_{q\in\mathbb{Q}}(B_n + q)$, it follows that $\bigcup_{i<\lambda} A_i$ is of measure zero. □

4.2. Can the same characterization be given for MA(Suslin)?

S. Todorčević and B. Veličković gave in [T-V] the following characterization of MA:

MA ⇔ Every poset satisfying the ccc of size $< 2^{\aleph_0}$ is σ-centered.

It is an open question whether the same characterization can be given for Suslin posets. i.e.,

MA(Suslin) $\Leftrightarrow \forall P$ Suslin ccc $\forall P' \subseteq P$ of size $< 2^{\aleph_0}$, P' is σ-centered?

The following Lemma proves the easy direction:

Lemma 4.2.1.

MA(Suslin) $\Rightarrow \forall P$ *Suslin ccc* $\forall P' \subseteq P$ *of size* $< 2^{\aleph_0}$, P' *is* σ*-centered.*

Proof. Let P be Suslin ccc and let $P' \subseteq P$ be of size $< 2^{\aleph_0}$. Let P_ω be the ω-product of P with finite support.

Claim 4.2.2. P_ω is Suslin ccc.

Proof of Claim. To see that it is Suslin, notice that:

$$p \perp_{P_\omega} q \text{ iff } \exists n \in \omega\ p(n) \perp_P q(n) \tag{1}$$

$$p \leq_{P_\omega} q \text{ iff } \forall n \in \omega\ p(n) \leq_P q(n) \tag{2}$$

Now, suppose $A \subseteq P_\omega$ is an uncountable antichain. Since the support is finite, by a Δ-system argument we can find $A' \subseteq A$ uncountable and $s \subseteq \omega$, s finite, such that for all $p, q \in A'$, $\text{support}(p) \cap \text{support}(q) = s$. Notice that since A' is an antichain, $s \neq \emptyset$. Hence, it would be enough to show that any finite product of Suslin ccc posets is ccc. But since the product of any two Suslin posets is a Suslin poset, we need only to show that the product of two Suslin ccc posets is ccc.

So, suppose P, Q are Suslin ccc posets. By Fact 1.1.4 above, $\Vdash_P$ "$\dot{Q}$ is ccc". Hence, $P * \dot{Q}$ is ccc. But since Q is Suslin, $Q^V \lessdot Q^{V^P}$. Hence, the map $(p, q) \longmapsto (p, \check{q})$ is an embedding of $P \times Q$ into $P * \dot{Q}$ which preserves incompatibility. Therefore, $P \times Q$ is ccc. □

Now, for each $p \in P$, let

$$D_p = \{q \in P_\omega : \exists n \in \omega\ q(n) = p\}$$

Since the support is finite, D_p is dense in P_ω. Also, since $|P'| < 2^{\aleph_0}$, $|\{D_p : p \in P'\}| < 2^{\aleph_0}$. Hence, we can apply MA(Suslin) to get a generic filter G for $\{D_p : p \in P'\}$. Clearly, $P' \subseteq \bigcup G$. For each $n \in \omega$, define $G_n = \{p(n) : p \in G\}$. Then, G_n is centered, since so is G. So, $\bigcup_{n\in\omega} G_n \cap P'$ gives a σ-centering partition of P'. □

References

[B] T. Bartoszyński, *Additivity of measure implies additivity of category*, Trans. Amer. Math. Soc. **281** (1984), no.1, 209-213.

[B-J] T. Bartoszyński and H. Judah, *Jumping with random reals*, Annals of Pure and Applied Logic, 48,1990, 197-213.

[E] R. Engelking, *General Topology*, PWN, Warszawa 1977.

[G-S] M. Gitik and S. Shelah, *Forcings with Ideals and Simple Forcing Notions*, Israel Journal of Mathematics, to appear.

[J] T. Jech, *Set Theory*, Academic Press, New York, 1978.

[J-S 1] J. Ihoda and S. Shelah, *Suslin Forcing*, The Journal of Symbolic Logic **53** (4) (Dec. 1988), 1188-1207.

[J-S 2] J. Ihoda and S. Shelah, *Martin's Axioms, Measurability and Equiconsistency Results*, The Journal of Symbolic Logic **54** (1) (March 1989), 78-94.

[J-S 3] J. Ihoda and S. Shelah, Δ^1_2 *sets of reals*, Annals of Pure and Applied Logic, 42 (1989), no.3, 207-223.

[K] K. Kunen, *Set Theory. An Introduction to Independence Proofs*, North-Holland, 1980.

[M-S] D. Martin and R. Solovay, *Internal Cohen extensions*, Annals of Mathematical Logic **2** (1970), 143–178.

[S] S. Shelah, *Can you take Solovay's inaccessible away?*, Israel Journal of Mathematics **48** (1) (1984), 1-47.

[So] R. Solovay, *A model of set-theory in which every set of reals is Lebesgue measurable*, Ann. of Math. (1970), 1–56.

[T1] S. Todorčević, *Two examples of Borel ccc posets*, unpublished notes (1989).

[T2] S. Todorčević, *Trees and Linearly Ordered Sets* in *Handbook of Set-Theoretic Topology, edited by K. Kunen and J.E. Vaughan*, Elsevier Science Publishers B.V., 1984.

[T-V] S. Todorčević and B.Veličković, *Martin's axiom and partitions*, Compositio Mathematica **63** 391–408 (1987), Martinus Nijhoff Publishers, Dodrecht.

[Tr] J. Truss, *Sets having calibre* $\aleph_1$, Logic Colloquium '76, 595-612. North-Holland Publishing Company.

[W] H. Woodin, *On the consistency strength of Projective Uniformization*, Proceedings of the Herbrand Symposium, Logic Colloquium '81, J. Stern (editor), North-Holland Publishing Company, 1982.

Department of Mathematics,
University of California, Berkeley CA 94720

Partially supported by the "Generalitat de Catalunya" (CIRIT).

Department of Mathematics, Bar-Ilan University, Ramat Gan, ISRAEL

MEASURE AND CATEGORY — FILTERS ON ω

TOMEK BARTOSZYNSKI AND HAIM JUDAH

ABSTRACT. We study measurability and Baire property of filters of ω.

The goal of this paper is to present several results about filters on ω in context of their topological and measure-theoretical properties. In other words we identify filters on ω with subsets of 2^ω via characteristic functions of their elements. In this way the question about measurability and Baire property makes sense.

In the first section we give combinatorial characterizations of filters which have Baire property and filters which are measurable.

In the second section we study intersections of filters. It turns out that the intersection of countably many filters without Baire property does not have Baire property. Same result holds for nonmeasurable filters. This symmetry vanishes if one considers intersections of uncountably many filters.

The third section concerns the relationship between Ramsey filters and Cohen reals and between p-points and unbounded reals.

Finally in section four we define Raisonnier's filter and examine its complexity under various assumptions. We show that it is a rapid filter.

In section five we construct a model where there are no rapid filters.

Through the paper we use standard set-theoretical notation.

1. MEASURABILITY AND BAIRE PROPERTY OF FILTERS

In this section we study those filters on ω which are measurable or have Baire property. Let us start with the following:

Definition 1.1. $\mathcal{F} \subset P(\omega)$ is a nonprincipal filter on ω if

(1) $\forall X_1, \ldots, X_n \in \mathcal{F}\ (X_1 \cap \ldots \cap X_n \in \mathcal{F})$,

(2) $\forall X, Y\ (X \subseteq Y\ \&\ X \in \mathcal{F} \rightarrow Y \in \mathcal{F})$,

(3) $\forall X\ (X \text{ is finite} \rightarrow \omega - X \in \mathcal{F})$.

$\mathcal{F}$ is called an ultrafilter if $\mathcal{F}$ is maximal.

Theorem 1.1 (Sierpinski). *Suppose that $\mathcal{F}$ is a filter on ω. Then $\mathcal{F}$ is null (meager) or is nonmeasurable (does not have Baire property). If $\mathcal{F}$ is an ultrafilter then $\mathcal{F}$ is nonmeasurable and does not have Baire property.*

Proof. Let us start with the following :

Definition 1.2. $A \subseteq 2^\omega$ is called a *tail-set* if for every $x, y \in 2^\omega$ if $\{n \in \omega : x(n) \neq y(n)\}$ is finite and $x \in A$ then $y \in A$.

The following lemma is well-known.

Lemma 1.2 ([O]). *Every measurable tail-set has measure 0 or 1. Every tail-set which has Baire property is meager or residual (co-meager).* □

It remains to show that no filter has measure 1 or is co-meager.

Consider function $F : 2^\omega \longrightarrow 2^\omega$ defined as $F(X)(n) = 1 - X(n)$ for $X \in 2^\omega, n \in \omega$. F is a homeomorphism preserving measure. Thus if $\mu(\mathcal{F}) = 1$ then $\mu(F(\mathcal{F})) = 1$ and there is $X \in \mathcal{F}$ such that $F(X) \in \mathcal{F}$ which is impossible. Same argument shows that a complement of a filter cannot be meager.

If $\mathcal{F}$ is an ultrafilter then $\mathcal{F} \cup F(\mathcal{F}) = 2^\omega$ which means that $\mathcal{F}$ cannot be measurable. □

Our first goal is to characterize those filters which do not have Baire property.

Definition 1.3. Let $\mathcal{F}$ be a filter on ω. For $X \in \mathcal{F}$ let $f_X \in \omega^\omega$ be an increasing enumeration of X. Let $\hat{\mathcal{F}} = \{f_X : X \in \mathcal{F}\}$. We say that filter $\mathcal{F}$ is unbounded if the family $\hat{\mathcal{F}}$ is unbounded in ω^ω.

Theorem 1.3 (Talagrand [T1]). *The following conditions are equivalent for any filter $\mathcal{F}$*

(1) *$\mathcal{F}$ does not have Baire property,*
(2) *$\hat{\mathcal{F}}$ is unbounded,*
(3) *For every partition of ω into finite sets, $\{I_n : n \in \omega\}$ there exists $X \in \mathcal{F}$ such that $X \cap I_n = \emptyset$ for infinitely many $n \in \omega$.*

Proof. $1 \to 2$ Suppose that $\hat{\mathcal{F}}$ is bounded by some function $f \in \omega^\omega$. Then $\mathcal{F} \subset \bigcup_{n\in\omega} A_n$ where

$$A_n = \{X \subseteq \omega : \forall k \geq n \ f_X(k) \leq f(k)\} \text{ for } n \in \omega\ .$$

It is easy to see that the sets A_n correspond to meager subsets of 2^ω.

$2 \to 3$ Suppose that there is a partition $\{I_n : n \in \omega\}$ of ω such that

$$\forall X \in \mathcal{F} \ \forall^\infty n \ X \cap I_n \neq \emptyset\ .$$

Define $f'(n) = \max\{I_n\}$ for $n \in \omega$. Let $f_k(n) = f'(n+k)$ for $n \in \omega$. Let $f \in \omega^\omega$ be any function dominating the family $\{f_k : k \in \omega\}$. It is easy to see that f dominates $\hat{\mathcal{F}}$.

$3 \to 1$ Let $F = \bigcup_{n\in\omega} F_n$ be any meager set of type F_σ. Fix some enumeration of $\omega^{<\omega}$.

Define by induction two sequences $\{k_n : n \in \omega\}$ and $\{s_n : n \in \omega\}$ as follows:

$$s_{n+1} = \min\{s \in \omega^{<\omega} : \ \forall t \in 2^{k_n} \ \forall i \leq n \ [t^\frown s] \cap F_i = \emptyset\}$$

and

$$k_{n+1} = k_n + \mathrm{lh}(s_{n+1}) \ .$$

Let $I_n = [k_n, k_{n+1})$ for $n \in \omega$. Find $X \in \mathcal{F}$ such that $X \cap I_n = \emptyset$ for infinitely many $n \in \omega$. Define $Y \in 2^\omega$ as follows

$$Y \restriction I_n = \begin{cases} X \restriction I_n & \text{if } X \cap I_n \neq \emptyset \\ s_n & \text{if } X \cap I_n = \emptyset \end{cases} \quad \text{for } n \in \omega \ .$$

Clearly $Y \supseteq X$ and $Y \notin \mathcal{F}$. $\square$

Notice that in particular we showed that every meager filter can be covered by an upwards closed meager set of type F_σ.

For measure the situation is a little more complicated but nevertheless we show that every null filter can be covered by an upwards closed null set of type G_δ.

Theorem 1.4 ([Ba2]). *For any filter $\mathcal{F}$ the following conditions are equivalent:*

(1) *$\mathcal{F}$ is measurable,*
(2) *there exists a family $\{\mathcal{A}_n : n \in \omega\}$ such that*
  (a) *$\mathcal{A}_n$ consists of finitely many finite subsets of ω for all $n \in \omega$,*
  (b) *$\bigcup \mathcal{A}_n \cap \bigcup \mathcal{A}_m = \emptyset$ whenever $n \neq m$,*
  (c) *$\sum_{n=1}^\infty \mu(\{X \subseteq \omega : \exists a \in \mathcal{A}_n \ a \subset X\}) < \infty$,*
  (d) *$\forall X \in \mathcal{F} \ \exists^\infty n \ \exists a \in \mathcal{A}_n \ a \subset X$.*

Proof. 2) $\to$ 1) This implication is obvious since by d) $\mathcal{F}$ is contained in the set $\{X \subseteq \omega : \exists^\infty n \ \exists a \in \mathcal{A}_n \ a \subset X\}$ which is null by c).

1) $\to$ 2). Let us start with the following classical fact.

Lemma 1.5 ([O]). *Suppose that $H \subset 2^\omega$ has measure zero. Then there exists a sequence $\{F_n : n \in \omega\}$ such that $F_n \subseteq 2^n$ for $n \in \omega$, $\sum_{n=1}^\infty |F_n| \cdot 2^{-n} < \infty$ and $H \subseteq \{x \in 2^\omega : \exists^\infty n \ x \restriction n \in F_n\}$.*

Proof. Since H has measure zero there are open sets $\{G_n : n \in \omega\}$ covering H such that $\mu(G_n) < \frac{1}{2^n}$ for $n \in \omega$. Represent each set G_n as a disjoint union of open basic intervals i.e.

$$G_n = \bigcup_{m=1}^{\infty} [s_m^n] \text{ for } n \in \omega .$$

Let $F_n = \{s \in 2^n : s = s_l^k \text{ for some } k, l \in \omega\}$ for $n \in \omega$. It is easy to check that it is the sequence we were looking for. □

The above lemma inspires the following definition:

Definition 1.4. Set $H \subseteq 2^\omega$ is called *small* if there exists a partition $\mathcal{A}$ of ω into pairwise disjoint, finite sets and a family $\mathcal{J} = \{J_a : a \in \mathcal{A}\}$ such that

(1) $J_a \subseteq 2^a$ for $a \in \mathcal{A}$,
(2) $H \subseteq \{x \in 2^\omega : \exists^\infty a \in \mathcal{A}\ x \restriction a \in J_a\}$,
(3) $\sum_{a \in \mathcal{A}} |J_a| \cdot 2^{-|a|} < \infty$.

Denote the set $\{x \in 2^\omega : \exists^\infty a \in \mathcal{A}\ x \restriction a \in J_a\}$ by $(\mathcal{A}, \mathcal{J})$. If $\mathcal{A} = \{I_n : n \in \omega\}$ and $\mathcal{J} = \{J_n : n \in \omega\}$ are two families defining a small set denote the set $(\mathcal{A}, \mathcal{J})$ by $(I_n, J_n)_{n=1}^\infty$.

Notice that by Borel-Cantelli lemma condition 3) is a necessary and sufficient condition for this set to have measure zero.

We will need several properties of small sets.

Lemma 1.6. *Suppose that $(\mathcal{A}_1, \mathcal{J}^1)$ and $(\mathcal{A}_2, \mathcal{J}^2)$ are two small sets. If $\mathcal{A}_1$ is a finer partition than $\mathcal{A}_2$ then $(\mathcal{A}_1, \mathcal{J}^1) \cup (\mathcal{A}_2, \mathcal{J}^2)$ is a small set.*

Proof. Define $\mathcal{A}_3 = \mathcal{A}_2$ and for $a \in \mathcal{A}_3$ let

$$J_a^3 = J_a^2 \cup \{s \in 2^a : \exists b \in \mathcal{A}_1 (a \cap b \neq \emptyset \ \&\ s \restriction b \in J_b^1\} .$$

It is easy to see that $(\mathcal{A}_3, \mathcal{J}^3) = (\mathcal{A}_1, \mathcal{J}^1) \cup (\mathcal{A}_2, \mathcal{J}^2)$. □

In particular if $(\mathcal{A}_1, \mathcal{J}^1)$ and $(\mathcal{A}_2, \mathcal{J}^2)$ are two small sets and there exists a partition which is coarser than both $\mathcal{A}_1$ and $\mathcal{A}_2$ then the union $(\mathcal{A}_1, \mathcal{J}^1) \cup (\mathcal{A}_2, \mathcal{J}^2)$ is small.

Lemma 1.7. *Suppose that $(\mathcal{A}_1, \mathcal{J}^1)$ and $(\mathcal{A}_2, \mathcal{J}^2)$ are two small sets and that $(\mathcal{A}_1, \mathcal{J}^1) \subset (\mathcal{A}_2, \mathcal{J}^2)$. Then for all but finitely many $a \in \mathcal{A}_1$ and for every $s \in J_a^1$ there exists $b \in \mathcal{A}_2$ such that $b \cap a \neq \emptyset$ and all extensions of $s \restriction b$ are in J_b^2.*

Proof. suppose that this is not true. Then there exists a sequence $\{a_n : n \in \omega\}$ of elements of $\mathcal{A}_1$ and a sequence $s_n \in J^1_{a_n}$ for $n \in \omega$ such that for all $b \in \mathcal{A}_2$ whenever $a_n \cap b \neq \emptyset$ then $s_n \restriction b$ has an extension which does not belong to J^2_b. For every $b \in \mathcal{A}_2$ choose one sequence $s_b \notin J^2_b$ and define $x \in 2^\omega$ as follows:

$x \restriction b = s_b$ if $b \cap \bigcup_{n\in\omega} a_n = \emptyset$ and

$x \restriction b =$ any extension of $s_n \restriction b$ which is not in J^2_b if $b \cap a_n \neq \emptyset$ and n is minimal like that.

It is obvious that $x \notin (\mathcal{A}_2, \mathcal{J}^2)$ since for all $b \in \mathcal{A}_2$, $x \restriction b \notin J^2_b$. On the other hand $x \restriction a_n = s_n$ for $n \in \omega$ which means that $x \in (\mathcal{A}_1, \mathcal{J}^1)$. Contradiction. □

As a corollary we get the following:

Lemma 1.8. *Suppose that $(\mathcal{A}_1, \mathcal{J}^1)$ and $(\mathcal{A}_2, \mathcal{J}^2)$ are two small sets and that $(\mathcal{A}_1, \mathcal{J}^1) \subset (\mathcal{A}_2, \mathcal{J}^2)$. Then there exists partition $\mathcal{A}_3$ finer than both $\mathcal{A}_1$ and $\mathcal{A}_2$ and a family $\mathcal{J}^3$ such that*

$$(\mathcal{A}_1, \mathcal{J}^1) \subset (\mathcal{A}_3, \mathcal{J}^3) \subset (\mathcal{A}_2, \mathcal{J}^2) \ .$$

Proof. Let $\mathcal{A}_3 = \mathcal{A}_1 \bigwedge \mathcal{A}_2 = \{a \cap b : a \in \mathcal{A}_1, b \in \mathcal{A}_2\}$. For $c = a \cap b \in \mathcal{A}_3$ define

$$J^3_c = \{s \in 2^c : \ \forall t \in 2^b \ (t \supset s \rightarrow t \in J^2_b)\} \ .$$

Notice that

$$\frac{|J_c|}{2^{|c|}} \leq 2^{|b-c|} \cdot \frac{|J_c|}{2^{|b|}} \leq \frac{|J_b|}{2^{|b|}}$$

which shows that $(\mathcal{A}_3, \mathcal{J}^3)$ is a small set. It is also easy to see that $(\mathcal{A}_3, \mathcal{J}^3) \subset (\mathcal{A}_2, \mathcal{J}^2)$.

Suppose that $x \in (\mathcal{A}_1, \mathcal{J}^1)$. By the definition there exists infinitely many $a \in \mathcal{A}_1$ such that $x \restriction a \in J^1_a$. By the previous lemma for all such a (except possibly finitely many) there exists $b \in \mathcal{A}_2$ such that such that $b \cap a \neq \emptyset$ and all extensions of $x \restriction b$ are in J^2_b. But that means that for $c = a \cap b$ we have $x \restriction c = J^3_c$. □

□

The next theorem shows that small sets are good approximations of null sets. Moreover it shows that we can assume that partitions used in the definition of small sets are partitions into intervals.

Theorem 1.9. *Every null set is a union of two small sets.*

Proof. Let $H \subseteq 2^\omega$ be a null set. By 1.5 we can assume that $H = \{x \in 2^\omega : \exists^\infty n\ x \restriction n \in F_n\}$ for some sequence $\{F_n : n \in \omega\}$.

Fix a sequence of positive reals $\{\varepsilon_n : n \in \omega\}$ such that $\sum_{n=1}^\infty \varepsilon_n < \infty$.

Define two sequences $\{n_k, m_k : k \in \omega\}$ as follows: $n_0 = 0$,

$$m_k = \min\{j \in \omega : 2^{n_k} \cdot \sum_{i=j}^{\infty} \frac{|F_i|}{2^i} < \varepsilon_k\} ,$$

and

$$n_{k+1} = \min\{j \in \omega : 2^{m_k} \cdot \sum_{i=j}^{\infty} \frac{|F_i|}{2^i} < \varepsilon_k\} \text{ for } k \in \omega .$$

Notice that we can assume that both sequences $\{n_k, m_k : k \in \omega\}$ are subsequences of any given increasing sequence.

Let $I_k = [n_k, n_{k+1})$ and $I'_k = [m_k, m_{k+1})$ for $k \in \omega$. We can assume that $n_k < m_k < n_{k+1} < m_{k+1}$ for $k \in \omega$. Define

$$s \in J_k \leftrightarrow s \in 2^{I_k} \ \&\ \exists i \in [m_k, n_{k+1})\ \exists t \in F_i\ s \restriction \mathrm{dom}(t) \cap \mathrm{dom}(s) = t \restriction \mathrm{dom}(t) \cap \mathrm{dom}(s) .$$

Similarly

$$s \in J'_k \leftrightarrow s \in 2^{I'_k} \ \&\ \exists i \in [n_{k+1}, m_{k+1})\ \exists t \in F_i\ s \restriction \mathrm{dom}(t) \cap \mathrm{dom}(s) = t \restriction \mathrm{dom}(t) \cap \mathrm{dom}(s) .$$

It remains to show that $(I_k, J_k)_{k=1}^\infty$ and $(I'_k, J'_k)_{k=1}^\infty$ are small sets and that their union covers H.

Consider the set $(I_k, J_k)_{k=1}^\infty$. Notice that for $k \in \omega$

$$\frac{|J_k|}{2^{I_k}} \leq 2^{n_k} \cdot \sum_{i=m_k}^{n_{k+1}} \frac{|F_i|}{2^i} \leq \varepsilon_k .$$

Since $\sum_{n=1}^\infty \varepsilon_n < \infty$ this shows that the set $(I_n, J_n)_{n=1}^\infty$ is null. Analogous argument works for the other set. Finally we have that

$$H \subseteq (I_n, J_n)_{n=1}^\infty \cup (I'_n, J'_n)_{n=1}^\infty .$$

To see this suppose that $x \in H$. Then the set $X = \{n \in \omega : x \restriction n \in F_n\}$ is infinite. Thus either

$$X \cap \bigcup_{k=1}^{\infty} [m_k, n_{k+1}) \text{ is infinite or}$$

$$X \cap \bigcup_{k=1}^{\infty} [n_{k+1}, m_{k+1}) \text{ is infinite} .$$

Without loss of generality we can assume that it is the first case. But it means that $x \in (I_n, J_n)_{n=1}^{\infty}$ because if $x \restriction n \in F_n$ and $n \in [m_k, n_{k+1})$ then by the definition there is $t \in J_k$ such that $x \restriction [n_k, n_{k+1}) = t$. We are done since it happens infinitely many times. □

¿From now on we will assume that partitions occurring in the definition of small set are always partitions into disjoint intervals.

Lemma 1.10. *Let $\mathcal{F}$ be a filter. Then $\mathcal{F}$ is a measurable filter iff $\mathcal{F}$ can be covered by a small set.*

Proof. $\leftarrow$ Trivial since every small set is null.

$\rightarrow$ Let $\mathcal{F}$ be a measurable filter. Fix a sequence $\{\varepsilon_n : n \in \omega\}$ of positive reals such that $\sum_{k=1}^{\infty} 2^k \cdot \varepsilon_k < \infty$.

By 1.1 we know that $\mathcal{F}$ can be covered by some null set $H \subseteq 2^\omega$. By applying 1.5 to the set H we get a sequence $\{F_k : k \in \omega\}$ such that

$$H \subseteq \{x \in 2^\omega : \exists^\infty k \; x \restriction k \in F_k\} .$$

Using the proof of 1.9 we can represent the set H as a union of two small sets. In other words we have the following: There exist two sequences of natural numbers $\{n_k, m_k : k \in \omega\}$ and a family $\{J_k, J'_k : k \in \omega\}$ such that:

(1) $n_k < m_k < n_{k+1} < m_{k+1}$ for $k \in \omega$,
(2) $J_k \subset 2^{[n_k, n_{k+1})}$, $J'_k \subset 2^{[m_k, m_{k+1})}$,
(3) $|J_k| \cdot 2^{n_k - n_{k+1}} < \varepsilon_k$, $|J'_k| \cdot 2^{m_k - m_{k+1}} < \varepsilon_k$ for $k \in \omega$,
(4) $H \subset ([n_k, n_{k+1}), J_k)_{k=1}^{\infty} \cup ([m_k, m_{k+1}), J'_k)_{k=1}^{\infty}$.

By the assumption $\mathcal{F} \subset ([n_k, n_{k+1}), J_k)_{k=1}^{\infty} \cup ([m_k, m_{k+1}), J'_k)_{k=1}^{\infty}$.

If $\mathcal{F} \subset ([n_k, n_{k+1}), J_k)_{k=1}^{\infty}$ or if $\mathcal{F} \subset ([m_k, m_{k+1}), J'_k)_{k=1}^{\infty}$ then we are done since both sets are small.

Therefore assume that neither set covers $\mathcal{F}$.

Define for $k \in \omega$

$$S_k = \{s \in 2^{[n_k, m_k)} : \; s \text{ has at least } 2^{n_{k+1} - m_k - k} \text{ extensions inside } J_k \} .$$

It is easy to check that

$$\frac{|S_n|}{2^{m_k - n_k}} \leq 2^k \cdot \varepsilon_k$$

holds for $k \in \omega$.

Similarly if we define

$$S'_k = \{s \in 2^{[n_k, m_m)} : s \text{ has at least } 2^{n_k - m_{k-1} - k} \text{ extensions inside } J'_k\}$$

then by the same argument we have that

$$\frac{|S'_k|}{2^{m_k-n_k}} \leq 2^k \cdot \varepsilon_k$$

for all $k \in \omega$.

Consider the set $([n_k, m_k), S_k \cup S'_k)_{k=1}^{\infty}$. This set is small since $\sum_{k=1}^{\infty} |S_k \cup S'_k| \cdot 2^{n_k - m_k} < \infty$.

Now we have three small sets

(1) $H_1 = ([n_k, n_{k+1}), J_k)_{k=1}^{\infty}$,
(2) $H_2 = ([m_k, m_{k+1}), J'_k)_{k=1}^{\infty}$,
(3) $H_3 = ([n_k, m_k), S_k \cup S'_k)_{k=1}^{\infty}$.

If $\mathcal{F} \subset H_2 \cup H_3$ we are done since by 1.6 $H_2 \cup H_3$ is a small set. Therefore assume that there exists $X \in \mathcal{F}$ such that $X \notin H_2 \cup H_3$. Since $\mathcal{F} \subset H_1 \cup H_2$ we get that $X \in H_1$. Therefore there exists an infinite sequence $\{k_u : u \in \omega\}$ such that

$$X \restriction [n_{k_u}, n_{k_u+1}) \in J_{k_u}$$

for $u \in \omega$.

Define for $u \in \omega$

$$\begin{aligned} I_u &= [m_{k_u+1}, n_{k_u+1}) \text{ and} \\ T_u &= \{s \in 2^{I_u} : \; X \restriction [n_{k_u}, m_{k_u+1})^\frown s \in J_{k_u} \text{ or} \\ &\qquad s^\frown X \restriction [n_{k_u+1}, m_{k_u+1}) \in J'_{k_u+1}\} \,. \end{aligned}$$

By the choice of X, $X \restriction [n_{k_u}, n_{k_u+1}) \in J_{k_u}$ but $X \restriction [n_{k_u}, n_{k_u+1}) \notin S_{k_u} \cup S'_{k_u}$ for sufficiently large $u \in \omega$. Thus $|T_u| \cdot 2^{-|I_u|} \leq 2^{-u}$ for all but finitely many $u \in \omega$.

Claim 1.11. $\mathcal{F} \subset (I_u, T_u)_{u=1}^{\infty}$.

Proof. Suppose that $\mathcal{F}$ is not contained in this set and let $Y \in \mathcal{F} - (I_u, T_u)_{u=1}^{\infty}$.

Define $Z \in 2^\omega$ as follows

$$Z(n) = \begin{cases} Y(n) & \text{if } n \in \bigcup_{u\in\omega} I_u \\ X(n) & \text{otherwise} \end{cases} \quad \text{for } n \in \omega \,.$$

Notice that $Z \in \mathcal{F}$ since $X \cap Y \subseteq Z$. We will show that $Z \notin H_1 \cup H_2$ which gives a contradiction. Consider an interval $I_m = [n_m, n_{m+1})$. If $m \neq k_u$ for every $u \in \omega$ then $I_m \cap \bigcup_{u\in\omega} I_u = \emptyset$ and $Z \restriction I_m \notin \dot{J}_m$ since $Z \restriction I_m = X \restriction I_m$ for such m's. On the other hand if $m = k_u$ for some $u \in \omega$ then $X \restriction I_m \in \dot{J}_m$ but by the choice of X, $Z \restriction [n_{k_u}, m_{k_u}) = X \restriction [n_{k_u}, m_{k_u})$ has only few extensions inside $J_{n_{k_u}}$ (since $X \notin H_3$). In fact if $Z \restriction I_m \in J_m$ then $Z \restriction I_u$ has to be an element of T_u. But this is impossible since $Z \restriction I_u = Y \restriction I_u \notin T_u$

for sufficiently large $u \in \omega$. Hence for all except finitely many $m \in \omega$, $Z \restriction I_m \notin J_m$ which means that $Z \notin H_1$. Similarly, using the second clause in the definition of H_3 we prove that $Z \notin H_2$. That finishes the proof since the set $(I_u, T_u)_{u=1}^{\infty}$ is small. □

Assume that $\mathcal{F}$ is a measurable filter. Then by the above lemma $\mathcal{F}$ can be covered by a set $\{x \in 2^\omega : \exists^\infty n\; x \restriction I_n \in J_n\}$ where sequences $\{I_n : n \in \omega\}$ and $\{J_n : n \in \omega\}$ satisfy the definition of a small set.

Define for $n \in \omega$

$$J'_n = \{s \in J_n : \forall u \in 2^{I_n}\; (s^{-1}(1) \subseteq u^{-1}(1) \to u \in J_n)\} .$$

Claim 1.12. $\mathcal{F} \subseteq (I_n, J'_n)_{n=1}^{\infty}$.

Proof. Suppose not. Let $X \in \mathcal{F} - \{x \in 2^\omega : \exists^\infty n\; x \restriction I_n \in J_n\}$. It is not very hard to see that there exists a set $X' \supseteq X$ which does not belong to $\{x \in 2^\omega : \exists^\infty n\; x \restriction I_n \in J_n\}$ Contradiction. □

Identify elements of J'_n with subsets of I_n and let

$$\mathcal{A}_n = \{a \subseteq I_n : a \text{ } is \text{ } \subseteq -\text{minimal element of } J'_n\} \text{ for } n \in \omega .$$

Obviously $\mathcal{F} \subseteq \{X \subseteq \omega : \exists^\infty n\; \exists a \in \mathcal{A}_n\; a \subset X\}$ and the family $\{\mathcal{A}_n : n \in \omega\}$ has properties $(a) - (d)$. □

As a corollary we get:

Theorem 1.13. *Every measurable filter extends to a measurable filter which does not have Baire property.*

Proof. Suppose that $\mathcal{F}$ is a measurable filter. Let $\mathcal{A} = \{\mathcal{A}_n : n \in \omega\}$ be a family from 1.4. For $X \subseteq \omega$ define

$$\text{supp}_{\mathcal{A}}(X) = \{a : \exists n \exists a \in \mathcal{A}_n\; a \subset X\} .$$

Notice that

$$\mathcal{F}^\star = \{\text{supp}_{\mathcal{A}}(X) : X \in \mathcal{F}\}$$

is a filter since $\text{supp}_{\mathcal{A}}(X) \cap \text{supp}_{\mathcal{A}}(Y) \supseteq \text{supp}_{\mathcal{A}}(X \cap Y)$ for any $X, Y \in \mathcal{F}$.

Let $\mathcal{H}$ be any ultrafilter containing $\mathcal{F}^\star$. Define

$$\mathcal{G} = \{X \subseteq \omega : \text{supp}_{\mathcal{A}}(X) \in \mathcal{H}\} .$$

It is clear that $\mathcal{G}$ is a filter; the fact that $\mathcal{G}$ does not have Baire property follows from 1.3. □

We do not know if every filter having Baire property extends to a non-measurable filter having Baire property. We only have:

Theorem 1.14 (Talagrand [T1]). *Assume MA. There exists a nonmeasurable filter which has Baire property.*

To prove this theorem we will need the following

Definition 1.5. For a set $X \subset \omega$ define density of X as

$$d(X) = \lim_{n\to\infty} \frac{|X \cap n|}{n}$$

if the above limit exists.

Let $\{K_\xi : \xi < 2^\omega\}$ be an enumeration of all closed sets of positive measure.

Lemma 1.15. *Assume MA. There exists a family $\{X_\xi : \xi < 2^\omega\}$ such that $X_\xi \in K_\xi$ for $\xi < 2^\omega$ and*

$$\forall n \in \omega\ \forall \xi_1, \dots, \xi_n\ d(X_{\xi_1} \cap \dots \cap X_{\xi_n}) = \frac{1}{2^n}\ .$$

Proof. We construct $\{X_\xi : \xi < 2^\omega\}$ by induction. Suppose that $\{X_\xi : \xi < \alpha\}$ are already constructed. Let $\kappa = |\alpha|$ and let $\{Y_\xi : \xi < \kappa\}$ be an enumeration of all finite intersections of elements of $\{X_\xi : \xi < \alpha\}$.

The following claim is an easy consequence of the fact that the family of subsets of ω having density $\frac{1}{2}$ has measure 1.

Claim 1.16. *Suppose that $d(Y) = c$. Then the set $\{X \subset \omega : d(X \cap Y) = \frac{c}{2}\}$ has measure 1.*

□Define for $\xi < \kappa$

$$H_\xi = \{X \subset \omega : d(X \cap Y_\xi) = \frac{d(Y_\xi)}{2}\}\ .$$

By the above claim all sets H_ξ have measure 1. By Martin Axiom $K_\alpha \cap \bigcap_{\xi<\kappa} H_\xi \neq \emptyset$. Let X_α be any element of this set. This finishes the construction and the proof of the lemma. □

Proof. 1.14 Let $\{X_\xi : \xi < 2^\omega\}$ be a family from the above lemma. Let $\mathcal{F}$ be a filter generated by this family. It is clear that $\mathcal{F}$ is a nonmeasurable filter as $\mathcal{F}$ intersects every set of positive measure. Let $I_n = [2^{n^2}, 2^{(n+1)^2})$ for $n \in \omega$. Notice that if $X \cap I_n = \emptyset$ for infinitely many $n \in \omega$ then

$$\liminf_{n\to\infty} \frac{|X \cap n|}{n} = 0$$

therefore $X \notin \mathcal{F}$ since $\mathcal{F}$ is generated by elements having positive density. □

Another application is related to the following.

Definition 1.6. $\mathcal{F}$ is called *rapid* if for every increasing function $f \in \omega^\omega$ there exists $X \in \mathcal{F}$ such that $|X \cap f(n)| \leq n$ for $n \in \omega$.

Theorem 1.17 (Mokobodzki). *Every rapid filter is nonmeasurable and does not have Baire property.*

Proof. It is clear from the definition that rapid filters are unbounded so by 1.3 they do not have Baire property.

Let $\mathcal{F}$ be a rapid filter. Suppose that $\mathcal{F}$ is covered by a set of form $\{X \subset \omega : \exists^\infty n \; \exists a \in \mathcal{A}_n \; a \subset X\}$ where $\{\mathcal{A}_n : n \in \omega\}$ is a family as in 1.4. Without losing generality we can assume that for all $n \in \omega$

$$\mu(\{X \subseteq \omega : \exists a \in \mathcal{A}_n \; a \subset X\}) < \frac{1}{2^{n+1}}$$

and that

$$\max\{\max(a) : a \in \mathcal{A}_n\} \geq \min\{\min(a) : a \in \mathcal{A}_m\} \; for \; n \geq m \; .$$

In particular it means that no set in $\mathcal{A}_n$ has less than $n+1$ elements. Define $f(n) = \max\{\max(a) : a \in \mathcal{A}_n\}$ for $n \in \omega$ and let $Z \in \mathcal{F}$ be such that $|Z \cap f(n)| \leq n$ for all $n \in \omega$. We immediately get that

$$Z \notin \{X \subset \omega : \exists^\infty n \; \exists a \in \mathcal{A}_n \; a \subset X\} \; .$$

Contradiction. □

2. Intersections of Filters

We start with the following:

Theorem 2.1 (Talagrand [T1]).

(1) *Intersection of countably many filters without Baire property is a filter without Baire property.*
(2) *Assume MA. Then intersection of $< 2^{\aleph_0}$ filters without Baire property is a filter without Baire property.*

Proof. Let $\{\mathcal{F}_\xi : \xi < \kappa < 2^{\aleph_0}\}$ be a family of filters without Baire property. Let $\mathcal{F} = \bigcap_{\xi<\kappa} \mathcal{F}_\xi$. Let $\{I_n : n \in \omega\}$ be a partition of ω into finite sets. By 1.3 it is enough to show that there exists $X \in \mathcal{F}$ such that $X \cap I_n = \emptyset$ for infinitely many $n \in \omega$. Define sequences $\{X_\xi : \xi < \kappa\}$ and $\{Y_\xi : \xi \leq \kappa\}$ such that

(1) $X_\xi \in \mathcal{F}_\xi$ for $\xi < \kappa$,
(2) $\forall \xi < \kappa \; \forall n \in Y_\xi \; X_\xi \cap I_n = \emptyset$,
(3) $Y_\xi - Y_\eta$ is finite for $\xi \geq \eta$.

Given sequences $\{X_\xi : \xi < \alpha\}$ and $\{Y_\xi : \xi \leq \alpha\}$ using MA find set Y'_α such that $Y'_\alpha - Y_\xi$ is finite for $\xi < \alpha$. Then using 1.3 find $Y_\alpha \subseteq Y_\alpha$ and X_α having properties 1) and 2). Finally let

$$X = \bigcup_{\xi<\kappa} (X_\xi - \bigcup_{n \in Y_\kappa - Y_\xi} I_n) .$$

Clearly $X \in \mathcal{F}$ and $X \cap I_n = \emptyset$ for $n \in Y_\kappa$. $\square$

For ultrafilters we have much stronger result:

Theorem 2.2 (Plewik [P]). *Intersection of $< 2^{\aleph_0}$ ultrafilters is a filter without Baire property.*

Proof. Let $\{\mathcal{F}_\xi : \xi < \kappa < 2^{\aleph_0}\}$ be a family of ultrafilters. Let $\mathcal{F} = \bigcap_{\xi<\kappa} \mathcal{F}_\xi$ and let $\{I_n : n \in \omega\}$ be a partition of ω into finite sets. By 1.3 it is enough to show that there exists $X \in \mathcal{F}$ such that $X \cap I_n = \emptyset$ for infinitely many $n \in \omega$.

Let $\{a_\xi : \xi < 2^{\aleph_0}\}$ be a family of almost disjoint subsets of ω. Consider sets $X_\xi = \bigcup_{n \in a_\xi} I_n$ for $\xi < 2^{\aleph_0}$. Only one of those sets can belong to a filter. Therefore there is $\eta < 2^{\aleph_0}$ such that $X_\eta \notin \mathcal{F}_\xi$ for $\xi < \kappa$. Thus $\omega - X_\eta \in \mathcal{F}$ and $(\omega - X_\eta) \cap I_n = \emptyset$ for $n \in a_\eta$. $\square$

Notice that in fact we showed that for every possible cover H of the intersection of filters we found a family of $2^{\aleph_0}$ almost disjoint sets $\{X_\xi : \xi < 2^{\aleph_0}\}$ such that neither X_ξ nor $\omega - X_\xi$ belongs to H for $\xi < 2^{\aleph_0}$.

The question whether the analog of the theorem above is true for measure is open. The following example shows why the same proof does not work for measure. Define

$I_n = [2^n, 2^{n+1})$ and $J_n = \{a \subset I_n : 2^{-n} \cdot |a| \geq \frac{3}{4}\}$ for $n \in \omega$. It is easy to check that families $\{I_n, J_n : n \in \omega\}$ satisfy conditions 1)- 4) of 1.4. On the other hand for every partition of ω into 4 sets the set $G = (I_n, J_n)_{n=1}^{\infty}$ will contain one of those sets or its complement.

For a countable case we have an analog of 2.1.

Theorem 2.3 (Talagrand [T1]). *Intersection of countably many nonmeasurable filters is a nonmeasurable filter.*

Proof. Let $\{\mathcal{F}_n : n \in \omega\}$ be a sequence of nonmeasurable filters. Denote $\mathcal{F} = \bigcap_{n\in\omega} \mathcal{F}_n$. By 1.4 it is enough to show that for every family $\{\mathcal{A}_n : n \in \omega\}$ satisfying a) - d) of 1.4 there is $X \in \mathcal{F}$ such that $X \not\supset a$ for $a \in \mathcal{A}_n$, $n \in \omega$.

Let $a \in (0, \frac{1}{2}]$ be a real number. Define a measure on 2^ω as $\mu_a = \prod_{n=1}^\infty \hat{\mu}_a$ where $\hat{\mu}_a$ is a measure on $\{0,1\}$ such that $\hat{\mu}_a(\{1\}) = a$ and $\hat{\mu}_a(\{0\}) = 1-a$. We will need the following:

Lemma 2.4. *Let $\mathcal{F}$ be a nonprincipal filter on ω. Then $\mathcal{F}$ is nonmeasurable iff $\mathcal{F}$ is μ_a-nonmeasurable for all $a \in (0, \frac{1}{2}]$.*

Proof. Find $n \in \omega$ and $c \in [0,1]$ such that

$$a = \frac{1}{2^n} + (1 - \frac{1}{2^n}) \cdot c$$

and define the function $F : (2^\omega)^n \times 2^\omega \longrightarrow 2^\omega$ as

$$F(X_1, \dots, X_n, Y) = Y \cup \bigcap_{k=1}^{n} X_k \ .$$

Notice that we identify here $X \in 2^\omega$ with $\{n \in \omega : X(n) = 1\}$.

Let ν be a measure on $(2^\omega)^n \times 2^\omega$ defined as $\mu \times \mu \times \ldots \times \mu \times \mu_c$.

Claim 2.5. *For every Borel set $A \subset 2^\omega$ $\mu_a(A) = \nu(F^{-1}(A))$.*

Proof. It is enough to check that it holds for sets $A_k = \{x \in 2^\omega : x(k) = 1\}$ for $k \in \omega$. Clearly $\mu_a(A_k) = a$ for $k \in \omega$. On the other hand

$$\langle X_1, \dots, X_n, Y\rangle \in F^{-1}(A_k) \leftrightarrow \forall i \leq n\ X_i(k) = 1$$

or

$$\exists i \leq n\ X_k = 0 \ \&\ Y(k) = 1 \ .$$

Therefore $\nu(F^{-1}(A_k)) = \frac{1}{2^n} + (1 - \frac{1}{2^n}) \cdot c = a$. □

Suppose that $\mathcal{F}$ is not measurable. Consider a set $B \subset 2^\omega$ such that $\mu_a(B) > 0$. Since the set $\mathcal{F} \times \mathcal{F} \times \ldots \times \mathcal{F} \times 2^\omega$ has outer measure 1 and mapping F preserves measure we can find $\langle X_1, \dots, X_n, Y\rangle \in F^{-1}(B) \cap \mathcal{F} \times \mathcal{F} \times \ldots \times \mathcal{F} \times 2^\omega$. Thus the set $\bigcap_{k=1}^n X_k \cap Y \in \mathcal{F} \cap B$ which finishes the proof.

Assume that $\mathcal{F}$ is measurable. Let $\{\mathcal{A}_n : n \in \omega\}$ be a family satisfying conditions a) - d) of 1.4. Let $H = \{X \in \omega : \exists^\infty n\ \exists a \in \mathcal{A}_n\ a \subset X\}$. Notice that if $\mu(H) = 0$ then $\mu_a(H) = 0$ for $a \in (0, \frac{1}{2}]$. □

We will use the following notation: if $\{\mathcal{A}_n : n \in \omega\}$ is a family satisfying conditions a) - d) of 1.4 then $(\mathcal{A}_n)_{n=1}^\infty$ denotes the set $\{X \subset \omega : \exists^\infty n\ \exists a \in \mathcal{A}_n\ a \subset X\}$. If $Y \subset \omega$ then $(\mathcal{A}_n - Y)_{n=1}^\infty$ denotes the set $\{X \subset \omega : \exists^\infty n\ \exists a \in \mathcal{A}_n\ a - Y \subset X\}$. Notice that if $Y \in (\mathcal{A}_n)_{n=1}^\infty$ then $(\mathcal{A}_n - Y)_{n=1}^\infty = 2^\omega$.

Let $\{p_n : n \in \omega\}$ be the sequence of reals defined as $p_0 = \frac{1}{2}$ and $p_{n+1} = 1 - \sqrt{1 - p_n}$ for $n \in \omega$.

Define a sequence $\{X_n : n \in \omega\}$ such that for $n \in \omega$

(1) $X_n \in \mathcal{F}_n$,
(2) $X_n \subset X_{n+1}$,
(3) $\mu_{p_n}((\mathcal{A}_k - X_n)_{k=1}^{\infty}) = 0$.

Suppose that X_n is already constructed for some $n \in \omega$. Consider the function $S : 2^\omega \times 2^\omega \longrightarrow 2^\omega$ defined as $S(X, Y) = X \cup Y$. In the same way as in 2.4 we show that for every Borel set $A \subset 2^\omega$

$$\mu_{p_{n+1}} \times \mu_{p_{n+1}}(S^{-1}(A)) = \mu_{p_n}(A) .$$

By the induction hypothesis $\mu_{p_n}((\mathcal{A}_k - X_n)_{k=1}^{\infty}) = 0$. Therefore

$$\mu_{p_{n+1}} \times \mu_{p_{n+1}}(S^{-1}(\mathcal{A}_k - X_n)_{k=1}^{\infty}) = 0 .$$

Since filter $\mathcal{F}_{n+1}$ is $\mu_{p_{n+1}}$-nonmeasurable using Fubini theorem we can find $Y \in \mathcal{F}_{n+1}$ such that

$$\mu_{p_{n+1}}(\{X : S(Y, X) \in (\mathcal{A}_k - X_n)_{k=1}^{\infty}\}) = 0 .$$

Let $X_{n+1} = X_n \cup Y$. It is clear that this set has desired properties.

Finally for every $n \in \omega$ find a natural number k_n such that $a \not\subset X_n$ for $a \in \mathcal{A}_m$ and $m \geq k_n$. Let $X_\omega = \bigcup_{n\in\omega}(X_n - k_n)$. Clearly $X_\omega \bigcap_{n\in\omega} \mathcal{F}_n$ and $X_\omega \notin (\mathcal{A}_k)_{k=1}^{\infty}$. $\square$

Surprisingly this theorem does not generalize to uncountable families of filters.

Theorem 2.6 (Fremlin [F]). *Assume MA & $\neg CH$. Then there exists a family $\{\mathcal{F}_\xi : \xi < 2^{\aleph_0}\}$ of nonmeasurable filters such that $\bigcap_{\xi\in I} \mathcal{F}_\xi$ is a measurable filter for every uncountable set $I \subset 2^{\aleph_0}$. In particular there exists a family of $\aleph_1$ nonmeasurable filters with measurable intersection.*

Proof. Let $\{I_\xi : \xi < 2^{\aleph_0}\}$ be a family of disjoint subsets of $2^{\aleph_0}$ of size $2^{\aleph_0}$. Let $\{K_\xi : \xi < 2^{\aleph_0}\}$ be an enumeration of closed sets of positive measure such that for every closed set of positive measure K and $\xi < 2^{\aleph_0}$ there exists $\eta \in I_\xi$ such that $K = K_\eta$. Let $\{X_\xi : \xi < 2^{\aleph_0}\}$ be a family from 1.15 constructed for the family $\{K_\xi : \xi < 2^{\aleph_0}\}$. Let $\mathcal{F}_\xi$ be the filter generated by the family $\{X_\eta : \eta \in I_\xi\}$ for $\xi < 2^{\aleph_0}$. It is clear that all those filters are nonmeasurable. Suppose that $I \subset 2^{\aleph_0}$ is uncountable. Let $X \in \bigcap_{\xi\in I} \mathcal{F}_\xi$. For each $\xi \in I$ there is a finite set $J_\xi \subset I_\xi$ such that $\bigcap_{\varrho\in J_\xi} X_\varrho \subset X$. Find an infinite set $I' \subset I$ and $k \in \omega$ such that $|J_\xi| = k$ for $\xi \in I'$. Let $Y_\xi = \bigcap_{\varrho\in J_\xi} X_\varrho$ for $\xi \in I'$. By the above remarks $d(Y_\xi) = 2^{-k}$ for $\xi \in I'$. Since all sets J_ξ are disjoint, for every finite set $I'' \subset I'$ we have

$$d(\bigcup_{\xi\in I''} Y_\xi) = 1 - \prod_{\xi\in I''}(1 - d(Y_\xi)) = 1 - (1 - \frac{1}{2^k})^{|I''|} .$$

Thus $d(\bigcup_{\xi\in I'} Y_\xi) = 1$ and therefore $d(X) = 1$. We conclude that

$$\bigcap_{\xi\in I} \mathcal{F}_\xi \subset \{X \subset \omega : d(X) = 1\} .$$

That finishes the proof since the family of sets having density 1 has measure zero. □

3. Ramsey Filters and p-Points

In this section we study the relationship between Ramsey filters and Cohen reals.

Definition 3.1. A filter $\mathcal{F}$ is called *p-point* if for every partition of ω, $\{Y_n : n \in \omega\}$ either there exists $n \in \omega$ such that $Y_n \in \mathcal{F}$ or there exists $X \in \mathcal{F}$ such that $X \cap Y_n$ is finite for $n \in \omega$.

$\mathcal{F}$ is called *Ramsey* if for every partition of ω $\{Y_n : n \in \omega\}$ either $Y_n \in \mathcal{F}$ for some $n \in \omega$ or there exists $X \in \mathcal{F}$ such that $|X \cap Y_n| \leq 1$ for $n \in \omega$.

$\mathcal{F}$ is called *q-point* if for every partition of ω into finite pieces $\{I_n : n \in \omega\}$ exists $X \in \mathcal{F}$ such that $|X \cap I_n| \leq 1$ for $n \in \omega$.

$\mathcal{F}$ is called *rapid* if for every increasing function $f \in \omega^\omega$ there exists $X \in \mathcal{F}$ such that $|X \cap f(n)| \leq n$ for all $n \in \omega$.

Let wD denote the sentence

$$\forall F \subset [\omega^\omega]^{<2^{\aleph_0}}\ \exists g \in \omega^\omega\ \forall f \in F\ \exists^\infty n\ f(n) < g(n) .$$

If we denote by d the size of the smallest dominating family then $wD \leftrightarrow d = 2^{\aleph_0}$.

Theorem 3.1 (Ketonen [K]). *wD iff every filter generated by $< 2^{\aleph_0}$ elements can be extended to a p-point.*

Proof. $\leftarrow$ Suppose that $F \subset \omega^\omega$ is a family of size $< 2^{\aleph_0}$. For $f \in F$ and $n \in \omega$ define $X_f = \{\langle n, k\rangle \in \omega \times \omega : k \geq f(n)\}$ and $X^n = \{\langle m, k\rangle \in \omega \times \omega : m \geq n\}$. It is easy to see that the family $\{X_f : f \in F\} \cup \{X^n : n \in \omega\}$ generates a proper filter. Consider a partition of $\omega\times\omega$ given by $Y_n = \{n\}\times\omega$ for $n \in \omega$. By the assumption there exists $X \subseteq \omega \times \omega$ such that $X \cap X_f$ is infinite for all $f \in F$ and $X \cap Y_n$ is finite for $n \in \omega$. Define the function $g(n) = \max\{k \in \omega : \langle n, k\rangle \in X \cap Y_n\}$ for $n \in \omega$. It is clear that the function g is defined on infinite subset of ω and is not dominated by any function $f \in F$.

$\rightarrow$ This is proved by induction. Single step looks as follows:

Suppose that $\mathcal{F}$ is a filter generated by $< 2^{\aleph_0}$ elements and let $\{Y_n : n \in \omega\}$ be a partition of ω.

CASE 1 There exists $X \in \mathcal{F}$ such that $\{n \in \omega : X \cap Y_n \neq \emptyset\}$ is finite.

We do nothing- every ultrafilter containing $\mathcal{F}$ will contain one piece Y_n for some $n \in \omega$.

CASE 2 For all $X \in \mathcal{F}$ $\{n \in \omega : X \cap Y_n \neq \emptyset\}$ is infinite.

In this case define for $X \in \mathcal{F}$

$$f_X(n) = \begin{cases} \min\{X \cap Y_n\} & \text{if } X \cap Y_n \neq \emptyset \\ \text{undefined} & \text{otherwise} \end{cases} \quad \text{for } n \in \omega .$$

Lemma 3.2. *d is equal to the smallest cardinal κ such that*

$$\exists F \in [\omega^\omega]^\kappa\ \exists G \in [[\omega]^\omega]^\kappa\ \forall g \in \omega^\omega \exists f \in F\ \exists X \in G\ \exists^\infty n \in X\ g(n) < f(n) .$$

Proof. Suppose that F and G are families of size κ having above property. We can assume that F consists of strictly increasing functions. For $f \in F$ and $X \in G$ define for $n \in \omega$, $f^X(n) = f(x_n)$ where x_n is n-th element of X. Easy computation shows that $\{f^X : f \in F,\ X \in G\}$ is a dominating family. □

Using wD and 3.2 find a function $f \in \omega^\omega$ such that

$$\forall X \in \mathcal{F}\ \exists^\infty n\ f_X(n) < f(n)$$

and define

$$X = \bigcup_{n \in \omega} Y_n \cap f(n) .$$

It is not hard to check that X has all the required properties. □

The next theorem was first proved by M. Canjar and independently by the authors.

Theorem 3.3. *The following conditions are equivalent:*

(1) *$\Re$ is not the union of $< 2^{\aleph_0}$ meager sets ,*
(2) *every filter generated by $< 2^{\aleph_0}$ elements can be extended to a Ramsey ultrafilter ,*
(3) *wD and every filter generated by $< 2^{\aleph_0}$ elements can be extended to a q-point.*

Proof. 1) → 2) Let $\mathcal{F}$ be a filter on ω generated by less than $2^{\aleph_0}$ elements. Ultrafilter we are looking for is constructed by induction with respect to all possible partitions of ω. We present a single induction step here. Let $\{Y_n : n \in \omega\}$ be a partition of ω. We have two cases:

CASE 1 There exists $X \in \mathcal{F}$ such that $\{n \in \omega : X \cap Y_n \neq \emptyset\}$ is finite.

In this case we do nothing - every ultrafilter containing $\mathcal{F}$ will contain exactly one set Y_n for some $n \in \omega$.

CASE 2 For every $X \in \mathcal{F}$ $\{n \in \omega : X \cap Y_n \neq \emptyset\}$ is infinite.

In this case we construct a desired selector as follows. Let $\mathcal{Y} = \prod_{n\in\omega} Y_n$. For every $X \in \mathcal{F}$ define

$$G_X = \{x \in Y : \exists^\infty n\ x(n) \in X \cap Y_n\}\ .$$

It is easy to verify that G_X's are dense G_δ subsets of $\mathcal{Y}$. Therefore by the assumption

$$\bigcap_{X\in\mathcal{F}} G_X \neq \emptyset\ .$$

Every element $x \in \bigcap_{X\in\mathcal{F}} G_X$ gives a set $Z = \{x(n) : n \in \omega\}$ which has infinite intersection with every element of $\mathcal{F}$ and selects one element out of every Y_n for $n \in \omega$.

2) $\rightarrow$ 3) Follows immediately from 3.1.

3) $\rightarrow$ 1) Let $\mathcal{A}$ be a family of size $< 2^{\aleph_0}$ which consists of closed, nowhere dense subsets of 2^ω . We have to show that $2^\omega - \bigcup \mathcal{A} \neq \emptyset$. Define a filter on $\omega \times 2^{<\omega}$ in the following way.

For $F \in \mathcal{A}$ let

$$X_F = \{\langle n, s\rangle \in \omega \times 2^{<\omega} :\ \forall t \in 2^{\leq n}\ [t^\frown s] \cap F = \emptyset\}$$

and for $n \in \omega$ let

$$X^n = \{\langle m, s\rangle \in \omega \times 2^{<\omega} : m \geq n\}\ .$$

Notice that $X_{F_1} \cap X_{F_2} \supseteq X_{F_1\cup F_2}$ hence the family $\{X_F : F \in \mathcal{A}\} \cup \{X^n : n \in \omega\}$ generates a proper filter.

For every $F \in \mathcal{A}$ define

$$f_F(n) = \min\{k \in \omega :\ \exists s \in 2^k\ \langle n, s\rangle \in X_F\} \text{ for } n \in \omega\ .$$

Using wD we can find a function $f \in \omega^\omega$ such that

$$\forall F \in \mathcal{A}\ \exists^\infty n\ f_F(n) \leq f(n)\ .$$

Let $X' = \{\langle n, s\rangle \in \omega \times 2^{<\omega} : s \in 2^{\leq f(n)}\}$. It is clear that $X' \cap X_F$ is infinite for all $F \in \mathcal{A}$. Let $\{k_n : n \in \omega\}$ be an increasing sequence of natural numbers such that $k_0 = 0$ and $\sum_{i\leq k_n} f(i) \leq k_{n+1}$ for $n \in \omega$.

Define

$$X'_1 = \{\langle n, s\rangle \in X' : n \in \bigcup_{j\in\omega} [k_{2j}, k_{2j+1})\}$$

and

$$X'_2 = \{\langle n, s\rangle \in X' : n \in \bigcup_{j\in\omega} [k_{2j+1}, k_{2j+2})\}\ .$$

Since $X' = X'_1 \cup X'_2$ one of those sets has infinite intersection with all sets X_F for $F \in \mathcal{A}$. Without loosing generality we can assume that it is X'_1.

Let $\mathcal{F}$ be a filter on the set X_1' generated by the family $\{X_1' \cap X_F : F \in \mathcal{A}\} \cup \{X_1' - U : U \in [X_1']^{<\omega}\}$. For $n \in \omega$ define

$$Y_n = \{\langle j, s\rangle \in X_1' : j \in [k_{2n}, k_{2n+1})\} .$$

Family $\{Y_n : n \in \omega\}$ defines a partition of X_1' so by the assumption there is a set $X \subseteq X_1'$ such that $|X \cap Y_n| \leq 1$ and $\mathcal{F} \cup \{X\}$ generates a proper filter. Suppose that $X = \{\langle u_n, s_n\rangle : n \in \omega\}$ where $u_n \in [k_{2n}, k_{2n+1})$ for $n \in \omega$.

Consider a point $x = s_1{}^\frown s_2{}^\frown \dots {}^\frown s_n{}^\frown \dots \in 2^\omega$. We show that $x \notin F$ for $F \in \mathcal{A}$. Fix $F \in \mathcal{A}$ and find $n \in \omega$ such that $\langle u_n, s_n\rangle \in X_F$. By the above construction

$$\mathrm{lh}(s_1 {}^\frown \dots {}^\frown s_{n-1}) \leq \sum_{i \leq k_{2n-1}} f(i) \leq k_{2n} .$$

Therefore by the definition of X_F, $[s_1 {}^\frown \dots s_{n-1} {}^\frown s_n] \cap F = \emptyset$. This finishes the proof since $x \in [s_1 {}^\frown \dots s_n]$. $\square$

It turns out that the condition 3) of 3.3 can be still weakened. Namely we have the following:

Theorem 3.4 (Fremlin [F1]). *The following conditions are equivalent:*

(1) *$\Re$ is not the union of $< 2^{\aleph_0}$ meager sets,*
(2) *wD and every filter generated by $< 2^{\aleph_0}$ elements can be extended to a rapid filter.*

Proof. $1 \to 2$ Follows immediately from 3.3.

$2 \leftarrow 1$ We will use the following result from [Ba1] (a simple proof can be found in [FM]).

Theorem 3.5. *The following conditions are equivalent:*

(1) *$\Re$ is not the union of $< 2^{\aleph_0}$ meager sets,*
(2) $\forall F \in [\omega^\omega]^{<2^{\aleph_0}}\ \exists g \in \omega^\omega\ \forall f \in F\ \exists^\infty n\ f(n) = g(n)$. $\square$

Let $\{f_\xi : \xi < \theta < 2^{\aleph_0}\}$ be a family of functions from ω^ω. Using wD find a function $f \in \omega^\omega$ such that the set $J_\xi = \{n : f_\xi(n) \leq f(n)\}$ is infinite for all $\xi < \theta$. Using wD again we can find a sequence $\{I_n : n \in \omega\}$ of pairwise disjoint, finite subsets of ω such that $\{n : |I_n \cap J_\xi| \geq n+1\}$ is infinite for all $\xi < \theta$. Define for $n \in \omega$

$$W_n = \bigcup\{I_k : n^2 \leq k < (n+1)^2\} ,$$
$$S_n = \{s : s \text{ is a function , } \mathrm{dom}(s) \subset W_n,\ s(j) \leq f(j) \text{ for } j \in \mathrm{dom}(s)\} .$$

Let $S = \bigcup_{n=1}^{\infty} S_n$ and let

$$X_\xi = \{s \in S : \ |\mathrm{dom}(s) \cap I_n| \geq n+1 \text{ and } s{\restriction}I_n \subset f_\xi \text{ for some } n \in \omega\} \text{ for } \xi < \theta \ .$$

It is easy to see that the family $\{X_\xi : \xi < \theta\}$ generates a proper filter. Let $\mathcal{F}$ be a rapid filter containing this family. Let $X \subset S$ be an element of $\mathcal{F}$ such that $|X \cap S_n| \leq n+1$ for all $n \in \omega$. Let g be a function obtained by diagonalizing the set $X \cap S_n$ over every set I_k for $n^2 \leq k < (n+1)^2$ and $n \in \omega$. This is possible since the above set contains only $\leq n+1$ functions domain of which has more than $n+1$ elements. Verification that

$$\forall \xi < \theta \ \exists^\infty n \ f_\xi(n) = g(n)$$

is straightforward. $\square$

4. Raisonnier's Filter

In this section we will define certain filter on ω called Raisonnier's filter. We prove that this filter is rapid assuming that all Σ^1_2 are Lebesgue measurable and unbounded assuming $MA_{\aleph_1}(\sigma - \text{centered})$.

Definition 4.1. Let $X \subseteq 2^\omega$ be an uncountable set of reals. Let $h : 2^\omega \times 2^\omega \longrightarrow \omega$ be the following function

$$h(x,y) = \min\{n \in \omega : \ x{\restriction}n \neq y{\restriction}n\} \ \ x,y \in 2^\omega \ .$$

For a relation $R \subset 2^\omega \times 2^\omega$ define

$$R_X = \{n \in \omega : \exists x,y \ \langle x,y\rangle \in R \,\&\, h(x,y) = n\} \ .$$

Define

$\mathcal{F}_X = \{R_X : R$ is a Borel equivalence relation with countably many equivalence classes $\}$.

Lemma 4.1. *$\mathcal{F}_X$ generates a proper, non-principal filter.*

Proof. Suppose that $X_1, X_2 \in \mathcal{F}_X$. Let R^1, R^2 be two relations such that $X_1 = R^1_X$ and $X_2 = R^2_X$. Define $R^3 = R^1 \cap R^2$. It is easy to see that $\mathcal{F}_X \ni R^3_X \subseteq R^1_X \cap R^2_X$.

Also every set of form R_X is infinite. To see that let R be a relation having countably many equivalence classes. Since the set X is uncountable at least one of those equivalence classes, call it Y, must be infinite. Function h maps $[Y]^2$ into R_X so if R_X is finite by Ramsey theorem we get an infinite h-homogeneous set. This is not possible since h does not have homogeneous sets of size > 2.

Suppose that $X_1 \in \mathcal{F}_X$ and $n \in \omega$. Let R be a relation such that $X_1 = R_X$. Define the relation R' as follows:
$(x,y) \in R'$ if $(x,y) \in R$ & $x \restriction n = y \restriction n$. Clearly $X_1 - n \subset R'_X$. $\square$

Lemma 4.2. *Let $X = L[a] \cap 2^\omega$ for some $a \in \Re$. Then $\mathcal{F}_X$ is a Σ^1_3 set.*

Proof. $b \in \mathcal{F}_X \leftrightarrow \exists R\, \exists x\ \varphi_1$ & φ_2 & φ_3 & φ_4 & φ_5 where

(1) $\varphi_1 \leftrightarrow$ R is a Borel relation (Π^1_1),
(2) $\varphi_2 \leftrightarrow$ R is an equivalence relation on $L[a]$ (Π^1_2),
(3) $\varphi_3 \leftrightarrow$ R has countably many equivalence classes (Π^1_2),
(4) $\varphi_4 \leftrightarrow \forall n \in \omega\ (n \notin b$ or $(\exists x_1, x_2 \in L[a]\ \langle x_1, x_2\rangle \in R$ & $h(x_1,x_2) = n))$ (Σ^1_2),
(5) $\varphi_5 \leftrightarrow \forall x_1, x_2\ (\langle x_1,x_2\rangle \notin L[a]$ or $h(x_1,x_2) \in b)$.

$\square$

Lemma 4.3 ([JS1]). *Assume $MA_{\aleph_1}$ and let $X = L[a] \cap 2^\omega$ for some $a \in \Re$. Then $\mathcal{F}_X$ is a Δ^1_3 set.*

Proof. We shall prove that the complement of $\mathcal{F}_X$ is Σ^1_3.

Claim 4.4. *Assume $MA_{\aleph_1}$. For a subset $b \subseteq \omega$ the following conditions are equivalent:*

(1) *There is no equivalence relation R on X having countably many equivalence classes such that*
$\forall x, y(\langle x,y\rangle \in R \to h(x,y) \in b)$,
(2) *If $P_b = \{f :\ dom(f) \in [X]^{<\omega}$ & $ran(f) \subset \omega$ & $f(x) = f(y) \to h(x,y) \in b\}$ then $(P_b, \subseteq)$ does not satisfy countable chain condition.*
(3) *There exists $\{\langle x^\xi_j : j \le n\rangle : \xi < \omega_1\} \subseteq [X]^n$ such that*
$\forall \xi < \eta\ \exists j\ h(x^\xi_j, x^\eta_j) \notin b$.

Proof. 1) $\to$ 2) is an immediate consequence of Martin Axiom.

2) $\to$ 3) By the assumption there exists $\{f_\xi : \xi < \omega_1\} \subset P_b$ such that $f_\xi \cup f_\eta \notin P_b$ for $\xi \neq \eta$.

Without loss of generality we can assume that there exists $k \in \omega$ such that

(1) $\mathrm{dom}(f_\xi) = \{x^\xi_1, \dots, x^\xi_m\}$ for $\xi < \omega_1$,
(2) $x^\xi_1 \restriction k, \dots, x^\xi_m \restriction k$ are all different for $\xi < \omega_1$,
(3) $x^\xi_1(l) = x^\eta_1(l), \dots, x^\xi_m(l) = x^\eta_m(l)$ for $\xi,\ \eta < \omega_1$ and $l \le m$,
(4) $\mathrm{dom}(f_\xi) \cap \mathrm{dom}(f_\eta) = \emptyset$ for $\xi \neq \eta$,
(5) For every $\xi < \eta < \omega_1$ there exists $l \le n$ such that $h(x^\xi_l, x^\eta_l) \notin b$.

This is proved by "thinning out" the original family several times. $\square$

3) $\to$ 1) Let $\{\langle x_j^\xi : j \le n\rangle : \xi < \omega_1\} \subseteq [X]^n$ be a family such that $\forall \xi < \eta\, \exists j\, h(x_j^\xi, x_j^\eta) \notin b$. Suppose that 1) is not true. Let R be a relation on X witnessing that. Define for $\xi < \omega_1$

$$F(\langle x_j^\xi : j \le n\rangle) = \langle [x_j^\xi]_R : j \le n\rangle\ .$$

As R has only countably many equivalence classes there are $\xi < \eta < \omega_1$ such that

$$\langle [x_j^\xi]_R : j \le n\rangle = \langle [x_j^\eta]_R : j \le n\rangle\ .$$

By the hypothesis there exists $l \le n$ such that

$$h(x_l^\xi, x_l^\eta) \notin b$$

which contradicts the choice of R. $\square$

Now we can conclude the proof of the lemma.

We have

$$b \notin \mathcal{F}_X \leftrightarrow \exists A \subseteq \omega_1\ \langle L_{\omega_1}[a],\ \in,\ b,\ A\rangle \models \varphi$$

where φ is the first order sentence expressing part 3) of the claim above. By having A to absorb a Skolem function for φ we can assume that φ is Π_1^1 . By $MA_{\aleph_1}$ every set $A \subseteq \omega_1$ can be coded by a real, say c and the encoding process is Δ_1 over $\langle L_{\omega_1}[a],\ \in,\ c\rangle$. Therefore

$$b \notin \mathcal{F}_X \ \leftrightarrow\ \exists c \subset \omega\ (c \textit{ codes } A \subseteq \omega_1\ \&\ \langle L_{\omega_1}[a],\ \in,\ b,\ A\rangle \models \varphi)\ .$$

The last expression is Σ_3^1. $\square$

Lemma 4.5 ([JS1]). *Assume $MA_{\aleph_1}$ and let $X = L[a] \cap 2^\omega$. If $\{a_n : n \in \omega\} \subset \mathcal{F}_X$ then there exists $a^\star = \{k_n : n \in \omega\} \in \mathcal{F}_X$ such that $k_n \in a_n$ for $n \in \omega$.*

Proof. Let $\{a_n : n \in \omega\} \subset \mathcal{F}$. Without loss of generality we can assume that $a_n \supseteq a_{n+1}$ for $n \in \omega$. Let R_n be an equivalence relation witnessing that $a_n \in \mathcal{F}$ for $n \in \omega$. Let Q be the following notion of forcing:

$\langle f,\ Y,\ \{k_l : l \le n\}\rangle \in Q$ if

(1) $Y \subset X$ and $f : Y \longrightarrow \omega$,
(2) $\forall y_1, y_2 \in Y\ (y_1 \ne y_2\ \&\ f(y_1) = f(y_2)) \to\ h(y_1, y_2) \in \{k_l : l \le n\}$,
(3) $k_l \in a_l$ for $l \le n$.

Elements of Q are ordered by reversed inclusion.

Claim 4.6. *Q satisfies countable chain condition.*

Proof. Let $\{\langle f^\xi, Y^\xi, \{k_l^\xi : l \le n^\xi\}\rangle : \xi < \omega_1\}$ be an uncountable subset of Q. By "thinning out" we can assume that there exist $j, n, m \in \omega$ such that for all $\xi, \eta < \omega_1$

(1) $Y^\xi = \{y_1^\xi, \dots, y_m^\xi\}$,
(2) $n^\xi = n$,
(3) $\{k_j^\xi : j \le n\} = \{k_j^\eta : j \le n\} \stackrel{def}{=} \{k_j : j \le n\}$,
(4) $\langle f^\xi(y_1^\xi), \dots, f^\xi(y_m^\xi)\rangle = \langle f^\eta(y_1^\eta), \dots, f^\eta(y_m^\eta)\rangle$,
(5) $y_1^\xi \restriction j, \dots, y_m^\xi \restriction j$ are all different,
(6) $\langle y_1^\xi \restriction j, \dots, y_m^\xi \restriction j\rangle = \langle y_1^\eta \restriction j, \dots, y_m^\eta \restriction j\rangle$,
(7) $y_1^\xi R_{n+m} y_1^\eta, \dots, y_m^\xi R_{n+m} y_m^\eta$.

Choose $\xi \ne \eta$ and let

$$Y = Y^\xi \cup Y^\eta \ , \ f = f^\xi \cup f^\eta$$

$$\{k_l : n \le l < n+m\} = \{h(y_l^\xi, y_l^\eta) : l \le m\} \ .$$

It is easy to verify that the condition $\langle f, Y, \{k_l : l \le n+m\}\rangle$ extends both $\langle f^\xi, Y^\xi, \{k_l^\xi : l \le n\}\rangle$ and $\langle f^\eta, Y^\eta, \{k_l^\eta : l \le n\}\rangle$. □

Now applying $MA_{\aleph_1}$ to forcing Q we get an element $a^\star \in \mathcal{F}_X$ having desired properties. □

Corollary 4.7 ([JS1]). *Assume $MA_{\aleph_1}$ & $\aleph_1$ is not an inaccessible cardinal in L. Then there exists Δ_3^1 rapid filter on ω.*

Proof. If $\aleph_1$ is not inaccessible in L then there is a real number a such that $L[a] \cap 2^\omega$ is uncountable. The rest follows immediately from 4.5. □

In fact we have the following.

Theorem 4.8 (Raisonnier [R]). *Suppose that Lebesgue measure is $\aleph_1$ additive and X is a well-ordered set of size $\aleph_1$. Then $\mathcal{F}_X$ is a rapid filter.*

5. A Model Where There Are No Rapid Filters

In this section we show that the existence of rapid filters is not provable in ZFC. Recall that a nonprincipal filter $\mathcal{F}$ is rapid if for every increasing sequence $\{n_k : k \in \omega\}$ there exists $X \in \mathcal{F}$ such that $|X \cap n_k| \le k$ for all $k \in \omega$.

In [Mi] Miller constructed a model where there are no rapid filters. Here we present a more general construction.

Theorem 5.1. *Con(ZFC) → Con(ZFC & there are no rapid filters).*

Proof. Let P_{ω_2} be the countable support iteration of Mathias forcing of length ω_2. Denote by B_κ a measure algebra adding κ many random reals. In other words consider standard product measure μ on 2^κ and let B_κ be the associated measure algebra.

Let V be a model satisfying GCH. Let $G \subset P_{\omega_2} \star B_\kappa$ be a V-generic filter. We will show that $V[G] \models$ " there are no rapid filters".

We need the following notation. Suppose that $\dot{a}$ is a B_κ-name for a subset of ω. Define a function $\hat{a} : \omega \longrightarrow [0,1]$ as $\hat{a}(n) = \mu([\![n \in \dot{a}]\!])$ for $n \in \omega$.

Lemma 5.2. *Let $\{n_k : k \in \omega\}$ be an increasing sequence of natural numbers. Suppose that $\dot{a}$ is a B_κ-name for a subset of ω such that $\Vdash_{B_\kappa}$ $\forall k\ |\dot{a} \cap n_k| \leq k$. Then for all $k \in \omega$ we have $\sum_{j=1}^{n_k} \hat{a}(j) \leq k$.*

Proof. Suppose that for some $k \in \omega$, $\sum_{j=1}^{n_k} \hat{a}(j) > k$. For $j \leq n_k$ let A_j be the set representing element $[\![j \in \dot{a}]\!]$. By the definition $\hat{a}(j) = \mu(A_j)$ for $j \leq n_k$. Let f_j be a characteristic function of the set A_j for $j \leq n_k$ and let $f = \sum_{j=1}^{n_k} f_j$. We have

$$\sum_{j=1}^{n_k} \hat{a}(j) = \sum_{j=1}^{n_k} \int f_j dx = \int f dx > k \ .$$

Therefore $p = \{x : f(x) > k\}$ has positive measure and clearly $p \Vdash |\dot{a} \cap n_k| > k$. Contradiction. □

Let $\{n_k : k \in \omega\}$ be the first Mathias real added by P_{ω_2}. Suppose that $a \in V[G]$ is a subset of ω such that $|a \cap n_k| \leq k$ for all $k \in \omega$. In the model $V[G \cap P_{\omega_2}]$ a has a B_κ-name $\dot{a}$. Therefore the function $\hat{a}$ defined as above belongs to the model $V[G \cap P_{\omega_2}]$ hence it has a name $f_a \in V$. From now on we will work with this name.

We will need the following technical lemma. Recall that for two conditions $p, q \in P_{\omega_2}$, a set $F \subset \omega_2$ and $n \in \omega$, $p \geq_{F,n} q$ if $p \geq q$ and $p \restriction \xi \Vdash p(\xi) \geq_n q(\xi)$ for $\xi \in F$.

Lemma 5.3. *Suppose that $p \in P_{\omega_2}$ and $\varepsilon > 0$. There exist sequence $\{F_n : n \in \omega\}$ of finite subsets of ω_2, sequence of conditions $\{p_n^1, p_n^2 : n \in \omega\} \subset P_{\omega_2}$, sequence of natural numbers $\{k_n : n \in \omega\}$ and sequence of finite subsets of ω , $\{B_n^1, B_n^2 : n \in \omega\}$ such that*

(1) $0 \in F_0 \subset F_1 \subset \ldots F_n \subset \ldots,$
(2) $\bigcup_{n\in\omega}(supp(p_n^1) \cup supp(p_n^2)) = \bigcup_{n\in\omega} F_n,$
(3) $p = p_0^i \leq_{F_0,0} p_1^i \leq_{F_1,1} p_2^i \leq_{F_2,2} \ldots$ *for* $i = 1, 2,$
(4) $B_0^i \subset B_1^i \subset B_2^i \ldots$ *for* $i = 1, 2$, $B_n^1 \cap B_n^2 \subset B_0^1 \cap B_0^2$ *for* $n \in \omega$ *and* $\bigcup_{n\in\omega} B^1 \cup B_n^2 = \omega,$

(5) *if n is even then for every $k > k_n$ there exists a condition $q \geq p_n^1$ such that*

(a) $q \Vdash \sum_{j=k_n}^{k} f_a(j) < \frac{\varepsilon}{10^n}$,

(b) $q \Vdash \sum_{j \notin B_n^1 \cap k_n} f_a(j) < \varepsilon \cdot (\frac{1}{10} + \frac{1}{10^2} + \dots \frac{1}{10^{n-1}})$,

(c) $p_n^2 \Vdash \sum_{j \notin B_n^2 \cap k_n} f_a(j) < \varepsilon \cdot (\frac{1}{10} + \frac{1}{10^2} + \dots \frac{1}{10^{n-1}})$.

(6) *if n is odd then for every $k > k_n$ there exists a condition $q \geq p_n^2$ such that*

(a) $q \Vdash \sum_{j=k_n}^{k} f_a(j) < \frac{\varepsilon}{10^n}$,

(b) $q \Vdash \sum_{j \notin B_n^2 \cap k_n} f_a(j) < \varepsilon \cdot (\frac{1}{10} + \frac{1}{10^2} + \dots \frac{1}{10^{n-1}})$,

(c) $p_n^1 \Vdash \sum_{j \notin B_n^1 \cap k_n} f_a(j) < \varepsilon \cdot (\frac{1}{10} + \frac{1}{10^2} + \dots \frac{1}{10^{n-1}})$.

Proof. We prove it by induction on n. Suppose that we succeeded in constructing first n elements of all those sequences. Since the last two conditions are symmetric we can assume that n is even.

For every $j \in \omega$ find a condition $p_{n,j}^2$ and family of functions $\{f_{j,l} : l < 2^{n \cdot |F_n|}\}$ such that

(1) $p_n^2 \leq_{F_n, n} p_{n,j}^2$ for $j \in \omega$,
(2) the $n+1$-th element of the infinite part of $p_{n,j}^2$ is bigger that j,
(3) $p_{n,j}^2 \Vdash \exists l < 2^{n \cdot |F_n|}\ \forall i \leq j\ |f_{j,l}(i) - f_a(i)| < \delta_n$ where δ_n is sufficiently small.

Using diagonalization argument we can easily show that there exists a sequence $\{j_m : m \in \omega\}$ such that the sequence $\{f_{j_m,l}(i) : m \in \omega\}$ converges for every $i \in \omega$ and $l < 2^{n \cdot |F_n|}$. Let

$$f^l(i) = \lim_{m \to \infty} f_{j_m,l}(i) \text{ for } i \in \omega \ .$$

Notice that $\sum_{i=1}^{\infty} f^l(i) \leq n+1$ for every $l < 2^{n \cdot |F_n|}$. Otherwise there exists $k \in \omega$ such that $\sum_{i=1}^{k} f^l(i) > n+1$. Therefore there is $j = j_m > k$ such that $\sum_{i=1}^{k} f_{j,l}(i) > n+1$. If δ_n is small enough then $p_{n,j}^2 \Vdash \sum_{i=1}^{k} f_a(i) > n+1$. This is impossible by 5.2 and the fact that the $n+1$-st element of the infinite part of $p_{n,j}^2$ is bigger that j.

Now let k_{n+1} be chosen such that

$$\sum_{i>k_{n+1}} f^l(i) \leq \frac{\varepsilon}{10^{n+2}} \text{ for all } l < 2^{n\cdot|F_n|} .$$

Also let $B^1_{n+1} = B^1_n \cup [k_n, k_{n+1})$, $B^2_{n+1} = B^2_n$ and $p^2_{n+1} = p^2_n$. By the construction p^1_n has an extension p^1_{n+1} such that

$$p^1_{n+1} \Vdash \sum_{j\notin B^1_n \cap k_n} f_a(j) < \varepsilon \cdot (\frac{1}{10} + \frac{1}{10^2} + \dots \frac{1}{10^{n-1}} + \frac{1}{10^n}) .$$

Finally let F_{n+1} be any finite set of ordinals containing F_n and first n elements of $\operatorname{supp}(p^i_j)$ for $i = 1, 2$ and $j \leq n$ (in some fixed enumeration of $\operatorname{supp}(p^i_j)$ in order type ω). □

Let $\dot{\mathcal{F}}$ be a P_{ω_2}-name for a B_κ-name for a rapid ultrafilter on ω. We have the following lemma.

Lemma 5.4. *For every $\varepsilon > 0$ and for every condition $p \in P_{\omega_2}$ there are conditions $p^1, p^2 \geq p$ and sets $B^1, B^2 \subset \omega$ such that*

(1) $p^1 \Vdash \mu([\![B^2 \in \dot{\mathcal{F}}]\!]) < \varepsilon$,
(2) $p^2 \Vdash \mu([\![B^1 \in \dot{\mathcal{F}}]\!]) < \varepsilon$,
(3) $B^1 \cup B^2 = \omega$ *and* $B^1 \cap B^2$ *is finite.*

Proof. Let $\{n_k : k \in \omega\}$ be the first Mathias real added by P_{ω_2}. Suppose that $a \in V[G]$ is an element of $\mathcal{F}$ such that $|a \cap n_k| \leq k$ for all $k \in \omega$. Since $\dot{\mathcal{F}}$ is a name for a rapid filter apply 5.3 to the function f_a and define $B^i = \bigcup_{n\in\omega} B^i_n$, $p^i = \lim_{n\to\infty} p^i_n$ for $i = 1, 2$. Also we have

$$p^i \Vdash \sum_{j\notin B^i} f_a(j) < \varepsilon \text{ for } i = 1, 2 .$$

Since the above sum estimates the "probability" that the set B^i is infinite and we assume that filter $\mathcal{F}$ is nonprincipal we immediately get 1) and 2). □

Next we have to show that in addition we can assume that the conditions p^1, p^2 from 5.3 are compatible. First notice that there exists $\delta < \omega_2$ such that $\operatorname{cf}(\delta) = \omega_1$ and 5.4 holds in $V[G \cap P_\delta]$. In other words using standard reflection argument using the fact that we force with proper forcing and that $V \models GCH$ we can see that there exists δ such that if $B^1, B^2 \in V[G \cap P_\delta]$ is a pair of almost disjoint sets covering ω and if there are conditions p^1, p^2 such that $p^1 \Vdash \mu([\![B^2 \in \dot{\mathcal{F}}]\!]) < \varepsilon$ and $p^2 \Vdash \mu([\![B^1 \in \dot{\mathcal{F}}]\!]) < \varepsilon$ then we can find those conditions in $V[G \cap P_\delta]$.

Fix δ as above. Since $P_{\omega_2} \cong P_{\omega_2} \backslash P_\delta$ 5.3 is true in $V[G \cap P_\delta]$ and we can find B^1, B^2, p^1, p^2 as in 5.3.

Since $\mathrm{cf}(\delta) = \omega_1$ there is $\alpha < \delta$ such that $B^1, B^2 \in V[G \cap P_\alpha]$. By the assumption about δ there are conditions $\hat{p}^1, \hat{p}^2 \in V[G \cap P_\delta]$ such that $\hat{p}^1 \Vdash \mu([\![B^2 \in \dot{\mathcal{F}}]\!]) < \varepsilon$ and $\hat{p}^2 \Vdash \mu([\![B^1 \in \dot{\mathcal{F}}]\!]) < \varepsilon$.

Consider $q = \hat{p}^1 \cdot p^2$. It is a nonzero condition since $\hat{p}^1$ and p^2 have disjoint supports. Without losing generality we can assume that $q \in G$. Therefore in $V[G \cap P_{\omega_2}]$ we have that

(1) $\mu([\![B^2 \in \dot{\mathcal{F}}[G \cap P_{\omega_2}]]\!]) < \varepsilon$,

(2) $\mu([\![B^1 \in \dot{\mathcal{F}}[G \cap P_{\omega_2}]]\!]) < \varepsilon$.

If $\varepsilon < \frac{1}{2}$ then there is a condition $q' \in B_\kappa$ which forces that neither B^1 nor B^2 belongs to $\mathcal{F}$. But this is a contradiction as $\mathcal{F}$ is assumed to be a nonprincipal ultrafilter. $\square$

References

[Ba] T. Bartoszynski, *On covering of the real line by null sets*, Pacific Journal of Mathematics **1** (1988).

[Ba1] T. Bartoszynski, *Combinatorial aspects of measure and category*, Fundamenta Mathematicæ (1987).

[Ba2] T. Bartoszynski, *On the structure of measurable filters on a countable set* (to appear).

[BJ] T. Bartoszynski and H. Judah, *Measure and category in Set theory - the asymmetry* (in preparation).

[C] M. Canjar, *On the generic existence of special ultrafilters* (to appear in Proc. AMS).

[F] D. Fremlin, *Note of Aug. 16, 1982.*

[F1] D. Fremlin, *Note of April 17, 1989.*

[FM] D. Fremlin, A. Miller, *On the properties of Hurewicz and Menger* (to appear).

[J] H. Judah, *Unbounded filters on ω*, Logic Colloquium (1989).

[JS] H. Judah and S. Shelah, *Q-sets, Sierpinski sets and rapid filters* (to appear in Journal of Symb. Logic).

[JS1] H. Judah and S. Shelah, *Martin's Axiom, measurability and equiconsistency results*, Journal of Symb. Logic (1989).

[K] J. Ketonen, *On the existence of p-points in Cech-Stone compactification of integers*, Fundamenta Mathematicæ (1976).

[Mi] A. Miller, *There are no Q-points in Laver's model for the Borel Conjecture*, Proc. AMS **178** (Jan. 1980).

[O] J. Oxtoby, "Measure and category", Springer Verlag.

[P] Sz. Plewik, *Intersections and unions of ultrafilters without Baire property*, Bull. of Polish Acad. of Sciences **35** (1987).

[R] J. Raisonnier, *A mathematical proof of Shelah's theorem*, Israel Journal of Math. (1984).

[T1] M. Talagrand, *Compacts de fonctions mesurables et filtres nonmesurables*, Studia Mathematica **T.LXVII** (1980).

[T2] M. Talagrand, *Filtres: mesurabilite, rapidite, propriete de Baire forte*, Studia Mathematica **T.T.LXXIV** (1982).

Department of Mathematics, Boise State University, Boise ID 83725

Department of Mathematics, Bar-Ilan University, Ramat-Gan, ISRAEL

UNIVERSALLY BAIRE SETS OF REALS

QI FENG, MENACHEM MAGIDOR AND HUGH WOODIN

ABSTRACT. We introduce a generalization of the Baire property for sets of reals via the notion that a set of reals is universally Baire. We show that the universally Baire sets can be characterized in terms of their possible Souslin representations and that in the presence of large cardinals every universally Baire set is determined. We also study the connections between large cardinals, generalizations of $\underset{\sim}{\Sigma}^1_2$ absoluteness with respect to set generic extensions, and various sets being universally Baire.

1. INTRODUCTION

We study in this paper a generalization of the property of Baire for sets of reals. Given a subset A of the set of reals, we say that A is universally Baire if for every topological space X and for every continuous function, $f: X \to \mathbb{R}$, the preimage of A under f, $f^{-1}[A]$, has the property of Baire in the space X. The main theorem we will prove states that a set A of reals is universally Baire if and only if A and its complement are projections of two class trees which have the property that they project to complements in every set generic extension of the universe. It follows that the universally Baire sets have the usual classical regularity properties of analytic and co-analytic sets. Thus it is perhaps more accurate to view the universally Baire sets as generalizations of the analytic and co-analytic sets. In fact we will show that in the presence of suitable large cardinals, every universally Baire set is determined. We will also show that the existence of large cardinals can be used to show that certain sets are universally Baire and conversely (at least in the sense of giving back inner models). For example every $\underset{\sim}{\Sigma}^1_2$ set of reals is universally Baire if and only if every set has a sharp. We shall also show that within the projective sets the property of being universally Baire has connections to the absoluteness of the theory of the reals under set forcing. More precisely, every $\underset{\sim}{\Delta}^1_2$ set of reals is universally Baire if and only if V is $\underset{\sim}{\Sigma}^1_3$ absolute with respect to every set generic extension. Further every $\underset{\sim}{\Sigma}^1_2$ set of reals is universally Baire if and only if every set generic extension of V is $\underset{\sim}{\Sigma}^1_3$ absolute with respect to all further set generic extensions. This is how we will prove that if every $\underset{\sim}{\Sigma}_2^1$ set is universally Baire then every set has a sharp.

Closely related to the universally Baire sets are the projections of weakly homogeneous trees, which for the sake of completeness we shall define below.

For our purpose, the set of reals, $\mathbb{R}$, is the set ω^ω of all functions $f : \omega \to \omega$, where ω is the set of nonnegative numbers. We let $\omega^{<\omega}$ denote the set of all finite sequences of elements of ω and for $s \in \omega^{<\omega}$ let N_s be the set

$$\{ f \in \omega^\omega \mid f \restriction \mathrm{lh}(s) = s \},$$

where $\mathrm{lh}(s)$ is the length of s. The set $\{N_s \mid s \in \omega^{<\omega} \}$ generates a topology on ω^ω; it is the product topology derived from the discrete topology on ω. Endowed with this topology, ω^ω is homeomorphic to the Euclidean space of irrationals.

Suppose X is a set. We denote by X^ω the set of all functions from ω to X and we denote by $X^{<\omega}$ the set of all finite sequences of elements of X. We adopt the usual convention that $X^{<\omega}$ is the set of all functions $f : dom(f) \to X$ such that $dom(f) \in \omega$ and if $s \in X^{<\omega}$ then $dom(s) = \mathrm{lh}(s)$ is the length of s. Suppose that λ is an ordinal larger than 0. A tree on $\omega \times \lambda$ is a subset $T \subseteq \omega^{<\omega} \times \lambda^{<\omega}$ such that for all pairs $(s,t) \in T$, $\mathrm{lh}(s) = \mathrm{lh}(t)$ and $(s\restriction i, t\restriction i) \in T$ for each $i \in \mathrm{lh}(s) \in \omega$. Suppose that T is a tree on $\omega \times \lambda$. For $s \in \omega^{<\omega}$ and for $x \in \omega^\omega$,

$$T_s = \{t \in \lambda^{<\omega} \mid (s,t) \in T \}$$

and

$$T_x = \bigcup \{ T_{x\restriction n} \mid n \in \omega \}.$$

For each $x \in \omega^\omega$, $T_x \subseteq \lambda^{<\omega}$ and is naturally viewed as a tree on λ. Let

$$[T] = \{ (x, f) \mid x \in \omega^\omega \ \wedge \ f \in \lambda^\omega \ \wedge \ \forall n \in \omega \, (x\restriction n, f\restriction n) \in T \}.$$

We also define

$$p[T] = \{ x \in \omega^\omega \mid \exists \, f \in \lambda^\omega \ (x,f) \in [T] \}.$$

Thus $p[T]$ is the *projection* of T, and for each $x \in \omega^\omega$, $x \in p[T]$ if and only if T_x is ill-founded.

Suppose that X is a nonempty set. We denote by $m(X)$ the set of countably complete ultrafilters on the Boolean algebra $P(X)$. μ is a *measure* on X if $\mu \in m(X)$. For $\mu \in m(X)$ and $A \subseteq X$ we write $\mu(A) = 1$ to indicate $A \in \mu$. Suppose $X = Y^{<\omega}$ and $\mu \in m(Y^{<\omega})$. Since μ is countably complete, there is a unique $n \in \omega$ such that $\mu(Y^n) = 1$. Suppose μ_1, μ_2 are in $m(Y^{<\omega})$, and $\mu_1(Y^{n_1}) = \mu_2(Y^{n_2}) = 1$. Then μ_1 *projects to* μ_2 if $n_1 < n_2$ and for all $A \subseteq Y^{n_1}$, $\mu_1(Y^{n_1}) = 1$ if and only if $\mu_2(\{ s \in Y^{n_2} \mid s \restriction n_1 \in A \}) = 1$.

For each $\mu \in m(X)$ there is a canonical elementary embedding $j_\mu : V \to M_\mu$ of the universe, V, into an inner model, M_μ, where M_μ is the transitive collapse of the ultrapower V^X/μ. Suppose that μ_1 projects to μ_2 are in $m(Y^{<\omega})$. Then there is also a canonical elementary embedding

$$j_{\mu_1\mu_2} : M_{\mu_1} \to M_{\mu_2}$$

such that

$$j_{\mu_2} = j_{\mu_1\mu_2} \circ j_{\mu_1}.$$

Suppose $\langle \mu_k \mid k \in \omega \rangle$ is a sequence of measures in $m(Y^{<\omega})$ such that for each $k \in \omega, \mu_k(Y^k) = 1$. The sequence $\langle \mu_k \mid k \in \omega \rangle$ is a *tower* if for all $n < k$ the measure μ_n projects to the measure μ_k. The tower $\langle \mu_k \mid k \in \omega \rangle$ is *countably complete* if for any sequence $\langle A_k \mid k \in \omega \rangle$ such that for all $k < \omega$, $A_k \subseteq Y^k$ and $\mu_k(A_k) = 1$, there exists $f \in Y^\omega$ such that $f \restriction k \in A_k$ for all $k \in \omega$. A tower $\langle \mu_k \mid k \in \omega \rangle$ of measures in $m(Y^{<\omega})$ is countably complete if and only if the direct limit of the sequence $\langle M_k \mid k \in \omega \rangle$ under the maps,

$$j_{nk} : M_n \to M_k$$

(where $n < k$) is well-founded.

The following is a standard reformulation of the definition of a weakly homogeneous tree (see [25]).

Definition. Suppose λ is an ordinal, $\lambda > 0$. A tree T on $\omega \times \lambda$ is *weakly homogeneous* if there is a countable set $\sigma \subseteq m(\lambda^{<\omega})$ such that for all $x \in \omega^\omega$, $x \in p[T]$ if and only if there exists a countably complete tower $\langle \mu_k \mid k \in \omega \rangle$ of measures in σ with $\mu_k(T_{x\restriction k}) = 1$ for all $k \in \omega$.

By standard results and Theorem 2.1 projections of weakly homogeneous trees are λ-universally Baire for some λ. Also it follows from the main theorem of this paper and results of Woodin [25] that if there is a supercompact (or Woodin) cardinal then every universally Baire set is the projection of a weakly homogeneous tree. A cardinal κ is *supercompact* if for every $\lambda > \kappa$ there is an elementary embedding $j : V \to M$ such that κ is the first ordinal moved by j and $j(\kappa) > \lambda$ and the inner model M is closed under λ sequences. A cardinal δ is a *Woodin cardinal* if for every function $f : \delta \to \delta$ there is an elementary embedding $j : V \to M$ such that, letting κ be the first ordinal moved by j, $\kappa < \delta$ and for every $\alpha < \kappa$ $f(\alpha) < \kappa$ and $V_{j(f)(\kappa)} \subseteq M$.

A topic related to the universally Baire sets, the absolutely $\underset{\sim}{\Delta}^1_2$ sets of the reals, has previously been studied by Solovay (unpublished) and by Fenstad and Norman [2]. Vaught (unpublished) and Schilling [16] studied the absolutely $\underset{\sim}{\Delta}^1_2$ sets in the context of Boolean operations. Suppose $A \subseteq P(\omega)$. One can associate to the set A an operation on the subsets of a topological space X as follows. Suppose $\langle B_i : i \in \omega \rangle$ is a sequence of subsets of X. Define a new set B^* by

$$B^* = \{ a \in X \mid \{i \mid a \in B_i\} \in A\}.$$

The main theorem of [16] is that this operation preserves the Baire property in any space X if the set A is absolutely $\underset{\sim}{\Delta}^1_2$. Define a set $A \subseteq P(\omega)$ to be universally Baire in the natural fashion by identifying $P(\omega)$ with $2^\omega \subseteq \omega^\omega$. It is not difficult to see that if the set A is universally Baire then this operation preserves the Baire property in any space X. In fact it can be shown that a set $A \subseteq P(\omega)$ is universally Baire if and only if for every $B \subseteq P(\omega)$ which is continuously reducible to A, the Boolean operation given by B preserves the Baire property in any space X.

Qi Feng would like to thank S. G. Simpson for communicating to him the results of Schilling and Vaught on absolutely $\underset{\sim}{\Delta}^1_2$ sets.

2. Universally Baire Sets

In this section, we study the universally Baire sets of reals. We show that the universally Baire sets can be characterized as those sets which have very nice Souslin representations. These representations are in some sense absolute for set generic extensions. Applying this characterization, we show that the universally Baire sets are Lebesgue measurable, Ramsey, and have the Bernstein property, etc.

Definition. Let A be a subset of the reals. Let λ be an infinite cardinal. A is *λ-universally Baire* if for every topological space X with a regular open basis of cardinality $\leq \lambda$, for every continuous function $f : X \to \omega^\omega$, $f^{-1}[A]$ has the property of Baire, i.e., there is an open set D such that $D \triangle f^{-1}[A]$ is meager. Where $X \triangle Y$ is the symmetric difference of X and Y and $f^{-1}[A] = \{ x \mid \exists\, a \in A\; a = f(x) \}$.

A is *universally Baire* if A is λ-universally Baire for every infinite cardinal λ.

Clearly, all the universally Baire subsets of the reals form a σ-algebra containing all the open sets. Hence every Borel subset of ω^ω is universally Baire. Also, if $f \colon \omega^\omega \to \omega^\omega$ is a continuous function, $A \subseteq \omega^\omega$ is universally

Baire, then $f^{-1}[A]$ is also universally Baire. In fact we shall see that every analytic set is universally Baire.

We are mainly interested in which sets of reals are universally Baire. Since each universally Baire set has the property of Baire, we will be primarily concerned with consistency results. By results of Solovay [21] and Shelah [18], the assertion that every projective set, or that every set of reals which is in $L(\mathbb{R})$ has the Baire property is not very strong in consistency strength. But we will see that even the assertion that every $\underset{\sim}{\Sigma}^1_2$ set is universally Baire is a relatively strong statement.

Theorem 2.1. *Let $A \subseteq \omega^\omega$. let λ be an infinite cardinal. Then the following are equivalent:*

(1) *There are two trees T, T^* such that*
  (a) $A = p[T]$, $\omega^\omega - A = p[T^*]$,
  (b) $\mathrm{Col}(\omega, \lambda) \Vdash p[T] \cup p[T^*] = \omega^\omega$.
(2) *There are trees T, T^* such that*
  (a) $A = p[T]$, $\omega^\omega - A = p[T^*]$,
  (b) $\mathbf{P} \Vdash p[T] \cup p[T^*] = \omega^\omega$ *for every forcing notion* $\mathbf{P}$ *of cardinality* $\leq \lambda$.
(3) *A is λ-universally Baire.*
(4) *For every continuous function $f\colon \lambda^\omega \to \omega^\omega$, $f^{-1}[A]$ has the property of Baire.*

Proof. (1) $\Rightarrow$ (2) By an absoluteness argument, noting that if $|\mathbf{P}| \leq \lambda$, then $\mathbf{P} \times \mathrm{Col}(\omega, \lambda)$ is isomorphic to $\mathrm{Col}(\omega, \lambda)$.

(2) $\Rightarrow$ (3) Let X be a topological space with a regular open basis of cardinality $\leq \lambda$. Let $\mathcal{B}$ be the complete Boolean algebra of the regular open sets in the space X. Let T, T^* be the two trees given by (2). Then we have that

$$\mathcal{B} \Vdash p[T] \cup p[T^*] = \omega^\omega.$$

Given a continuous function $f\colon X \to \omega^\omega$, we want to show that $f^{-1}[A]$ has the property of Baire.

Take κ to be a sufficiently large regular cardinal such that all the relevant objects are in H_κ, the set of sets whose transitive closure have cardinality smaller than κ.

If $G \subseteq \mathcal{B}$ is any generic, we denote the unique real x, by $f(G)$, such that

$$\forall n\, f^{-1}[N_{x\restriction n}] \in G.$$

Define

$$B_0 = [\![f(\dot{G}) \in p[T]]\!],$$

and

$$B_1 = [\![f(\dot{G}) \in p[T^*]]\!].$$

Claim. $B_0 \triangle f^{-1}[A]$ is meager.

To prove this claim, we use the following Banach-Mazur game on X and apply a theorem of Oxtoby [15].

Given $D \subseteq X$, the Banach-Mazur game $\mathcal{G}(D)$ is defined as follows:

$$\begin{array}{lccc} \text{I} & D_0 & D_2 & \cdots \\ \text{II} & D_1 & D_3 & \cdots \end{array}$$

Two players play in turn nonempty open sets so that $D_{n+1} \subseteq D_n$. After ω many steps, I wins if and only if $\bigcap_{n<\omega} D_n \subseteq D$ and II wins otherwise.

A winning strategy for either player has the standard meaning (cf. [4]).

We apply the following theorem of Oxtoby [15] which is easy to prove.

Theorem (Oxtoby). *If I has a winning strategy σ in the game $\mathcal{G}(D)$, and if D_0 is the first move of I according to the strategy σ, then $D_0 - D$ is meager.*

To prove the claim, it suffices to show that $B_0 - f^{-1}[A]$ is meager and $B_1 - f^{-1}[\omega^\omega - A]$ is meager.

By the symmetry, we may assume that $B_0 \neq \emptyset$, and we only prove that $B_0 - f^{-1}[A]$ is meager.

First, let us observe the following fact.

Fact. *If $M \prec H_\kappa$ is a countable elementary submodel containing all the relevant objects, and if G is a $\mathcal{B}$-generic filter over M, then*

$$f(G) \in p[T] \iff B_0 \in G.$$

To see this, assume $B_0 \in G$. Then $B_0 \Vdash f(\dot{G}) \in p[T]$. Hence

$$M \models B_0 \Vdash f(\dot{G}) \in p[T].$$

So $M[G] \models f(G) \in p[T]$. By upward absoluteness, $f(G) \in p[T]$.

If $B_0 \notin G$, then $B_1 \in G$. Similar arguments show that $f(G) \in p[T^*]$.

Now we are ready to show that $B_0 - f^{-1}[A]$ is meager. To do this, we show that I has a winning strategy σ in the game $\mathcal{G}(f^{-1}[A])$ such that B_0 is the first move according to σ.

For bookkeeping, fix a bijective function $\pi: \omega \to \omega \times \omega$ such that if $\pi(n) = (k, l)$, then $l < n$.

Let $D_0 = B_0$ be the first move of I. Let D_1 be the response of II. Now I chooses a countable elementary submodel $M_0 \prec H_\kappa$ with all relevant things in M_0. Let $\langle C_{n0} \mid n < \omega \rangle$ be an enumeration of all dense subsets of $\mathcal{B}$ in M_0.

Let G_0 be a $\mathcal{B}$-generic over M_0 such that $D_1 \in G_0$. Let $x_0 = f(G_0)$. By the fact above, $x_0 \in p[T]$. Let k_0 be so large that there is $D \in C_{00}$ such that

$$f^{-1}[N_{x_0 \restriction k_0}] \subseteq D \cap D_1.$$

I then responds with $D_2 = f^{-1}[N_{x_0 \restriction k_0}]$.

Inductively, let D_{2n+1} be played by II. I chooses a countable elementary submodel $M_n \prec H_\kappa$ such that $D_{2n+1} \in M_n$ and $M_{n-1} \subseteq M_n$. Then let $\langle C_{in} \mid i < \omega \rangle$ be an enumeration of all dense subsets of $\mathcal{B}$ in M_n.

Let G_n be a $\mathcal{B}$-generic over M_n such that $D_{2n+1} \in G_n$. Let $x_n = f(G_n)$. Then $x_n \in p[T]$ and $x_n \restriction k_{n-1} = x_{n-1} \restriction k_{n-1}$. Let k_n be so large that there is $D \in C_{\pi(n)}$ such that

$$f^{-1}[N_{x_n \restriction k_n}] \subseteq D \cap D_{2n+1}.$$

Then I plays with $D_{2n+2} = f^{-1}[N_{x_n \restriction k_n}]$.

This defines a strategy for I. We show that it is a winning strategy for I.

Let $M_\omega = \bigcup_{n<\omega} M_n$. Then $M_\omega \prec H_\kappa$. Now the filter G generated by

$$\{ D_n \mid n < \omega \}$$

of the play is $\mathcal{B}$-generic over M_ω. Then we have

$$\bigcap_{n<\omega} D_n \subseteq f^{-1}[A].$$

This finishes the proof of the claim, and hence (2) $\Rightarrow$ (3) is proved.

(3) $\Rightarrow$ (4) is trivial.

(4) $\Rightarrow$ (1) By (4), we have that for every continuous function $f\colon \lambda^\omega \to \omega^\omega$, $f^{-1}[A]$, $f^{-1}[\omega^\omega - A]$ have the property of Baire, i.e., we can find two open sets B_0, B_1, and dense open sets $\langle D_n \mid n < \omega \rangle$ such that

$$f^{-1}[A] \cap \bigcap_{n<\omega} D_n = B_0 \cap \bigcap_{n<\omega} D_n$$

and

$$f^{-1}[\omega^\omega - A] \cap \bigcap_{n<\omega} D_n = B_1 \cap \bigcap_{n<\omega} D_n.$$

Consider the forcing $\mathcal{B} = \mathrm{Col}(\omega, \lambda)$. Identify each $\mathcal{B}$-name $\dot{x}$ for a real with

$$\{ \langle p, \langle n, m \rangle \rangle \mid p \Vdash \langle n, m \rangle \in \dot{x} \}.$$

Let N be the set of such names.

We consider the following complete metric space S:

$$S = \{ f \mid f(0) \in N, \ f(n+1) \in \lambda \text{ for } n < \omega \} = N \times \lambda^\omega.$$

Suppose $e \colon S \to \omega^\omega$ is a continuous function. Then it follows that $e^{-1}[A]$ and $e^{-1}[\omega^\omega - A]$ each have the property of Baire. To see this note that $\{B_a \mid a \in N\}$ is a discrete partition (the union of any subset is clopen) of the space S into clopen sets each of which is homeomorphic to the space λ^ω where for each $a \in N$, $B_a = \{f \mid f(0) = a\}$.

For $f \in S$, define

$$e(f) = \{ \langle n, m \rangle \mid \exists k \, \langle (f(1), \cdots, f(k)), (n, m) \rangle \in f(0) \}.$$

Notice that on a dense G_δ set $S' \subseteq S$ we have $e : S' \to \omega^\omega$ is continuous. By a theorem of Stone [23] S' is homeomorphic to S.

Thus by the remarks above both $e^{-1}[A]$ and $e^{-1}[\omega^\omega - A]$ have the property of Baire.

Let B_0, B_1 be two open sets, and let $\langle D_n \mid n < \omega \rangle$ be dense open such that

$$e^{-1}[A] \cap \bigcap_{n<\omega} D_n = B_0 \cap \bigcap_{n<\omega} D_n$$

and

$$e^{-1}[\omega^\omega - A] \cap \bigcap_{n<\omega} D_n = B_1 \cap \bigcap_{n<\omega} D_n.$$

Notice that we have the following:

(a) $e[e^{-1}[A] \cap \bigcap_{n<\omega} D_n] = A$,
(b) $e[e^{-1}[\omega^\omega - A] \cap \bigcap_{n<\omega} D_n] = \omega^\omega - A$.
(c) $B_0 \cup B_1$ is dense open.

Let us see why (a) is true. Let $x \in A$. Let

$$\tau = \{ (p, (n, m)) \mid p \in \lambda^{<\omega}, \ (n, m) \in x \}.$$

Then τ is the canonical name for x. Look at the clopen set

$$B_x = \{ f \in S \mid f(0) = \tau \}.$$

We have for $f \in B_x$, $e(f) = x$, and

$$\{ f \in B_x \mid f \in \bigcap_{n<\omega} D_n \}$$

is dense in B_x by the Baire category theorem.

Now we define the trees T, T^* as follows.

Define that $(\sigma, (\tau, p_0, \cdots, p_k)) \in T$ if and only if $\tau \in N$ is a canonical name for a real, and $\sigma \in \omega^{<\omega}$, $\mathrm{lh}(\sigma) = k$, and for each $i \leq k$, we have $\widehat{p_0 \cdots p_i} \Vdash \tau\restriction i = \sigma\restriction i$, and for some $s \in \lambda^{<\omega}$ we have $\forall f \in S$ if $\tau\widehat{p_0 \cdots p_k}s \subseteq f$, then $f \in B_0 \cap \bigcap_{j\leq k} D_j$.

Similarly define T^*, replacing B_0 by B_1.

We claim that the following hold:

(a) $A = p[T]$, $\omega^\omega - A = p[T^*]$,

(b) $\mathrm{Col}(\omega, \lambda) \Vdash\ p[T] \cup p[T^*] = \omega^\omega$.

(a) follows from that $e[B_0 \cap \bigcap_{n<\omega} D_n] = A$ and $e[B_1 \cap \bigcap_{n<\omega} D_n] = \omega^\omega - A$.

To see (b), let $\hat{x}$ be a canonical name for a real. Let

$$B_{\hat{x}} = \{ f \in S \mid f(0) = \hat{x} \}.$$

It is a clopen set. By the Baire category theorem,

$$B_{\hat{x}} \cap (B_0 \cup B_1) \cap \bigcap_{n<\omega} D_n$$

is comeager in $B_{\hat{x}}$.

If G is a $\mathrm{Col}(\omega, \lambda)$-generic over V, let

$$g(0) = \hat{x},\ \ g(n+1) = G(n) \text{ for } n < \omega.$$

Then in $V[G]$,

$$g \in B_{\hat{x}} \cap (B_0 \cup B_1) \cap \bigcap_{n<\omega} D_n.$$

We may assume that

$$g \in B_{\hat{x}} \cap B_0 \cap \bigcap_{n<\omega} D_n.$$

Then g witnesses $x = \hat{x}/G \in p[T]$ in $V[G]$.

This finishes the proof of the theorem. $\square$

As an immediate consequence we obtain the following corollary.

Corollary 2.1. *Let $A \subseteq \omega^\omega$. The following are equivalent:*

(1) *A is universally Baire.*
(2) *For every infinite cardinal λ, for every continuous function $f\colon \lambda^\omega \to \omega^\omega$, $f^{-1}[A]$ has the property of Baire.*
(3) *For every poset $\mathbf{P}$ there are two trees T, T^* such that*

$$A = p[T], \quad \omega^\omega - A = p[T^*]$$

and

$$P \Vdash \; p[T] \cup p[T^*] = \omega^\omega. \qquad \square$$

Corollary 2.2. *Every analytic $A \subseteq \omega^\omega$ is universally Baire.*

Proof. Suppose that A is analytic. There are class trees T, T^* such that $A = p[T]$ and such that in any set generic extension of V the trees T, T^* project to complements. Therefore by Corollary 2.1 the set A is universally Baire. $\square$

Using the structural characterization given in Corollary 2.1 we prove that the universally Baire sets assume all the regular properties which analytic sets and coanalytic sets share.

Suppose X is a topological space which is not meager. One could define a set $A \subseteq \omega^\omega$ to be X-universally Baire if for any function $f\colon X \to \omega^\omega$ which is continuous on a comeager set, $f^{-1}[A]$ has the property of Baire. With this definition A is λ^ω-universally Baire if and only if A is λ-universally Baire (in the previous sense). Theorem 2.1 could then be reformulated as follows. The following are equivalent:

(1) A is X-universally Baire.
(2) There exist trees T, T^* such that $A = p[T]$ and

$$RO(X) \Vdash \; p[T] = \omega^\omega - p[T^*].$$

Slight strengthenings of the classical regularity properties are in effect assertions that a set is X-universally Baire for the appropriate space X. For example a set $A \subseteq \omega^\omega$ is universally measurable if and only if A is X-universally Baire where X is the maximal ideal space of the measure algebra.

We recall that a subset $A \subseteq \omega^\omega$ is Ramsey, if there is an infinite $a \subseteq \omega$ such that either $[a]^\omega \subseteq A$ or $[a]^\omega \subseteq \omega^\omega - A$, where $[a]^\omega$ is the set of all increasing functions from ω to a.

Theorem 2.2. *If $A \subseteq \omega^\omega$ is universally Baire, then A is Lebesgue measurable and A is Ramsey.*

Proof. Let $\lambda \geq 2^{\aleph_0}$ be regular. Let T, T^* be two trees such that $A = p[T]$ and $\omega^\omega - A = p[T^*]$ and for every forcing notion of size at most λ

$$\Vdash \ p[T] \cup p[T^*] = \omega^\omega.$$

Let $\mathcal{B}$ be the measure algebra. We have

$$\mathcal{B} \Vdash \ p[T] \cup p[T^*] = \omega^\omega.$$

Let $\dot{x}_G$ be the canonical name for a Random real. Let

$$B_0 = [\![\dot{x}_G \in p[T]]\!]$$

and

$$B_1 = [\![\dot{x}_G \in p[T^*]]\!].$$

Claim. $B_0 \triangle A$ is of measure zero.

To see this, let κ be a regular cardinal large enough such that every thing relevant is in H_κ. Let $M \prec H_\kappa$ be a countable elementary submodel containing all the relevant objects.

Subclaim. If x is Random over M, then

$$x \in p[T] \iff x \in B_0.$$

If $x \in B_0$, then $B_0 \Vdash \ \dot{x}_G \in p[T]$. Hence

$$M \models B_0 \Vdash \ \dot{x}_G \in p[T].$$

Then $M[x] \models \ x \in p[T]$. Hence $x \in p[T] = A$.

If $x \notin B_0$, then $x \in B_1$. So $B_1 \Vdash \ \dot{x}_G \in p[T^*]$. By a similar argument, $x \in p[T^*]$.

Since M is countable, there are only countably many maximal antichains of $\mathcal{B}$ in M. Since the measure algebra satisfies the countable chain condition, we have that $B_0 \triangle A$ is of measure zero.

Therefore, A is Lebesgue measurable.

To see that A is Ramsey, one uses a similar argument with Mathias forcing replacing the measure algebra.

Recall that Mathias forcing $\mathcal{B}$ is the following. Each condition p is a pair $p = (t, s)$ with $t \in [\omega]^{<\omega}$ and $s \in [\omega]^\omega$. The ordering is defined as

$(t,s) \leq (t',s')$ if and only if $t' \subseteq t,\ s \subseteq s'$,and $t - t' \subseteq s'$. For properties of the Mathias forcing, see [14,24].

Let $\dot{G}$ be the canonical name for the generic filter of $\mathcal{B}$. Let $\dot{f}_G$ be the Mathias real. Let T, T^* be two trees such that

$$p[T] = A, \;\; p[T^*] = \omega^\omega - A$$

and

$$\mathcal{B} \Vdash \;\; p[T] \cup p[T^*] = \omega^\omega.$$

We may assume without loss of generality that

$$(\emptyset, s) \Vdash \;\; \dot{f}_G \in p[T].$$

Let κ be sufficiently large and $M \;\prec\; H_\kappa$ be a countable elementary submodel containing the objects of interest. Let $g \in [s]^\omega$ be Mathias generic over M. Then

$$M[g] \;\models\; f_g \in p[T]$$

and if $g' \in [g]^\omega$, then $M[g'] \;\models\; f_{g'} \in p[T]$.

Therefore, $\forall\, x \in [g]^\omega,\;\; f_x \in p[T]$.

This finishes the proof. $\square$

We now consider the perfect subset property. Since there might be some uncountable $\underset{\sim}{\Pi}^1_1$ set that does not contain a perfect subset, one can not hope to prove that every uncountable universally Baire set contains a perfect subset. However, it is true that every universally Baire set has the Bernstein property and that if every $\underset{\sim}{\Delta}^1_2$ set is universally Baire then every uncountable $\underset{\sim}{\Sigma}^1_2$ set does contains a perfect subset.

A subset $X \subseteq \mathbb{R} = 2^\omega$ has the *Bernstein property* if for every perfect set P either $X \cap P$ contains a perfect set or $(\mathbb{R} - X) \cap P$ contains a perfect set.

All subsets of 2^ω with the Bernstein property form a σ-algebra, containing all the closed sets.

Theorem 2.3. *If $A \subseteq 2^\omega$ is universally Baire, then A has the Bernstein property.*

Remark. From this theorem, if a $\underset{\sim}{\Pi}^1_1$ set does not contain a perfect subset, then it must be an S_0-set (i.e., for each perfect set P there is a perfect subset $Q \subseteq P$ disjoint from it).

Proof. Let $A \subseteq 2^\omega$ be a universally Baire set. Let T, T^* be two trees such that $A = p[T]$ and $2^\omega - A = p[T^*]$ and

$$\mathbb{P} \Vdash \;\; p[T] \cup p[T^*] = 2^\omega$$

where $\mathbb{P}$ is Sacks forcing, i.e., conditions in $\mathbb{P}$ are the perfect subtrees of $2^{<\omega}$, ordered by inclusion.

Let x be a name for a Sacks real. Let p be a perfect tree. By symmetry, it suffices to prove the following claim.

Claim. If $p \Vdash x \in p[T]$, then there is a perfect tree $q \leq p$ such that $[q] \subseteq p[T]$.

To prove the claim, we consider the following game:

$$\begin{array}{llll} I & p_0 & p_1 & \cdots \\ \mathbb{I} & \quad q_0 & \quad q_1 & \quad \cdots \end{array}$$

Rules: $p_n,\ q_n \in \mathbb{P}$ and $p_{n+1} \leq q_n \leq p_n$.

I wins if and only if $\bigcap_{n<\omega} [p_n] \subseteq A$.

Subclaim. I has a winning strategy σ such that p is the first move according to σ.

Given this subclaim, we show the claim holds.

Fix a winning strategy σ for I in the game with p as the first move. For each $\tau \in 2^{<\omega}$, we associate with τ a perfect tree $p_\tau \leq p$ so that for any $f \in 2^\omega$, the sequence $\langle p_{f\restriction n},\ q_{f\restriction n+1} \mid n < \omega\rangle$ is a play with I playing according to σ, and $\bigcap_{n<\omega} [p_{f\restriction n}]$ is a singleton, denoted by f^*, and the mapping $f \to f^*$ is a one-to-one continuous mapping.

We do this by induction. $p_{\langle\rangle} = p$.

For $\tau \in 2^{<\omega}$, assume that p_τ, q_τ are defined and satisfying the obvious requirements. Let $s \in p_\tau$ be the first branching point of p_τ. Then define

$$q_{\widehat{\tau\langle 0\rangle}} = p_\tau\restriction \widehat{s\langle 0\rangle},\quad q_{\widehat{\tau\langle 1\rangle}} = p_\tau\restriction \widehat{s\langle 1\rangle}$$

$$p_{\widehat{\tau\langle 0\rangle}} = \sigma(p_{\langle\rangle}, q_{\tau\restriction 1}, p_{\tau\restriction 1}, \cdots, p_\tau, q_{\widehat{\tau\langle 0\rangle}}),$$

$$p_{\widehat{\tau\langle 1\rangle}} = \sigma(p_{\langle\rangle}, q_{\tau\restriction 1}, p_{\tau\restriction 1}, \cdots, p_\tau, q_{\widehat{\tau\langle 1\rangle}}).$$

Clearly this gives us what we want.

We need to prove the subclaim.

Fix a bijective function $\pi\colon \omega \to \omega \times \omega$ such that if $\pi(n) = (k, l)$, then $l < n$.

Let $p_0 = p$ be I's first move. Let q_0 be $\mathbb{I}$'s response. Pick a countable elementary submodel $M_0 \prec H_\kappa$ with all relevant things in M_0. Let $\langle C_{n0} \mid n < \omega\rangle$ be an enumeration of all dense subsets of $\mathbb{P}$ in M_0.

Let $q_0 \in G_0$ be a $\mathbb{P}$-generic over M_0. Let $x_0 = x/G_0$. Then we have $x_0 \in p[T]$. Then let $p_1 \in C_{00}$ be stronger than q_0 such that $p_1 \Vdash x(0) = x_0(0)$. I then plays this p_1.

Inductively, let q_n be played by $\mathbb{I}$. I chooses a countable elementary submodel $M_n \prec H_\kappa$ such that $q_n \in M_n$ and $M_{n-1} \subseteq M_n$. Then let $\langle C_{in} \mid i < \omega \rangle$ be an enumeration of all dense subsets of $\mathbb{P}$ in M_n.

Let G_n be a $\mathbb{P}$-generic over M_n such that $q_n \in G_n$. Let $x_n = x/G_n$. Then $x_n \in p[T]$ and $x_n \restriction n = x_{n-1} \restriction n$. Let $p_n \in C_{\pi(n)}$ such that $p_n \leq q_n$ and $p_n \Vdash x(n) = x_n(n)$. Then I plays this p_n.

This defines a strategy for I. We show that it is a winning strategy for I.

Let $M_\omega = \bigcup_{n<\omega} M_n$. Then $M_\omega \prec H_\kappa$. Now the filter G generated by $\{ p_n \mid n < \omega \}$ of the play is $\mathbb{P}$-generic over M_ω. Then we have

$$\bigcap_{n<\omega} [p_n] = \{x/G\} \subseteq p[T].$$

□

To end this section, we state the following theorem below, whose proof will be given in the next section.

Theorem 2.4. *If every $\underset{\sim}{\Delta}^1_2$ set is universally Baire, then every uncountable $\underset{\sim}{\Sigma}^1_2$ set contains a perfect subset.*

3. Absoluteness

One of the consequences of Shoenfield absoluteness [19] is that the first order $\underset{\sim}{\Sigma}^1_2$ theory of the reals can never be changed by forcing, i.e., in any forcing extension of the universe with the same ordinals, the old reals form an $\underset{\sim}{\Sigma}^1_2$ elementary submodel of the reals in the generic extension. By Martin-Solovay absoluteness [11], no forcing notion of size smaller than or equal to any measurable cardinal can change the $\underset{\sim}{\Sigma}^1_3$ theory of the reals. On the other hand, by results of Levy [8], Silver [20], and Martin-Solovay [11], these are the best possible. It is relatively consistent with ZFC that forcing can change the Σ^1_3 theory of the reals; and it is relatively consistent with ZFC + *There is a measurable cardinal*, that small forcing can change the Σ^1_4 theory of the reals. In fact, one can show [24] that $\underset{\sim}{\Sigma}^1_3$ absoluteness between V,$V[G_P]$ and $V[G_P * G_Q]$ for every iteration $P * Q$ is equivalent to every set has a sharp.

In [24], Woodin showed that if one assumes that every projective set has the property of Baire and every projective relation is projectively uniformizable, then one can not change the theory of the reals by adding Cohen reals. A similar statement is true for Random real forcing.

Also Woodin showed that when the supremum of ω many strong cardinals is Levy collapsed to ω, then no further set forcing can affect the first order theory of the reals in a strong sense.

To illustrate some of the connections between absoluteness and sets being universally Baire, we prove the following theorem from which it follows that every $\underset{\sim}{\Delta}^1_2$ set is universally Baire if and only if V is $\underset{\sim}{\Sigma}^1_3$ absolute with respect to every set generic extension.

Given two models $\mathbb{M} \subseteq \mathbb{N}$ of set theory, we say that $\mathbb{N}$ is $\underset{\sim}{\Sigma}^1_n$ absolute with respect to $\mathbb{M}$ if $\mathbb{R} \cap \mathbb{M}$ is a Σ_n elementary submodel of $\mathbb{R} \cap \mathbb{N}$. And $\mathbb{N}$ is absolute with respect to $\mathbb{M}$ if it is $\underset{\sim}{\Sigma}^1_n$ absolute for every n.

Theorem 3.1. *Let λ be an infinite cardinal. Then the following are equivalent:*

(1) *If $\mathcal{B}$ is a forcing notion of size $\leq \lambda$, and $\phi(x_1, \cdots, x_n)$ is a Π^1_3 formula with free variables shown, and $a_1, \cdots, a_n$ are reals, then*

$$\phi(a_1, \cdots, a_n) \iff \mathcal{B} \Vdash \phi(a_1, \cdots, a_n).$$

(2) *Every $\underset{\sim}{\Delta}^1_2$ subset $A \subseteq \omega^\omega$ is λ-universally Baire.*

Proof. First, if $(1')$ is the statement (1) of the theorem replacing $\mathcal{B}$ by $\mathrm{Col}(\omega, \lambda)$, then (1) $\iff$ $(1')$. This is because if $|\mathcal{B}| \leq \lambda$, then $\mathcal{B} \times \mathrm{Col}(\omega, \lambda)$ is isomorphic to $\mathrm{Col}(\omega, \lambda)$, and the Σ^1_3 statements are upward absolute.

(1) $\Rightarrow$ (2) Let $A \subseteq \omega^\omega$ be $\underset{\sim}{\Delta}^1_2$. That is, there are two formulas ϕ, φ, both are Π^1_1 with some real parameters, such that

$$A = \{\, x \in \omega^\omega \mid \exists\, y\, \phi(x, y) \,\}$$

and

$$\omega^\omega - A = \{\, x \in \omega^\omega \mid \exists\, y\, \varphi(x, y) \,\}.$$

There exist trees T, T^* such that $A = p[T]$, $\omega^\omega - A = p[T^*]$ and

$$\mathrm{Col}(\omega, \lambda) \Vdash p[T] = \{\, x \in \omega^\omega \mid \exists\, y\, \phi(x, y) \,\}$$

and

$$\mathrm{Col}(\omega, \lambda) \Vdash p[T^*] = \{\, x \in \omega^\omega \mid \exists\, y\, \varphi(x, y) \,\}.$$

Since $\forall\, x\, (\exists\, y\, \phi(x, y)$ or $\exists\, y\, \varphi(x, y))$ is a $\underset{\sim}{\Pi}^1_3$ statement, by (1),

$$\mathrm{Col}(\omega, \lambda) \Vdash \forall\, x\, (\exists\, y\, \phi(x, y) \text{ or } \exists\, y\, \varphi(x, y)).$$

Hence

$$\mathrm{Col}(\omega, \lambda) \Vdash p[T] \cup p[T^*] = \omega^\omega.$$

By the main theorem of section one, A is λ-universally Baire.

To see (2) $\Rightarrow$ (1′), we first prove that (2) implies (3).

(3) For each continuous function $f:\lambda^\omega \to \omega^\omega$, for every $\underset{\sim}{\Pi}^1_1$ function $g:\omega^\omega \to \omega^\omega$, there is a comeager set $A \subseteq \lambda^\omega$, there is a continuous function $h: A \to \omega^\omega$ such that

$$\forall\, x\, (\, x \in A \Rightarrow h(x) = g(f(x))).$$

Let $f:\lambda^\omega \to \omega^\omega$ be continuous and $g:\omega^\omega \to \omega^\omega$ be a $\underset{\sim}{\Pi}^1_1$ function. Then for each $s \in \omega^{<\omega}$, $g^{-1}[N_s]$ is a $\underset{\sim}{\Delta}^1_2$ set. Hence there is an open set D_s such that

$$B_s = D_s \triangle f^{-1}[g^{-1}[N_s]]$$

is meager.

Let $A = \lambda^\omega - \bigcup\{\, B_s \mid s \in \omega^{<\omega}\,\}$. Then A is a comeager set, and the function $h(x) = g(f(x))$ is continuous on A.

We now proceed to show that (3) $\Rightarrow$ (1′).

Assume that $V[G] \models \exists\, x\, \forall\, y\, \varphi(x,y)$. Where φ is the negation of a Π^1_1 formula ϕ with parameters from $V \cap \omega^\omega$. Let $\dot{x}$ be a canonical name such that

$$V[G] \models \forall\, y\, \varphi(\dot{x}/G, y).$$

We can find a function $f:\lambda^\omega \to \omega^\omega$ such that f is continuous on a G_δ comeager set and in $V[G]$ we have $f(G) = \dot{x}/G$.

We assume for a contradiction that $V \models \forall\, x\, \exists\, y\, \phi(x,y)$. By the Addison-Kondo theorem, we can find a $\underset{\sim}{\Pi}^1_1$ function $g:\omega^\omega \to \omega^\omega$ such that $\forall\, x\, \phi(x, g(x))$. By a theorem of Stone [23], every G_δ comeager subset of λ^ω is homeomorphic to λ^ω. Hence by (3), we can have a G_δ comeager set $A = \bigcap_{n<\omega} D_n$, where each D_n is open dense, such that f is continuous on A, and $g(f(x))$ is continuous on A. Let $F(x) = (f(x), g(f(x)))$. Then $F(x)$ is continuous on A.

Let T be a tree on $\omega \times \omega \times \omega$ such that

$$p[T] = \{\, (x,y) \mid \varphi(x,y)\,\}$$

and

$$\mathrm{Col}(\omega,\lambda) \Vdash p[T] = \{\, (x,y) \mid \varphi(x,y)\,\}.$$

We have such a tree since $\varphi(x,y)$ is $\underset{\sim}{\Sigma}^1_1$.

Define a tree T^* as follows: $(\sigma, \tau_1, \tau_2, \tau_3) \in T^* \iff$

(a) $(\tau_1, \tau_2, \tau_3) \in T$ & $\mathrm{lh}(\sigma) = \mathrm{lh}(\tau_1)$ and

(b) $\exists\, t \in \lambda^{<\omega}\; N_{\sigma \frown t} \subseteq \bigcap_{i \le \mathrm{lh}(\sigma)} D_i$ & $F[N_{\sigma \frown t} - \bigcup_{n<\omega}(\lambda^\omega - D_n)] \subseteq N_{(\tau_1,\tau_2)}$.

Since $V[G] \models \varphi(f(G), g(f(G)))$, and $G \in \bigcap_{n<\omega} D_n$ in $V[G]$, we have that $T_{F(G)}$ is ill-founded. Hence, in $V[G]$, G witnesses that T^* is ill-founded. By absoluteness, in V, T^* is ill-founded. But any branch in $[T^*]$ would imply that $\exists\, x \in A$ such that $T_{F(x)}$ is ill-founded, which contradicts that

$$\forall\, x\ (\ x \in A \Rightarrow \phi(f(x), g(f(x)))).$$

This contradiction shows that $V \models \exists\, x\, \forall\, y\, \varphi(x, y)$.

This proves (1′). And hence the theorem is proved. □

Remarks.

(a) We note that (1) of the theorem above is equivalent to that every total $\underset{\sim}{\Pi}^1_1$ function, $f\colon \omega^\omega \to \omega^\omega$, is forced to be total by any forcing of size at most λ.

(b) When $\lambda = \omega$, then (2) of the theorem is equivalent to that every $\underset{\sim}{\Delta}^1_2$ subset of the reals has the property of Baire, which in turn is equivalent to the claim that every $\underset{\sim}{\Pi}^1_1$ total function, $f\colon \omega^\omega \to \omega^\omega$, is continuous on a comeager set. This special case was first proved by Bagaria [1].

(c) Similar statements can be proved characterizing when every $\underset{\sim}{\Delta}^1_2$ set is Lebesgue measurable by replacing Cohen forcing with Random forcing. See [1].

Corollary 3.1. *The following are equivalent:*

(1) *Every $\underset{\sim}{\Delta}^1_2$ set is universally Baire.*

(2) *V is $\underset{\sim}{\Sigma}^1_3$ absolute with respect to every set generic extension.* □

The following theorem, which underlies the proof of Theorem 3.1, will be useful in the proofs of several theorems in this section which generalize Theorem 3.1. It can be proved by an elementary analysis of terms. This theorem shows that under certain circumstances forcing arguments can be done *pointwise*.

Theorem 3.2. *Suppose that A is universally Baire and that T, T^* are class trees which witness this with $A = p[T]$. Suppose λ is an infinite cardinal.*

(1) *For each term τ for a real in $V^{\mathrm{Col}(\omega,\lambda)}$ there corresponds a partial function $f_\tau\colon \lambda^\omega \to \omega^\omega$ which is defined and continuous on a comeager subset of λ^ω (and conversely).*

(2) *If τ is a term for a real with corresponding function f_τ then for any condition $p \in \mathrm{Col}(\omega, \lambda)$,*

$$p \Vdash \tau \in p[T]$$

if and only if

$$\{a \mid a \in O_p \text{ and } f_\tau(a) \in A\}$$

is comeager in O_p where O_p is the basic open subset of λ^ω defined by p. $\square$

While this theorem is useful for certain arguments, it should be used with some care. Difficulties can arise when the tree T does not project in $V[G]$ to the intended set from a semantical point of view. For example we shall see in Theorem 3.8 that it is possible that every $\underset{\sim}{\Delta}^1_3$ be universally Baire and that $\underset{\sim}{\Sigma}^1_4$ absoluteness fail between V and $V[G]$. Thus while one may have that every $\underset{\sim}{\Delta}^1_3$ set is universally Baire there can exist a $\underset{\sim}{\Delta}^1_3$ set A defined by $\underset{\sim}{\Sigma}^1_3$ formulas $\varphi_1(x)$, $\varphi_2(x)$ such that if the trees T, T^* witness that A is universally Baire then in some generic extension of V these trees do not project to the sets defined by these formulas (the formulas while necessarily defining disjoint sets in the extension may not define complements).

An immediate corollary to Theorem 3.1 is that if every $\underset{\sim}{\Delta}^1_2$ set is universally Baire then ω_1 is inaccessible in $L[x]$ for every real x since by Theorem 3.1 V is $\underset{\sim}{\Sigma}^1_3$ absolute relative to $V[G]$ for any set generic extension of V. Thus if every $\underset{\sim}{\Delta}^1_2$ set is universally Baire it follows that every uncountable $\underset{\sim}{\Sigma}^1_2$ set contains a perfect subset. This proves the theorem stated at the end of the previous section.

The following theorem gives the consistency strength of the statement that every $\underset{\sim}{\Delta}^1_2$ set is universally Baire.

Theorem 3.3. *The following are equiconsistent:*

(1) *ZFC + Every $\underset{\sim}{\Delta}^1_2$ set is universally Baire.*
(2) *ZFC + There exists an inaccessible cardinal κ such that,*

$$V_\kappa \prec_{\Sigma_2} V$$

Proof. Assume every $\underset{\sim}{\Delta}^1_2$ set is universally Baire. Then by Theorem 3.1, V is $\underset{\sim}{\Sigma}^1_3$ absolute with respect to $V[G]$ for any set generic extension $V[G]$. Thus it follows that κ is inaccessible in L and that

$$L_\kappa \prec_{\Sigma_2} L$$

where $\kappa = \omega_1^V$.

Conversely suppose κ is strongly inaccessible and that

$$V_\kappa \prec_{\Sigma_2} V.$$

Let G be V-generic for $\mathrm{Col}(\omega, < \kappa)$. It follows that if H is set generic over $V[G]$ then $V[G]$ is $\underset{\sim}{\Sigma}^1_3$ absolute with respect to $V[G][H]$. $\square$

Note that the consistency strength of (2) is less than that of a Mahlo cardinal.

The following theorem gives a characterization of when every $\underset{\sim}{\Sigma}^1_2$ set is universally Baire. In particular it follows that every $\underset{\sim}{\Sigma}^1_2$ set is universally Baire if and only if for any two step forcing iteration, the extensions are iteratively $\underset{\sim}{\Sigma}^1_3$ absolute.

Theorem 3.4. *The following are equivalent:*

(1) *Every $\underset{\sim}{\Sigma}^1_2$ subset of the reals is universally Baire.*
(2) *For any set forcing $\mathcal{B}$, if G is any $\mathcal{B}$-generic over V, then in $V[G]$, every $\underset{\sim}{\Delta}^1_2$ subset of the reals is universally Baire.*
(3) *For every set x, $x^\#$ exists.*

Proof. We first prove the following claim which is a generalization of (3) within the proof of Theorem 3.1. Assume every $\underset{\sim}{\Sigma}^1_2$ set is universally Baire.

Claim. Suppose that $f\colon \lambda^\omega \to \omega^\omega$ is a continuous function and that $g\colon \omega^\omega \to \omega^\omega$ is a partial function which is $\underset{\sim}{\Pi}^1_1$. Let $h\colon \lambda^\omega \to \omega^\omega$ be the partial function given by the composition of g and f. Then there exists an open set $O \subseteq \lambda^\omega$ and a comeager set $A \subseteq \lambda^\omega$ such that

$$A \cap O \;=\; A \cap dom(h) \;\; \textit{and } h \textit{ is continuous on } A \cap O.$$

Let $f\colon \lambda^\omega \to \omega^\omega$ be continuous and let $g\colon \omega^\omega \to \omega^\omega$ be a $\underset{\sim}{\Pi}^1_1$ partial function. Then for each $s \in \omega^{<\omega}$, $g^{-1}[N_s]$ is a $\underset{\sim}{\Sigma}^1_2$ set. Hence there is an open set D_s such that

$$B_s = D_s \triangle f^{-1}[g^{-1}[N_s]]$$

is meager.

Let $A = \lambda^\omega - \cup\{\, B_s \mid s \in \omega^{<\omega} \,\}$ and let $O = \cup\{\, D_s \mid s \in \omega^{<\omega} \,\}$. Then A is comeager in λ^ω, O is open, $A \cap O \;=\; A \cap dom(h)$ and h is continuous on $A \cap O$. This proves the claim.

The claim has the following corollary.

(i) Assume every $\underset{\sim}{\Sigma}^1_2$ set is universally Baire. Suppose T is a tree such that $p[T] = A$ where A is a $\underset{\sim}{\Sigma}^1_2$ set given by a $\underset{\sim}{\Sigma}^1_2$ formula $\varphi(x)$. Then in any set generic extension of V, $p[T] \subseteq A$ where A is defined in $V[G]$ using the same formula $\varphi(x)$.

To see this let $\varphi(x) = \exists\, y\ \phi(x,y)$ where $\phi(x,y)$ is a $\underset{\sim}{\Pi}^1_1$ formula. Let B be the subset of ω^ω defined by $\phi(x,y)$ and let $g\colon \omega^\omega \to \omega^\omega$ be a $\underset{\sim}{\Pi}^1_1$ partial function that uniformizes B. Fix a cardinal λ and suppose τ is a term for a real such that

$$\mathrm{Col}(\omega,\lambda) \Vdash\ \tau \in p[T]$$

The term τ defines in a canonical fashion a partial function $f\colon \lambda^\omega \to \omega^\omega$ which is defined and continuous on a comeager set, with range in $p[T]$. Every dense G_δ subset of λ^ω is homeomorphic to λ^ω. Therefore we can apply the claim to f and g to get that the function h given by the composition of g with f is continuous on a comeager set. The function h being continuous on a comeager set defines a term σ such that

$$\mathrm{Col}(\omega,\lambda) \Vdash\ \phi[\tau,\sigma]$$

and so

$$\mathrm{Col}(\omega,\lambda) \Vdash\ \tau \in A$$

(ii) Suppose that A is a $\underset{\sim}{\Sigma}^1_2$ set and that T, T^* are class trees which witness that A is universally Baire (with $A = p[T]$). Let $V[G]$ be a set generic extension of V. Then in $V[G]$,

$$A \subseteq p[T],$$

where A is defined in $V[G]$ using the same formula as used in V.

If this were to fail in $V[G]$ then in $V[G]$; $p[S]$, $p[T^*]$ have nonempty intersection where S is the (class) Shoenfield tree for A. By absoluteness the intersection is nonempty when computed in V, a contradiction.

Combining (i) and (ii) we have now proved the following. Assume every $\underset{\sim}{\Sigma}^1_2$ set is universally Baire. Suppose that A is a $\underset{\sim}{\Sigma}^1_2$ set defined by a $\underset{\sim}{\Sigma}^1_2$ formula $\varphi(x)$. Suppose that T, T^* are class trees witnessing that A is universally Baire with $A = p[T]$. Then for any set generic extension of V, $A = p[T]$ in $V[G]$ where A is defined in $V[G]$ using the formula $\varphi(x)$. Thus there is a class tree in V whose projection in any set generic extension of V is the universal $\underset{\sim}{\Sigma}^1_3$ set. Therefore for any forcing iteration $\mathcal{B}_1 * \mathcal{B}_2$, forcing with $\mathcal{B}_1$ over V cannot change the $\underset{\sim}{\Sigma}^1_3$ theory of the reals in V, and forcing with $\mathcal{B}_2$ over $V^{\mathcal{B}_1}$ cannot change the $\underset{\sim}{\Sigma}^1_3$ theory of the reals in $V^{\mathcal{B}_1}$. So (2) follows.

Now (2) $\Rightarrow$ (3) follows from the results in [24] (cf. Lemma 1 [24]). By the results of Martin-Solovay [11] and Corollary 2.1 it follows that (3) $\Rightarrow$ (1). $\square$

Corollary 3.2. *The following are equivalent:*

1. *Every $\underset{\sim}{\Sigma}^1_2$ set is universally Baire.*
2. *In every set generic extension of V, every $\underset{\sim}{\Sigma}^1_2$ set is universally Baire.*
3. *In every set generic extension of V, every $\underset{\sim}{\Sigma}^1_2$ set has the property of Baire.*
4. *In every set generic extension of V, every $\underset{\sim}{\Delta}^1_2$ set is Lebesgue measurable.*
5. *In every set generic extension of V, every $\underset{\sim}{\Sigma}^1_2$ set is Lebesgue measurable.*
6. *In every set generic extension of V, every uncountable $\underset{\sim}{\Pi}^1_1$ ($\underset{\sim}{\Sigma}^1_2$) set has a perfect subset.*

Proof. That (1) is equivalent to (2) is immediate from Theorem 3.4. To finish we prove the following claim.

Claim. Suppose either that in every set generic extension of V, every $\underset{\sim}{\Sigma}^1_2$ set has the property of Baire, or that in every set generic extension of V, every $\underset{\sim}{\Delta}^1_2$ set is Lebesgue measurable. Then (6) holds.

This claim gives the corollary, since it is an immediate consequence of Jensen's covering lemma that (6) is equivalent to that every set has a sharp.

Suppose that (6) does not hold. Then there is a set generic extension of V, $V[G]$ in which there is a real x such that

$$\omega_1^{L[x]} = \omega_1^{V[G]}.$$

By Lemma 4 of [24] if c is a Cohen real over $V[G]$, then in $V[G][c]$ there is no random real over $L[x][c]$ and so by results of Judah and Shelah [5] it follows that there exists in $V[G][c]$ a $\underset{\sim}{\Delta}^1_2$ set which is not Lebesgue measurable. This proves the claim in the case of Lebesgue measurability.

Notice that in this case the only generic extensions one need consider are those of generic collapses.

For the case of the Baire property it suffices to show that if there exists a real x such that $\omega_1^{L(x)} = \omega_1$ then there is a generic extension of V in which there exists a real y such that the set of Cohen reals over $L(y)$ is not comeager. We may assume that CH holds (otherwise force it). Fix a sequence $\langle a_\alpha : \alpha < \omega_1\rangle$ of almost disjoint subsets of ω such that the sequence is in $L(x)$. Fix an enumeration $\langle \sigma_\alpha : \alpha < \omega_1\rangle$ of all the subsets of ω. Suppose $C \subseteq \omega_1$ is closed and unbounded, and contains only limit ordinals. Let $\pi_C : \omega_1 \to C$ be the canonical isomorphism. Define a subset $A_C \subseteq \omega_1$ as follows.

$$A_C = \{\pi_C(\alpha) + k \mid k \in \sigma_\alpha,\ \alpha < \omega_1\}$$

The point is the following. Suppose C is a reasonably fast club. (For example suppose that for each $\alpha \in C$, there exists an elementary substructure, $X \prec V_{\omega_1+1}$, containing x,α, the two sequences, the function π_C and such that $\pi_C(\alpha+1) > X \cap \omega_1$.) Suppose that g is generic for almost disjoint coding $\omega_1 - A_C$ relative to $\langle a_\alpha : \alpha < \omega_1 \rangle$. Then in $V[g]$ the set of Cohen reals over V is not comeager. Hence in $V[g]$ the set of Cohen reals over $L[x,g]$ is not comeager.

This gives the claim. The corollary follows. $\square$

Corollary 3.2 suggests that the following might be true. Suppose that in every set generic extension of V, every projective set is Lebesgue measurable. Then every projective set is universally Baire. Actually the proof of the corollary suggests that one need only assume that for every cardinal λ,

$V^{\mathrm{Col}(\omega,\lambda)} \models$ Every projective set is Lebesgue measurable.

However this weaker condition cannot be sufficient.

Theorem. *Assume that there are $\omega+1$ Woodin cardinals. Then there is a transitive model of ZFC satisfying;*

(1) *For every poset, $\mathbb{P}$, which is σ-centered or forces the collapse of ω_1,*

$$V^{\mathbb{P}} \models \textit{Every projective set is Lebesgue measurable, has the property of Baire etc.}$$

and

(2) *Every universally Baire set is $\underset{\sim}{\Delta}^1_3$.*

The proof uses an argument of [26] for obtaining the consistency of,

$$\text{ZFC+There exists a Woodin cardinal + Every weakly homogeneously Souslin set is } \underset{\sim}{\Sigma}^1_2,$$

together with the following. Suppose $A \subseteq \omega^\omega$ and $\mathbb{P}$ is a poset that is ω-closed. Then A is universally Baire in $V^{\mathbb{P}}$ iff A is universally Baire in V.

Remark. The following is true probably:

> Suppose that for every cardinal λ, $V^{\mathrm{Col}(\omega,\lambda)} \models$ Every set in $L(\mathbb{R})$ is Lebesgue measurable. Then in every set generic extension of V, $AD^{L(\mathbb{R})}$ holds.

We shall prove in section 4 that if in every set generic extension of V, $AD^{L(\mathbb{R})}$ holds, then every set in $L(\mathbb{R})$ is universally Baire.

Theorem 3.5. *Assume that V is $\utilde{\Sigma}^1_4$ absolute with respect to every set generic extension. Then every $\utilde{\Delta}^1_3$ set is universally Baire.*

Proof. Notice that there is a Σ^1_4 formula $\varphi(x)$ which (provably) expresses:

$$\omega_1 \textit{ is a successor cardinal in } L[x].$$

It now follows by covering that for every real x, $x^\#$ exists. However this is expressible by a Π^1_4 sentence and so by one more application of the absoluteness of V with respect to every set generic extension, it follows that for every set x, $x^\#$ exists.

Suppose $\phi_1(x)$ and $\phi_2(x)$ are $\utilde{\Sigma}^1_3$ formulas which define a $\utilde{\Delta}^1_3$ set. By the Σ^1_4 absoluteness of V with respect to every set generic extension, it follows that $\phi_1(x)$ and $\phi_2(x)$ define a $\utilde{\Delta}^1_3$ set in every set generic extension. Since every set has a sharp it follows from the results of Martin-Solovay [11] that there is a class tree definable in V whose projection in every set generic extension of V is the universal $\utilde{\Sigma}^1_3$ set. From this tree one can easily define class trees which witness that the $\utilde{\Delta}^1_3$ set defined by $\phi_1(x)$ and $\phi_2(x)$ is universally Baire. □

In analogy with Theorem 3.1 one would expect the converse of this theorem to be true. While we cannot prove this (it is false) we can prove an approximation to the converse.

Theorem 3.6. *Suppose that every $\utilde{\Delta}^1_3$ set is universally Baire and that V is not $\utilde{\Sigma}^1_4$ absolute with respect to some set generic extension. Then for every real x, $x^\dagger$ exists.*

Proof. We first prove the following claim. Assume every $\utilde{\Delta}^1_3$ set is universally Baire.

Claim. Suppose that $f\colon \lambda^\omega \to \omega^\omega$ is a continuous function and that $g\colon \omega^\omega \to \omega^\omega$ is a function which is $\utilde{\Sigma}^1_3$. Let $h\colon \lambda^\omega \to \omega^\omega$ be the function given by the composition of g and f. Then h is continuous on a comeager set.

Let f and g be given. Then for each $s \in \omega^{<\omega}$, $g^{-1}[N_s]$ is a $\utilde{\Delta}^1_3$ set. Hence there is an open set D_s such that

$$B_s = D_s \triangle f^{-1}[g^{-1}[N_s]]$$

is meager.

Let $A = \lambda^\omega - \cup\{\, B_s \mid s \in \omega^{<\omega}\,\}$. Then A is comeager in λ^ω and h is continuous on A. This proves the claim.

Assume that for some real x_0, $x_0^\dagger$ does not exist. Since every $\utilde{\Delta}^1_3$ set is universally Baire clearly every $\utilde{\Sigma}^1_2$ set is universally Baire and so by

Theorem 3.3 every set has a sharp. Let z be any real with $x_0 \in L[z]$ and let K_z be the Jensen-Dodd core model constructed relative to the real z. Since $x_0^\dagger$ does not exist it follows that $z^\dagger$ does not exist and so by Jensen's absoluteness theorem, K_z is $\underset{\sim}{\Sigma}^1_3$ absolute with respect to V. Therefore because of the uniformity of the definition of K_z with respect to z, it follows that every $\underset{\sim}{\Sigma}^1_3$ subset of $\omega^\omega \times \omega^\omega$ can be uniformized by a $\underset{\sim}{\Sigma}^1_3$ function.

Suppose that $\varphi(x,y)$ is a $\underset{\sim}{\Pi}^1_2$ formula and that $\forall\, x\, \exists\, y\, \varphi(x,y)$ is true in V. Fix a cardinal λ. It suffices to prove

$$\mathrm{Col}(\omega,\lambda) \Vdash \forall\, x\, \exists\, y\, \varphi(x,y).$$

Fix a $\underset{\sim}{\Sigma}^1_3$ function $g\colon \omega^\omega \to \omega^\omega$ such that for all $a \in \omega^\omega$, $\varphi[a, g(a)]$. Suppose τ is a term for a real. The term τ defines in a canonical fashion a partial function $f\colon \lambda^\omega \to \omega^\omega$ which is defined and continuous on a comeager set. Again, every dense G_δ subset of λ^ω is homeomorphic to λ^ω and therefore we can apply the claim to get that the composition of g with f is continuous on a comeager set. Therefore the composition defines a term σ such that

$$\mathrm{Col}(\omega,\lambda) \Vdash \varphi[\tau,\sigma].$$

This proves

$$\mathrm{Col}(\omega,\lambda) \Vdash \forall\, x\, \exists\, y\, \varphi(x,y)$$

using Theorem 3.2 and the observation that since every set has a sharp there are class trees T, T^* which witness that the set defined by $\varphi(x,y)$ is universally Baire and such that the tree T projects to the set defined by $\varphi(x,y)$ in every set generic extension of V. $\Box$

We can use Theorem 3.6 to compute the consistency strength of the assertion that every $\underset{\sim}{\Delta}^1_3$ set is universally Baire.

Theorem 3.7. *The following are equiconsistent:*

1. *ZFC + Every $\underset{\sim}{\Delta}^1_3$ set is universally Baire.*
2. *ZFC + For every set x, $x^\#$ exists, and there exists an inaccessible cardinal κ such that,*

$$V_\kappa \prec_{\Sigma_3} V.$$

Proof. Assume every $\underset{\sim}{\Delta}^1_3$ set is universally Baire. By Theorem 3.6, either V is $\underset{\sim}{\Sigma}^1_4$ absolute with respect to $V[G]$ for any set generic extension $V[G]$, or $0^\dagger$ exists. In either case it follows that κ is inaccessible in $L^\#$ and that

$$L^\#_\kappa \prec_{\Sigma_3} L^\#$$

where $\kappa = \omega_1^V$ and $L^\#$ is the smallest transitive inner model of ZFC closed under the sharp operation and containing the ordinals.

Conversely suppose every set has a sharp, κ is strongly inaccessible, and that

$$V_\kappa \prec_{\Sigma_3} V.$$

Let G be V-generic for $\mathrm{Col}(\omega, <\kappa)$. It follows that if H is set generic over $V[G]$ then $V[G]$ is $\underset{\sim}{\Sigma}^1_4$ absolute with respect to $V[G][H]$. □

A version of the converse of Theorem 3.5 follows assuming the appropriate generalization of Jensen's absoluteness theorem: Assume that x is a real and that there is no transitive inner model of ZFC + *There is a Woodin cardinal* containing x and the ordinals. Assume that for every real z, $z^\#$ exists. Let K_x be the core model for 1 Woodin cardinal in the sense of Steel [22] constructed relative to x. Then K_x is $\underset{\sim}{\Sigma}^1_3$ absolute with respect to V.

Given this one can prove the following are equivalent:

(1) In every set generic extension of V every $\underset{\sim}{\Delta}^1_3$ set is universally Baire.
(2) If $V[G_P] \subseteq V[G_P][G_Q]$ are set generic extensions of V then $V[G_P]$ is $\underset{\sim}{\Sigma}^1_4$ absolute with respect to $V[G_P][G_Q]$.

The proof splits into two cases depending on whether or not $\underset{\sim}{\Delta}^1_2$ sets are determined, see [26] for more details on this kind of argument.

This is essentially the best one can hope for.

Theorem 3.8. *Assume that every $\underset{\sim}{\Delta}^1_2$ set is determined and that every $\underset{\sim}{\Sigma}^1_4$ sequence of distinct reals is countable. Then there is a transitive model of ZFC satisfying:*

(1) *Every $\underset{\sim}{\Delta}^1_3$ set is universally Baire*
(2) *V is not $\underset{\sim}{\Sigma}^1_4$ absolute with respect to $V^{\mathrm{Col}(\omega,\omega_1)}$.*

Proof. (sketch) Let C_4 be the largest countable Σ^1_4 set. Let M be the smallest transitive set such that;

(1) $C_4 \subseteq M$, $\omega_1 \subseteq M$ and
(2) For each $a \in M$, $Q_3(a) \subseteq M$.

Here the operation $Q_3(a)$ is generalized to countable transitive sets in the natural fashion see [26].

M can also be defined as following

$$M = \bigcap \{N \mid C_4 \subseteq N, \omega_1 \subseteq N, and\ N \models \text{ ZFC}^{\text{-powerset}} + \text{Det}\, \underset{\sim}{\Delta}^1_2\}$$

Thus $M \models$ ZFC, M has height ω_1 and $C_4 = \omega^\omega \cap M$. Therefore M is $\underset{\sim}{\Sigma}^1_4$ absolute with respect to V. Suppose $\varphi_1(x)$ and $\varphi_2(x)$ are $\underset{\sim}{\Sigma}^1_3$ formulas with parameters from M, that define in M a $\underset{\sim}{\Delta}^1_3$ set. Hence the formulas define a $\underset{\sim}{\Delta}^1_3$ set in V. Suppose G is M-generic for a poset in M (with $G \in V$). Therefore $M[G]$ is closed under the Q_3 operation and further the Q_3 operation is definable in $M[G]$. Let U be a (reasonable) Π^1_3 set which is complete ($U \subseteq \omega^\omega \times \omega^\omega$). Therefore by the Generalized Spector-Gandy Theorem, $U \cap M[G] \in M[G]$ and further $U \cap M[G]$ is definable in $M[G]$. Thus since Π^1_3 has the scale property it follows that there is a tree $T_G \in M[G]$, definable in $M[G]$, such that $U \cap M[G] \;=\; p[T_G] \cap M[G]$. Finally it follows that there are definable (within $M[G]$) trees $T_G^{\varphi_1}$, $T_G^{\varphi_2}$ in $M[G]$ such that

$$p[T_G^{\varphi_1}] \cap M[G] \;=\; A_1 \cap M[G]$$

and

$$p[T_G^{\varphi_2}] \cap M[G] \;=\; A_2 \cap M[G]$$

where A_1 is the set defined in V by φ_1 and A_2 is the set defined in V by φ_2. These trees are uniformly definable (independent of the poset for which G is generic) hence the sets defined in M by φ_1 and φ_2 are universally Baire in M.

This proves that M satisfies every $\underset{\sim}{\Delta}^1_3$ set is universally Baire. Suppose G is M-generic for $\mathrm{Col}(\omega, \kappa)$ where $\kappa = \omega_1^M$. Then Det $\underset{\sim}{\Pi}^1_2$ fails in $M[G]$ since there exists a real $a \in M[G]$ (any real coding C_4, G) such that $Q_3(a) = \omega^\omega \cap M[G]$. Therefore M is not $\underset{\sim}{\Sigma}^1_4$ absolute with respect to $M[G]$. This completes the proof. $\square$

Our next theorem gives a characterization for every set of reals which is in $L(\mathbb{R})$ to be universally Baire.

Theorem 3.9. *Every set $A \subseteq \omega^\omega$ which is in $L(\mathbb{R})$ is universally Baire if and only if $\mathbb{R}^\#$ exists and $\mathbb{R}^\#$ is universally Baire.*

Proof. ($\Rightarrow$) Since each $\underset{\sim}{\Sigma}^1_2$ set is universally Baire, by the previous theorem, $\mathbb{R}^\#$ exists. Since $\mathbb{R}^\# = \bigcup_{n<\omega} A_n$, where each A_n is in $L(\mathbb{R})$, we have $\mathbb{R}^\#$ is universally Baire, being a countable union of universally Baire sets.

($\Leftarrow$) Since each $A \in L(\mathbb{R})$ is continuously reducible to $\mathbb{R}^\#$, i.e., there is a continuous function $f\colon \omega^\omega \to \omega^\omega$ such that for each $x \in \omega^\omega$, $x \in A$ iff $f(x) \in \mathbb{R}^\#$. The universally Baire sets are closed under continuous preimages hence A is universally Baire.

This completes the proof. $\square$

4. Universally Baire Sets and Large Cardinals

In last section we have shown that every $\underset{\sim}{\Sigma}^1_2$ set is universally Baire if and only if for every set x, $x^\#$ exists. In particular, if there is a strong cardinal, then every $\utilde{\Sigma}^1_2$ set is universally Baire. In this section, we will show that if there is a supercompact cardinal, then every set in $L(\mathbb{R})$ is universally Baire. Also we show that this conclusion is very strong.

Theorem 4.1. *If there is a supercompact cardinal, then every subset of the reals which is in $L(\mathbb{R})$ is universally Baire.*

Proof. By Theorem 3.9, we need only to show that $\mathbb{R}^\#$ exits and $\mathbb{R}^\#$ is universally Baire. Notice that if κ is a strong cardinal, then for each $A \subseteq \omega^\omega$, A is universally Baire if and only if for each $\lambda < \kappa$ A is λ-universally Baire.

Let κ be a supercompact cardinal. By a theorem of Woodin [25] there are trees T and T^* on $\omega \times \kappa$ such that for any partial order $\mathbb{P} \in V_\kappa$, if $G \subseteq \mathbb{P}$ is a V-generic, then

$$V[G] \models p[T] = \mathbb{R}^\# \ \wedge \ p[T] \cup p[T^*] = \omega^\omega.$$

Since κ must be strong, we are done. □

Remark. It follows from results of Woodin that the conclusion can be proved from a much weaker large cardinal hypothesis or from strong compactness.

Let WO be a canonical set of reals which codes countable ordinals. More specifically, let $\langle t_i \mid i < \omega \rangle = \omega^{<\omega}$ be a recursive 1-1 enumeration satisfying $t_0 = \emptyset$, $t_j \subseteq t_i \neq t_j \Rightarrow j < i$, and so $\mathrm{lh}(t_i) \leq i$.

We say that $s \in \omega^{<\omega} - \{\emptyset\}$ codes $t \in \omega^{<\omega} - \{\emptyset\}$ if

$$\forall j \, (\emptyset \neq t_j \subseteq t \Rightarrow s(j) = 0).$$

For each $s \in \omega^{<\omega} - \{\emptyset\}$, define an order $<_s$ on $\mathrm{lh}(s)$ by letting $i <_s j$ if and only if $j \neq i \neq 0$ and exactly one of the following conditions is satisfied:

(1) $j = 0$, or
(2) s does not code any of the t_i and t_j, and $i < j$, or
(3) s codes t_j, but s does not code t_i, or
(4) s codes both t_i and t_j, and $\exists\, n < \mathrm{lh}(t_i)\ t_j = t_i \restriction n$, or
(5) s codes both t_i and t_j, and $t_i(m) < t_j(m)$, where m is the least $t_i(m) \neq t_j(m)$.

Notice that if $s \subseteq t$ then $<_s \subseteq <_t$. For each $x \in \omega^\omega$, let

$$<_x = \bigcup_{n<\omega} <_{x\restriction n}.$$

This defines a linear ordering of ω for each $x \in \omega^\omega$. We then define that $x \in WO$ if and only if $<_x$ is a well ordering. For $x \in WO$, let $\|x\|$ be the rank of 0 in $<_x$, i.e., $\|x\| + 1$ is the length of $<_x$.

Let WO^ω be defined as $y \in WO^\omega \iff \forall\, n < \omega\ (y)_n \in WO$ for each real y. We then define $\pi : WO^\omega \to [\omega_1]^{\leq\omega}$ by

$$\pi(y) = \{\|(y)_n\| \mid n < \omega\}.$$

Given $A \subseteq \omega_1$, we define a set A^* of reals to code A as follows.

$$(x, y) \in A^* \iff x \in WO \ \wedge\ y \in WO^\omega\ \wedge \pi(y) = A \cap \|x\|.$$

The following theorem is a reformulation of the theorem of Kechris in [7] that if $\aleph_1$ is measurable then for every subset $A \subseteq \omega_1$, A is constructible from a real if and only if A^* is Souslin (and hence co-Souslin).

Theorem. *Assume that there is a measurable cardinal. Let A be a subset of ω_1. Then A^* is universally Baire if and only if A is constructible from a real.*

Proof. One direction is easy. If $A \subseteq \omega_1$ is constructible from a real then A^* is $\underset{\sim}{\Pi}^1_1$.

Main Fact. If $A \subseteq \omega_1$, A^* is universally Baire, then A is constructible from a real.

Let κ be measurable.

We define the tree S on $\omega \times \kappa$ so that $WO = p[S]$ as follows.

For $(s, u) \in \omega^{<\omega} \times \kappa^{<\omega}$, define that $(s, u) \in S$ if and only if $u : \mathrm{lh}(s) \to \kappa$ is a $<_s$-order preserving function. Then $WO = p[S]$ and for $(s, u) \in S$, $0 < i < \mathrm{lh}(s)$, we have $u(i) < u(0)$.

Also we fix a tree F on $\omega \times \omega \times \kappa$ such that

$$p[F] = \{(x, y) \mid x \in WO\ \wedge y \in WO\ \wedge \|x\| \leq \|y\|\}.$$

To prove the main fact, we are going to play the following Solovay game.

Let $A \in P(\omega_1) \cap L(\mathbb{R})$. A game $\mathcal{G}_A$ is defined as follows. Player I and Player $\mathbb{I}$ play natural numbers in turn producing two reals x and y respectively. $\mathbb{I}$ wins if and only if x does not code an ordinal or else $y \in WO^\omega$ and there is some ordinal $\alpha > \|x\|$ such that $\pi(y) = A \cap \alpha$.

Fact 1. $\mathbb{I}$ has a winning strategy in the game $\mathcal{G}_A$.

Being universally Baire, we can have two trees on $\omega \times \omega \times \lambda$ for some $\lambda \geq \kappa$ such that

$$A^* = p[T],\ \omega^\omega \times \omega^\omega - A^* = p[T^*]$$

and in any generic extension of the universe

$$\omega^\omega \times \omega^\omega = p[T] \cup p[T^*].$$

Fix such two trees T, T^* as above until the end of the section.

We play a game $\mathcal{G}_T$ as follows.

I	$\mathbb{I}$
x, f	y, z, g, h

$\mathbb{I}$ wins if and only if $(x, f) \in [S]$ implies that $(x, y, g) \in [F]$ and $(y, z, h) \in [T]$.

Since the game $\mathcal{G}_T$ is played along trees, it is a closed game. Hence one of the player must have a winning strategy. We want to show that $\mathbb{I}$ has a winning strategy.

We will play the same game in a generic extension, where we can show that $\mathbb{I}$ has a winning strategy. Then using an absoluteness argument we conclude that $\mathbb{I}$ has a winning strategy.

To proceed, let σ be a strategy of I in the game $\mathcal{G}_T$. Call a sequence s a correct partial play according to σ if it is a partial play of the game $\mathcal{G}_T$ such that neither player has lost the play so far and I has played according to σ.

Define a tree P_σ to be the set of all such correct partial plays according to σ. Then σ is a winning strategy of I in the game $\mathcal{G}_T$ if and only if P_σ is well founded.

Let $\mathrm{Col}(\omega, < \kappa)$ be the partial order for the Levy collapse of everything $< \kappa$ to ω so that κ becomes ω_1. Let $G \subseteq \mathrm{Col}(\omega, < \kappa)$ be a generic over V.

Now working in $V[G]$, consider the game $\mathcal{G}_T$ played in the extension, call it $\mathcal{G}_T^*$.

Fact 2. If I has a winning strategy σ in $\mathcal{G}_T$ in V, then σ is a winning strategy for I in the game $\mathcal{G}_T^*$ in $V[G]$.

This is because all the partial plays are the same both in V and in $V[G]$. So a strategy of I in the ground model remains to be a strategy of I in the extension. Therefore, in $V[G]$, we still have that σ is a winning strategy for I in the game $\mathcal{G}_T^*$ if and only if P_σ is well founded. Hence Fact 2 follows from absoluteness.

Fact 3. $\mathbb{I}$ has a winning strategy in $\mathcal{G}_T^*$ in $V[G]$. Hence $\mathbb{I}$ has a winning strategy in the game $\mathcal{G}_T$ in V.

Proof of Fact 3.

Claim. There is $B \subseteq \kappa = \omega_1$ such that $B^* = p[T]$.

We show first the following hold.

(1) $(x, y) \in p[T] \Rightarrow x \in WO \wedge y \in WO^\omega$,
(2) $(x, y) \in p[T],\ \pi(z) = \pi(y) \Rightarrow (x, z) \in p[T]$,
(3) $(x, y) \in p[T],\ (x', y') \in p[T],\ \|x\| \le \|x'\| \Rightarrow \pi(y) = \|x\| \cap \pi(y')$,
(4) $(u, z) \in p[T],\ x \in WO,\ y \in WO^\omega,\ \pi(y) = \|x\| \cap \pi(z) \Rightarrow (x, y) \in p[T]$.

Given (1)–(4), let

$$B = \bigcup \{\pi(y) \mid \exists\, x\ (x, y) \in p[T]\}.$$

Then $B \subseteq \kappa = \omega_1$ and $B^* = p[T]$.

To see (1), notice that $\{(x, y) \mid (x, y) \notin WO \times WO^\omega\}$ is a $\underset{\sim}{\Sigma}^1_1$ set. So there is a tree Q on κ representing this set. We then merge the two trees T and Q to get a tree $T * Q$ as follows.

$$(s, t, u, v) \in T * Q \iff (s, t, u) \in T\ \wedge\ (s, t, v) \in Q.$$

Since $T * Q$ is well founded in V, it is well founded in $V[G]$. So (1) holds.

For (2), take a tree Q on κ in V such that

$$p[Q] = \{(x, y) \mid \pi(x) = \pi(y)\}$$

both in V and in $V[G]$.

Now merge T, T^*, and Q as follows.

$$(s, t, u, v, w, r) \in T{*}T^*{*}Q \iff (s, t, v) \in T \wedge (s, u, w) \in T^* \wedge (t, u, r) \in Q.$$

Since $T * T^* * Q$ is well founded in V, it must be well founded in $V[G]$. If for some $(x, y) \in p[T]$, $\pi(z) = \pi(y)$, $(x, z) \notin p[T]$, then $(x, z) \in p[T^*]$. But then $T * T^* * Q$ is ill-founded in $V[G]$.

For (3), take a tree Q on κ in V such that

$$p[Q] = \{(x, y, z) \mid x \in WO,\ y,\ z \in WO^\omega,\ \pi(y) \ne \|x\| \cap \pi(z)\}$$

both in V and in $V[G]$.

Define a tree $T * F * Q$ by

$$(s, t, u, v, w_0, w_1, w_2, w_3) \in T * F * Q \iff$$
$$(s, t, w_0) \in T \ \&\ (u, v, w_1) \in T \ \&\ (s, u, w_2) \in F \ \&\ (s, t, v, w_3) \in Q.$$

Again, $T * F * Q$ is well founded in V. So any counterexample to (3) in $V[G]$ would give the ill-foundedness of $T * F * Q$.

For (4), let Q be a tree on κ in V such that

$$p[Q] = \{(x, y, z) \mid x \in WO,\ y, z \in WO^\omega,\ \pi(y) = \|x\| \cap \pi(z)\}$$

both in V and in $V[G]$.

Then merge the trees T, T^* and Q to get a tree $T * T^* * Q$ as follows.

$$(s, t, u, v, w_0, w_1, w_2) \in T * T^* * Q \iff$$
$$(s, v, w_0) \in T \ \&\ (t, u, w_1) \in T^* \ \&\ (u, t, v, w_2) \in Q.$$

From the definition of A^*, we see that in V the tree $T * T^* * Q$ is well founded. Hence in $V[G]$ it is well founded. So there is no counterexample to (4) in $V[G]$.

This establishes (1)–(4), hence the claim.

Now we proceed to prove that $\mathbb{I}$ has a winning strategy for the game $\mathcal{G}_T^*$ in $V[G]$.

Actually the winning strategy is very simple. First notice that we still have $WO = p[S]$ in the extension and if $(x, f) \in [S]$, then $\|x\| \leq f(0)$. So after $x(0), f(0)$ are played, I lost the game. Namely, let $\alpha < \kappa$ be such that $\alpha > f(0)$. Let $y \in WO$ be such that $\|y\| = \omega + \alpha$, and pick $z \in WO^\omega$ such that $\pi(z) = B \cap \|y\|$. Then we have $(y, z) \in p[T]$. $\mathbb{I}$ simply plays them and the needed witnesses to against the play by I. This certainly wins.

This finishes the proof of Fact 3. $\square$

Proof of Fact 1. First, let us consider the following auxiliary game $\mathcal{G}_A^*$.

$$\begin{array}{cc} \text{I} & \mathbb{I} \\ x, f & y \end{array}$$

where $f(i) \leq f(0) < \kappa$. $\mathbb{I}$ wins if and only if $(x, f) \in [S] \Rightarrow \exists\ \alpha > \|x\| \ \&\ \pi(y) = A \cap \alpha$.

Since $\mathbb{I}$ has a winning strategy in the game $\mathcal{G}_T$, $\mathbb{I}$ has a winning strategy for $\mathcal{G}_A^*$ by consulting the winning strategy for the game $\mathcal{G}_T$ and hiding the witnesses.

Now we can translate a winning strategy for $\mathbb{I}$ in the game $\mathcal{G}_A^*$ to a winning strategy in the game $\mathcal{G}_A$ via the measures associated with the tree S in a standard way.

Specifically, let U be a normal ultrafilter on κ. Inductively define ultrafilters U_n on $[\kappa]^n$ for $n \geq 1$ as follows.

For $X \subseteq [\kappa]^{n+1}$, let $X \in U_{n+1}$ if and only if

$$\{\alpha < \kappa \mid \{t \in [\kappa]^n \mid \alpha < min(t) \;\&\; \{\alpha\} \cup t \in X\} \in U_n\} \in U.$$

For $i < \omega$, let $U_{\langle i \rangle} = U_1$. For $s \in \omega^{<\omega}$, $\mathrm{lh}(s) \geq 2$, let $\pi_s : \mathrm{lh}(s) \to ls(s)$ be the permutation such that

$$i <_s j \iff \pi_s(i) < \pi_s(j).$$

Let $n = \mathrm{lh}(s)$. For $t \in [\kappa]^n$, $t = \{t(0), t(1), \cdots, t(n-1)\}_<$, let

$$s^*(t) = \langle t(\pi_s(0)), t(\pi_s(1)), \cdots, t(\pi_s(n-1))\rangle.$$

Then define U_s on κ^n by letting for $X \subseteq \kappa^n$, $X \in U_s$ if and only if

$$\{\, t \in [\kappa]^n \mid s^*(t) \in X\} \in U_n.$$

Then for each $s \in \omega^{<\omega} - \{\emptyset\}$, U_s is a κ-complete ultrafilter on $\kappa^{\mathrm{lh}(s)}$ and there is $X \in U_s$ such that for each $t \in X$ we have $(s,t) \in S$.

If τ^* is a winning strategy for $\mathbb{I}$ in the game $\mathcal{G}^*_A$, then define a strategy τ for $\mathbb{I}$ in the game $\mathcal{G}_A$ as follows: letting $\tau(x \restriction n+1) = y(n)$ if and only if there is an $X \in U_{x\restriction n+1}$ such that for all $t \in X$ $\tau^*(x\restriction n+1, t) = y(n)$. Then τ is a well defined strategy and is a winning strategy for $\mathbb{I}$. $\square$

Corollary 4.1. *If there is a measurable cardinal, and every subset of the reals which is in $L(\mathbb{R})$ is universally Baire, then $\aleph_1$ is measurable in $L(\mathbb{R})$.*

Corollary 4.2. *If there is a measurable cardinal and every projective set is universally Baire, then every subset of ω_1 which is projective in codes is constructible from a real.*

We end this section with the following theorem which shows that some additional hypothesis is necessary for the conclusion of Theorem 4.2 to hold. Recall that a cardinal κ is an Erdős cardinal if $\kappa \to (\omega)^{<\omega}_2$.

Theorem 4.3. *Assume $V = K$, for every set a, $a^\#$ exists, $0^\dagger$ does not exist and there are no Erdős cardinals. Then there is a subset $A \subseteq \omega_1$ such that A^* is universally Baire but A is not constructible from a real.*

Proof. Let F be the function given by $F(a) = a^\#$ where a is an arbitrary set. Define the set A as follows. $\alpha \in A$ if there exists a transitive model M closed under F such that $M \models$ 'ZFC$^{\text{-replacement}}$ $+ V = K$', $\alpha < \kappa^M$ and α is an infinite successor cardinal of M where κ^M is the least Erdős cardinal of M or the height of M if none exist in M. The key is the following claim.

Claim. Suppose M is a transitive set,

$$M \models \text{`ZFC}^{\text{-replacement}} + V = K\text{'}$$

and M is closed under F. Suppose $\alpha \in M$ and $\alpha < \kappa^M$. Then either M is a witness for $\alpha \in A$ or there is a witness for $\alpha \in A$ which is an element of M or $\alpha \notin A$.

To prove the claim suppose N is a witness for $\alpha \in A$, $\alpha \in M$ and that M is not a witness for $\alpha \in A$. Since $0^\dagger$ does not exist and since the transitive sets M, N are closed under F it follows that for each $\beta < \alpha$, $P(\beta)^M \subset P(\beta)^N$ or $P(\beta)^N \subset P(\beta)^M$. There are two cases depending on whether or not α is a cardinal of M. First suppose α is a cardinal of M. Then for some $\beta < \alpha$, $P(\beta)^N \not\subset M$. However α is an uncountable cardinal of N and so there exists $a \in N$ such that $a \subset \beta$ for some $\beta < \alpha$, $a \notin M$ and $L[a] \models$ '$V = K$'. Therefore $M \subset L[a]$. However $a^\# \in N$ hence α is an indiscernible for M and so α is an Erdős cardinal in M, a contradiction. Now suppose α is not a cardinal of M. Arguing as above it follows that there exists a set $a \subset \alpha, a \in M$ such that $L[a] \models$ '$V = K$' and such that $N \subset L[a]$. Let γ be an indiscernible of $L[a]$ above α with $\gamma \in M$. It follows that N_γ witnesses $\alpha \in A$. This proves the claim.

For every set a, $a^\#$ exists and so there is a definable class tree which projects to the graph of F in any set generic extension of V. Therefore by the claim the set A^* is universally Baire. It remains to show that A is not constructible from a real. Suppose $x \in \mathbb{R}$ and $A \in L[x]$. We may assume that $L[x] \models$ '$V = K$'. Further we may also assume that A contains all the indiscernibles of $L[x]$ below ω_1 since A must contain a tail of them. Let M be the smallest transitive set closed under F such that $x \in M$ and $M \models \text{ZFC}^{\text{-replacement}}$. Therefore $M \models$ '$V = K$'. Let α be the second uniform indiscernible of M. α is an indiscernible of $L[x]$ and so $\alpha \in A$. M is not a witness for $\alpha \in A$ and so by the claim there is a witness N with $N \in M$. By the choice of M, $x \notin N$. Therefore $N \subset L[x]$ a contradiction since α is an indiscernible of $L[x]$ and yet is a successor cardinal of N. □

The previous theorem is quite general. For example assume $AD^{L(\mathbb{R})}$ and let γ be the least Erdős cardinal of $HOD^{L(\mathbb{R})}$. Then in $HOD^{L(\mathbb{R})} \cap V_\gamma$ there is a set A for which A^* is universally Baire and A is not constructible from a real.

5. Universally Baire Sets and Determinacy

In this section we consider the relationships between determinacy and the universally Baire sets. We will prove in this section that if $AD^{L(\mathbb{R})}$

holds in every set generic extension of V, then every subset of the reals which is in $L(\mathbb{R})$ is universally Baire. We conjecture that the converse is also true. By the results of Martin-Steel and Woodin [13,26], we conclude that if there are two Woodin cardinals then every universally Baire set is determined. So in particular, if there are two Woodin cardinals and every subset of the reals which is in $L(\mathbb{R})$ is universally Baire, then $AD^{L(\mathbb{R})}$ holds. Further the theory of $L(\mathbb{R})$ is absolute for forcing extensions by posets of size less than the second Woodin cardinal.

Theorem 3.4, which characterizes when every $\utilde{\Sigma}^1_2$ set is universally Baire can be reformulated as follows.

Theorem 5.1. *The following are equivalent:*

(1) *Every $\utilde{\Sigma}^1_2$ set of the reals is universally Baire.*
(2) Det $\utilde{\Pi}^1_1$ *holds in every set generic extension of the universe.*

Theorem 5.2. *If $AD^{L(\mathbb{R})}$ holds in every set generic extension of V, then every $A \subseteq \mathbb{R}$ which is in $L(\mathbb{R})$ is universally Baire.*

Proof. The proof depends on the following theorem due to Solovay, see [10,12].

Theorem (Solovay). *Assume that $AD^{L(\mathbb{R})}$ holds and $\mathbb{R}^\#$ exists. If $A \subseteq \mathbb{R}$, $A \in L(\mathbb{R})$, and A is definable over $L(\mathbb{R})$ from finitely many Silver indiscernibles for $L(\mathbb{R})$, then there is a definable tree T such that A is the projection of the tree T.*

So it follows from the theorem that under the hypothesis of the theorem, there are two definable trees T and T^* such that $\mathbb{R}^\# = p[T]$ and $\mathbb{R} - \mathbb{R}^\# = p[T^*]$. Since $\mathbb{R}^\#$ is a definable countable union of sets in $L(\mathbb{R})$ which are definable over $L(\mathbb{R})$ from finitely many Silver indiscernibles for $L(\mathbb{R})$, one can merge countably many trees in a definable way to get the desired tree.

We now proceed to prove the theorem.

Notice that under the hypothesis of the theorem it follows that in every set generic extension of V, $\mathbb{R}^\#$ exists. We need only show that $\mathbb{R}^\#$ is universally Baire.

Let κ be an infinite cardinal. Let $\mathrm{Col}(\omega, \kappa)$ be the partial order for the Levy collapse of κ to ω. Let G be a $\mathrm{Col}(\omega, \kappa)$-generic over V.

Then in $V[G]$, $\mathbb{R}^\#$ exists and $AD^{L(\mathbb{R})}$ holds. Applying the quoted theorem above and the remark following it, let T, T^* be two ground model trees such that

$$\mathbb{R}^\# = p[T], \ \& \ \mathbb{R} - \mathbb{R}^\# = p[T^*].$$

Then the following lemma finishes the proof.

Lemma. *Assume that T is a tree in V_1 and $V_1 \subseteq V_2$ and both satisfy ZFC. If in V_2, $p[T] = \mathbb{R}^\#$ then in V_1, $p[T] = \mathbb{R}^\#$.*

To see this, we show that in V_1 the projection of the tree T satisfies the properties of being a sharp of the set of the reals. Then by the uniqueness we have $\mathbb{R}^\# = p[T]$.

Since in V_2 the projection $p[T]$ is the sharp of the set of the reals, it follows easily that in V_1, $p[T]$ is a well-founded, complete, consistent $\mathbb{R}^\#$-like theory. The only potential problem is the witness condition.

So let us check this.

To simplify notation let $a \in p[T]$ be a code for $\exists\ x\ \varphi(x, r_0, c_0)$, where r_0 is a real parameter and c_0 is the constant for (the least) Silver indiscernible. We show that there is some $b \in p[T]$ which is a code for $\varphi(t(r_0, r_1, c_0, \cdots, c_m), r_0, c_0)$ where t is a term, r_1 is an additional real parameter and $c_1, \cdots, c_m$ are additional constants for Silver indiscernibles.

Look at the set A of all such codes. Since the coding is done in a uniform Borel way, there is a tree S such that both in V_1 and V_2 the projection $p[S]$ of this tree S is the set of all such codes in the respective models. Now merge the two trees S and T to get $T * S$ so that

$$p[T * S] = p[T] \cap p[S].$$

Then in V_2, $T * S$ is ill-founded. Hence in V_1, the tree $T * S$ is ill-founded. We are done. $\square$

The following theorem offers some evidence that if every set in $L(\mathbb{R})$ is universally Baire then $AD^{L(\mathbb{R})}$ holds in every set generic extension of V.

Theorem 5.3. *Suppose that every set in $L(\mathbb{R})$ is universally Baire and that $AD^{L(\mathbb{R})}$ holds. Then $AD^{L(\mathbb{R})}$ holds in every set generic extension of V.*

Proof. (sketch) Suppose $AD^{L(\mathbb{R})}$ holds and that $\mathbb{R}^\#$ exists. For each $k < \omega$ let Γ_k be the pointclass of sets in $L(\mathbb{R})$ which can be defined by a Σ_1 formula in $L(\mathbb{R})$ using k indiscernibles as parameters. Solovay's theorem (cf. the proof of Theorem 5.2) that every set in $L(\mathbb{R})$ is Souslin actually states that for each k every set $A \in \Gamma_k$ admits in a canonical fashion a scale each norm of which is in $\cup\{\Gamma_j \mid j \in \omega\}$.

For each k let $G^k \subseteq \omega \times \omega^\omega \times \omega^\omega$ be the canonical universal Γ_k set. For each $j \in \omega$ and $x \in \omega^\omega$, let $G^k_{j,x}$ be the set,

$$\{y \in \omega^\omega \mid (j, x, y) \in G^k\}.$$

It follows from the nature of the scales that exist that for all $k \in \omega$ and for all $j \in \omega, x \in \omega^\omega$ if $G^k_{j,x} \neq \emptyset$ then there exists $y \in G^k_{j,x}$ such that for all $l \in \omega$, $G^l_{n(k,j,l),x} = \{y_l\}$, where $y_l(i) = y(i)$ if $i \leq l$ and $y_l(i) = 0$ otherwise. Here $n: \omega \times \omega \times \omega \to \omega$ is a (recursive) function which depends on the actual (cooperative) choice of the scales. For each $k \in \omega$ let T_k, T^*_k be trees witnessing that G^k is universally Baire with $G^k = p[T_k]$.

Since every set in $L(\mathbb{R})$ is universally Baire we have that $\mathbb{R}^\#$ is universally Baire. Let T, T^* be trees that witness $\mathbb{R}^\#$ is universally Baire with $\mathbb{R}^\# = p[T]$. Suppose that $V[G]$ is a set generic extension of V with $G \subseteq \mathbb{P}$. It suffices to show that $p[T] = \mathbb{R}^\#$ in $V[G]$. Again by absoluteness $p[T]^{V[G]}$ is an $\mathbb{R}^\#$ like theory. We need verify the witness condition. To verify the witness condition return to V. Fix a cardinal δ with $\mathbb{P} \in V_\delta$ and such that for each $k \in \omega$,

$$\mathbb{P} \Vdash p[T] = p[S]$$

$$\mathbb{P} \Vdash p[T_k] = p[S_k], p[T^*_k] = p[S^*_k]$$

where $S = T \cap V_\delta$, $S_k = T_k \cap V_\delta$, etc. Now choose a countable set $X \prec V_{\delta+1}$ such that $\{\mathbb{P}, S_k, S^*_k\} \subseteq X$. Let $g \subseteq X \cap \mathbb{P}$ be X-generic and let M be the transitive collapse of X.

Note that

$$\langle S_k, S^*_k \mid k, i \in \omega \rangle \in X$$

and so

$$\langle G^k \cap M[g] \mid k \in \omega \rangle \in M[g].$$

We shall show the following. Suppose N is a transitive model of ZFC$^{\text{-replacement}}$ and that

$$\langle G^k \cap N \mid k \in \omega \rangle \in N.$$

Then $(\mathbb{R}^\#)^N = N \cap \mathbb{R}^\#$. From this it follows that

$$(\mathbb{R}^\#)^{M[g]} = \mathbb{R}^\# \cap M[g] = p[S \cap X] \cap M[g].$$

Now suppose $x \in N \cap \omega^\omega$ and that $G^k_{j,x} \neq \emptyset$. By the remarks above there exists $y \in G^k_{j,x}$ such that for all $l \in \omega$, $G^l_{n(j,k,l),x} = \{y_l\}$ where y_l is defined as above. Finally for each $l \in \omega$, $y_l \in N$. Further the function n is recursive and so it follows that $y \in N$. Thus for each $k \in \omega$ and for each $j \in \omega, x \in N \cap \omega^\omega$ if $G^k_{j,x} \neq \emptyset$ then $G^k_{j,x} \cap N \neq \emptyset$. From this it follows that $(\mathbb{R}^\#)^N = \mathbb{R}^\# \cap N$. This completes the proof. $\square$

The previous theorem is really quite general. For example the version for the projective sets is true: Suppose every projective set is determined and is universally Baire. Then projective determinacy holds in every set generic extension of V.

As we have indicated in the presence of large cardinals every universally Baire set is determined. In fact even more is true, every universally Baire set is homogeneously Souslin.

Theorem 5.4. *Assume there are two Woodin cardinals. Then every universally Baire set is homogeneously Souslin and (therefore) determined.*

This theorem follows from the following theorems of Martin-Steel [13] and Woodin [26] together with Theorem 2.1.

Theorem 5.5 (Martin-Steel). *Assume δ is a Woodin cardinal. If T is a tree which is δ^+ weakly homogeneous then the set $\omega^\omega - p[T]$ is homogeneously Souslin.*

Theorem 5.6 (Woodin). *Assume δ is a Woodin cardinal. Suppose T, T^* are trees such that,*

$$\mathrm{Col}(\omega, \delta) \Vdash p[T] = \omega^\omega - p[T^*].$$

Then both trees T, T^* are $< \delta$ weakly homogeneous.

Corollary 5.1. *Assume there is a proper class of Woodin cardinals. Then a set $A \subseteq \omega^\omega$ is universally Baire if and only if the set A is ∞-homogeneously Souslin.*

Schilling and Vaught [17] associate to every Borel set $A \in \omega^\omega$ an operation, G_A, on subsets of a topological space using a game quantifier. They show using Borel determinacy that this operation preserves the Baire property in any topological space. Using the previous corollary one can generalize their results to any universally Baire set (assuming there is a proper class of Woodin cardinals) and to more complicated operations.

If there exists a Woodin cardinal then *any* tree can be forced to be weakly homogeneous.

Theorem 5.7 (Woodin). *Assume that δ is a Woodin cardinal. If T is a tree, then there exists some $\lambda < \delta$ such that for each generic $G \subseteq \mathrm{Col}(\omega, \lambda)$, T is $< \delta$ weakly homogeneous in $V[G]$.*

Corollary (Woodin). *Assume that δ is a Woodin cardinal. Then there exists $\kappa < \delta$ such that if $G \subseteq \mathrm{Col}(\omega, \kappa)$ is generic, then in $V[G]$ every projective set is $< \delta$ weakly homogeneous Souslin.*

Remark. The corollary can be proved from a much weaker hypothesis.

The following theorem is an unpublished result of Woodin.

Theorem 5.8 (Woodin). *Assume there are infinitely many strong cardinals below κ. Suppose G is generic for $\mathrm{Col}(\omega, \kappa)$. Then for every projective formula $\varphi(x)$ there is a class tree $T_\varphi \subseteq (\omega \times Ord)^{<\omega}$ such that*

$$p[T_\varphi] = \{x \in \mathbb{R} \mid \varphi(x)\}$$

in every set generic extension of $V[G]$.

Corollary. *Assume that there are infinitely many strong cardinals below κ. If $G \subseteq \mathrm{Col}(\omega, \kappa)$ is V-generic, then in $V[G]$, every projective set is universally Baire.*

6. Open Questions

In this section, we list seven questions which we think are interesting.

1. Assume that every projective set is universally Baire. Is it the case that every projective sentence is absolute with respect to every set generic extension? In fact, is it the case that for each projective formula φ there is a class tree which represents φ in every set generic extension?

2. Assume that every set of reals which is in $L(\mathbb{R})$ is universally Baire. Is $\mathbb{R}^\#$ invariant under set forcing? Is there a class tree which projects to $\mathbb{R}^\#$ in every set generic extension of the universe?

3. Assume that A is a set of reals. Assume that every set of reals which is projective in A is universally Baire. Let B be a set of reals which is projective in A. Is there a class tree T such that T projects to B in every set generic extension of V (in the obvious sense)?

4. Assume that $(A, \mathbb{R})^\#$ is universally Baire. Does there exist a class tree which projects to $(A, \mathbb{R})^\#$ in every set generic extension (again in the obvious sense)?

5. Assume that A is a subset of ω_1 and that B is universally Baire where B is any set projective in A^*. Is A constructible from a real? (cf. Theorem 4.2).

6. Assume that $\mathbb{R}^\#$ is universally Baire. Does $AD^{L(\mathbb{R})}$ hold? Let A be a set of reals. Assume that $(A, \mathbb{R})^\#$ is universally Baire. Does $AD^{L(A,\mathbb{R})}$ hold?

7. Assume that V is (projectively) absolute with respect to every set generic extension. Or even weaker simply assume that in every set generic extension of V, every projective set has the property of Baire. Is every projective set universally Baire?

By Theorem 5.3 the answer to (2) is yes if one assumes in addition that $AD^{L(\mathbb{R})}$ holds. A positive answer to (5) would likely yield a positive answer to (2) in the strong sense that if $\mathbb{R}^{\#}$ is universally Baire then $AD^{L(\mathbb{R})}$ holds and so a positive answer to (5) would likely give a partial answer to (6).

One can show that a positive answer to (3) implies a positive answer to (4). By the results indicated in the previous section one cannot hope to prove that if every projective set is universally Baire then every projective set is determined. The pointclass of the projective sets is simply not sufficiently closed. For more on this see [26].

Acknowledgments. This paper was written during the year 1989-90 when Magidor and Woodin were at MSRI, 1000 Centennial Drive, Berkeley CA 94720, and Feng was visiting the Department of Mathematics, UC Berkeley. Feng would like to thank the Department for its hospitality and to thank Professor Hugh Woodin for his support.

REFERENCES

1. J. Bagaria, *Definable Forcing and Regularity Properties of Projective Sets of Reals*, Ph. D. Thesis, UC Berkeley (1991).
2. J. Fenstad and D. Norman, *On absolutely measurable sets*, Fund. Math. **81** (1974), 91–98.
3. L. Harrington, *Analytic determinacy and* $0^{\#}$, J. S. L. **43** (1978), 685–693.
4. T. Jech, *Set Theory*, Academic Press, 1978.
5. H. Judah and S. Shelah, $\underset{\sim}{\Delta}^1_2$ *sets of reals*, Ann. Pure and Applied Logic **42** (1989), 207–223.
6. A. Kechris, *Homogeneous trees and projective scales*, Cabal Seminar 77-79 (A. Kechris, D. A. Martin, Y. N. Moschovakis, eds.), Lecture Notes in Mathematics, vol. 839, Springer-Verlag, 1981, pp. 33–74.
7. A. Kechris, *Subsets of* $\aleph_1$ *constructible from a real*, Cabal Seminar 81-85 (A. Kechris, D. A. Martin, J. Steel, eds.), Lecture Notes in Mathematics, vol. 1333, Springer-Verlag, 1988, pp. 110–116.
8. A. Levy, *Definability in axiomatic set theory*, "Mathematical Logic and Foundations of Set Theory" (Y. Bar-Hilled, ed.), North-Holland, Amsterdam, 1970.
9. D. A. Martin, *Measurable Cardinals and analytic games*, Fund. Math. **66** (1970), 287–291.
10. D. A. Martin, *The real game quantifier propagates scales*, Cabal Seminar 79-81, Lecture Notes in Mathematics, vol. 1019, Springer-Verlag, 1983, pp. 157–171.
11. D. A. Martin and R. Solovay, *A basis theorem for* Σ^1_3 *sets of reals*, Ann. Math. **89** (1969), 138–159.
12. D. A. Martin and J. Steel, *The extend of scales in* $L(\mathbb{R})$, Cabal Seminar 79-81, Lecture Notes in Mathematics, vol. 1019, Springer-Verlag, 1983, pp. 93–106.
13. D. A. Martin and J. Steel, *A proof of projective determinacy*, J. AMS **2** (1989), 71–125.
14. A. Mathias, *Happy Families*, Ann. Math. Logic **12** (1977), 59–111.
15. J. Oxtoby, *The Banach-Mazur game and Banach category theorem*, in Contributions to the theory of games jour Ann. Math. Studies **39** (1957), 159–163.

16. K. Schilling, *On Absolutely* $\underset{\sim}{\Delta}^1_2$ *operations*, Fund. Math. **121 (3)** (1984 pages 239–250).
17. K. Schilling and R. Vaught, *Borel Games and the Baire Property*, Trans. AMS **279**, 411–428.
18. S. Shelah, *Can you take Solovay's inaccessible away?*, 48 (1984), 1–47.
19. J. Shoenfield, *The problem of predicativity*, Essays on the Foundations of Mathematics (Y. Bar-Hilled, ed.), The Magnes Press, Jerusalem, 1961, pp. 132–142.
20. J. Silver, *Measurable cardinals and* Δ^1_3 *well–orderings*, Ann. Math. **94** (1971), 414–446.
21. R. Solovay, *A model of set theory in which every set of reals is Lebesgue measurable*, Ann. Math. (1970), 1–56.
22. J. Steel, *The Core Model Iterability Problem*, Manuscript (June 1990).
23. A. Stone, *Nonseparable Borel sets*, Rozprawy Mathematyczne, 1962.
24. H. Woodin, *On the strength of projective uniformization*, Logic Colloquium '81 (J. Stern, ed.), pp. 365–383.
25. H. Woodin, *Supercompact cardinals, sets of reals, and weakly homogeneous trees*, Proc. Natl. Acad. Sci., USA **85** (1988), 6587–6589.
26. H. Woodin, *Large cardinals and Determinacy*, in preparation.

Department of Mathematics, National University of Singapore, SINGAPORE 0511

Department of Mathematics, Hebrew University of Jerusalem, Jerusalem, ISRAEL

Department of Mathematics, University of California, Berkeley CA 94720

THE SINGULAR CARDINAL HYPOTHESIS REVISITED

Moti Gitik and Menachem Magidor

Introduction

Cardinal arithmetic had been one of the central themes in set theory, but in the late 60's and early 70's, it seemed that there were actually very few theorems that could be proved about cardinal arithmetic. For example, except for some trivial facts, the behavior of cardinal exponentiation (which is the only non-trivial operation in cardinal arithmetic) is almost completely arbitrary, and the accepted system of axioms for set theory, ZFC, does not yield any structure theory for cardinal arithemetic.

The most clear formulation of the lack of any deep structure is Eastons's result [Ea]; namely that for regular cardinals, the only theorems one can prove in ZFC are the trivial fact of monotonicity of exponentiation ($\alpha < \beta$ implies $2^\alpha \leq 2^\beta$) and the Zermelo-König inequality (the cofinality of $2^\alpha > \alpha$).

In [Ea] it is shown that for every reasonable "function" F, from cardinals to cardinals, which satisfies the above requirements there exists a model of set theory in which for *regular* α, $2^\alpha = F(\alpha)$.

In the models constructed by Easton there was a very simple rule that determines the exponents of singular α, 2^α for α singular was the smallest cardinal having cofinality $> \alpha$ and not smaller than 2^β for $\beta < \alpha$. (Hence for instance if α is strong limit, i.e. $\beta < \alpha$ implies $2^\beta < \alpha$, then $2^\alpha = \alpha^+$.) More formally, for singular α

$$\alpha^{cf(\alpha)} = \max(2^{cf(\alpha)}, \alpha^+).$$

The above assumption became known as the Singular Cardinals Hypothesis (SCH). Is SCH a theorem of ZFC? If it were, then the study of cardinal arithmetic would be completely finished and we would have a very simple and complete classification of all possible behaviors of the function $\alpha \to 2^\alpha$. It would mean that we know all there is to know about cardinal arithmetic. Fortunately, for the career of the authors, but probably unfortunately for mathematics, the situation turned out to be much more complicated.

The main difficulty in getting a model which is a counterexample to SCH is that the forcing notion used to increase the size of the power set of a cardinal α is nicely behaving when α is regular (for instance it introduces no bounded subset of α), but it has disasterous effects when α is singular; and typically when using it, α ceases to be a cardinal.

The saving idea was to start from a model in which α is regular, blow up 2^α to any desired value, keeping α strong limit, and then make α singular, without changing the fact that α is strong limit and without collapsing cardinals. Having a forcing notion that keeps α a cardinal while making α singular requires some special properties of α, and the standard assumption is that α is a measureable cardinal. Making this assumption, we have the forcing notion introduced by Prikry, [Pr], which starts from a measurable cardinal κ and makes it singular of cofinality ω, by creating what bacame known as a Prikry sequence for κ—complete ultrafilter U on κ. (A sequence $\langle \alpha_n | n < \omega \rangle$ is called a Priky sequence for U if the sequence is eventually included in every $A \in U$.) Prikry's forcing does not add any bounded subsets to κ, and it satisfies the κ^+—c.c. It follows that no cardinals are collapsed. We get that κ is still a strong limit cardinal in the extension. So we are now faced with the problem of getting a model with a measurable cardinal violating the GCH. In 1971 Silver was able to get such a model, starting from the strong assumption of having a 1–extendible cardinal. Combining the results of Prikry and Silver one gets a model which violates SCH; so SCH is not a theorem, unless the large cardinals used by Silver are inconsistent with ZFC. In the model produced by Silver and Prikry the violation of SCH occurs at a very large cardinal. In [Ma1] a model was constructed in which the smallest singular cardinal, i.e. $\aleph_\omega$ violates SCH.

The results of Prikry, Silver and [Ma1] seemed to be weaker in several senses than the results of Easton for regular cardinals. First, the consistency assumption made was stronger than the natural assumption of the consistency of ZFC. Second, it was not clear to what extent one has the same freedom in determining the powers of singular cardinals as one has for regular cardinals. For instance, the fact that one starts with a measurable cardinal κ violating the GCH immediately implies, by the usual reflection properties of measurable cardinals, that there are unboundedly many cardinals below κ that voilate GCH. So if one uses Prikry's forcing in the extension, κ which is the counterexample to SCH is not the first cardinal violating GCH. Straightforward forcing notions to rearrange GCH below κ collapse cardinals above κ in such a way that κ is not a counterexample to SCH anymore. Similar obstacles were encountered in the construction of [Ma1] where the method for collapsing cardinals below κ in order to make it $\aleph_\omega$ did require leaving some cardinals untouched, and they were exactly the cardinals which left an unbounded sequence of cardinals below κ at which GCH was violated. So naturally there arises the problem: "Can a singular cardinal be the first counterexample to GCH?"

In the early 70's it was generally believed by set theorists that the above weakness of the proofs are only artifacts and, for singular cardinals, one should expect the complete analogues of the results for regular cardinals. The situation was dramatically changed in 1975 when Silver proved what became known as "Silver's singular Cardinals Theorem" claiming that a singular cardinal of cofinality $> \aleph_0$ can not be the first cardinal violating the GCH. So here is a non-trivial theorem about cardinal arithmetic which applies only to singular cardinals. Further results followed, for instance the Theorem of Galvin and Hajnal [GH] giving a bound for powers of singular cardinals of uncountable cofinality.

Silver's theorem triggered a striking series of results, which became the cornerstone of Inner Models Thoery. They were Jensen's Covering Theorem for L, the Jensen-Dodd Covering Theorem for the Core Model K, and the Mitchell Core Model with its weak covering properties. These results showed that the use made of large cardinals assumptions in the construction of the models of SCH was really necessary.

Shelah has proved many further deep theorems, extending the Galvin-Hajnal bound also to singular cardinals of cofinality $\aleph_0$ and (in many cases) improving them. His most recent result shows that $\aleph_\omega^{\aleph_0} < \max(2^{\aleph_0}, \aleph_{\omega_4})$. One cannot avoid comparing these deep results with the absence of any deep theorems restricting the powers of regular cardinals.

One is now faced with the problem of classifying all the possible behaviors of powers of singular cardinals. Natural test problems are the problem like the problem mentioned above "Can a singular cardinal be the first cardinal violating GCH?" (In view of Silver's result it must have cofinality $\aleph_0$.) and the problems like "Is it possible that every cardinal violates the GCH?" "Assuming that $2^{\aleph_0} < \aleph_\omega$, how large can $\aleph_\omega^{\aleph_0}$ be?" (Is Shelah's bound the best possible?)

The first test problem was handled in [Ma2], where starting from stronger large cardinals ("huge") a model was constructed in which GCH holds below $\aleph_\omega$ and $2^{\aleph_0} = \aleph_{\omega+2}$. Concerning the third problem: the original construction of [Ma1] gave as a possible value for $\aleph_\omega$ any cardinal of the form $\aleph_{\omega+\alpha+1}$ for $\alpha \leq \omega$. Shelah in [Sh1] showed that a change in the construction can give as possible value for $\aleph_\omega^{\aleph_0}$ any $\aleph_{\omega+\alpha+1}$ for $\alpha < \omega_1$. So the best possible upper bound for $\aleph_\omega^{\aleph_0}$ is $\aleph_{\omega_1}$ (and it is still open whether this is actually an upper bound). Since the constructions used by Shelah in [Sh1] followed [Ma1], GCH was failing below $\aleph_\omega$. The methods of Shelah and [Ma2] were combined by the second author (unpublished) to get models for every given $\alpha < \omega_1$ in which GCH holds below $\aleph_\omega$ and $2^{\aleph_\omega} = \aleph_{\omega+\alpha+1}$. Again very large cardinals were used in these proofs. The second problem was solved by Foreman and Woodin [FW], who constructed a model in which every cardinal violates the GCH. The construction of models in which a singular cardinal is the first counterexample to GCH required association of the cardinals between κ and 2^κ to cardinals below κ. Actually each $\kappa < \alpha < 2^\kappa$ had to have its set of associates below κ such that for different α's the corresponding sets were disjoint. This meant that these methods could not have produced a model in which κ is singular, GCH holds below κ and $2^\kappa > \kappa^{+\kappa}$. Was there some hidden theorem?

Another class of problems comes up naturally. The results of Jensen, Jensen-Dodd and Mitchell showed that some large cardinals are needed for the failure of SCH. More formally, if SCH holds then some inner model has some large cardinals. The definitions of large cardinals form a natural hierarchy. It is very desirable to pin down, if possible, the exact large cardinal notion equiconsistent with the statement under study. The linear scale of large cardinals is used to measure the degree of independence of the statement, or dually, what risk of inconsistency is involved in assuming the truth of this statement. In case such an exact equiconsistency result is not available, the alternative is to give as tight consistency bounds as possible, a lower bound (namely a large cardinal notion whose consistency is implied by the statement under consideration), and an upper bound (namely a large cardinal notion whose consistency implies the consistency of the statement). The closer these two bounds are, the better the result.

Until 1988, the results of Mitchell gave the best lower bounds; namely, $\neg$SCH implies the existence of a sequence of measurable cardinals κ_n such that $o(\kappa_n) \geq \kappa_{n-1}$. For $\neg$SCH at a singular cardinal of cofinality $> \aleph_0$ Mitchell proved the much better lower bound $o(\kappa) = \kappa^{++}$. (Mitchell in [Mi1] introduces an heirarchy of measurable cardinals, which are determined by their order: $o(\kappa) = 1$ means simply being measurable. $o(\kappa) = \kappa^{++}$ is the largest possible order. The hierarchy can be generalized to hypermeasurables and so we can talk about $o(\kappa) = \lambda$ for $\lambda = \kappa^{++}$. The notion of a strong cardinal is equivalent to $o(\kappa) = \infty$). This lower bound was much weaker than the large cardinals used in [Ma1] and [Sh1], and even more so for the cardinals used in [Ma2]. So pinning down the exact consistency stength of $\neg$SCH became an important research problem. The major step towards solving this problem was made by Woodin in [Wo] (These results will be included in the forthcoming book [C-Wo]. See also [Ca]).

He lowered the upper bound substantially by getting models for $\neg$SCH and also for the failure of GCH at a measurable cardinal, starting from hypermeasureable cardinals. His final result was one that can get $\neg$SCH from a cardinal κ having the following property: "There exists an elementary embedding $j : V \to M$, where M is transitive, $M^\kappa \subset M$, for κ the critical point of j and $\kappa^{+2} = j(f)(\kappa)$ for some $f : \kappa \to \kappa$." He also showed that like in [Ma1] one can get the failure of SCH at $\aleph_\omega$. The assumption above together with GCH was needed for getting $2^{\aleph_\omega} = \aleph_{\omega+2}$. For getting $2^{\aleph_\omega} = \aleph_{\omega+\alpha+1}$ for $\alpha < \omega_1$, one needed the obvious strengthening of the above assumption to $\kappa^{+\alpha} = j(f)(\kappa)$. He later showed that one could dispense with the function f and simply require $j(\kappa) \geq \kappa^{+2}$. Gitik in [Gi] proved that Woodin's condition can be forced starting from $o(\kappa) = \kappa^{++}$. So the upper bound became $o(\kappa) = \kappa^{++}$. In recent work Gitik combined Mitchell's methods with the *pcf* theory of Shelah (see [Sh2]) to improve the lower bound to $o(\kappa) = \kappa^{++}$. Thus the consistency strength of $\neg$SCH was finally pinned down at $o(\kappa) = \kappa^{++}$. If κ was a singular strong limit cardinal such that $2^\kappa = \kappa^{+n}$, then Gitik's lower bound was $o(\kappa) = \kappa^{+n}$, provided the Mitchell's Covering Theorem can be generalized to the higher core models. (The analoguous problem for $2^\kappa = \lambda^+$ where $\lambda \geq \kappa^{+\omega}$ is still open.) The constructions of Woodin described above were along the lines of [Ma1], hence they did not produce models with GCH below κ (where κ is the first counterexample to SCH). So there remained the (unlikely) possibility that the consistency strength of "A singular cardinal is the first violating GCH" could be much higher than that of $\neg$SCH. But Woodin modified his construction, and assuming GCH and the existence of an elementary embedding $j : V \to M$ such that $M^\kappa \subset M$ and $j(\kappa) \geq \kappa^{+2}$, where κ is the critial point of j, he constructed a model in which GCH holds below $\aleph_\omega$ and $2^{\aleph_\omega} = \aleph_{\omega+2}$. The method of proof involved collapsing cardinals in a special way so as to get rid of the cardinals below κ initially violating GCH and some essential ingredients of it seemed not to work if one wanted higher values for $2^{\aleph_\omega}$ while keeping GCH below $\aleph_\omega$; so again one was left with the possibility that the consistency strength of "GCH below $\aleph_\omega$ and $2^{\aleph_\omega} = \aleph_{\omega+3}$" could be much higher than that of the same statement with $\aleph_{\omega+3}$ replaced by $\aleph_{\omega+2}$.

In this paper we present a different method of constructing models violating SCH. The main merit of this new method is that we eliminate the need to blow up 2^κ while κ is still regular, hence we are not forced to have an unbounded set of cardinals below 2^κ violating the GCH while κ is still regular, and so it is much easier to have GCH below the counterexample to SCH. The idea is to start with some large cardinal κ (of course we have to use some large cardinal!) and then add simultaneously many new ω–sequences to κ, such that κ becomes singular of cofinality ω and 2^κ becomes large. Thus the two steps essential to all the previous proofs become merged into one step. The construction adds no new bounded subsets κ and it satisfies the κ^{++}–chain condition (so if we started from a model of GCH, we still have GCH below κ). No cardinals are collapsed (one needs a special argument for κ^+). The idea is that if for a given λ we want to introduce λ many new ω–sequences into κ, we might try to add Prikry sequences for a system $\langle U_\alpha | \alpha < \lambda \rangle$ of λ many κ–complete ultrafilters on κ. (Of course if $\lambda > \kappa^{++}$ we must have many repetitions among the U_α's). A typical condition for adding a Prikry sequence for an ultrafilter U has the form, $\langle \alpha_0, \ldots, \alpha_n, T \rangle$ where $\langle \alpha_0, \ldots, \alpha_n \rangle$ is a finite sequence (giving an initial segment of the Prikry sequence $\langle \alpha_n | n < \omega \rangle$ for U) and T is a tree of possible continuations of the sequence $\langle \alpha_0, \ldots, \alpha_n \rangle$ such that if $p \subset q \in T$ then the set $\{\alpha | q \frown \{\alpha\} \in T\}$ is a set in the ultrafilter U. (In the case U is normal one can replace T by one set $A \in U$). For a fixed U the arguments of Prikry show that no new bounded subsets of κ are introduced. However, if one tries to do the same construction λ many times (even for different U's) in a straightforward way, then it is not true that no now bounded subsets of κ are introduced. For instance, if one uses the product forcing of the two Prikry forcings for U_1 and U_2, getting the two Prikry sequences $\langle \alpha_n | n < \omega \rangle$ and $\langle \beta_n | n < \omega \rangle$, the set $\{n | \beta_n < \alpha_n\}$ is a new subset of ω. Typically, cardinals are collapsed. In order to block the above example one has to assume some "coupling" between the different Prikry sequences, so that the relation between two different ones will not generate a new bounded subset of κ. The most natural coupling can be created if in the ground model we have $f : \kappa \to \kappa$ such that one Prikry sequence is obtained from the other one by applying f to it's members. One can easily verify that if $\langle \alpha_n | n < \omega \rangle$ is a Prikry sequence for U_1, thus $\langle \beta_n | n < \omega \rangle$ is a Prikry sequence for U_2, and for all n (or only for sufficiently large n's) $f(\alpha_n) = \beta_n$ then f is a Rudin-Keisler projection of U_1 to U_2. (Namely f reduces the problem of membership in U_2 to that of membership in U_1, i.e. $A \in U_2 \Leftrightarrow f^{-1}(A) \in U_1$.) So we assume that $\langle U_\alpha | \alpha < \lambda \rangle$ form a directed system under Rudin-Keisler reductions. Namely there is a partial order $\prec$ on λ, such that if $\alpha \prec \beta$, there is a Rudin-Keisler reduction $f_{\beta\alpha}$ of U_β to U_α. It is also natural to require that this system is commutative, i.e. if $\alpha \prec \beta \prec \gamma$ then for some A in U_γ, and for all $x \in A$ the equality $f_{\beta\alpha}(f_{\gamma\beta}(x)) = f_{\gamma\alpha}(x)$ holds. So now the idea is to introduce for each U_α from our system a Prikry sequence $\langle \beta_n(\alpha) | n < \omega \rangle$, such that if $\alpha \prec \gamma$, then $\beta_n(\alpha) = f_{\gamma\alpha}(\beta_n(\gamma))$ for large enough n. In order to guarantee that the sequences will all be different (remember that the U_α's can be the same for different α's), one should require that if $\alpha \prec \gamma$, $\beta \prec \gamma$, then for some $A \in U_\gamma$ and all $x \in A$ the inequality $f_{\gamma\alpha}(x) \neq f_{\gamma\beta}(x)$ holds. There are further technical condition on the system of ultrafilters $\langle U_\alpha | \alpha < \lambda \rangle$ which are all incorporated in the notion of a "nice system of ultrafilters."

In Section 1 it is shown that if one has a nice system of ultrafilters on κ of length λ and one forces with the forcing notion intended to introduce the Prikry sequences as described above, then $2^\kappa \geq \lambda$, no new bounded subsets of κ are introduced and the forcing satisfies κ^{++}–c.c. A special argument shows that κ^+ is not collapsed. So if $\lambda > \kappa^+$ we get a model in which κ violates SCH and GCH holds below κ. How does one get a nice system of ultrafilters of a given length λ? In Section 1 it is shown that having an elementary embedding $j : V \to M$ such that $V_{\kappa+\lambda} \subseteq M$ is enough. This is a hyper-measurable assumption, much weaker than an assumption used in [Ma2] or its generalizations. Also note that if κ is strong, then λ is arbitrary and for every λ we get a model in which κ is the first cardinal violating GCH, it is singular and $2^\kappa \geq \lambda$. So no bound can be proved in general about powers of singular cardinals even if one assumes GCH below κ. The assumption used can be somewhat weakened; in a forthcoming paper it will be shown how to start from $o(\kappa) = \lambda$ and force a nice system of ultrafilters of length λ.

In Section 2 it is shown how to get the singular cardinal κ of Section 1 to be $\aleph_\omega$, at least in the case $2^\kappa = \kappa^{+m}$ for finite m. So using what seems to be the exact consistency strength needed (which is much weaker than what was used before for the case $m > 2$) one gets a model of "GCH holds below $\aleph_\omega$ and $2^{\aleph_\omega} = \aleph_{\omega+m+2}$."

In Section 3 we merge the methods of Section 2 with Shelah's [Sh1] to get (for each $\alpha < \omega_1$) a model of "GCH below $\aleph_\omega$ and $2^{\aleph_\omega} = \aleph_{\omega+\alpha+1}$." Again this is a major improvement in the strength of the large cardinals used.

"Old problems never die, they just fade away." The singular cardinals problem is an example. In spite of all the progress, some very interesting open problems are still left. We already mentioned the problem of finding out whether Shelah's bound for $\aleph_\omega^{\aleph_0}$ is the best possible. The first cardinal which according to the present knowledge, cannot be ruled out as an improvement of Shelah's bound is $\aleph_{\omega_1}$. So a subproblem is: "Can one get a model in which $\aleph_\omega$ is a strong limit and $2^{\aleph_\omega} \geq \aleph_{\omega_1}$?" It seems that completely new methods are called for in solving this problem. It seems plausible that, the consistency strength needed for getting a model in which $\aleph_\omega$ is a strong limit and $2^{\aleph_\omega} \geq \aleph_{\omega_1}$ is much higher than what was sufficient for the statements considered in this paper.

Similar problems concern the first cardinal fixed point, i.e. the first κ such that $\kappa = \aleph_\kappa$. Shelah proved many important bounds for powers of cardinal fixed points, but no such bound was proved for the first cardinal fixed point. The methods of this paper can be used to get for each $\alpha < \omega_1$ a model in which κ is the first cardinal fixed point; GCH holds below κ and 2^κ has α many cardinal fixed points below it. However, it is not known how to get a similar model, in which 2^κ has ω_1 many fixed points below it. Arguments similar to the $\aleph_\omega$–case indicate that the consistency strength of the statment "If κ is the first cardinal fixed point, then 2^κ has ω_1 fixed point below it and κ is strong limit" is much larger than the cardinals used in this paper. The first singular cardinal for which we know that no bound on its power set can be proved is the first cardinal fixed point of order ω ([Sh1], see Def. 4.13 below). For small fixed points of countable cofinality the problem is open. It seems that cardinal arithmetic still has some surprises for us in stock.

We tried to keep the notation of this paper rather standard. A fair amount of acquaintance with forcing techniques is expected. We shall refer to many notions of large cardinals, not all of them properly documented in the published literature, but the paper should be understandable even without knowing the exact definitions.

1. Making κ^ω Large

Given κ which is an appropriate large cardinal and $\lambda > \kappa$, we shall present in this section a forcing notion that will make κ singular of cofinality ω while simultaneously introducing λ many ω sequences in κ. Our forcing notion will not introduce any bounded subsets to κ and will not collapse any cardinals. The exact assumption on the ground model we need is that κ carries a "nice system of ultrafilters".In order to motivate this, rather technical, definition we shall start from the the assumption that κ is the critical point of an appropriate elementary embedding and we shall define a certain sequence of ultrafilters defined from this embedding. Absracting the properties of this sequence of ultrafilters gives the definition of "nice sequence of ultrafilters".

Suppose $j : V \to M$ is an elementary embedding of V into a transitive model M with the critical point κ. Let λ be a successor ordinal or a cardinal of cofinality $> \kappa^+$. Assume that (a) $V_{\kappa+\lambda} \subseteq M$ and (b) for a function $f_\lambda : \kappa \to \kappa$ $j(f_\lambda)(\kappa) = \lambda$. Suppose for simplicity that V satisfies GCH. Note that under the above assumptions we have ${}^{\kappa^+}V_{\kappa+\lambda} \subseteq M$.

Let us define now an extender which catches M up to $V_{\kappa+\lambda}$ and forms

a κ^{++}-directed system. For a technical reason we would like also to have the normal measure generated by j, i.e. $U = \{X \subseteq \kappa \mid \kappa \in j(X)\}$ to be "covered" by every measure of the extender. Fix some well-ordering $\prec$ of V_κ so that for every inaccessible cardinal $\alpha < \kappa$ $\prec |\alpha^{+f_\lambda(\alpha)} : \alpha^{+f_\lambda(\alpha)} \leftrightarrow [\alpha^{+f_{\lambda(\alpha)}}]^{\leq \alpha^+}$ and $\prec (\alpha) = \{\alpha\}$. Consider a set $\mathcal{A} = \{\delta < \kappa^{+\lambda} | j(\prec)(\delta)$ is a subset
of $\kappa^{+\lambda}$ of cardinality $\leq \kappa^+$ with the minimal element κ and for every $\gamma \in j(\prec)(\delta)\, j(\prec)(\gamma) \cap (\kappa^{+\lambda} \setminus \kappa) \subseteq j(\prec)(\delta)\}$. Define for $\delta_1, \delta_2 \in \mathcal{A}\; \delta_1 <_{\mathcal{A}} \delta_2$ iff $\delta_1 \in j(\prec)(\delta_2)$. Then $\langle \mathcal{A}, \leq_{\mathcal{A}} \rangle$ will be a partial ordered set which is κ^{++}-directed and has the minimal element κ.

For every $\delta \in A$ let us define an ultrafilter U_δ over κ as follows:

$$X \in U_\delta \quad \Longleftrightarrow \quad \delta \in j(X)\ .$$

If $\delta_2 {}_{\mathcal{A}}\!\geq \delta_1$, then U_{δ_2} can be naturally projected onto U_{δ_1}. Let us define the projections $\langle \pi_{\delta_2\delta_1} \mid \delta_2 {}_{\mathcal{A}}\!\geq \delta_1 \rangle$. Proceed as follows. Set $\pi_{\delta\delta} = id$ for every $\delta \in \mathcal{A}$. Set $\pi_{\delta_2\delta_1}(0) = 0$ for every $\delta_2 {}_{\mathcal{A}}\!\geq \delta_1$. Let $\pi_{\delta\kappa}(\alpha) = \min(\prec (\alpha) \cap On)$ for every $\delta \in \mathcal{A}, 0 < \alpha < \kappa$. Let now $\delta_2 {}_{\mathcal{A}}\!> \delta_1 \neq \kappa$. Consider the following commutative diagram:

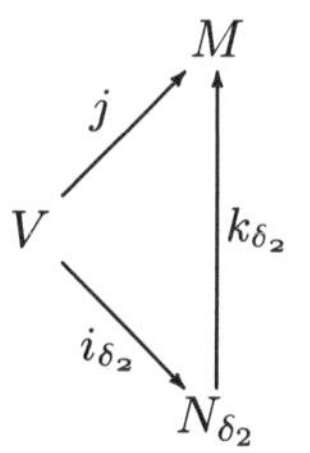

where $N_{\delta_2} \simeq Ult(V, U_{\delta_2}), i_{\delta_2}$ is the corresponding elementary embedding and $k_{\delta_2}\ ([f]_{U_{\delta_2}}) = j(f)(\delta_2)$.

The critical point of k_{δ_2} is $(\kappa^{++})^{N_{\delta_2}} > \kappa^+$. $i_{\delta_2}(\prec)([id]_{U_{\delta_2}})$ is mapped by k_{δ_2} to $j(\prec)(\delta_2)$. Since the cardinality of the last set is $\leq \kappa^+$ in M, the same is true in N_{δ_2}. So $j(\prec)(\delta_2) = k''_{\delta_2}(i_{\delta_2}(\prec)([id]_{U_{\delta_2}})$. Pick δ_1' to be the element of $i_{\delta_2}(\prec)([id]_{U_{\delta_2}})$ which is mapped by k_{δ_2} on δ_1. Now, any function representing δ_1' in N_{δ_2} will project U_{δ_2} onto U_{δ_1}. Let $t : \kappa \to \kappa$ be such a function. Find a set $X \in U_{\delta_2}$ so that for every $\nu \in X$ $\pi_{\delta_1\kappa}(t(\nu)) = \pi_{\delta_2\kappa}(\nu)$. Define

$$\pi_{\delta_2\delta_1}(\nu) = \begin{cases} t(\nu), & \text{if } \nu \in X \setminus \{0\} \\ \nu, & \text{otherwise} \end{cases}$$

Denote by $i_\delta : V \to N_\delta \simeq Ult(V, U_\delta)$ and let $k_{\delta_1\delta_2} : N_{\delta_1} \to N_{\delta_2}$ be defined for $\delta_1 <_{\mathcal{A}} \delta_2$ as follows $k_{\delta_1\delta_2}([f]_{U_{\delta_1}}) = [g]_{U_{\delta_2}}$, where $g(\alpha) = f(\pi_{\delta_2\delta_1}(\alpha))$. One can easily show that under these definitions for $\delta_1, \delta_2 \in \mathcal{A}$ we have

$\pi_{\delta_1\kappa}(\nu) = \pi_{\delta_2\kappa}(\nu)$ for all $\nu \in \kappa$. Also if $\delta_1 \leq_{\mathcal{A}} \delta_2$ then $\pi_{\delta_2\delta_1}(\pi_{\delta_1\kappa}(\nu)) = \pi_{\delta_2\kappa}(\nu)$.

Then $\langle < N_\delta \mid \delta \in \mathcal{A}\rangle, \langle k_{\delta_1\delta_2} \mid \delta_1 <_{\mathcal{A}} \delta_2 >\rangle$ is a κ^{++}-directed commutative system. Further we shall identify it with $\langle U_\delta, \pi_{\delta_1\delta_2} \mid \delta \in \mathcal{A}, \delta_1 <_{\mathcal{A}} \delta_2\rangle$. Notice that each U_δ is a P-point ultrafilter since δ which defines it sits between κ and $\kappa^{+\lambda} = \kappa^{+j(f_\lambda)(\kappa)}$.

In the sequel we shall use an arbitrary sequence of ultrafilters, having the properties of the sequence defined from j. Let us list the properties we need. (Note that in the definitions below 0 plays the role of the minimal member of $\mathcal{A}$; this role was played in the sequence defined above by κ.)

Let κ be a fixed measurable cardinal.

Suppose that $\langle \mathcal{A}, <_{\mathcal{A}}\rangle$ is a κ^{++}-directed partial ordered set.

Definition 1.0. A sequence $\langle U_\alpha \mid \alpha \in \mathcal{A}\rangle$ of κ-complete ultrafilter over sets of cardinality κ is called a Rudin-Keisler directed commutative, iff there exists a sequence $\langle \pi_{\alpha\beta} \mid \alpha, \beta \in \mathcal{A}$ and $\alpha \geq_{\mathcal{A}} \beta\rangle$ of projections so that

(1) $\pi_{\alpha\beta}$ projects U_α onto U_β, i.e.

$$X \in U_\beta \iff \pi_{\alpha\beta}^{-1''}(X) \in U_\alpha$$

(2) $\pi_{\alpha\alpha}$ is the identity, for every $\alpha \in \mathcal{A}$.

(3) (commutativity) for every $\alpha >_{\mathcal{A}} \beta >_{\mathcal{A}} \gamma$ there is $X \in U_\alpha$ so that for every $\nu \in X$

$$\pi_{\alpha\gamma}(\nu) = \pi_{\beta\gamma}(\pi_{\alpha\beta}(\nu))$$

(4) for every $\alpha \neq \beta$ and γ in $\mathcal{A}$, if $\gamma >_{\mathcal{A}} \alpha, \beta$ then

$$\{\nu < \kappa \mid \pi_{\gamma,\alpha}(\nu) \neq \pi_{\gamma,\beta}(\nu)\} \in U_\gamma\ .$$

Definition 1.1. A set $U = \ll \langle U_\alpha \mid \alpha \in \mathcal{A}\rangle, \langle \pi_{\alpha\beta} \mid \alpha, \beta \in \mathcal{A}$ and $\alpha \geq_{\mathcal{A}} \beta\rangle \gg$ is called a nice system if

(1) $\mathcal{A}$ has the least element 0

(2) $\langle U_\alpha \mid \alpha \in \mathcal{A}\rangle$ is a Rudin-Keisler directed commutative sequence

(3) U_0 is a normal measure over κ.

(4) for every $\alpha \in \mathcal{A}$ U_α is an ultrafilter over κ

(5) $\langle \pi_{\alpha\beta} \mid \alpha, \beta \in \mathcal{A}$ and $\alpha \geq_{\mathcal{A}} \beta\rangle$ satisfies conditions (1)–(4) of Definition 1.0

(6) (full commutativity at 0) for every $\alpha \geq_{\mathcal{A}} \beta$, $\nu < \kappa$ $\pi_{\alpha 0}(\nu) = \pi_{\beta 0}(\pi_{\alpha\beta}(\nu))$

(7) (independence of the choice of projection to zero) for every $\alpha, \beta \in \mathcal{A}\backslash\{0\}$, $\nu < \kappa$ $\pi_{\alpha 0}(\nu) = \pi_{\beta 0}(\nu)$

(8) for every $\alpha \in \mathcal{A}$ U_α is a P-point ultrafilter, i.e. for every $f \in {}^\kappa\kappa$, if f is not constant mod U_α, then there exists $X \in U_\alpha$ such that for every $\nu < \kappa$ $| X \cap f^{-1''}\{\nu\}| < \kappa$.

Let us call $|\mathcal{A}|$ the length of U.

Let us point out the following.

Proposition 1.1.1. *Let $\langle \mathcal{A}, <_\mathcal{A}\rangle$ be a κ^{++}-directed partial order and $\langle U_\alpha \mid \alpha \in \mathcal{A}\rangle$ a Rudin-Keisler directed commutative sequence of P-points. Then there exists a nice system of the length $|\mathcal{A}|$.*

Proof. Let $\alpha \in \mathcal{A}$. Consider $i : V \to N$ where N is the directed limit of the directed system of the structures of the form $Ult(V, U_\alpha)$ where $\alpha \in \mathcal{A}$. Define a normal ultrafilter U over κ to be the set of all $X \subseteq \kappa$ so that $\kappa \in i(X)$. We like to add U to the sequence as its least element. Let $i_\alpha : V \to N_\alpha$ be the natural embedding of V into $N_\alpha = Ult(V, U_\alpha)$. Define U^*_α by $X \in U^*_\alpha iff \kappa \in i_\alpha(X)$.Note that for α which is large enough (according to $<_\mathcal{A}$) we have $U^*_\alpha = U$.Define $\mathcal{A}' = \{\alpha \in \mathcal{A} \mid \alpha \neq 0$ and $U^*_\alpha = U\}$. Let $<_{\mathcal{A}'} = <_\mathcal{A} \restriction \mathcal{A}'$. Define $\mathcal{A}^* = \mathcal{A}' \cup \{0\}$ let $<_{\mathcal{A}^*} \restriction \mathcal{A}' = <_{\mathcal{A}'}$ and $0 <_{\mathcal{A}^*} \alpha$ for every $\alpha \in \mathcal{A}'$. Clearly $\mathcal{A}^*$ is still κ^{++}-directed and $|\mathcal{A}^*| = |\mathcal{A}|$. Denote U by U_0. Consider $\langle U_\alpha \mid \alpha \in \mathcal{A}^*\rangle$. Obviously, it is a Rudin-Keisler directed commutative sequence of P-points W.l. of g. we can assume that all U_α's are concentrating on κ.

For every $\alpha \in \mathcal{A}^*$ let us replace each U_α by an isomorphic ultrafilter U'_α over κ^2. Set $U'_0 = U_0$. If $\alpha \in \mathcal{A}^*$ $\alpha \neq 0$, then set $X \in U'_\alpha \iff \langle \kappa, [id]_{U_\alpha}\rangle \in i_\alpha(X)$, where $i_\alpha : V \to N_\alpha \simeq Ult(V, U_\alpha)$. Then the function $\delta : \kappa^2 \to \kappa$ defined by $\delta(\alpha, \beta) = \alpha$ will project each U'_α $(\alpha \in \mathcal{A}^*\backslash\{0\})$ onto U'_0. Fix some $\sigma : \kappa^2 \leftrightarrow \kappa$ a coding of pairs such that $\sigma\restriction\nu^2 : \nu^2 \leftrightarrow \nu$ for every cardinal ν. Let U^*_α be the isomorphic image of U'_α by σ, for every $\alpha \in \mathcal{A}\backslash\{0\}$. Set $U^*_0 = U_0$. Then $\langle U^*_\alpha \mid \alpha \in \mathcal{A}^*\rangle$ is still a Rudin-Keisler directed commutative sequence of P-points. Let $\langle \pi_{\alpha\beta} \mid \alpha \leq_{\mathcal{A}^*} \beta\rangle$ be a sequence of projection witnessing this. We shall use it to define a new sequence of projections $\langle \pi^*_{\alpha\beta} \mid \alpha <_{\mathcal{A}^*} \beta\rangle$ which will satisfy conditions (5),(6) and (7) of Definition 1.1. Set $\pi^*_{\alpha\alpha} = id$ for every $\alpha \in \mathcal{A}^*$. For $\alpha \in \mathcal{A}^*\backslash\{0\}$, $\nu < \kappa$ let $\pi^*_{\alpha 0}(\nu) = \delta(\sigma^{-1}(\nu))$, i.e. the first coordinate of the pair coded by ν. Clearly $\pi^*_{\alpha,0}$ projects U^*_α onto U^*_0 and it does not depend on α. Suppose now we have $\alpha, \beta \in \mathcal{A}^*\backslash\{0\}$, $\alpha >_{\mathcal{A}^*} \beta$. Define always $\pi^*_{\alpha,\beta}(0) = 0$. Pick a set $X \in U_\alpha$ so that for every $\nu \in X$ $\pi^*_{\alpha 0}(\nu) = \pi^*_{\beta 0}(\pi_{\alpha\beta}(\nu))$. Such X exists since $[\pi^*_{\alpha 0}]_{U^*_\alpha} = [\pi_{\alpha 0}]_{U^*_\alpha}$ $[\pi^*_{\beta 0}]_{U^*_\beta} = [\pi_{\beta 0}]_{U^*_\beta}$ and $\pi_{\alpha 0}(\nu) = \pi_{\beta 0}(\pi_{\alpha\beta}(\nu))$ on a set of ν's in U_α. Define

$$\pi^*_{\alpha\beta}(\nu) = \begin{cases} \pi_{\alpha\beta}(\nu) & \text{if } \nu \in X\backslash\{0\} \\ 0\,, & \text{otherwise.} \end{cases}$$

Then $\pi^*_{\alpha\beta}$ projects U^*_α onto U^*_β and for every $\nu < \kappa$ $\pi^*_{\alpha 0}(\nu) = \pi^*_{\beta 0}(\pi^*_{\alpha\beta}(\nu))$. So, $\mathbf{U} = \ll U^*_\alpha \mid \alpha \in \mathcal{A}^* >, < \pi^*_{\alpha\beta} \mid \alpha, \beta \in \mathcal{A}^*, \alpha <_{\mathcal{A}^*} \beta \gg$ is as desired. □

Remark 1.2. The system constructed above from an embedding $j : V \to M$, $M \supseteq V_{\kappa+\lambda}$ is a nice system of the length $\kappa^{+\lambda}$. The strength of this assumption is $o(\kappa) = \kappa^{+\lambda} + 1$. But for a nice system alone, the existence of a Rudin-Keisler directed commutative sequence of P-points is sufficient, by Proposition 1.1.1. It will be shown in [G-M] that $o(\kappa) = \kappa^{+\lambda}$ for λ a successor ordinal or $\kappa < cf\lambda < \lambda$ is enough for a Rudin-Keisler directed commutative sequence of P-points to exist in a generic extension. On the other hand, by Mitchell [Mi4] for $\lambda = 2$ and by [G4], modulo the weak covering lemma for hypermeasures, $o(\kappa) = \kappa^{+\lambda}$ looks also necessary for this.

Let $\mathbf{U}$ be some fixed nice system.

For $\nu < \kappa, 0 < \delta \in \mathcal{A}$ let us denote $\pi_{\delta,0}(\nu)$ by ν^0. By °-increasing sequence of ordinals we mean a sequence $\langle \nu_1, \ldots, \nu_n \rangle$ of ordinals below κ so that

$$\nu^0_1 < \nu^0_2 < \cdots < \nu^0_n\,.$$

For every $\delta \in \mathcal{A}$ by $X \in U_\delta$ we shall always mean that X for $\nu_1, \nu_2 \in X$ if $\nu^0_1 < \nu^0_2$ then $|\{\alpha \in X \mid \alpha^0 = \nu^0_1\}| < \nu^0_2$. Since U_δ is a P-point, most of its sets satisfy this condition. Also the following weak version of normality holds: if $X_i \in U_\delta (i < \kappa)$ then also $X = \Delta^*_{i<\kappa} X_i = \{\nu \mid \forall i < \nu^0\ \nu \in X_i\} \in U_\delta$.

Let $\nu < \kappa$ and $\langle \nu_1, \ldots, \nu_n \rangle$ be a finite sequence of ordinals below κ. Then ν is called permitted for $\langle \nu_1, \ldots, \nu_n \rangle$ if $\nu^0 > \max\{\nu^0_i \mid 1 \le i \le n\}$.

Let us now define a forcing notion for adding $|\mathcal{A}|$ ω-sequences to κ.

Definition 1.3. The set of forcing conditions $\mathcal{P}$ consists of all the elements p of the form $\{\langle \gamma, p^\gamma \rangle \mid \gamma \in g \backslash \{\max g\} \cup \{\langle \max g, p^{\max g}, T \rangle\}$, where

(1) $g \subseteq \mathcal{A}$ of cardinality $\le \kappa$ which has a maximal element (i.e. ${}_{\mathcal{A}}\ge$ than every element of g) and $0 \in g$. Further let us denote g by supp(p), $\max(g)$ by $mc(p)$, T by T^p and $p^{\max(g)}$ by p^{mc} (mc for the maximal coordinate).

(2) for $\gamma \in g$ p^γ is a finite °-increasing sequence of ordinals $< \kappa$.

(3) T is a tree with a trunk p^{mc} consisting of °-increasing sequences. All the splittings in T are required to be on sets in $U_{mc(p)}$, i.e. for every $\eta \in T$, if $\eta_T \ge p^{mc}$ then the set

$$\mathrm{Suc}_T(\eta) = \{\nu < \kappa \mid \eta^\frown \nu \in T\} \in U_{mc(p)}\,.$$

Also assume that for $\eta_{1T} \geq \eta_{2T} \geq p^{mc}$

$$\mathrm{Suc}_T(\eta_1) \subseteq \mathrm{Suc}_T(\eta_2) .$$

(4) for every $\gamma \in g$, $\pi_{mc(p),\gamma}(\max(p^{mc}))$ is not permitted for p^γ
(5) For every $\nu \in \mathrm{Suc}_T(p^{mc})$

$$|\{\gamma \in g \mid \nu \quad \text{is permitted for} \quad p^\gamma\}| \leq \nu^0$$

(6) $\pi_{mc(p),0}$ projects p^{mc} onto p^0, in particular, p^{mc} and p^0 are of the same length.

Let us give some intuitive motivation for the definition of forcing conditions. We like to add a Prikry sequence for every $U_\delta (\delta \in \mathcal{A})$. The finite sequences p^γ ($\gamma \in \mathrm{supp}\, p$) are initial segments of such sequences. The support of p has two distinguished coordinates. The first is the 0-coordinate of p and the second is its maximal coordinate. The 0-coordinate or more precisely the Prikry sequence for the normal measure will be used further in order to push the present construction to $\aleph_\omega$. Also condition (6) will be used only for this purpose. The maximal coordinate of p is responsible for extending the Prikry sequences for γ's in the support of p. The tree T^p is a set of possible candidates for extending p^{mc} and by using the projections map $\pi_{mc(p),\gamma}$ ($\gamma \in \mathrm{supp}\, p$) it becomes also the set of candidates for extending p^γ's. Instead of working with a tree, it is possible using the diagonal intersection Δ^* to replace it by a single set. Condition (4) means that the information carried by $\max(p^{mc})$ is impossible to project down. The reasons for such a condition are technical. Condition (5) is desired to allow the use of the diagonal intersections.

Definition 1.4. Let $p, q \in \mathcal{P}$. We say that p extends $q (p \geq q)$ if

(1) $\mathrm{supp}(p) \supseteq \mathrm{supp}(q)$
(2) for every $\gamma \in \mathrm{supp}(q)$ p^γ is an endextension of q^γ
(3) $p^{mc(q)} \in T^q$
(4) for every $\gamma \in \mathrm{supp}(q)$
$p^\gamma \backslash q^\gamma = \pi_{mc(q),\gamma}$ "$((p^{mc(q)} \backslash q^{mc(q)}) \restriction (\text{length } (p^{mc}) \backslash (i+1))$ where $i \in \mathrm{dom}\, p^{mc(q)}$ is the largest such that $p^{mc(q)}(i)$ is not permitted for q^γ.
(5) $\pi_{mc(p),mc(q)}$ projects $T^p_{p^{mc}}$ into $T^q_{q^{mc}}$
(6) for every $\gamma \in \mathrm{supp}\, q$, for every $\nu \in \mathrm{Suc}_{T^p}(p^{mc})$, if ν is permitted for p^γ. then

$$\pi_{mc(p),\gamma}(\nu) = \pi_{mc(q),\gamma}(\pi_{mc(p),mc(q)}(\nu)) .$$

In clause (5) above we use denote, for tree T which is a tree of finite sequences, $\eta \in T$, by T_η the subtree above η, namely all the finite sequences μ such that $\eta^\frown\mu$ is in T.

Intuitively, we are allowing to add almost everything on the new coordinates and restrict ourselves to choosing extensions from the sets of measure one on the old coordinates. Actually here we are really extending only the maximal old coordinate and then we are using the projection map. This idea goes back to [G1] and further to Mitchell [Mi1].

Definition 1.5. Let $p, q \in \mathcal{P}$. We say that p is a direct or Prikry extension of $q(p \geq^* q)$ if

(1) $p \geq q$
(2) for every $\gamma \in \text{supp}(q)$ $p^\gamma = q^\gamma$.

Our strategy would be to show that $\langle \mathcal{P}, \leq, \leq^* \rangle$ is a κ-weakly closed forcing satisfying the Prikry condition and κ^{++}-c.c. Where κ-weakly closedness means that $\langle \mathcal{P}, \leq^* \rangle$ is κ-closed and the Prikry condition means the following: for every statement σ of the forcing language and for every $q \in \mathcal{P}$ there is $p^* \geq q$ deciding σ.

The Prikry condition together with κ-weak closedness insure that no new bounded subsets of κ are added. κ^{++}-c.c. takes care of cardinals $\geq \kappa^{++}$. Since κ will change its cofinality to $\aleph_0$, an argument similar to those of [M2], 4.2 will be used to show that κ^+ is preserved. Condition (6) of the definition of nice system insures that at least $|\mathcal{A}|$-many ω-sequences will be added to κ.

Lemma 1.6. *The relation $\leq$ is a partial order.*

Proof. Let us check the transitivity of $\leq$. Suppose that $r \leq q$ and $q \leq p$. Let us show that $r \leq p$. Conditions (1) and (2) of Definition 1.4 are obviously satisfied. Let us check (3), i.e. let us show that $p^{mc(r)} \in T^r$. Since $p \geq q \geq r$, $mc(r) \in \text{supp}(q), q^{mc(r)} \in T^r$ and

$$p^{mc(r)} \backslash q^{mc(r)} = \pi''_{mc(q),mc(r)}(p^{mc(q)} \backslash q^{mc(q)}) \ .$$

Also $p^{mc(q)} \in T^q$. By (5) of 1.4 (for q and r) $\pi_{mc(q),mc(r)}$ projects $T^q_{q^{mc}}$ into subtree of $T^r_{q^{mc(r)}}$. Hence $p^{mc(r)} \in T^r$ and, so condition (3) is satisfied.

Let us check condition (4). Suppose that $\gamma \in \text{supp}(r)$. We need to show that $p^\gamma \backslash r^\gamma = \pi''_{mc(r),\gamma}(p^{mc(r)} \backslash r^{mc(r)})$. In order to simplify the notation, we are assuming here that every element of $p^{mc(r)} \backslash r^{mc(r)}$ is permitted for r^γ. Since $q \geq r, q^\gamma \backslash r^\gamma = \pi''_{mc(r),\gamma}\left(q^{mc(r)} \backslash r^{mc(r)}\right)$. So, we need to show only that $p^\gamma \backslash q^\gamma = \pi''_{mc(r),\gamma}\left(p^{mc(r)} \backslash q^{mc(r)}\right)$. Since $p \geq q, p^{mc(q)} \in T^q$ and

$p^\gamma\backslash q^\gamma = \pi''_{mc(q),\gamma}\left(p^{mc(q)}\backslash q^{mc(q)}\right)$. Using condition (6) of 1.4 for $q \geq r$ and the elements of $p^{mc(q)}\backslash q^{mc(q)}$, we obtain the following

$$\begin{aligned} p^\gamma\backslash q^\gamma &= \pi''_{mc(q),\gamma}\left(p^{mc(q)}\backslash q^{mc(q)}\right) \\ &= \pi_{mc(r),\gamma}\left(\pi''_{mc(q),mc(r)}(p^{mc(q)}\backslash q^{mc(q)})\right) \\ &= \pi''_{mc(r),\gamma}\left(p^{mc(r)}\backslash q^{mc(r)}\right). \end{aligned}$$

The last equality holds by condition (4) of 1.4 used for p and q.

Let us check condition (5), i.e. $\pi_{mc(p),mc(r)}$ projects $T^p_{p^{mc}}$ into $T^r_{p^{mc(r)}}$. Since $p \geq q$, $T^p_{p^{mc}}$ is projected by $\pi_{mc(p),mc(q)}$ into $T^q_{q^{mc}}$. Since $q \geq r, \pi_{mc(q),mc(r)}$ projects $T^q_{q^{mc}}$ into $T^r_{q^{mc(r)}}$. Now, using condition (6) for p and q with $\gamma = mc(r)$, we obtain condition (5) for p and r.

Finally, let us check condition (6). Let $\gamma \in \mathrm{supp}(r)$, $\nu \in \mathrm{Suc}_{T^p}(p^{mc})$ and suppose that ν is permitted for p^γ. Using condition (5) for p and q, we obtain that $\pi_{mc(p),mc(q)}(\nu) \in \mathrm{Suc}_{T^q}(q^{mc})$. Recall, that it was required in the definition of a condition that each splitting contains splitting below it in the tree. Denote $\pi_{mc(p),mc(q)}(\nu)$ by δ. By condition (6) for q and r $\pi_{mc(q),\gamma}(\delta) = \pi_{mc(r),\gamma}(\pi_{mc(q),mc(r)}(\delta))$. Using (6) for p and q, we obtain

$$\begin{aligned} \pi_{mc(p),\gamma}(\nu) &= \pi_{mc(q),\gamma}(\pi_{mc(p),mc(q)}(\nu)) = \\ =&\pi_{mc(q),\gamma}(\delta) = \pi_{mc(r),\gamma}(\pi_{mc(q),mc(r)}(\delta)) \ . \end{aligned}$$

Once more using (6) for p and q,

$$\pi_{mc(q),mc(r)}(\pi_{mc(p),mc(q)}(\nu)) = \pi_{mc(p),mc(r)}(\nu) \ .$$

This completes the checking of (6) and also the proof of the lemma. □

The main point of the proof appears in the next lemma.

Lemma 1.7. *Let $q \in \mathcal{P}$ and $\alpha \in \mathcal{A}$ then there is $p \geq^* q$ so that $\alpha \in \mathrm{supp}(p)$.*

Proof. If $\alpha <_{\mathcal{A}} mc(q)$, then it is obvious. Thus, if $\alpha \in \mathrm{supp}(q)$, then we can take $p = q$. Otherwise add to q a pair $\langle \alpha, t\rangle$ where t is any $^\circ$-increasing sequence so that $\max(q^{mc})$ is not permitted for t.

Suppose now that $\alpha \not\leq_{\mathcal{A}} mc(q)$. Pick some $\beta \in \mathcal{A}$ so that $\beta \geq_{\mathcal{A}} \alpha$ and $\beta \geq_{\mathcal{A}} mc(q)$. W.l. of g let us assume that $\beta = \alpha$. We shall define p to be of the form

$$q' \cup \{\langle \alpha, t, T\rangle\}$$

where q' is constructed from q by removing T^q from the triple $\langle mc(q), q^{mc}, T^q\rangle$, t is an $^\circ$-increasing sequence which projects onto q^0 by $\pi_{\alpha 0}$ and the tree T will be defined below.

Consider first the tree T_0 which is the inverse image of $T^q_{q^{mc}}$ by $\pi_{\alpha,mc(q)}$, with t added as the trunk. Then $p_0 = q' \cup \{\langle \alpha, t, T_0\rangle\}$ is a condition in $\mathcal{P}$ which is "almost" an extension and even a direct extension of q. The only problematic thing is that condition (6) of Definition 1.4 may not be satisfied by p_0 and q. In order to repair this, let us shrink the tree T_0 a little.

Denote $\mathrm{Suc}_{T_0}(t)$ by A. For $\nu \in A$ set $B_\nu = \{\gamma \in \mathrm{supp}(q) \mid \gamma \neq mc(q)$ and ν is permitted for $q^\gamma\}$. Then $|B_\nu| \leq \nu^0$, since $\pi_{\alpha,mc(q)}(\nu) \in \mathrm{Suc}_{T^q}(q^{mc}), \nu^0 = \pi_{\alpha 0}(\nu) = \pi_{mc(q),0}\ (\pi_{\alpha mc(q)}(\nu))$ and q being in $\mathcal{P}$, satisfies condition (5) of Definition 1.3. Clearly, for $\nu, \delta \in A$, if $\nu^0 = \delta^0$ then $B_\nu = B_\delta$, and if $\nu^0 > \delta^0$ then $B_\nu \supseteq B_\delta$. Also, if $\nu \in A$ and ν^0 is a limit point of $\{\delta^0 \mid \delta \in A\}$, then $B_\nu = \cup\{B_\delta \mid \delta \in A \text{ and } \delta^0 < \nu^0\}$. So the sequence $\langle B_\nu \mid \nu \in A\rangle$ is increasing and continuous (according to $\nu^0 - s$). Obviously $\cup\{B_\nu \mid \nu \in A\} = \mathrm{supp}(q)\backslash\{mc(q)\}$. Let $\langle \xi_i \mid i < \kappa\rangle$ be an enumeration of $\mathrm{supp}(q)\backslash mc(q)$ such that for every $\nu \in A$

$$B_\nu \subseteq \{\xi_i \mid i < \nu^0\}\ .$$

Pick now for every $i \in A$ a set $C_i \subseteq A, C_i \in U_\alpha$ so that for every $\nu \in C_i$ $\pi_{\alpha\xi_i}(\nu) = \pi_{mc(q),\xi_i}(\pi_{\alpha,mc(q)}(\nu))$. Let $C = A \cap \Delta^*_{i<\kappa} C_i = \{\nu \in A \mid \forall_{i<\nu^0} \nu \in C_i\}$. Then $C \in U_\alpha$.

Define now T to be the tree obtained from T_0 by intersecting every level of T_0 with C. Let us show that condition (6) of Definition 1.4 is now satisfied. Suppose $\gamma \in \mathrm{supp}(q)$. If $\gamma = mc(q)$, then everything is trivial. Assume that $\gamma \in \mathrm{supp}(q)\backslash\{mc(q)\}$. Then for some $i_0 < \kappa$ $\gamma = \xi_{i_0}$. Suppose that some $\nu \in C$ is permitted for q^γ. Then $\xi_{i_0} = \gamma \in B_\nu$. Since $B_\nu \subseteq \{\xi_i \mid i < \nu^0\}$, $i_0 < \nu^0$. Then $\nu \in C_{i_0}$. Hence

$$\pi_{\alpha\xi_{i_0}}(\nu) = \pi_{mc(q),\xi_{i_0}}(\pi_{\alpha,mc(q)}(\nu))\ .$$

So condition (6) is satisfied by p. It means that $p^* \geq q$. $\square$.

Lemma 1.8.

(a) $\langle \mathcal{P}, \leq\rangle$ *satisfies* κ^{++}*-c.c.*

(b) $\langle \mathcal{P}, \leq^*\rangle$ *is* κ*-closed.*

Proof of (a). Let $\langle p_\alpha \mid \alpha < \kappa^{++}\rangle$ be a set of forcing conditions. W.l. of g. let us assume their supports form a Δ-system and are contained in κ^{++}. Also assume that there are s and $\langle t, T\rangle$ so that for every $\alpha < \kappa^{++}$ $p_\alpha \restriction \alpha = s$

and $\langle p_\alpha^{mc}, T^{p_\alpha}\rangle = \langle t, T\rangle$. Let us show then that p_α and p_β are compatible for every $\alpha, \beta < \kappa^{++}$.

Let $\alpha, \beta < \kappa^{++}$ be fixed.

We would like simply to take the union $p_\alpha \cup p_\beta$ and to show that this is a condition stronger than both p_α and p_β. The first problem is that $p_\alpha \cup p_\beta$ may not be in $\mathcal{P}$, since $\text{supp}(p_\alpha \cup p_\beta) = \text{supp}(p_\alpha) \cup \text{supp}(p_\beta)$ may not have a maximal element. In order to fix this, let us add say to p_α some new coordinate δ so that $\delta_{\mathcal{A}} \geq mc(p_\alpha), mc(p_\beta)$. Let p_α^* be the extension of p_α defined in the previous lemma by adding δ as a new coordinate to p_α. Then $p_\alpha^* \cup p_\beta \in \mathcal{P}$. But we do need a condition stronger than both p_α and p_β. The condition $p_\alpha^* \cup p_\beta$ is a good candidate for it. The only problematic things here are (5) and (6) of Definition 1.4. Actually, (5) can be easily satisfied by intersecting $T^{p_\alpha^*}_{(p_\alpha^*)^{mc}}$ with $\pi^{-1}_{\delta, mc(p_\beta)}$ "$(T^{p_\beta}_{p_\beta^{mc}})$. In order to satisfy (6), we need to shrink $T^{p_\alpha^*}$ more. The argument of the previous lemma can be used for this. □

Proof of (b). Let $\delta < \kappa$ and let $\langle p_i \mid i < \delta\rangle$ be an $\leq^*$-increasing sequence of elements of $\mathcal{P}$. Pick $\alpha \in \mathcal{A}$ above $\{mc(p_i) \mid i < \delta\}$. Let p be the union of p_i's with T^{p_i} removed. Set $T = \bigcap_{i<\delta} \pi^{-1}_{\alpha, mc(p_i)}$ "(T^{p_i}). Also remove all τ's with $\tau^0 \leq \delta$ from this tree. Let t be a °-increasing sequence so that $\pi''_{\alpha 0}(t) = p_0^0$. Consider $p \cup \{\langle \alpha, t, T\rangle\}$. Clearly, it belongs to P. Now, as in Lemma 1.6, shrink T to a tree T^i so that $p \cup \{\langle \alpha, t, T^i\rangle\} {}^*\!\geq p_i$, where $i < \delta$. Let $T^* = \bigcap_{i<\delta} T^i$ and consider $r = p \cup \{\langle \alpha, t, T^*\rangle\}$. Then $r {}^*\!\geq p_i$ for every $i < \delta$. □

Lemma 1.9. *$\langle \mathcal{P}, \leq, \leq^*\rangle$ satisfies the Prikry condition, i.e. for every statement σ of the forcing language, for every $q \in \mathcal{P}$ there exists $p \geq^* q$ deciding σ.*

Proof. Let σ be a statement and $q \in \mathcal{P}$. In order to simplify the notation we are assuming that $q = \phi$. Pick an elementary submodel N of V_μ for μ large enough containing all the relevant information of cardinality κ^+ and closed under κ-sequences of its elements. Pick $\alpha \in \mathcal{A}$ which is above all the elements of $N \cap \mathcal{A}$. Let T be a tree so that $\{\langle \alpha, \phi, T\rangle\} \in \mathcal{P}$. More precisely, we should write $\{\langle 0, \phi\rangle\} \cup \{\langle \alpha, \phi, T\rangle\}$. But let us omit the least coordinate when the meaning is clear. If there is $p \in N$ so that $p \cup \{\langle \alpha, \phi, T'\rangle\} \in \mathcal{P}$ and decides σ, for some $T' \subseteq T$, then we are done. Suppose otherwise. Denote $\text{Suc}_T(\langle\rangle)$ by A. We shall define by induction sequences $\langle p_\nu \mid \nu \in A\rangle$ and $\langle T^\nu \mid \nu \in A\rangle$.

For this purpose fix some well ordering $\prec$ of A so that $\nu_1^0 < \nu_2^0$ implies $\nu_1 \prec \nu_2$. We are assuming that A is just a subset of κ and $\prec$ is the usual well-ordering of ordinals.

Let $\nu = \min A$. Consider $\{< \alpha, \langle\nu\rangle, T_{\langle\nu\rangle} >\}$. If there is no $p \in N$ and $T' \subseteq T_{\langle\nu\rangle}$ such that $p \cup \{\langle\alpha, < \nu >, T'\rangle\}$ is in $\mathcal{P}$ and decides σ, then set $p_\nu = \phi$ and $T^\nu = T_{\langle\nu\rangle}$. Otherwise, pick some p and $T' \subseteq T_{\langle\nu\rangle}$ so that $p \cup \{\langle\alpha, < \nu >, T'\rangle\}$ is in $\mathcal{P}$ and decides σ. Set $p_\nu = p$ and $T^\nu = T'$.

Suppose now that p_ξ and T^ξ are defined for every $\xi < \nu$ in A. We shall define p_ν and T^ν. But let us first define p'_ν and p''_ν. Define p''_ν to be the union of all p_ξ's with $\xi \in A \cap \nu$. Let $p'_\nu = \{\langle\gamma, p'^{\gamma}_\nu\rangle \mid \gamma \in \text{supp}(p''_\nu)\}$, where for $\gamma \in \text{supp}(p''_\nu)$ $p'^{\gamma}_\nu = p''^{\gamma}_\nu$ unless ν is permitted for p''^{γ} and then $p'^{\gamma}_\nu = p''^{\gamma}_\nu \cap < \pi_{\alpha\gamma}(\nu) >$. If there is no $p \in N$ and T' so that $q = p \cup \{\langle\alpha, < \nu >, T'\rangle\} \in \mathcal{P}$, $q \,{}^{*}\!\!\geq p'_\nu \cup \{\langle\alpha, < \nu >, T_{\langle\nu\rangle}\rangle\}$ and $q \| \sigma$, then set $p_\nu = p''_\nu$ and $T^\nu = T_{\langle\nu\rangle}$. Suppose otherwise. Let p, T' be witnessing this. Then set $T^\nu = T'$ and $p_\nu = p''_\nu \cup (p \backslash p'_\nu)$.

This completes the inductive definition. Set $p = \bigcup_{\nu \in A} p_\nu$. For $i < \kappa$ let

$$C_i = \begin{cases} A, & \text{if there is no } \delta \in A \text{ such that } \delta^0 = i \\ \cap \{\text{Suc}_{T^\delta}(\langle\delta\rangle) \mid \delta \in A \text{ and } \delta^0 = i\}, & \text{otherwise} \end{cases}$$

Note that always $C_i \in U_\alpha$ since $A \in U_\alpha$ and this means by our agreement that for $\nu_1, \nu_2 \in A$ if $\nu_1^0 < \nu_2^0$ then $|\{\gamma \in X \mid \gamma^0 = \nu_1^0\}| < \nu_2^0$. Set $A^* = A \cap \Delta^*_{i<\kappa} C_i$. Then for every $\nu \in A^*$ for every $\delta \in A$ if $\delta^0 < \nu^0$ then $\nu \in \text{Suc}_{T^\delta}(\langle\delta\rangle)$. Let S be the tree obtained from T by first replacing $T_{\langle\nu\rangle}$ by T^ν for every $\nu \in A^*$ and then intersecting all levels of it with A^*.

Claim 1.9.1. $p \cup \{\langle\alpha, \phi, S\rangle\}$ belongs to $\mathcal{P}$.

Proof. The only nontrivial point here is to show that $p \cup \{\langle\alpha, \phi, S\rangle\}$ satisfies condition (5) of the definition of $\mathcal{P}$. So let $\nu \in \text{Suc}_S(\langle\rangle)$. By definition of $S, \text{Suc}_S(\langle\rangle) = A^*$. Consider the set

$$B_\nu = \{\gamma \in \text{supp}(p) \mid \nu \text{ is permitted for } p^\gamma\} .$$

For every $\delta \in A$ let $B_{\nu,\delta} = \{\gamma \in \mathit{supp}(p_\delta) \mid \nu \text{ is permitted for } p^\gamma_\delta\}$. Then $B_\nu = \bigcup_{\delta \in A} B_{\nu,\delta}$. But, actually the definition of the sequence $\langle p_\delta \mid \delta \in A\rangle$ implies that $B_\nu = \cup\{B_{\nu,\delta} \mid \delta \in A \text{ and } \delta^0 < \nu^0\}$. The number of δ's in A with $\delta^0 < \nu^0$ is $\leq \nu^0$, since $A \in U_\alpha$ and it means in particular, that for every $\xi < \nu^0$ $|\{\delta \in A \mid \delta^0 = \xi\}| < \nu^0$. So it is enough to show that for every $\delta \in A, \delta^0 < \nu^0$ implies $|B_{\nu,\delta}| \leq \nu^0$. Fix some $\delta \in A$ such that $\delta^0 < \nu^0$. Since $\nu \in A^*$ and $\delta^0 < \nu^0$, $\nu \in \text{Suc}_{T^\delta}(\langle\delta\rangle)$. But $p_\delta \cup \{\langle\alpha, < \delta >, T^\delta\rangle\} \in \mathcal{P}$. So, by the definition of $\mathcal{P}, |B_{\nu,\delta}| \leq \nu^0$.
□ of the claim.

Then, clearly, $p \cup \{\langle\alpha, \phi, S\rangle\} \,{}^{*}\!\!\geq \{\langle\alpha, \phi, T\rangle\}$.

For $\delta \in \mathrm{Suc}_S(<>) = A^*$ let us denote by $(p \cup \{\langle \alpha, \phi, S\rangle\{)_\delta$ the sequence $\{\langle \gamma, (p^\gamma)_{\pi_{\alpha\gamma}(\delta)}\rangle \mid \gamma \in \operatorname{supp} p\} \cup \{\langle \alpha, < \delta >, S_{\langle \delta \rangle}\rangle\}$, where

$$(p^\gamma)_{\pi_{\alpha\gamma}(\delta)} = \begin{cases} p^\gamma \cap \pi_{\alpha\gamma}(\delta), & \text{if } \delta \text{ is permitted for } p^\gamma \\ p^\gamma, & \text{otherwise} \end{cases} .$$

Note that $(p \cup \{\langle \alpha, \phi, S\rangle\})_\delta$ is a condition and $\pi_{\alpha\gamma}(\delta)$ is added only for γ's which appear in the support of some p_ξ with $\xi^0 < \delta^0$ and hence, with $\xi < \delta$. Also $(p \cup \{\langle \alpha, \phi, S\rangle\})_\delta{}^* \geq p_\delta \cup \{\langle \alpha, < \delta >, T^\delta\rangle$.

Claim 1.9.2. For every $\delta \in \mathrm{Suc}_{<>} S$ if for some $q, R \in N$ $(p \cup \{\langle \alpha, \phi, S\rangle\})_\delta \leq^* q \cup \{\langle \alpha, < \delta >, R\rangle\}$ and $q \cup \{\langle \alpha, < \delta >, R\rangle \Vdash \sigma$ (or $\neg\sigma$), then $(p \cup \{\langle \alpha, \phi, S\rangle\})_\delta \Vdash \sigma$ (or $\neg\sigma$).

Proof. Note that such $q \cup \{\langle \alpha, < \delta >, R\rangle\}$ is a direct extension of $p_\delta \cup \{\langle \alpha, < \delta >, T^\delta\rangle\}$. By the choice of p_δ and T^δ, then $p_\delta \cup \{\langle \alpha, < \delta >, T^\delta\rangle\}$ forces σ (or $\neg\sigma$). But $(p \cup \{\langle \alpha, \phi, S\rangle\})_\delta{}^* \geq p_\delta \cup \{\langle \alpha, < \delta >, T^\delta\rangle\}$.
□ of the claim.

Let us shrink now the first level of S in order to insure that for every δ_1 and δ_2 in the new first level

$$(p \cup \{\langle \alpha, \phi, S\rangle\})_{\delta_1} \Vdash \sigma \quad (\text{or } \neg\sigma)$$

iff

$$(p \cup \{\langle \alpha, \phi, S\rangle\})_{\delta_2} \Vdash \sigma \quad (\text{or } \neg\sigma) .$$

Let us denote such shrunken tree by the same letter.

Claim 1.9.3. For every $\delta \in \mathrm{Suc}_{<>} S$ $(p \cup \{\langle \alpha, \phi S\rangle\})_\delta \nVdash \sigma$.

Proof. Suppose otherwise. Then every δ in $\mathrm{Suc}_{<>} S$ will force the same truth value of σ. Suppose, for example, that σ is forced. Then $p \cup \{\langle \alpha, \phi, S\rangle\}$ will force σ. Since every $q \geq p \cup \{\langle \alpha, \phi, S\rangle\}$ is compatible with one of $(p \cup \{\langle \alpha, \phi, S\rangle\})_\delta$ for $\delta \in \mathrm{Suc}_{<>} S$. This contradicts the initial assumption.
□ of the claim.

Now, climbing up level by level in the fashion described above for the first level, construct a direct extension $p^* \cup \{\langle \alpha, \phi, S^*\rangle\}$ of $p \cup \{\langle \alpha, \phi, S\rangle\}$ so that

(a) for every $\eta \in S^*$, if for some $q, R \in N$ $(p^* \cup \{\langle \alpha, \phi, S^*\rangle\})_\eta \leq^* q \cup \{\langle \alpha, \eta, R\rangle\}$ and $q \cup \{\langle \alpha, \eta, R\rangle\} \Vdash \sigma$ (or $\neg\sigma$), then $(p^* \cup \{\langle \alpha, \phi, S^*\rangle\})_\eta \Vdash \sigma$ (or $\neg\sigma$)

(b) If $\eta_1, \eta_2 \in S^*$ are of the same length then

$$(p^* \cup \{\langle \alpha, \phi, S^* \rangle\})_{\eta_1} \Vdash \sigma \quad (\text{or } \neg\sigma)$$

iff

$$(p^* \cup \{\langle \alpha, \phi, S^* \rangle\})_{\eta_2} \Vdash \sigma \quad (\text{or } \neg\sigma)$$

As in Claim 1.9.3, it is impossible to have $\eta \in S^*$ so that $(p^* \cup \{\langle \alpha, \phi, S^* \rangle\})_\eta$ decides σ. Combining this with (a) we obtain the following.

Claim 1.9.4. For every $q, R, t \in N$, if $q \cup \{\langle \alpha, t, R \rangle\} \geq p^* \cup \{\langle \alpha, \phi, S^* \rangle\}$ then $q \cup \{\langle \alpha, t, R \rangle\}$ does not decide σ.

Proof. Just note that $q \cup \{\langle \alpha, t, R \rangle\} ^* \geq (p^* \cup \{\langle \alpha, \phi, S^* \rangle\})_t$ and use (a).
□ of the claim.

Pick some $\beta \in N \cap \mathcal{A}$ which is above every element of $\text{supp}(p^*)$. It is possible since $\text{supp}(p^*) \in N$. Shrink S^* to a tree S^{**}, as in Lemma 1.7 in order to insure the following:

for every $\nu \in \text{Suc}_{S^{**}}(<>)$, for every $\gamma \in \text{supp}(p^*)$, if ν is permitted for $(p^*)^\gamma$, then $\pi_{\alpha\gamma}(\nu) = \pi_{\beta\gamma}(\pi_{\alpha\beta}(\nu))$.

Let S^{***} be the projection of S^{**} to β via $\pi_{\alpha\beta}$. Denote $p^* \cup \{\langle \beta, \phi, S^{***} \rangle\}$ by p^{**}. Then $p^{**} \in N$ and $p^{**} \cup \{\langle \alpha, \phi, S^{**} \rangle\}^* \geq p^* \cup \{\langle \alpha, \phi, S^* \rangle\}$. Since N is an elementary submodel there is some $q \in N$ $q \geq p^{**}$ deciding σ. Let, for example, $q \Vdash \sigma$. Pick some $t \in S^{**}$ so that $\pi''_{\alpha\beta}(t) = q^\beta$. Such t exists, since by Definition 1.4 q^β belongs to S^{***} which is the image of S^{**} under $\pi_{\alpha\beta}$. Note also that $mc(q) <_{\mathcal{A}} \alpha$, by the choice of N. Let R be a tree obtained from S_t^{**} by intersecting S_t^{**} with $\pi_{\alpha,mc(q)}^{-1}(T^q)$ and shrinking, if necessary, as in Lemma 1.7 in order to insure the equality of projections $\pi_{\alpha\gamma}$ and $\pi_{mc(q),\gamma} \circ \pi_{\alpha,mc(q)}$ for permitted γ's in $\text{supp}(q)$. Then $q \cup \{\langle \alpha, t, R \rangle\}$ will be a condition stronger than q. Hence, it forces σ. But this contradicts Claim 1.9.4, since $q \cup \{\langle \alpha, t, R \rangle\} \geq p^* \cup \{\langle \alpha, \phi, S^* \rangle\}$.
Contradiction. □

Remark. It is possible to replace κ^{++}-directness by κ^+-directness. For this instead of working with one fixed α as the maximal element of supports, an increasing sequence of the length κ of α's should be used. The proof becomes more complicated, but essentially no new ideas are needed.

Let G be a generic subset of $\mathcal{P}$. By Lemma 1.7, for every $\alpha \in \mathcal{A}$ there is $p \in G$ with $\alpha \in \text{supp}(p)$. Let us denote $\cup\{p^\alpha \mid p \in G\}$ by G^α.

Lemma 1.10.

(a) *For every $\alpha \in \mathcal{A}, G^\alpha$ is a Prikry sequence for U_α,* i.e. *an ω-sequence s.t. for every $X \in U_\alpha$ it is almost contained in X.*

(b) G^0 *is an* ω*-sequence unbounded in* κ.

(c) *If* $\alpha \neq \beta$ *are in* $\mathcal{A}$ *then* $G^\alpha \neq G^\beta$.

Proof. (a) follows from the definition of $\mathcal{P}$. (b) is a trivial consequence of (a). For (c) note that there is $\gamma \in \mathcal{A}\gamma \geq_{\mathcal{A}} \alpha, \beta$. Condition (6) of the definition of a nice system requires that $\{\nu < \kappa \mid \pi_{\gamma\alpha}(\nu) \neq \pi_{\gamma\beta}(\nu)\} \in U_\gamma$. This together with the definition of $\mathcal{P}$ implies that $G^\alpha \neq G^\beta$. □

Lemma 1.11. κ^+ *remains a cardinal in* $V[G]$.

Proof. Suppose otherwise. Then it changes its cofinality to some $\mu < \kappa$. Let $g : \mu \to (\kappa^+)^V$ be unbounded in $(\kappa^+)^V$. Pick $p \in G$ forcing this. Suppose for simplicity that $\phi \Vdash \underset{\sim}{g}:\check{\mu} \to \check{\kappa}^+$ unbounded. Pick an elementary submodel N as in Lemma 1.10. Let $\alpha \in \mathcal{A}$ be above every element of $N \cap \mathcal{A}$. Pick a tree T so that $\{\langle \alpha, \phi, T \rangle\} \in \mathcal{P}$. As in Lemma 1.9, define by induction an $\leq^*$ increasing sequence of direct extensions of $\{\langle \alpha, \phi, T \rangle\}$ $\langle q_i \cup \{\langle \alpha, \phi, S^i \rangle\} \mid i < \mu \rangle$ so that

(a) $q_i \in N$

(b) If for some $q, R, t \in N$, some $j < \kappa^+$ $q \cup \{\langle \alpha, t, R \rangle\} \geq q_i \cup \{\langle \alpha, \phi, S^i \rangle\}$ and $q \cup \{\langle \alpha, t, R \rangle\} \Vdash \underset{\sim}{g}\,(\check{i}) = \check{j}$, then

$$(q_i \cup \{\langle \alpha, \phi, S^i \rangle\})_t \Vdash \underset{\sim}{g}\,(\check{i}) = \check{j} .$$

Using Lemma 1.8, find S so that $\bigcup_{i<\mu} q_i \cup \{\langle \alpha, \phi, S \rangle\} ^* \geq q_i \cup \{\langle \alpha, \phi, S^i \rangle\}$ for every $i < \mu$. Denote $\bigcup_{i<\mu} q_i$ by p. As in Lemma 1.9, pick $\beta \in N \cap \mathcal{A}$ above $\operatorname{supp}(p)$ and project S to β using $\pi_{\alpha\beta}$. Denote the projection by S^*. Let $p^* = p \cup \{\langle \beta, \phi, S^* \rangle\}$. Then $p^* \in N$ and $p^* \cup \{\langle \alpha, \phi, S \rangle\} ^* \geq p \cup \{\langle \alpha, \phi, S \rangle\}$. Since N is an elementary submodel, for every $i < \mu$ there will be $q \in N, q \geq p^*$ forcing a value of $\underset{\sim}{g}\,(i)$. Then, using (b), as in Lemma 1.9, for some $t \in S$ $(p \cup \{\langle \beta, \phi, S \rangle\})_t$ will force the same value of $\underset{\sim}{g}\,(i)$. But $|S| = \kappa$. So, all such values are bounded in κ^+ by some ordinal δ. Which is impossible, since $N \supseteq \kappa^+$ and $N \models (\phi \Vdash (\underset{\sim}{g}:\check{\mu} \to \check{\kappa}^+$ unbounded)).

Contradiction. □

Now combining the lemmas together, we obtain the following.

Theorem 1.12.

(a) $V[G]$ *is a cardinal preserving extension of* V.

(b) *No new bounded subsets are added to* κ.

(c) $cf\kappa = \aleph_0$
(d) $\kappa^{\aleph_0} \geq |\mathcal{A}|$.

If κ is a strong cardinal, then for every λ a nice system of a length $\geq \lambda$ can be constructed. The Solovay arguments [So-Re-Ka] for producing a function $f : \kappa \to \kappa$ and $j : V \to M$ so that $j(f)(\kappa) > \lambda$ for supercompact κ, can work without changes also for strong κ. Now, having f and j we can use a nice system defined from them in the beginning. So, the following holds.

Theorem 1.13. *Let V be a model of GCH, κ be a strong cardinal. Then for every λ there exists a cardinal preserving set generic extension $V[G]$ of V so that*

(a) *no new bounded subsets are added to κ.*
(b) *κ changes its cofinality to $\aleph_0$.*
(c) $2^\kappa \geq \lambda$.

2. Down to $\aleph_\omega$, a Finite Gap

In this section we shall define a forcing notion which will combine the forcing of Section 1 with collapsing of cardinals in order to construct a model satisfying GCH below $\aleph_\omega$ and $2^{\aleph_\omega} = \aleph_{\omega+m}$ for any m, $1 < m < \omega$. The ideas of this construction are going back to M. Magidor [M1,2] and to H. Woodin, see[Ca,G2]. The consistency of $2^{\aleph_n} = \aleph_{n+1}$ $(n < \omega) + 2^{\aleph_\omega} = \aleph_{\alpha+1}$ for every $\alpha < \omega_1$ was shown by M. Magidor using huge cardinal, for $\alpha = 1$ H.Woodin, see [Ca], constructed such a model from $o(\kappa) = \kappa^{++}$.

Let $1 \leq m < \omega$ be fixed. We are going to construct a model satisfying "$2^{\aleph_n} = \aleph_{n+1}$ for every $n < \omega$ and $2^{\aleph_\omega} = \aleph_{\omega+m}$". The initial assumption will be the existence of a $\mathcal{P}^m(\kappa)$-hypermeasurable cardinal, whose strength is $o(\kappa) = \kappa^{+m} + 1$.

Actually, what will be used here is the existence of a nice system $\mathcal{A}$ of the length κ^{+m} so that the function $f(\alpha) = \alpha^{+m}$ represents κ^{+m} in the direct limit of the system. By [G2], see also [G-M], $o(\kappa) = \kappa^{+m}$ is sufficient for this, but in a model of $\neg$GCH, i.e. $o(\kappa) = \kappa^{+m}$ is sufficient for constructing a model satisfying "$\aleph_\omega$ is a strong limit cardinal and $2^{\aleph_\omega} = \aleph_{\omega+m}$". But it seems that $o(\kappa) = \kappa^{+m} + 1$ is needed for getting $2^{\aleph_\omega} = \aleph_{\omega+m}$ with G.C.H. below $\aleph_\omega$.

Let $j : V \to M$ be the embedding of $\mathcal{P}^m(\kappa)$-hypermeasurable cardinal, i.e, $crit(j) = \kappa, {}^\kappa M \subseteq M$ and $V_{\kappa+m} \subseteq M$. Assume that $V \models$ GCH. It is not a restrictive assumption, since by W. Mitchell [Mi2], GCH holds in the inner model for $\mathcal{P}^m(\kappa)$-hypermeasurable.

Denote κ^{+m} by λ. Clearly $f_\lambda : \kappa \to \kappa$ defined by $f_\lambda(\alpha) = \alpha^{+m}$ represents λ in M, i.e. $j(f)(\kappa) = \lambda$. Let $\mathbf{U} = \langle < U_\alpha \mid \alpha \in \mathcal{A} \rangle, \langle \pi_{\alpha\beta} \mid \alpha_{\mathcal{A}} \geq \beta > \rangle$ be a nice system of ultrafilters defined from j, as in Section 1. Consider the following commutative diagram

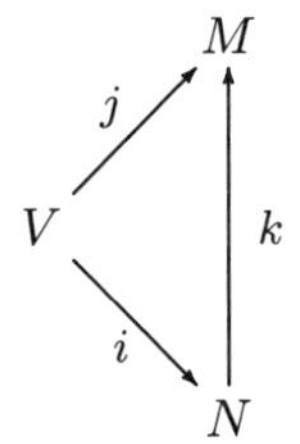

where $i : V \to N \simeq Ult(V, U_0), k([f]_{U_0}) = j(f)(\kappa)$.

We would like to have in V an M-generic subset of $(\mathrm{Col}(\lambda^+, j(\kappa)))^M$. The way of obtaining it was pointed out by H. Woodin. Proceed as follows. Pick $H_0 \subseteq (\mathrm{Col}(\kappa^{+m+1})^N, i(\kappa)))^N$, which is possible since both $(\kappa^{+m+1})^N$ and $i(\kappa)$ are of cardinality κ^+ in V. Then let H be generated by $k''(H_0)$. If $D \in M$ is a dense open subset of $\mathrm{Col}(\lambda^+, j(\kappa))^M$, then for some ordinals $\kappa_1, \dots, \kappa_n, \lambda > \kappa_n > \cdots > \kappa_1 > \kappa$ and a function $g : [\kappa]^{n+1} \to V_\kappa$ $D = j(g)(\kappa, \kappa_1, \dots, \kappa_n)$. The set $A = \{\alpha < \kappa \mid$ there exist ordinals $\alpha_1, \dots, \alpha_n, \alpha < \alpha_1 < \cdots < \alpha_n < \alpha^{+m}$ $g(\alpha, \alpha_1, \dots, \alpha_n)$ is a dense open subset of $\mathrm{Col}(\alpha^{+m+1}, \kappa)\}$ is in U_0. Define $\overline{g} : A \to V_\kappa$ as follows $\overline{g}(\alpha) = \cap\{g(\alpha, \alpha_1, \dots, \alpha_n) \mid \alpha < \alpha_1 < \cdots < \alpha_n < \alpha^{+m}$ and $g(\alpha, \alpha_1, \dots, \alpha_n)$ is a dense open subset of $\mathrm{Col}(\alpha^{+m+1}, \kappa)\}$.

Then $j(\overline{g})(\kappa)$ is a dense open subset of D. But, also $i(\overline{g})(\kappa)$ is dense in $(\mathrm{Col}((\kappa^{+m+1})^N, i(\kappa)))^N$. Hence $i(\overline{g})(\kappa) \cap H_0 \neq \phi$. It implies $H \cap D \neq \phi$.

Now we are ready to define the forcing conditions.

Definition 2.1. The set of forcing conditions $\mathcal{P}$ consists of all elements p of the form $\{\langle 0, \langle \tau_1, \dots, \tau_n \rangle,$
$\langle f_0, \dots, f_n \rangle, F \rangle\} \cup \{\langle \gamma, p^\gamma, b(p,\gamma) \rangle \mid \gamma \in g \backslash \{\max g, 0\}\} \cup \{\langle \max g, p^{\max g}, T \rangle\}$, where

(1) $\{\langle 0, \langle \tau_1, \dots, \tau_n \rangle\rangle\} \cup \{\langle \gamma, p^\gamma \rangle \mid \gamma \in g \backslash \{\max g, 0\}\} \cup \{\langle \max g, p^{\max g}, T \rangle\}$

is as in Definition 1.3. Let us use the notations introduced there. So, we denote g by $\mathrm{supp}(p)$ $\max(g)$ by $mc(p), T$ by T^p and $p^{\max(g)}$ by p^{mc}. Also let us denote further $\langle \tau_1, \dots, \tau_n \rangle$ by p^0, $\langle f_0, \dots, f_n \rangle$ by f^p, for $i < n$ f_i by f_i^p, n by n^p and F by F^p.

(2) $b(p,\gamma)$, the bound over γ, is either 0 or cardinal below κ and above $\sup\left(\bigcup_{0 \leq i \leq n} f_i\right)$. If $b(p,\gamma) = 0$, then we shall omit it.

The new meaning of "permitted for" will be used further. Thus ν is called permitted for p^γ iff $\nu^0 > \max\{(p^\gamma)^0, b(p,\gamma)\}$. We require that condition (6) of Definition 1.3 holds now in this new sense.

(3) $f_0 \in \mathrm{Col}(\omega, \tau_1)$, $f_i \in \mathrm{Col}(\tau_i^{+m+1}, \tau_{i+1})$ for $0 < i < n$ and $f_n \in \mathrm{Col}(\tau_n^{+m+1}, \kappa)$.

(4) F is a function on the projection of $T_{p^{mc}}$ by $\pi_{mc(p),0}$ so that

$$F(\langle \nu_0, \dots, \nu_{i-1}\rangle) \in \mathrm{Col}(\nu_{i-1}^{+m+1}, \kappa)\ .$$

Let us denote the projection of T by π_{mc0} by $T^{p,0}$.

(5) For every $\eta \in T^{p,0}_{p^0}$ let F_η be defined by $F_\eta(\nu) = F(\eta^\frown\nu)$. Then $j(F_\eta)(\kappa)$ belongs to H.

Let us call $\{\langle p^0, \overline{f}^p, \rangle\}$ the lower part of p.

Intuitively, the forcing $\mathcal{P}$ is intended to turn κ to $\aleph_\omega$ simultaneously blowing up its power to κ^{+m+1}. The part of $\mathcal{P}$, which is responsible for blowing up the power of κ is the forcing used in Section 1. The additions to that forcing notion made here are responsible for the collapsing. Basically, $\mathcal{P}$ is modeled after the forcing of [Ma1] and its reduction to hypermeasures made by H. Woodin see [Ca], [G2], or [C-Wo]. The function $f_0, \dots, f_{n-1}$ provide partial information about collapsing already known elements of the Prikry sequence for the normal measure U_0. F is a set of possible candidates for collapsing between further, still unknown elements of this sequence. Condition (5) is desired to insure that these candidates are compatible at least modU_0. This is crucial for proving that the forcing satisfies the Prikry condition. Condition (2), namely the bound $b(p,\gamma)$ is also needed for this purpose. Since here we shall diagonalize over collapsing functions of unknown yet collapse. Note, that for $i < n$ we are starting the collapse with τ_i^{+m+1}, i.e. we intend to preserve all $\tau_i, \tau_i^+, \dots, \tau_i^{+m+1}$. The reason for this, as it appears in the proof, is that $H \subseteq \mathrm{Col}(\kappa^{+m+1}, j(\kappa))$ is M-generic and belongs to V. We were able to construct such H using the fact that the hypermeasure producing M has all the generators below κ^{+m}. The reason looks technical, but the recent work of S. Shelah [Sh3] and [G5] suggest that it is more or less necessary to leave the gap of $m+1$ cardinals below in order to have the gap of the same width between κ and 2^κ.

Definition 2.2. Let $p, q \in \mathcal{P}$. We say that p extends $q (p \geq q)$. If

(1) $\{\langle 0, p^0\rangle\} \cup \{\langle \gamma, p^\gamma, b(p,\gamma)\rangle \mid \gamma \in \mathrm{supp}(p)\backslash\{mc(p), 0\}\} \cup \{\langle mc(p), p^{mc}, T^p\rangle\}$ extends $\{\langle 0, q^0\rangle\} \cup \{\langle \gamma, q^\gamma, b(q,\gamma)\rangle \mid \gamma \in \mathrm{supp}(q)\backslash\{mc(q), 0\}\} \cup \{\langle mc(q), q^{mc}, T^q\rangle\}$ in sense of Definition 1.4.

(2) for every $\gamma \in \mathrm{supp}\, q\backslash\{0, mc(q)\}$ if $\sup\left(\bigcup_{i \leq n^p} f_i^p\right) \geq b(q,\gamma)$ or $p^\gamma \neq$

q^γ, then $b(p,\gamma)=0$ otherwise $b(p,\gamma)=b(q,\gamma)$. If $mc(p)\neq mc(q)$, then $b(p,mc(q))=0$.

Intuitively, this means that if the bound on the old γ is overcome either by increasing the collapsing parts of the condition or by adding new elements to q^γ (which are certainly above $b(q,\gamma)$) it is impossible to set a new bound.

(3) for every $i <$ length $(q^0)=n^q$, $f_i^p \geq f_i^q$
(4) for every $\eta \in T_{p^0}^{p,0}$, $F^p(\eta)\supseteq F^q(\eta)$
(5) for every $i, n^q \leq i < n^p$

$$f_i^p \supseteq F^q((p^0\backslash q^0)\restriction i+1)$$

(6) $\min(p^0\backslash q^0) > \sup(rngf_{n^q})$

Definition 2.3. Let $p,q\in\mathcal{P}$. We say that p is a direct extension (or a Prikry extension) of q ($p^*\!\geq q$) if

(a) $p\geq q$
(b) for every $\gamma\in\operatorname{supp}(q)$ $p^\gamma=q^\gamma$.

The following lemmas are analogous to the corresponding lemmas of the previous section and they have the same proof.

Lemma 2.4. *The relation $\leq$ is a partial order.*

Lemma 2.5. *Let $q\in\mathcal{P}$ and $\alpha\in\mathcal{A}$. Then there is $p^*\!\geq q$ so that $\alpha\in\operatorname{supp}(p)$.*

Lemma 2.6. *$\langle\mathcal{P},\leq\rangle$ satisfies κ^{++}-c.c.*

If $p\in\mathcal{P}$ and $\tau\in p^0$, then the set $\mathcal{P}/p$ of all extensions of p in $\mathcal{P}$ can be split in the obvious fashion into two parts: one everything above τ and the second everything below τ. Denote them by $(\mathcal{P}/p)^{\geq\tau}$ and $(\mathcal{P}/p)^{<\tau}$. Then $\mathcal{P}/p$ can be viewed as $(\mathcal{P}/p)^{\geq\tau}\times(\mathcal{P}/p)^{<\tau}$. The part $(\mathcal{P}/p)^{<\tau}$ consists of finitely many Levy collapses and the part $(\mathcal{P}/p)^{\geq\tau}$ is similar to $\mathcal{P}$ but has a slight advantage, namely the Levy collapses used in it are τ^{+m+1}-closed. Using this observation, one can show the following analog of Lemma 1.8:

Lemma 2.7. *Let $p\in\mathcal{P}$ and $\tau\in p^0$ then $\langle(\mathcal{P}/p)^{\geq\tau},\leq^*\rangle$ is τ^{+m+1}-closed.*

Let us turn now to the Prikry condition.

Lemma 2.8. *$\langle\mathcal{P},\leq,\leq^*\rangle$ satisfies the Prikry condition.*

Proof. Let σ be a statement of the forcing language and $q\in\mathcal{P}$. We shall find $p^*\!\geq q$ deciding σ. In order to simplify notation, assume that $q=\phi$.

Pick an elementary submodel N $\alpha \in \mathcal{A}$ and T as in Lemma 1.9. Consider condition $\{\langle \alpha, \phi, T\rangle\}$. More precisely, we should write $\{\langle 0, \phi, \phi, \phi\rangle\} \cup \{\langle \alpha, \phi, T\rangle\}$. But when the meaning is clear we shall omit $\{\langle 0, \phi, \phi, \phi\rangle\}$. Also for function F as in Definition 2.1 of the forcing condition, we shall relax sometimes condition (3) of 2.1, allowing $\operatorname{dom} F$ to be bigger than just the projection $(T_{p^{mc}})^0$ of $T_{p^{mc}}$ to the zero coordinate. Still the relevant values will come from $(T_{p^{mc}})^0$. If for some $p \in N$ $\{\langle 0, \phi, f, F\rangle\} \cup p \cup \{\langle \alpha, \phi, T'\rangle\} \in \mathcal{P}$ and decides σ, for some $T' \subseteq T, f$ and F, then we are done. Suppose otherwise.

Claim 2.8.1. There are p, F and S in N so that

(a) $\{\langle 0, \phi, \phi, F\rangle\} \cup p \cup \{\langle \alpha, \phi, S\rangle\}^* \geq \{\langle \alpha, \phi, T\rangle\}$

(b) if for some $q \in N, q^0, q^\alpha,\ F', T'$ and $\vec{f}$,

$$\{\langle 0, q^0, \vec{f}, F'\rangle\} \cup q \cup \{\langle \alpha, q^\alpha, T'\rangle\}$$

is a direct extension of $\{\langle 0, \phi, \phi, F\rangle\} \cup p \cup \{\langle \alpha, \phi, T^*\rangle\}$ and forces σ (or $\neg\sigma$) then also

$$\{\langle 0, q^0, \vec{f}, F\rangle\} \cup p \cup \{\langle \alpha, q^\alpha, S_{q^\alpha}\rangle\}$$

forces the same.

Proof. Let A denote $\mathrm{Suc}_T(\langle\, \rangle)$. Assume that $A \subseteq \kappa$ and for $\nu_1, \nu_2 \in A$ $\nu_1 < \nu_2$ implies $\nu_1^0 < \nu_2^0$. Let $\{\langle q_i^0, \vec{f_i}, q_i^\alpha\rangle \mid i < \kappa\}$ be an enumeration of $[\kappa]^{<\omega} \times \bigcup_{\omega \leq \delta < \kappa} \mathrm{Col}(\delta, \nu^0) \times [\kappa]^{<\omega}$. W.l. of g. let us assume that for every $\nu \in A$ $\{\langle q_i^0, \vec{f_i}, q_i^\alpha\rangle \mid i < \nu^0\}$ enumerates $[\nu_0]^{<\omega} \times \bigcup_{\omega \leq \delta < \nu^0} \mathrm{Col}(\delta, \nu^0) \times [\nu_0]^{<\omega}$.

Define by induction sequences $\langle p_i \mid i < \kappa\rangle, \langle T^i \mid i < \kappa\rangle$ and $\langle F^i \mid i < \kappa\rangle$. Set $p_0 = \phi, T^0 = T$ and $F^0 = \phi$.

Suppose that p_j, T^j and F^j are defined for every $j < i$. Define p_i, T^i and F^i.

Set first $p_i'' = \bigcup_{j<i} p_j$. Let $p_i' = \{\langle \gamma, p_i'^\gamma\rangle \mid \gamma \in \mathrm{supp}(p_i'')\}$, where for $\gamma \in \mathrm{supp}(p_i'')$ $p_i'^\gamma = p_i''^\gamma$ unless there is $\nu \in q_i^\alpha$ permitted for $p_i''^\gamma$ and then $p_i'^\gamma = p_i''^\gamma \cup$ the maximal final segment of $\pi''_{\alpha\gamma}(q_i^\alpha)$ permitted for $p_i''^\gamma$.

If $\{\langle 0, q_i^0, \vec{f_i}, \phi\rangle\} \cup p_i' \cup \{\langle \alpha, q_i^\alpha, T_{q_i^\alpha}\rangle\} \notin \mathcal{P}$ or it belongs $\mathcal{P}$ and there is no $p \in N, T'$ and F so that $\{\langle 0, q_i^0, \vec{f_i}, F\rangle\} \cup p \cup \{\langle \alpha, q_i^\alpha, T'\rangle\} \in \mathcal{P}$ extends $\{\langle 0, q_i^0, \vec{f_i}, \phi\rangle\} \cup p_i' \cup \{\langle \alpha, q_i^\alpha, T_{q_i^\alpha}\rangle\}$ and decides σ or $\neg\sigma$, then set $p_i = p_i''$, $T^i = T_{q_i^\alpha}$ and $F^i = \phi$. Otherwise, pick some p, T' and F witnessing this. Define then $T^i = T', F^i = F$. Set $p_i = p_i'' \cup p^*$, where $\mathrm{supp}(p^*) = \mathrm{supp}(p \backslash p_i')$

and for every $\gamma \in \operatorname{supp} p^*$ $\langle \gamma, p^\gamma, \max\{b(p,\tau), \cup(\cup \vec{f_i})\}\rangle \in p^*$. This means that over the new γ's (i.e. $\gamma \notin \operatorname{supp} p_i''$) nothing below $\max \cup \vec{f_i}$ is not permitted.

This completes the inductive definition. Set $p = \bigcup_{i<\kappa} p_i$. Define now a subtree S of T by putting together all T_i's ($i < \kappa$). The definition is level by level. Thus, if S is defined up to level n and t sits in S on this level, then set $\mathrm{Suc}_S(t) = \{\nu \in A \mid \nu^0 > \max t$ and for every $i < \nu^0$ $\nu \in \mathrm{Suc}_{T^i}(\langle\ \rangle)$ and if $t \in T_i$ then $\nu \in \mathrm{Suc}_{T^i}(t)\}$. So $\mathrm{Suc}_S(t) \in U_\alpha$.

Let us now put together all F_i's and define a function F on a subtree of $(S)^0$. It is not hard to do since for every $i < \kappa$ $[F_i]_{U_0} \in H_0$, the ultrapower $N \simeq Ult(V, U_0)$ is closed under κ sequences and the forcing $\mathrm{Col}^N(\kappa^{+m+1}, j(\kappa))$ is closed enough. Pick $[F]_{U_0}$ to be a condition stronger than all $[F_i]_{U_0}$, $i < \kappa$. Shrink S to tree so that every $\eta^\frown\nu$ in it $F(\eta^\frown\nu) \in \mathrm{Col}(\nu^{+m+1}, \kappa)$. Let us denote this tree still by S. More precisely, the definition of F should be carried level by level, i.e. we should define F_η putting together all F_i so that $\eta = q_i^\alpha$.

Subclaim 2.8.2. For every $i < \kappa$, if $q_i^\alpha \in S$ then

$$\{\langle 0, q_i^0, \vec{f_i}, F\rangle\} \cup (p)_i \cup \{\langle \alpha, q_i^\alpha,\ S\rangle\}$$

belongs to $\mathcal{P}$ and it is a direct extension of $\{\langle 0, q_i^0, \vec{f}, F_i\rangle\} \cup (p)_i \cup \{\langle \alpha, q_i^\alpha, T_i\rangle\}$, where $(p)_i$ is obtained from p by extending p^γ's using $\pi''_{\alpha\gamma}(q_i^\alpha)$ and correcting $b(p,\gamma)$'s according to $\sup \cup \vec{f_i}$.

The proof is similar to those of Claim 1.9.1. The bounds $b(p,\gamma)$'s are used in order to show that for every $\nu \in \mathrm{Suc}_S(q_i^\alpha)$ $\mid \{\gamma \in \operatorname{supp} p \mid \nu$ is permitted for $p^\gamma\} \mid \leq \nu^0$. Namely the problem (in the simplest setting) is due to the fact that $|\mathrm{Col}(\omega, <\kappa)| = \kappa$. So there will be κ indexes i such that $f_i \in \mathrm{Col}(\omega, <\kappa)$ and $q_i^0 = q_i^\alpha = \emptyset$. Hence long unions of conditions with $p_i^0 = p_i^\alpha = \emptyset$ should be taken. But the supports of p_i's may increase. The bounds $b(p,\gamma)$'s ($\gamma \in \operatorname{supp}(p)$) where introduced in order to keep the number of ν's permitted for p^γ small.

The rest of the proof is as in Lemma 1.9.

□of the claim.

As in Lemma 1.9, it is possible to show that the assumption "$q \in N$" is not really restrictive. Briefly, if there is some q outside of N which is used to decide σ, then there exists one also inside N.

So the following claim will provide the desired contradiction.

Claim 2.8.3. There exists F^* and S^* so that

(a) $r^* = \{\langle 0, \phi, F^* \rangle\} \cup p \cup \{\langle \alpha, \phi, S^* \rangle\}^* \geq r$

(b) for every $q \in N, R, G, \vec{f}$ and $\vec{\eta}$, if $\{\langle 0, \vec{\eta}^{\,0}, \vec{f}, G \rangle\} \cup q \cup \{\langle \alpha, \vec{\eta}, R \rangle\} \geq r^*$ and $\{\langle 0, \vec{\eta}^{\,0}, \vec{f}, G \rangle\} \cup q \cup \{\langle \alpha, \vec{\eta}, R \rangle\} \Vdash \sigma$ (or $\neg\sigma$) then

$$\{\langle 0, \phi, \vec{f}(0), F^* \rangle\} \cup p \cup \{\langle \alpha, \phi, S^* \rangle\}$$

forces the same, where $\vec{f}(0)$ is the first function in the sequence $\vec{f}$, i.e. a member of $\mathrm{Col}(\omega, \kappa)$.

Proof. Instead of dealing with $\vec{\eta}, \vec{f}$ of arbitrary length, let us concentrate on the case of $|\vec{\eta}| = 1$, $|\vec{f}| = 1$. In this case the notation are much simpler and it contains all the techniques needed for the general one. In order to obtain the general case the argument below should be applied level by level through the tree S^*.

So we like to show the following:

$(*)$ There exist F^* and S^* so that

(a) $r^* = \{\langle 0, \phi, F^* \rangle\} \cup p \cup \{\langle \alpha, \phi, S^* \rangle\}^* \geq r$

(b) for every $q \in N, R, G, f_0, f_1$ and ν, if

$$\{\langle 0, <\nu^0>, f_0, f_1, G \rangle\} \cup q \cup \{\langle \alpha, <\nu>, R \rangle\} \geq r^*$$

and $\{\langle 0, <\nu^0>, f_0, f_1, G \rangle\} \cup q \cup \{\langle \alpha, <\nu>, R \rangle\} \Vdash \sigma$ (or $\neg\sigma$)then $\{\langle 0, \phi, f_0, F^* \rangle\} \cup p \cup \{\langle \alpha, \phi, S^* \rangle\}$ forces the same.

Let $\langle f_{0i} \mid i < \kappa \rangle$ be an enumeration of $\mathrm{Col}(\omega, \kappa)$.

Define by induction sequences $\langle S_i \mid i < \kappa \rangle$ and $\langle F_i \mid i < \kappa \rangle$.

Stage O. Consider in M the following two sets

$$D = \{b \in \mathrm{Col}(\kappa^{+m+1}, j(\kappa)) \mid \{\langle 0, <\kappa>, f_{00}, b, \\ j(F) \rangle\} \cup j(p) \cup \{\langle j(\alpha), <\alpha>, (j(S))_{<\alpha>} \rangle\} \Vdash_{j(\mathcal{P})} j(\sigma)\}$$

$D^* = \{b \mid b \in D$ or there is no element of D stronger than $b\}$.

Then D^* is a dense subset of $\mathrm{Col}(\kappa^{+m+1}, j(\kappa))$ in M. Pick F_0' to be a function on the projection $(S)^0$ of S to the 0-coordinate so that $j(F_0')(\kappa) \in H \cap D^*$ and $\{\langle 0, \phi, F_0' \rangle\} \cup p \cup \{\langle \alpha, \phi, S \rangle\} \geq r$. Set $S_0' = S$.

Now consider in V the following three sets.

$$X_i = \{\nu \in \mathrm{Suc}_{S_0'}(<>) \mid \nu^0 > \sup(rng f_{00}) \ ,$$

for some $g_\nu \supseteq F_0'(\nu^0)$

$$\{\langle 0, \langle \nu^0 \rangle, f_{00}, g_\nu, F_0' \rangle\} \cup p \cup \{\langle \alpha, <\nu>, (S_0')_{<\nu>} \rangle \Vdash^i \sigma\}$$

where $i \in 2, {}^0\sigma = \sigma, {}^1\sigma = \neg\sigma$, $X_2 = \mathrm{Suc}_{S_0'}(<>)\backslash(X_0 \cup X_1)$. For some $i \in 3$ $X_i \in U_\alpha$. Set $S_0 =$ the tree obtained from S_0' by intersecting all of its levels with X_i. Let $F_0 = F_0' \restriction (S_0)^0$.

Note, that if for some $q \in N$, R, G, ν and g_ν, $\{\langle 0, <\nu^0>, f_{00}, g_\nu, G\rangle\} \cup q \cup \{\langle \alpha, <\nu>, R\rangle\}$ is a condition stronger than $r_0 = \{\langle 0, \phi, F_0\rangle\} \cup p \cup \{\langle \alpha, \phi, S_0\rangle\}$ forcing σ (or $\neg\sigma$) then $\{\langle 0, <\nu^0>, f_{00}, g_\nu, F_0\rangle \cup p \cup \{\langle \alpha, <\nu>, S_{0<\nu>}\rangle\} \Vdash \sigma$ (or $\neg\sigma$).

Then $\mathrm{Suc}_{<>} S_0 = X_0$ since $\nu \in \mathrm{Suc}_{<>} S_0$. Hence for every $\nu \in \mathrm{Suc}_{<>} S_0$ there exists $g_\nu \supseteq F_0(\nu^0)$,

$$\{\langle 0, <\nu^0>, f_{00}, g_\nu, F_0\rangle\} \cup p \cup \{\langle \alpha, <\nu>, S_{0<\nu>}\rangle\} \Vdash \sigma \ .$$

Set $g(\nu) = g_\nu$ for $\nu \in \mathrm{Suc}_{<>} S_0$. Then, in M $j(g)(\alpha) \geq j(F_0)(\kappa) = j(F_0')(\kappa)$ and

$$\{\langle 0, <\kappa>, f_{00}, j(g)(\alpha), j(F_0)\rangle\} \cup j(p) \cup \{\langle j(\alpha), <\alpha>, j(S_0)_{<\alpha>}\rangle\} \Vdash_{j(\mathcal{P})} j(\sigma) \ .$$

By the choice of p, F, and S, then also

$$\{\langle 0, <\kappa>, f_{00}, j(g)(\alpha), j(F)\rangle\} \cup j(p) \cup \{\langle j(\alpha), <\alpha>, j(S)_{<\alpha>}\rangle\} \Vdash_{j(\mathcal{P})} j(\sigma) \ .$$

Hence $j(g)(\alpha) \in D$. But then $j(F_0)(\kappa) \in D$. So

$$\{\langle 0, \kappa, f_{00}, j(F_0)(\kappa)), j(F_0)\rangle\} \cup j(p) \cup \{\langle j(\alpha), <\alpha>, j(S_0)_{<\alpha>}\rangle\} \Vdash j(\sigma) \ .$$

Hence

$$C = \{\nu \in \mathrm{Suc}_{<>} S_0 \mid \{\langle 0, \nu^0, f_{00}, F_0(\nu^0), F_0\rangle\} \cup p \cup \{\langle \alpha, <\nu>, (S_0)_{<\nu>}\rangle\} \Vdash \sigma\} \in U_\alpha \ .$$

Restricting now everything to C, we obtain a condition of the form $\{< 0, \phi, f_{00}, G\rangle\} \cup p \cup \{\langle \alpha, \phi, R\rangle\}$ forcing σ. Then also $r_0 = \{\langle 0, \phi, f_{00}, F_0\rangle\} \cup p \cup \{\langle \alpha, \phi, S_0\rangle\}$ forces σ.

Stage i. Choose S_i^* and F_i^* so that

$$\{\langle 0,\phi,F_i^*\rangle\}\cup p\cup\{\langle\alpha,\phi,S_i^*\rangle\}\geq\{\langle 0,\phi,F_i^*\rangle\}\cup p\cup\{\langle\alpha,\phi,S_{i'}\rangle\}$$

for every $i'<i$. Define F_i' as F_0' above, just replace f_{00} by f_{0i} and require $\{\langle 0,\phi,F_i'\rangle\}\cup p\cup\{\langle\alpha,\phi,S_i^*\rangle$ to be stronger than $\{\langle 0,\phi,F_i^*\rangle\}\cup p\cup\{\langle\alpha,\phi,S_i^*\rangle\}$. Set $S_i'=S_i^*$. Define F_i,S_i from F_i', S_i' as above. Then $r_i=\{\langle 0,\phi,F_i\rangle\}\cup p\cup\{\langle\alpha,\phi,S_i\rangle\}$ will satisfy the following:

(**) If for some $q\in N,R,G,\nu$ and g_ν, $\{\langle 0,<\nu^0>,f_{0i},g_\nu,G\rangle\}\cup q\cup\{\langle\alpha,<\nu>,R\rangle\}\geq r_i$ and forces σ (or $\neg\sigma$) then $\{\langle 0,\phi,f_{0i},F_i\rangle\}\cup p\cup\{\langle\alpha,\phi,S_i\rangle\}\Vdash\sigma$ (or $\neg\sigma$).

This completes the construction of $\langle F_i\mid i<\kappa\rangle,\langle S_i\mid i<\kappa\rangle$.

Let us combine now $\langle S_i\mid i<\kappa\rangle$ into one tree. Proceed as follows. Let $A=\mathrm{Suc}_{<>}S$. Shrink A to a set $A'\in U_\alpha$ so that for every $\nu\in A'$

(i) if $i<\nu^0$ then $\sup(rng f_{0i})<\nu^0$;
(ii) if $f\in Col(\omega,\nu^0)$ then for some $i<\nu^0$ $f=f_{0i}$.

Set $A^*=\{\nu\in A'\mid\forall i<\nu^0\ \nu\in\mathrm{Suc}_{<>}S_i\}$. Then $A^*\in U_\alpha$. Define $\mathrm{Suc}_{<>}S^*$ to be A^*. Let $S_i\restriction A^*$ be the tree obtained from S_i by intersecting it level by level with A^*. For every $\nu\in A^*$ set $S^*_{<\nu>}=\cap\{(S_i)_{<\nu>}\mid i<\nu^0\}$. Now, for $\langle\nu_1,\dots,\nu_n\rangle\in S^*$ let $F^*(\langle\nu_1^0,\dots,\nu_n^0\rangle)=\cup\{F_i(\langle\nu_1^0,\dots,\nu_n^0\rangle)\mid i<\nu_1^0\}$.

It is easy to see that

$$r^*=\{\langle 0,\phi,F^*\rangle\}\cup p\cup\{\langle\alpha,\phi,S^*\rangle\}\in\mathcal{P}$$

and it is stronger than r.

Suppose now that $q\in N,R,G,f_0,f_1$ and ν are as in $(*)_{(b)}$. Then $\nu\in\mathrm{Suc}_{<>}S^*=A^*, f_0\in\mathrm{Col}(\omega,\nu^0)$. So for some $i<\nu^0$ $f_0=f_{0i}$. Also $\nu\in\mathrm{Suc}_{S_i}(<>), S^*_{<\nu>}\subseteq(S_i)_{<\nu>}$ and $F^*(\vec\eta)\supseteq F_i(\vec\eta)$ for every $\vec\eta\in S^*$ with the first element ν. Hence

$$\{\langle 0,<\nu^0>,f_{0i},f_1,G\rangle\}\cup q\cup\{\langle\alpha,<\nu>,R\rangle\}\geq r_i\ .$$

Then, by (**), $\{\langle 0,\phi,f_{0i},F_i\rangle\}\cup p\cup\{\langle\alpha,\phi,S_i\rangle\}\Vdash\sigma$ (or $\neg\sigma$).

By the choice of r, then $\{\langle 0,\phi,f_{0i},F\rangle\}\cup p\cup\{\langle\alpha,\phi,S\rangle\}\Vdash\sigma$ (or $\neg\sigma$). But $r^*\geq r$. So $\{\langle 0,\phi,f_{0i},F^*\rangle\}\cup p\cup\{\langle\alpha,\phi,S^*\rangle\}\Vdash\sigma$ (or $\neg\sigma$).

□ of the claim. This completes the proof of the lemma. □

Using Lemma 2.8 as replacement of Lemma 1.10 the arguments of 1.11 show the following.

Lemma 2.9. *κ^+ remains a cardinal in $V^{\mathcal{P}}$.*

Lemma 1.10 transfers directly to the present forcing notion. For G a generic subset of $\mathcal{P}$, $\alpha \in \mathcal{A}$ defines as in Section 1, G^α to be $\cup\{p^\alpha \mid p \in G\}$. Let $G^0 = \langle \kappa_0, \kappa_1, \ldots, \kappa_n, \ldots \rangle$.

Lemma 2.10.

(a) *For every $\alpha \in \mathcal{A}, G^\alpha$ is a Prikry sequence for U_α.*
(b) *G^0 is an ω-sequence unbounded in κ.*
(c) *If $\alpha \neq \beta$ are in $\mathcal{A}$ then $G^\alpha \neq G^\beta$.*

The next lemma is obvious.

Lemma 2.11. *If $\tau < \kappa, \tau > \aleph_0$ and τ remains a cardinal in $V[G]$, then for some n $\tau = \kappa_n$ or for some $m' \leq m$ $\tau = \kappa_n^{+m'+1}$.*

Combining now all the lemmas, we obtain the following.

Theorem 2.12. *In a generic extension $V[G]$ $2^{\aleph_n} = \aleph_{n+1}$ for every $n < \omega$ and $2^{\aleph_\omega} = \aleph_{\omega+m}$.*

3. Down to $\aleph_\omega$, an Infinite Gap

In this section we shall modify the construction of Section 2 in order to obtain a model satisfying GCH below $\aleph_\omega$ and $2^{\aleph_\omega} = \aleph_{\xi+1}$ for any ξ, $\omega < \xi < \omega_1$. The crucial tool will be the method of S. Shelah [Sh2] allowing to construct models with a countable gap between $\aleph_\omega$ and $2^{\aleph_\omega}$.

Fix an ordinal ξ, $\omega < \xi < \omega_1$. Suppose that κ is $\kappa + \xi + 1$-strong, i.e. there is an elementary embedding $j : V \to M$ with κ a critical point and $M \supseteq V_{\kappa+\xi+1}$.

Pick an increasing sequence of finite sets $\langle D_n \mid n < \omega \rangle$ so that $(\xi + 1) \backslash 1 = \bigcup_{n<\omega} D_n$. For each $n < \omega$, we would like to be able to collapse all the cardinals between κ^{++} and $\kappa^{+\xi+1}$, with exceptions for κ^{+i+1} for $i \in D_n$, preserving enough of strongness of κ. It can be easily achieved by making the right preparation forcing below κ. Actually, the models with indestructible κ as of R. Laver [L] or [G-Sh] can be used. But in order to simplify the further arguments, we would like rather to use the direct construction for this particular case.

We define $\mathrm{Col}_n(\delta)$ to be the combination of Levy collapses which are intended to preserve only the cardinals of the form δ^{+i+1} for $i \in D_n$ between δ^{++} and $\delta^{+\xi+1}$, where δ is an inaccessible cardinal. Thus

$$\mathrm{Col}_n(\delta) = \prod_{m<|D_n|} \mathrm{Col}(\delta^{+i_m+1}, < \delta^{+i_{m+1}})$$

where $\langle i_m \mid m < |D_n|\rangle$ is an increasing enumeration of D_n. Let Add $= \{f \mid f : 1 \to \omega\}$ be an atomic forcing notion. It would decide generically for which $n < \omega$ to use $\mathrm{Col}_n(\delta)$ on stage δ.

Define now on iteration $\langle P_\alpha, \underset{\sim}{Q}_\alpha \mid \alpha \le \kappa\rangle$. Set $P_0 = \phi$. $\underset{\sim}{Q}_\alpha = \phi$ unless α is an inaccessible. Then define in V_α^P $\underset{\sim}{Q}_\alpha =$ Add$*$ $\underset{\sim}{\mathrm{Col}}\ _{\underset{\sim}{f}_{\alpha(0)}}(\alpha)$, where $\underset{\sim}{f}_\alpha$ is the name of the generic choice made by Add on stage α. Use on a limit stage α the direct limit for inaccessible α and the inverse limit otherwise. Then, for every $n, j : V \to M$ extends in $V^{P_\kappa * \mathrm{Col}_n(\kappa)}$ to an embedding $j_n : V^{P_\kappa * \mathrm{Col}_n(\kappa)} \to M^{P_\kappa * \mathrm{Col}_n(\kappa) * P'}$ where $P' = j(P_\kappa)/P_\kappa * \{\langle 0, n\rangle\} * \mathrm{Col}_n(\kappa)$. Denote $M^{P_\kappa * \mathrm{Col}_n(\kappa) * P'}$ by M_n and $V^{P_\kappa * \mathrm{Col}_n(\kappa)}$ by V_n.

Using j, define in V a nice system $\mathbf{U} = \ll U_\alpha \mid \alpha \in \mathcal{A}\rangle, \langle \pi_{\alpha\beta} \mid \alpha_{\mathcal{A}} \ge \beta \gg$ of the length $\kappa^{+\xi+1}$, as in Section 1. Now, in V_n using j_n we define a nice system $\underset{\sim}{\mathbf{U}}_n = \ll \underset{\sim}{U}_{n\alpha} \mid \alpha \in \mathcal{A} >, < \underset{\sim}{\pi}_{n\alpha\beta} \mid \alpha_{\mathcal{A}} \ge \beta \gg$ so that $\underset{\sim}{U}_{n\alpha} \supseteq U_\alpha$ for every $\alpha \in \mathcal{A}$. Then $\underset{\sim}{\pi}_{n\alpha\beta}$ can be chosen to be $\pi_{\alpha\beta}$.

As in section 2, fix some $\underset{\sim}{H}_n \in V^{P_\kappa * \mathrm{Col}_n(\kappa)}$ which is M_n-generic subset of $(\mathrm{Col}((\kappa^{+\xi+2})^V, j(\kappa)))^{M_n}$. Note that $(\kappa^{+\xi+2})$ is $\kappa^{+2+|D_n|}$ in M_n.

Let $G_\kappa \subseteq P_\kappa$ be a generic subset. We shall work in $V[G_\kappa]$. Let B_n be the complete Boolean algebra of regular open sets of $\mathrm{Col}_n(\kappa)$. Denote by $\sigma_{k,n}$ the natural projection of B_k onto B_n for $\omega > k \ge n$.

We are ready now to define the main forcing notion for turning κ to $\aleph_\omega$.

Definition 3.1. A set $\mathcal{P}$ of forcing conditions consists of all elements p of the form $\{r\} \cup \{\langle 0, \langle \tau_1, \dots, \tau_n\rangle, \langle f_0, \dots, f_n\rangle, F\rangle\} \cup \{\langle \gamma, p^\gamma, b(p,\gamma)\rangle \mid \gamma \in g\backslash\{\max g, 0\}\} \cup \{\langle \max g, p^{\max g}, T\rangle$, where

(1) $\{\langle 0, \langle \tau_1, \dots, \tau_n\rangle, \langle f_0, \dots, f_n\rangle\} \cup \{\langle \gamma, p^\gamma, b(p,\gamma)\rangle \mid \gamma \in g\backslash\{0\}\}$ is as in Definition 2.1.
We shall use further the notation introduced there.

(2) T is a tree with a trunk p^{mc} consisting of $^\circ$-increasing sequences.

(3) F is a function on $(T_{p^{mc}})^0$ so that for $\eta \in T, \eta_T > p^{mc}$, $F(\eta^0) \in \mathrm{Col}((\max(\eta^0))^{+2+m}, \kappa)$, where $m = |B_{|\eta|-1}|$, i.e. the collapsing starts with $((\max(\eta))^{+\xi+2})^V$.

(4) $r \in B_n$.
Denote it further by $p(\mathrm{col})$.

(5) r forces the following "for every $\eta \in \check{T}$ $\mathrm{Suc}_{\check{T}}(\eta) \in \underset{\sim}{U}_{|\eta|, mc(p)}$ and $j_{|\eta|}(F_{\eta^0})(\kappa) \in \underset{\sim}{H}_{|\eta|}$, where $F_{\eta^0}(\nu^0) = F(\eta^{0\frown}\nu^0)$ for every $\nu \in \mathrm{Suc}_T(\eta)$".

Explanation. The set of conditions defined above is similar to those of Section 2. The difference is that we like to preserve all the cardinals between κ and $\kappa^{+\xi+1}$, but simultaneously to collapse all but finitely many cardinals between τ and $\tau^{+\xi+1}$ for each element τ of the zero coordinate. The idea of S. Shelah [Sh2] for doing this, is to leave more and more cardinals climbing up to κ along the zero coordinate, and to give only small pieces of information about collapses between κ and $\kappa^{+\xi+1}$, which finally would not produce a real collapse. $r = p(\mathrm{col})$ is such a piece.

Definition 3.2. Let $p, q \in \mathcal{P}$. We say that p extends $q (p \geq q)$, if

(1) $p \backslash \{p(\mathrm{col})\}$ extends $q \backslash \{q(\mathrm{col})\}$ in sense of Definition 2.2.

(2) $p(\mathrm{col})$ is stronger than $\sigma_{mn}(q(\mathrm{col}))$ in the forcing with B_n, where $m =$ length (q^0), $n =$ length (p^0).

Definition 3.3. Let $p, q \in \mathcal{P}, p^* \geq q$ if

(a) $p \geq q$

(b) for every $\gamma \in \mathrm{supp}(q)$ $p^\gamma = q^\gamma$.

The proofs of the following lemmas are more or less routine translations of the proofs of corresponding lemmas of the previous sections.

Lemma 3.4. *The relation $\leq$ is a partial order.*

Lemma 3.5. *Let $q \in \mathcal{P}$ and $\alpha \in \mathcal{A}$. Then there is $p^* \geq q$. so that $\alpha \in \mathrm{supp}(p)$.*

Lemma 3.6. *Let $p \in \mathcal{P}$ and $\tau \in p^0$. Then $\langle (\mathcal{P}/p)^{\geq \tau}, \leq^* \rangle$ is $(\tau^{+\xi+1})^V$-closed, where $(\mathcal{P}/p)^{\geq \tau}$ is defined as in Section 2.*

Lemma 3.7. *$\langle \mathcal{P}, \leq, \leq^* \rangle$ satisfies the Prikry condition.*

Lemma 3.8. *κ^+ remains a cardinal in $V^{\mathcal{P}}$.*

Lemma 3.9. *(a) $G^0 = \langle \kappa_0, \ldots, \kappa_n \ldots \rangle$ is an ω-sequence unbounded in κ.*

(b) If $\alpha \neq \beta$ are in $\mathcal{A}$ then $G^\alpha \neq G^\beta$, where G^α's are defined as in Section 2.

Lemma 3.10. *If $\tau, \aleph_0 < \tau < \kappa$ and τ remains a cardinal in $V^{P_\kappa * \mathcal{P}}$, then there exists n such that $\kappa_n \leq \tau < \kappa_{n+1}$ and for some $m \leq |D_n|+2$ $\tau = \kappa_n^{+m}$.*

The new point here is to show that all the cardinals between κ^+ and $\kappa^{+\xi+1}$ are preserved. Note that $\mathcal{P}$ satisfies now only $\kappa^{+\xi+2}$-c.c.

Lemma 3.11. *Let $\delta, 1 < \delta \leq \xi$ be an ordinal. Then $\kappa^{+\delta+1}$ is preserved in $V^{P_\kappa * \mathcal{P}}$.*

Proof. Suppose first that δ is a successor ordinal. Let $n < \omega$ be the least so that $\delta - 1, \delta \in D_n$. Let $q \in \mathcal{P}$. Extend q to p so that $|p^0| > n$. Consider the forcing notion $\mathcal{P}/p = \{t \in \mathcal{P} \mid t \geq p\}$. For every $t \in \mathcal{P}/p$, $t(\mathrm{col}) \in B_m$ for some $m \geq n$. Recall that then B_m is a complete subalgebra of B_n. Also, $\kappa^{+\delta+1}$ is preserved by forcing with B_n. Namely B_n splits into $B_{n,1}$

$\times B_{n,2}$ so that $B_{n,1}$ is of cardinality $\leq \kappa^{+\delta}$ and $B_{n,2}$ is δ-closed. Force with $B_{n,2}$. Let $G(B_{n,2})$ be a $V[G_\kappa]$-generic subset of $B_{n,2}$, where G_κ is V-generic subset of P_κ. Consider in $V^* = V[G_\kappa, G(B_{n,2})]$

$$\mathcal{P}_n^* = \{t \in \mathcal{P}/p \mid t(\text{col}) \cap B_{n,2} \in G(B_{n,2})\} .$$

Then, as in Lemma 2.6, $\mathcal{P}_n^*$ satisfies $\kappa^{+\delta}$-c.c. So, $\kappa^{+\delta+1}$ is preserved by $B_{n,2} * \mathcal{P}_n^*$.

But it is not hard to see that the forcing $\mathcal{P}/p$ can be completely embedded into $B_{n,2} * \mathcal{P}_n^*$. Hence $\mathcal{P}/p$ cannot collapse $\kappa^{+\delta+1}$.

Suppose now that δ is a limit ordinal. Let us use the notation introduced in the previous case. The problem now is that $B_{n,1}$ and, hence $\mathcal{P}^*$ may fail to satisfy $\kappa^{+\delta+1}$-c.c. Using an appropriate inductive assumption we can assume w.l. of g. that for every $\gamma < \delta$, $\kappa^{+\gamma}$ remains a cardinal in $V^{P_\kappa * \mathcal{P}}$. So, if $\kappa^{+\delta+1}$ is collapsed then it changes its cofinality to some $\kappa^{+\gamma+1} < \kappa^{+\delta}$. Pick some $\overline{n} \geq n$ so that $\gamma, \gamma+1, \gamma+2 \in D_{\overline{n}}$. The previous argument gives the splitting $B_{\overline{n},2} * \mathcal{P}_{\overline{n}}^*$ so that $\mathcal{P}_{\overline{n}}^*$ satisfies $\kappa^{+\gamma+2}$-c.c. and $B_{\overline{n},2}$ is $\kappa^{+\gamma+2}$-closed. Clearly, $B_{\overline{n},2} * \mathcal{P}_{\overline{n}}^*$ cannot change the cofinality of $\kappa^{+\delta+1}$ to $\kappa^{+\gamma+1}$. Then, the same is true for $\mathcal{P}/p$, since it is completely imbeddible into $B_{\overline{n},2} * \mathcal{P}_{\overline{n}}^*$.

Hence $\kappa^{+\delta+1}$ is always preserved.

$\square$ of the lemma.

So the following holds:

Theorem 3.12. In $V^{P_\kappa * \mathcal{P}}$ GCH is true below $\aleph_\omega = \kappa$ and $2^{\aleph_\omega} = \aleph_{\xi+1}$.

S. Shelah [Sh2] showed that the power of the least fixed point of order ω of the aleph function can take any reasonable value below an inaccessible. A supercompact cardinal was used by him for this result. Let us indicate how to obtain this result from a strong cardinal.

Definition 3.13. (Shelah [Sh2]) Let $C^0 =$ the class of all infinite cardinals. $C^{n+1} = \{\lambda \in C^n \mid C^n \cap \lambda \text{ has order type } \lambda\}$ and $C^\omega = \bigcap_{n<\omega} C^n$. The order of a cardinal κ is the maximal $n \leq \omega$ such that $\kappa \in C^n$. Thus, if κ is of order 1, then it is a fixed point of the $\aleph$ function.

Theorem 3.14. *Suppose that V is a model of GCH and κ is a strong cardinal in V without inaccessible above it then for every cardinal μ, $\kappa^+ \leq \mu$ there exists a generic extension V^* of V so that in V^* the following holds*

(1) *κ is the first element of C^ω (i.e. the first fixed point of $\aleph$-function of the order ω)*

(2) $2^\kappa = \mu^+$

(3) *all the cardinals and its cofinalities above* κ *are preserved*
(4) *GCH holds below* κ

4. Sketch of the Proof

By S. Shelah [Sh2], Lemma 2.5 there exists an increasing sequence $\langle D_n \mid n < \omega\rangle$ so that $\bigcup_{n<\omega} D_n = \{\chi \mid \chi \text{ is a cardinal, } \kappa^{++} < \chi \le \mu^+\}$, and for every $n < \omega$ there is no elements of C^n between κ and μ^+ in the generic extension V_n of V obtained by preserving only elements of D_n as cardinals between κ^{++} and μ^+. Now, using a strongness of κ, find $j : V \to M$, so that

(a) $M \supseteq V_{\mu^+}$, ${}^{\kappa}M \subseteq M$ and
(b) for some $f : \kappa \to \kappa, d : \kappa \to V_\kappa$ $j(f)(\kappa) = \mu^+, j(d)(\kappa) = \langle D_n \mid n < \omega\rangle$.

Using this j define a nice system $\mathbf{U}$ of the length μ^+. Define $\mathrm{Col}_n(\delta)$ to be the product of the Levy collapses which preserves between δ^{++} and $(f(\delta))^+$ only elements of $d(\delta)(n)$.

Now we continue exactly as in the previous construction. In the final model κ will the least element of C^ω, $2^\kappa = \mu^+$, all the cardinals above κ will be preserved and GCH will hold below κ. □

Acknowledgments. This work was done while both authors were visitors at MSRI during the logic year there. They would like to thank the organizers for giving them this rare opportunity to work together.

The second author is partially supported by the U.S.-Israel Binational Science Foundation, Grant No. 87-00040.

REFERENCES

[B] S. Baldwin, *Between strong and superstrong*, J. Sym. Logic **51** (1986), 547–449.

[Ca] J. Cummings, A model in which GCH holds at successors but fails at limits; (to appear).

[C-Wo] J. Cummings and H. Woodin, *Applications of Radin's Forcing*, a forthcoming book.

[De-Jen] K. Devlin and R. Jensen, *Marginalia to a Theorem of Silver*, vol. 499, Logic Conference, Kiel 1974, Lec. Notes in Math., Springer-Verlag, Berlin and New York 1975.

[D] A. Dodd, *The Core Model*, London Mathematical Society Lecture Notes Series **61** (1982), Cambridge Univ. Press.

[Ea] W. Easton, *Powers of regular cardinals*, Ann. Math. Logic **1**, 139–178.

[F-W] M. Foreman and H. Woodin, *GCH can fail everywhere*, Ann. of Math.

[Ga-H] F. Galvin and A. Hajnal, Ann. Math. **101**, 491–498.

[G1] M. Gitik, *Changing cofinalities and the nonstationary ideal*, Israel J. Math. **56** (3), 280–314.

[G2] M. Gitik, *The negation of SCH from $o(\kappa) = \kappa^{++}$*, Ann. of Pure and Appl. Logic **43**, 209–234.

[G3] M. Gitik, *The strength of the failure of SCH*, Ann. of Pure and Appl. Logic **51**, 215–240.

[G4] M. Gitik, *On measurable cardinals violating GCH*, to appear.

[G5] M. Gitik, Handwritten notes, UCLA 1991.

[G-M] M. Gitik and M. Magidor, *Extender Based Forcing Notions.*

[G-Sh] M. Gitik and S. Shelah, *On Certain Indestructibility of Strong Cardinals and a Question of Hajnal*, Arch. Mth. Logic **28**, 35–42.

[J] T. Jech, *Set Theory*, Academic Press, New York.

[K-M] A. Kanamori and M. Magidor, *Müller and Scott, eds., Lecture Note in Math. 699, Higher Set Theory*, (Springer, Berlin).

[L] R. Laver, *Making the supercompactness of κ indestructible under κ-directed closed forcing*, Israel J. Math. **29**, 385–388.

[Ma1] M. Magidor, *On the singular cardinals problem, I*, Israel. J. Math. **28** (1) (1977), 1–31.

[Ma2] M. Magidor, *On the singular cardinal problem, II*, Ann. of Math. **106** (1977), 517–647.

[Mi1] W. Mitchell, *Indiscernibles, skies and ideals*, Contemporary Math. **31** (1984), 161–182.

[Mi2] W. Mitchell, Hypermeasurable cardinals, *Boffa (eds) Logic Coll.*, vol. 78, North Holland, 1979.

[Mi3] W. Mitchell, *The Core Model for sequences of measures I*, Math. Proc. Cambridge Phil. Soc. **95** (1984), 229–260.

[Mi4] W. Mitchell, *The Core Model for sequences of measures II.*

[Mi5] W. Mitchell, *On the SCH*, Trans A.M.S.

[P] K. Prikry, *Changing measurable into accessible cardinals*, Diss. Math. **68** (1970), 5–52.

[Sh1] S. Shelah, *Proper forcing*, vol. 940, Springer-Verlag, Lecture Notes in Math., 1982.

[Sh2] S. Shelah, *The Singular Cardinals Problem. Independence Results*, Mathias (ed.), vol. 87, London Math. Soc. Lecture Note Series, Surveys in Set Theory, pp. 116–133.

[Sh3] S. Shelah, *Cardinal Arithmetic*, in preparation.

[Si] J. Silver, *On the singular cardinals problem*, Proc. of the International Congress of Math.; Vancouver, vol. 1, 1974, pp. 265–268.

[So-Re-Ka] R. Solovay, W. Reinhardt and A. Kanomori, *Strong axioms of infinity and elementary embeddings*, Ann. Math. Logic **13** (1) (1978), 73–116.

[W] H. Woodin; private communication.

School of Mathematical Sciences, Raymond and Beverly Sackler Faculty of Exact Sciences, Tel Aviv University, Ramat Aviv, Tel Aviv 69978, ISRAEL

Department of Mathematics, Hebrew University of Jerusalem, Jerusalem, ISRAEL

A WEAK VERSION OF AT FROM OCA

WINFRIED JUST

ABSTRACT. We use ideas of Veličković to derive from the Open Coloring Axiom a number of statements which were originally proved by lengthy and difficult forcing arguments.

Throughout this note, "ideal" means a proper ideal $\mathcal{I}$ in the Boolean algebra $\mathcal{P}(\omega)$ that contains Fin—the ideal of finite subsets of ω. We often identify a subset $a \subseteq \omega$ with its characteristic function. Thus $\mathcal{P}(\omega)$ inherits the product topology on 2^ω, and whenever we consider topological notions (like "Borel ideal") we have this topology in mind.

In [J] and [J1], I formulated statements which I abbreviated CSP and AT, and proved their consistency relative to ZFC by rather lengthy and involved forcing arguments. Various consequences of AT have been derived in [J1]-[J4].

Recently Boban Veličković proposed an alternative approach: Instead of using forcing, derive these consequences, and possible AT itself, from the Open Coloring Axiom OCA (see Definition 0 below). He succeeded in establishing that OCA + MA implies that $\mathcal{P}(\omega)$ Fin has no nontrivial automorphisms (see [V]). This was originally proved consistent by Shelah using the oracle chain condition (see chapter 4 of [S]). It is still open whether AT is also a consequence of OCA + MA.

In the present note we use the ideas of [V] to establish that a weak version of AT, abbreviated WAT (see definition 10 below), does indeed follow from OCA. We also show that all consequences derived from AT in [J1]-[J4] follow from WAT, and thus from OCA.

I would like to thank Krzysztof Mazur for pointing out an error in a previous version of this note.

0. Definition. By *OCA* we abbreviate the following statement: "For every separable metric space $\mathcal{X}$ and every partition $[\mathcal{X}]^2 = \mathcal{K}_0 \cup \mathcal{K}_1$ such that $\mathcal{K}_0$ is open in $[\mathcal{X}]^2$

- *either* $\exists \mathcal{Y} \subseteq \mathcal{X}$ $\mathcal{Y}$ is uncountable and $[\mathcal{Y}]^2 \subseteq \mathcal{K}_0$
- *or* $\mathcal{X} = \bigcup_n \mathcal{X}_n$, where $[\mathcal{X}_n]^2 \subseteq \mathcal{K}_1$ for all $n < \omega$."

It is known that if ZFC is consistent, then so is the theory ZFC + OCA + MA (see [T]).

1. Definition. Let $\mathcal{I}$ be an ideal, $\mathcal{M}$ a subset of $P(\omega)$. We say that $\mathcal{M}$ is an approximation of $\mathcal{I}$, iff the following hold:

(i) $\mathcal{M}$ is downward closed, i.e., $\forall a \in \mathcal{M} \forall b \subset a\ b \in \mathcal{M}$.

(ii) $\forall a \in \mathcal{I} \exists n \in \omega\ a - n \in \mathcal{M}$.

$\mathcal{M}$ will be called a closed approximation of $\mathcal{I}$ if $\mathcal{M}$ is a closed subset of $\mathcal{P}(\omega)$.

2. Examples.

(a) Every ideal $\mathcal{I}$ is an approximation of itself.

(b) If $\mathcal{I}$ is an F_σ-ideal, then there exists a closed approximation $\mathcal{M}$ of $\mathcal{I}$ such that

$$\mathcal{I} = \{a \subset \omega : \exists b \in \mathrm{Fin}\ a \triangle b \in \mathcal{M}\}.$$

(see [M]).

(c) In particular, if $\mathcal{I} = \mathrm{Fin}$, then in (b) we can take $\mathcal{M} = \{\emptyset\}$.

(d) Let $h : \mathcal{P}(\omega) \to \mathbf{R}^+$ and define

$$\mathcal{I}_h = \{a \subset \omega : \limsup_{n \to \infty} \frac{\sum_{m \in a \cap n} h(m)}{\sum_{m<n} h(m)} = 0\}.$$

For a large class of functions h, called *EU-functions* in [JK], the family $\mathcal{I}_h$ is an ideal. (More precisely, h is an EU-function, if $\sum_{n \in \omega} h(n) = \infty$ and $\lim_{m \to \infty}(h(m)/\sum_{n \leq m} h(n)) = 0$.) If $f(n) = 1$ and $g(n) = \frac{1}{n+1}$ for all $n \in \omega$, then $\mathcal{I}_f$ and $\mathcal{I}_g$ are called *the ideal of sets of density zero* and *the ideal of sets of logarithmic density zero* respectively.

If h is an EU-function, and $\mathcal{I}_h$ is the corresponding ideal, then let for $\varepsilon > 0$

$\mathcal{M}(h, \varepsilon) = \{a \subset \omega : \sup_{n \in \omega} \frac{\sum_{k \in a \cap n} h(k)}{\sum_{k<n} h(k)} \leq \varepsilon\}$.

Clearly, $\mathcal{M}(h, \varepsilon)$ is a closed approximation of $\mathcal{I}_h$.

(e) Let $\mathcal{I}$ be an ideal, $\mathcal{M}$, $\mathcal{N}$ be approximations of $\mathcal{I}$. Denote:

$\mathcal{M} \oplus \mathcal{N} = \{a \subset \omega : \exists b \in \mathcal{M} \exists c \in \mathcal{N}\ \ a = b \cup c\}$. Then $\mathcal{M} \oplus \mathcal{N}$ is also an approximation of $\mathcal{I}$. Moreover, if both $\mathcal{M}$ and $\mathcal{N}$ are closed, then so is $\mathcal{M} \oplus \mathcal{N}$. Sometimes we shall write $2\mathcal{M}$ instead of $\mathcal{M} \oplus \mathcal{M}$.

3. Definition. Let $\mathcal{I}$ be an ideal, $\mathcal{M}$ an approximation of $\mathcal{I}$, and let $F : \mathcal{P}(a) \to \mathcal{P}(\omega)$ (where $a \subseteq \omega$). We say that F is $\mathcal{M}$-precise, iff there are Borel functions $G_n : \mathcal{P}(a) \to \mathcal{P}(\omega)$ for $n < \omega$ such that $\forall b \subseteq a\ \exists n < \omega\ \ F(b) \triangle G_n(b) \in \mathcal{M}$.

F is called semi-$\mathcal{M}$-precise iff $\exists k \exists a_0, \ldots, a_k\ a = a_0 \cup \ldots \cup a_k \& \ F|a_i$ is $\mathcal{M}$-precise for every $i \leq k$.

F is called $\mathcal{M}$-sharp ($\mathcal{M}$-trivial) iff there exists a Baire measurable (continuous) function $G : \mathcal{P}(a) \to \mathcal{P}(\omega)$ such that $F(b)\Delta G(b) \in \mathcal{M}$ for every $b \subseteq a$.

The notions of semi-$\mathcal{M}$-sharp and semi-$\mathcal{M}$-trivial functions are defined in an analogous way as semi-$\mathcal{M}$-precise functions.

4. Definition. A function F preserves intersections (unions) mod $\mathcal{M}$ iff $(F(a) \cap F(b))\Delta F(a \cap b) \in \mathcal{M}$ (*resp.* $(F(a) \cup F(b))\Delta F(a \cup b) \in \mathcal{M}$.)

F is called Fin-invariant mod $\mathcal{M}$ iff $F(a)\Delta F(b) \in \mathcal{M}$ whenever $a\Delta b \in$ Fin.

5. Fact. (a) *Suppose $a \subseteq \omega$, $F : \mathcal{P}(a) \to \mathcal{P}(\omega)$ preserves intersections mod some ideal $\mathcal{I}$ and is $\mathcal{M}$-sharp for some approximation $\mathcal{M}$ of $\mathcal{I}$. Then F is semi-$\mathcal{I} \oplus \mathcal{M}$-trivial.*

(b) *If moreover F preserves unions mod $\mathcal{I}$, then F is $\mathcal{I} \oplus 2\mathcal{M}$-trivial.*

(c) *Suppose $F : \mathcal{P}(\omega) \to \mathcal{P}(\omega)$ is* Fin-*invariant mod $\mathcal{I}$ and $a, b \subseteq \omega$ are such that $a\Delta b \in$* Fin. *Then $F|\mathcal{P}(a)$ is semi-$\mathcal{I} \oplus \mathcal{M}$-trivial (-sharp) iff $F|\mathcal{P}(b)$ is.*

Proof. Suppose the function G witnesses sharpness of F. Since G is Baire measurable, there is some comeagre subfamily $\mathcal{A} \subseteq \mathcal{P}(a)$ such that $G|\mathcal{A}$ is continuous. Find a decomposition $a = a_0 \cup a_1 \cup a_2 \cup a_3$ of a into four pairwise disjoint sets such that for all $x \subseteq a_0 \cup a_1$ and for all $y \subseteq a_2 \cup a_3$ we have $x \cup a_2 \in \mathcal{A}$ and $a_0 \cup y \in \mathcal{A}$. Define for $i \in \{0, 1\}$ and $b \subseteq a_{2i} \cup a_{2i+1}$: $G_i(b) = G(a_{2-2i} \cup b) \cap F(a_{2i} \cup a_{2i+1})$.

Since F preserves intersections mod $\mathcal{I}$, $F(b)\Delta G_i(b) \in \mathcal{I} \oplus \mathcal{M}$. This proves (a).

To prove (b), let for $b \subseteq a$: $G_2(b) = G_0(b \cap (a_0 \cup a_1)) \cup G_1(b \cap (a_2 \cup a_3))$. Since $\mathcal{I} \oplus \mathcal{I} = \mathcal{I}$, we are done.

(c) is obvious. $\square$

The following lemma generalizes Theorem 2 of [V].

6. Lemma. (a) *Suppose $a \subseteq \omega$; $F : \mathcal{P}(a) \to \mathcal{P}(\omega)$ preserves intersections mod some Σ^1_1-ideal $\mathcal{I}$ and is $\mathcal{M}$-precise for some Σ^1_1-approximation $\mathcal{M}$ of $\mathcal{I}$. Suppose furthermore that $\mathcal{A}$ is an infinite family of pairwise almost disjoint subsets of a.*

Then the set $\{b \in \mathcal{A} : F|\mathcal{P}(b)$ is not semi-$\mathcal{I} \oplus \mathcal{M}$-trivial$\}$ is finite.

(b) *Suppose moreover that $\mathcal{A}$ is contained in some comeagre subfamily $\mathcal{C}$ of $\mathcal{P}(a)$. Then we can find $b \in \mathcal{A}$ and $c \subset b$ so that $c \in \mathcal{C}$ and $F|\mathcal{P}(c)$ is $\mathcal{I} \oplus \mathcal{M}$-sharp.*

Proof. Let G_n, $n < \omega$ be functions that witness the $\mathcal{M}$-precision of F. The proofs of (a) and (b) are similar: In both cases we assume there is a witness $B = \{b_k : k \in \omega\} \subseteq \mathcal{A}$ to the contrary. (In the case of (a) that means $F|\mathcal{P}(b_k)$ is not semi-$\mathcal{I} \oplus \mathcal{M}$-trivial, in the case of (b), $F|\mathcal{P}(b)$ is not $\mathcal{I} \oplus \mathcal{M}$-sharp whenever $b \in \mathcal{P}(b_k) \cap \mathcal{C}$. In both cases we may without loss of generality assume that the b_k's are pairwise disjoint.)

Then we build inductively disjoint sets a_n and x_n and families $B^n = \{b^n_k : k \in \omega\}$ for $n < \omega$ so that for all n :

(1) $x_n \subseteq a_n \subseteq a$,
(2) $\bigcup B^n \subseteq a - \bigcup_{i \leq n} a_i$,
(3) for every $x \subseteq a - \bigcup_{i \leq n} a_i$ $\quad (G_n(\bigcup_{i \leq n} x_i \cup x) \cap F(a_n)) \Delta F(x_n) \notin \mathcal{I} \oplus \mathcal{M}$.
(4) for every $k \in \omega$:
$F|\mathcal{P}(b^n_k)$ is not semi-$\mathcal{I} \oplus \mathcal{M}$-trivial (for the proof of (a))
resp. $b^n_k \in \mathcal{P}(b_j) \cap \mathcal{C}$ for some j (for the proof of (b)).

If we succeed then we have reached a contradiction:

Let $x = \bigcup_{n<\omega} x_n$. It follows that $(G_n(x) \cap F(a_n)) \Delta F(x_n) \notin \mathcal{I} \oplus \mathcal{M}$ for every $n < \omega$. But this is impossible, since $(F(x) \cap F(a_n)) \Delta F(x_n) \in \mathcal{I}$ and $G_n(x) \Delta F(x) \in \mathcal{M}$ for some $n < \omega$.

It remains to show that under our assumptions the inductive construction can be carried out. Set $B^{-1} = B$. Suppose $\langle a_i : i < n \rangle$, $\langle x_i : i < n \rangle$ and B^{n-1} have been constructed and satisfy (1) to (4). Denote $c_n = a - (\bigcup_{i<n} a_i \cup b^{n-1}_0)$ and $z_n = \bigcup_{i<n} x_i$.

For $y \subseteq b^{n-1}_0$ define:

$$H_n(y) = \{x \subseteq c_n : \ (G_n(z_n \cup y \cup x) \cap F(b^{n-1}_0)) \Delta F(y) \in \mathcal{I} \oplus \mathcal{M}\}.$$

Notice that H_n is the inverse image of an analytic set under a Borel function, and is thus analytic. In particular, H_n has the Baire property for every n.

7. Claim. *$\exists y \subseteq b^{n-1}_0$ $H_n(y)$ is not comeagre.*

Proof. If $H_n(y)$ is comeagre for every $y \subseteq b^{n-1}_0$, then the graph of $F|\mathcal{P}(b^{n-1}_0)$ is contained in the set

$$\{\langle y, u \rangle \in \mathcal{P}(b^{n-1}_0) \times \mathcal{P}(\omega) :$$
$$\{x \subseteq c_n : (G_n(z_n \cup y \cup x) \cap F(b^{n-1}_0)) \Delta u \in \mathcal{I} \oplus \mathcal{M}\} \text{ is comeagre}\}. \quad \square$$

8. Fact. *Whenever $\mathcal{X}$ and $\mathcal{Y}$ are Polish spaces, and $\mathcal{Z} \subseteq \mathcal{X} \times \mathcal{Y}$ is Σ^1_1, then the set*

$$\{x \in \mathcal{X} : \{y : \langle x, y \rangle \in \mathcal{Z}\} \text{ is comeagre in } \mathcal{Y}\}$$

is also of class Σ^1_1.

Proof. See [K] or [Mo], page 262. □

Since every total multifunction of class Σ^1_1 has a Baire measurable uniformization, $F|\mathcal{P}(b_0^{n-1})$ is $\mathcal{I} \oplus \mathcal{M}$-sharp, and thus by Fact 5(a) semi-$\mathcal{I} \oplus \mathcal{M}$-trivial, which contradicts both the (a) and (b) versions of (4). This proves claim 7. □

Back to the proof of 6, fix $y \subseteq b_0^{n-1}$ and a basic clopen set $[s] \in \mathcal{P}(c_n)$ such that $H_n(y)$ is meagre in $[s]$. Let $u_0 = s^{-1}\{0\}$, $u_1 = s^{-1}\{1\}$, and $u = u_0 \cup u_1$. Find a decomposition $c_n = c_n^0 \cup c_n^1$ and subsets $t_0 \subset c_n^0$, $t_1 \subset c_n^1$ such that $u_1 \cup x \cup t_{1-i} - u_0 \notin H_n(y)$ for every $i \in \{0, 1\}$ and $x \subseteq c_n^i$. In the case of the proof of (b), we require additionally that $c_n^i \cap b_j^{n-1} - u \in \mathcal{C}$ for $i \in \{0, 1\}, j > 0$.

Now, for the proof of (a), observe that by inductive assumption (4), there is $i \in \{0, 1\}$ so that for infinitely many k the function $F|\mathcal{P}(b_k^{n-1} \cap c_n^i)$ is not semi-$\mathcal{I} \oplus \mathcal{M}$-trivial. Let us assume, for concreteness, that this is true for $i = 0$. Then set $a_n = b_0^{n-1} \cup u \cup c_n^1$, $x_n = y \cup u_1 \cup t_1$, and

$B^n = \{b_k^{n-1} \cap c_n^0 : \mathit{the\ function\ F\ restricted\ to\ this\ set\ is\ not\ semi} - \mathcal{I} \oplus \mathcal{M} - \mathit{trivial}\}$.

For the proof of (b), we can define a_n, x_n as in the proof of (a), and set $B^n = \{b_k^{n-1} \cap c_n^0 - u : k > 0\}$. This completes the inductive construction, and thus the proof of 6.

Let $\mathcal{I}$ be an ideal. By AT($\mathcal{I}$) we abbreviate the following statement: "Let $F : \mathcal{P}(\omega) \to \mathcal{P}(\omega)$ be a function that is both Fin-invariant mod $\mathcal{I}$ and preserves intersections mod $\mathcal{I}$, and let $\mathcal{A}$ be an uncountable family of pairwise almost disjoint subsets of $\mathcal{P}(\omega)$. Then $F|\mathcal{P}(a)$ semi-$\mathcal{I}$-trivial for all but countably many $a \in \mathcal{A}$."

By AT we abbreviate the assertion that AT($\mathcal{I}$) holds for every Σ^1_1-ideal $\mathcal{I}$.

9. Definition. A family $\mathcal{A}$ of almost disjoint subsets of ω is neat if there is a 1-1 map $e : \omega \to 2^{<\omega}$ such that if $a \in \mathcal{A}$ and $n, m \in a$ then $e(n) \subseteq e(m)$ or $e(m) \subseteq e(n)$. (In other words, e is such that $\bigcup e''a$ is an infinite branch through $2^{<\omega}$ for every $a \in \mathcal{A}$).

10. Definition. By WAT we abbreviate the following statement:

"Let $\mathcal{I}$ be an ideal, and $\mathcal{M}$ a closed approximation of $\mathcal{I}$. Furthermore, let $\mathcal{A}$ be an uncountable neat family of pairwise almost disjoint subsets of ω, and let $F : \mathcal{P}(\omega) \to \mathcal{P}(\omega)$ be a function that preserves intersections mod $\mathcal{I}$. Then $F|\mathcal{P}(a)$ is semi-$\mathcal{M}\oplus\mathcal{M}$-precise for all but countably many $a \in \mathcal{A}$."

11. Theorem. *OCA $\Rightarrow$ WAT.*

Proof. The proof of Theorem 11 goes very much along the lines of the proof of Lemma 2 in [V]. In what follows we assume OCA. Moreover, we fix $\mathcal{I}, \mathcal{M}, F$ and $\mathcal{A}$ as in the statement WAT.

Let $e : \omega \to 2^{<\omega}$ witness that $\mathcal{A}$ is neat. Let $\mathcal{X}$ be the set of all pairs $\langle a, b\rangle$ of subsets of ω such that there exists $c \in \mathcal{A}$ such that $b \subseteq a \subseteq c$, and define the partition:

$$[\mathcal{X}]^2 = \mathcal{K}_0 \cup \mathcal{K}_1$$

by $\{\langle a_0, b_0\rangle, \langle a_1, b_1\rangle\} \in \mathcal{K}_0$ iff

(a) $\bigcup e''a_0 \neq \bigcup e''a_1$,

(b) $a_0 \cap b_1 = a_1 \cap b_0$,

(c) $(F(a_0) \cap F(b_1))\Delta(F(a_1) \cap F(b_0)) \notin \mathcal{M} \oplus \mathcal{M}$.

Then $\mathcal{K}_0$ is open in the product of the separable metric topology τ on $\mathcal{X}$ obtained by identifying $\langle a, b\rangle$ with $\langle a, b, F(a), F(b)\rangle$.

Notice that if $\mathcal{I} = \text{Fin}$ and $\mathcal{M} = \{\emptyset\}$, then we get exactly the partition from [V].

12. Claim. *There are no uncountable 0-homogeneous subsets of $\mathcal{X}$.*

Proof. Suppose $\mathcal{Y}$ is an uncountable 0-homogeneous set. Let c be the union of all b such that for some a the pair $\langle a, b\rangle$ belongs to $\mathcal{Y}$. Let $\langle a, b\rangle$ be such a pair. By (b) in the definition of $\mathcal{K}_0$ it follows that $c \cap a = b$ and hence $(F(c) \cap F(a))\Delta F(b) \in \mathcal{I}$. Now by 1(ii) we can find an uncountable $\mathcal{Z} \subset \mathcal{Y}$ and $n < \omega$ such that $((F(c) \cap F(a))\Delta F(b)) - n \in \mathcal{M}$ for all $\langle a, b\rangle \in \mathcal{Z}$. Then there are distinct $\langle a_0, b_0\rangle$ and $\langle a_1, b_1\rangle$ in $\mathcal{Z}$ such that $F(a_0) \cap n = F(a_1) \cap n$ and $F(b_0) \cap n = F(b_1) \cap n$. It follows that if we denote $F(a_0) \cap F(b_1) = d$ and $F(a_1) \cap F(b_0) = e$, then $d\Delta e \subseteq \omega - n$. It is easy to check that $d\Delta e \subset ((F(a_0) \cap F(c))\Delta F(b_0)) \cup (F(a_1) \cap F(c))\Delta F(b_1))$. Since $\mathcal{M} \oplus \mathcal{M}$ is downward closed, it follows that $d\Delta e \in \mathcal{M} \oplus \mathcal{M}$, contradicting point (c) of the definition of $\mathcal{K}_0$.

Now, by OCA, we can find a decomposition $\mathcal{X} = \bigcup_{n<\omega} \mathcal{X}_n$, where $\mathcal{X}_n$ is 1-homogeneous for all n. Fix for each n a countable subset $\mathcal{D}_n$ of $\mathcal{X}_n$ which is dense in $\mathcal{X}_n$ in the sense of τ. For each $\langle a, b\rangle \in \mathcal{X}$, pick $\sigma(a) \in \mathcal{A}$ such that $b \subseteq a \subseteq \sigma(a)$. Let $\mathcal{B} = \{\sigma(a) : \langle a, b\rangle \in \mathcal{D}_n\ ;\ n < \omega\}$. We shall show that F is semi-$\mathcal{M} \oplus \mathcal{M}$-precise on every $c \in \mathcal{A} - \mathcal{B}$.

Thus, fix any such c and decompose it into two disjoint sets $c = c_0 \cup c_1$ such that for every $i \in \{0,1\}$, $n < \omega$ and $\langle a_0, b_0\rangle \in \mathcal{X}_n$, if $a_0 \subseteq c_i$ then for every $m < \omega$ there exists $\langle a_1, b_1\rangle \in \mathcal{D}_n$ such that: $a_0 \cap b_1 = a_1 \cap b_0$, $a_0 \cap m = a_1 \cap m$, $b_0 \cap m = b_1 \cap m$, $F(a_0) \cap m = F(a_1) \cap m$, and $F(b_0) \cap m = F(b_1) \cap m$.

This is done as follows. An increasing sequence $\langle n_i : i < \omega\rangle$ is constructed by induction. Let $n_0 = 0$. Suppose $\langle n_i : i \le k\rangle$ has been defined. Then, n_{k+1} is chosen sufficiently large such that for every $x, y, u, v \subseteq n_k$ and every $i \le k$ if there exist $\langle a, b\rangle \in \mathcal{X}_i$ such that $a \cap n_k = x$, $b \cap n_k = y$, $F(a) \cap n_k = u$, and $F(b) \cap n_k = v$, then there exists $\langle a, b\rangle \in \mathcal{D}_i$ with the same property such that in addition $a \cap c \subseteq n_{k+1}$. This is possible since a is almost disjoint from c whenever there is b such that $\langle a, b\rangle \in \mathcal{D}_n$. Finally, let

$c_0 = \bigcup\{c \cap [n_k, n_{k+1}) : k \text{ is even}\}$, and let $c_1 = c - c_0$.

For $n < \omega$, $i \in \{0,1\}$, define a tree $T_{n,i} \subset (2^{<\omega})^4$ as follows:

$\langle s, t, u, v\rangle \in T_{n,i}$ iff

$\exists m \in \omega\ \ s, t, u, v \in 2^m$,

$\exists \langle a, b\rangle \in \mathcal{D}_n\ \ a \cap c_i \subseteq m$

$\langle \chi_a|m, \chi_b|m, \chi_{F(a)}|m, \chi_{F(b)}|m\rangle = \langle s, t, \chi_{F(c_i)|m}, v\rangle$.

By $B_{n,i}$ we denote the set of infinite branches through $T_{n,i}$. If $\mathbf{d} \in B_{n,i}$, then $\mathbf{d}$ is of the form $\mathbf{d} = \langle (d)_0, (d)_1, (d)_2, (d)_3\rangle$.

13. Fact. *If $b \subseteq c_i$ and $\langle c_i, b\rangle \in \mathcal{X}_n$, then $\langle c_i, b, F(c_i), F(b)\rangle \in B_{n,i}$.*

Proof. This follows immediately from the fact that $\mathcal{D}_n$ is dense in $\mathcal{X}_n$ and from the choice of the c_i's. □

14. Fact. *Let i, n and $b \subset c_i$ be such that $\langle c_i, b\rangle \in \mathcal{X}_n$. Suppose $\langle c_i, b, F(c_i), d\rangle \in B_{n,i}$. Then $F(c_i) \cap (d \Delta F(b)) \in \mathcal{M} \oplus \mathcal{M}$.*

Proof. Suppose $\mathbf{b} = \langle c_i, b, F(c_i), F(b)\rangle$, $\mathbf{d} = \langle c_i, b, F(c_i), d\rangle$ witness the contrary. Since $\mathcal{M} \oplus \mathcal{M}$ is topologically closed and closed under subsets, there is an $m \in \omega$ such that $F(c_i) \cap (d \Delta F(b)) \cap m \notin \mathcal{M} \oplus \mathcal{M}$. We may without loss of generality assume that $\mathbf{b}|m, \mathbf{d}|m \in T_{n,i}$.

But now recall the reason why $\mathbf{d}|m$ was put into $T_{n,i}$:

there are $\langle a_0, b_0\rangle \in \mathcal{D}_n$ such that

$a_0 \cap m = c_i \cap m = a_0 \cap c_i$

$b_0 \cap m = b \cap m = b_0 \cap b$

$F(a_0) \cap m = F(c_i) \cap m$

$F(b_0) \cap m = d \cap m$.

It follows from the choice of d, m that $F(c_i) \cap (F(b) \Delta F(b_0)) \cap m \notin \mathcal{M} \oplus \mathcal{M}$.

On the other hand, $\langle\langle c_i, b\rangle, \langle a_0, b_0\rangle\rangle \in \mathcal{K}_1$; and since (a),(b) in the definition of $\mathcal{K}_0$ are clearly satisfied, we must have $\neg$(c), i.e.,

$(F(a_0) \cap F(b))\Delta(F(c_i) \cap F(b_0)) \in \mathcal{M} \oplus \mathcal{M}$.

But since $F(a_0) \cap m = F(c_i) \cap m$, we have $(F(c_i) \cap (F(b)\Delta F(b_0)) \cap m \subseteq (F(a_0) \cap F(b))\Delta(F(c_i) \cap F(b_0))$. This yields a contradiction, since $\mathcal{M} \oplus \mathcal{M}$ is closed under subsets.

For $n \in \omega$, $i \in \{0,1\}$ let $K_{n,i} = \{b : \exists d \in B_{n,i} \ (d)_0 = c_i \ \& \ (d)_1 = b\}$.

15. Fact. *For $n \in \omega$, $i \in \{0,1\}$ there exists a Borel function $G_{n,i} : \mathcal{P}(c_i) \to \mathcal{P}(\omega)$ such that $\forall b \in K_{n,i} \ \langle c_i, b, F(c_i), G_{n,i}(b)\rangle \in B_{n,i}$.*

Proof. This follows from the fact that every Borel set with compact sections has a Borel uniformization (see [Mo], page 254). □

Theorem 11 is an easy consequence of Facts 13–15.

Let us now look at some consequences of WAT.

16. Lemma. *WAT + MA implies AT($\mathcal{I}$), where $\mathcal{I}$ is any F_σ-ideal. In particular, WAT + MA implies AT(*Fin*).*

Proof. Assume WAT and MA, let $\mathcal{I}$ be an F_σ-ideal, and let $F, \mathcal{A}$ be as in AT($\mathcal{I}$). By 2(b), there exists a closed approximation $\mathcal{M}$ of $\mathcal{I}$ such that $\mathcal{M} \subset \mathcal{I}$. In this case, $\mathcal{I} \oplus \mathcal{M} \oplus \mathcal{M} = \mathcal{I}$. Fix such $\mathcal{M}$.

Use MA to find an uncountable family $\mathcal{B}$ of pairwise almost disjoint subsets of ω so that for all $b \in \mathcal{B}$ there exist infinitely many $a \in \mathcal{A}$ such that $a - b \in$ Fin.

By [V, Lemma 3] and MA there is an uncountable $\mathcal{C} \subseteq \mathcal{B}$ and for every $c \in \mathcal{C}$ a decomposition $c = c_0 \cup c_1$ such that the family $\mathcal{C}_i = \{c_i : c \in \mathcal{C}\}$ is neat for $i \in \{0,1\}$.

By WAT we find $b \in \mathcal{C}$ such that $F|\mathcal{P}(b)$ is semi-$\mathcal{M} \oplus \mathcal{M}$-precise. Fix such b, and fix a decomposition $b = b_0 \cup ... \cup b_k$ so that $F|\mathcal{P}(b_i)$ is $\mathcal{M} \oplus \mathcal{M}$-precise for $i \leq k$. Let $\langle a_n : n < \omega\rangle$ be a sequence of elements of $\mathcal{A}$ which are almost contained in b. By applying Lemma 6(a) $k+1$ times we find n so that $F|\mathcal{P}(a_n \cap b_i)$ is semi-$\mathcal{I} \oplus \mathcal{M} \oplus \mathcal{M}$-trivial for $i \leq k$. Recall that $\mathcal{I} \oplus \mathcal{M} \oplus \mathcal{M} = \mathcal{I}$. But this means that $F|\mathcal{P}(b \cap a_n)$ is semi-$\mathcal{I}$-trivial. By Fact 5(c) we are done. □

We now conclude that several interesting consequences of AT(Fin) also follow from OCA + MA. (In fact, MA can be dropped from the assumptions of 17-19. This is immediate from the proofs quoted below).

17. Corollary. *OCA + MA implies that for every $n < \omega$, the topological space $(\omega^*)^{n+1}$ is not a continuous image of $(\omega^*)^n$.*

Proof. See [J2] for definitions and a proof.

18. Corollary. *OCA + MA implies that no nowhere dense P-subset of ω^* is homeomorphic to ω^* itself.*

Proof. See [J3] for definitions and a proof.

19. Corollary. *Suppose OCA and MA hold and every Σ^1_{n+2}-set of reals has the property of Baire. If $\mathcal{I}$ is an ideal of class Σ^1_n, and if the quotient algebra $\mathcal{P}(\omega)/\mathcal{I}$ can be isomorphically embedded into $\mathcal{P}(\omega)/\mathrm{Fin}$, then $\mathcal{I}$ is generated over* Fin *by at most one set.*

Proof. See [J1, Theorem 0.6].

We conclude this note with an application of WAT to a problem of Erdős and Ulam. Let f and g be as in example 1(d). Erdős and Ulam asked whether the Boolean algebras $\mathcal{P}(\omega)/\mathcal{I}_f$ and $\mathcal{P}(\omega)/\mathcal{I}_g$ are isomorphic (see [E]). It was shown in [JK] that in the presence of CH these algebras are indeed isomorphic, and later I showed that $\mathrm{AT}(\mathcal{I}_g)$ implies that they are not isomorphic (see [J], [J4]). Here we show that WAT suffices for the latter result.

20. Theorem. *WAT $\Rightarrow \mathcal{P}(\omega)/\mathcal{I}_f \not\cong \mathcal{P}(\omega)/\mathcal{I}_g$.*

Proof. The argument in [J4] goes as follows: Suppose $F : \mathcal{P}(\omega) \to \mathcal{P}(\omega)$ induces an isomorphism from $\mathcal{P}(\omega)/\mathcal{I}_f$ onto $\mathcal{P}(\omega)/\mathcal{I}_g$. Choose an uncountable family $\mathcal{A}$ of pairwise almost disjoint subsets of ω that are "large" in the sense that every $a \in \mathcal{A}$ contains infinitely many intervals of ω of the form $[k, k \cdot 2^k)$. Then use $\mathrm{AT}(\mathcal{I}_g)$ and a version of Fact 5 to find $a \in \mathcal{A}$ so that F is $\mathcal{I}_g$-trivial on a. The remainder of the proof is a ZFC argument that F cannot both induce an isomorphism and be $\mathcal{I}_g$-trivial on a large set.

In a WAT-setting we may argue as follows: Suppose F is as above, and let $\mathcal{A}$ be an uncountable neat family of pairwise almost disjoint sets which are large in the above sense. By WAT, for every $\varepsilon > 0$ there are at most countably many $a \in \mathcal{A}$ so that $F|\mathcal{P}(a)$ is *not* semi-$\mathcal{M}(g,\varepsilon)\oplus\mathcal{M}(g,\varepsilon)$-precise. This allows us to pick $a \in \mathcal{A}$ so that $F|\mathcal{P}(a)$ is semi-$\mathcal{M}(g,2^{-n})\oplus\mathcal{M}(g,2^{-n})$-precise for all $n < \omega$.

Since a is large, we can find a family $\mathcal{B}$ of infinitely many pairwise disjoint large subsets of a.

Since the family of large subsets of a is comeagre in $\mathcal{P}(a)$, by Lemma 6(b), there is some $b \in \mathcal{B}$ so that $F|\mathcal{P}(b)$ is $\mathcal{I}_g \oplus \mathcal{M}(g, 2^{-1})$-sharp. Now iterate this construction to produce a $\subset$-decreasing sequence $\langle b^k : k < \omega\rangle$ of large sets such that $F|\mathcal{P}(b^k)$ is $\mathcal{I}_g \oplus \mathcal{M}(g, 2^{-k-1})$-sharp and $b^\omega = \bigcap_{k\in\omega} b^k$ is large. Since F is Fin-invariant mod $\mathcal{I}_g$, the latter can easily be arranged by fixing a finite portion of b^ω at every step. Clearly, $F|\mathcal{P}(b^\omega)$ is $\mathcal{I}_g \oplus \mathcal{M}(g, \varepsilon)$-sharp for every $\varepsilon > 0$. Since $2\mathcal{M}(g, \varepsilon) \subseteq \mathcal{M}(g, 2\varepsilon)$, we infer from 5(a),(b) that $F|\mathcal{P}(b^\omega)$ is $\mathcal{I}_g \oplus \mathcal{M}(g, \varepsilon)$-trivial for every $\varepsilon > 0$.

In order to get to the starting point of the ZFC-part of the argument in [J4], it suffices now to prove the following:

21. Claim. *If $F|\mathcal{P}(a)$ as above is $\mathcal{I}_g \oplus \mathcal{M}(g, \varepsilon)$-trivial for every ε, then $F|\mathcal{P}(a)$ is $\mathcal{I}_g$-trivial.*

Proof. For $c, d \subseteq \omega$ denote:

$\rho(c, d) = \limsup_{n\to\infty} \frac{\sum_{m\in n\cap(c\Delta d)} g(m)}{\sum_{m<n} g(m)}$.

Notice that ρ satisfies the triangle inequality, and that $\rho(c, d) = 0$ iff $c\Delta d \in \mathcal{I}_g$. Thus ρ can be considered a metric on $\mathcal{P}(\omega)/\mathcal{I}_g$; and it is not hard to show that the metric space $(\mathcal{P}(\omega)/\mathcal{I}_g, \rho)$ is complete.

Suppose the functions $F_k : \mathcal{P}(a) \to \mathcal{P}(\omega)$ witness $\mathcal{I}_g \oplus \mathcal{M}(g, 2^{-k})$-triviality of $F|\mathcal{P}(a)$, i.e., $\rho(F(b), F_k(b)) \leq 2^{-k}$ for all $k < \omega$. Define

$$\Gamma = \{\langle b, c\rangle : b \subseteq a \;\&\; c \subseteq \omega \;\&\; \lim_{k\to\infty} \rho(F_k(b), c) = 0\}$$

Since the sequence $\mathcal{F} = \langle F_k : k \in \omega\rangle$ can be encoded as a closed subset of $\mathcal{P}(a) \times \mathcal{P}(\omega)^\omega$, and since Γ can be defined from $\mathcal{F}$ by quantifying over natural numbers only, the set Γ is Borel.

Since $\langle F_k(b);\ k \in \omega\rangle$ is a Cauchy sequence in the sense of ρ, and since ρ is complete, for every $b \subseteq a$ there is some c so that $\langle b, c\rangle \in \Gamma$.

Moreover, if $\langle b, c\rangle \in \Gamma$, then $c\Delta F(b) \in \mathcal{I}_g$.

We conclude that there is a Baire measurable uniformization G of Γ, and this suffices by Fact 5(a),(b) applied to $\mathcal{M} = \{\emptyset\}$ to conclude that there is a continuous one as well.

22. Corollary. *OCA* $\Rightarrow \mathcal{P}(\omega)/\mathcal{I}_f \not\simeq \mathcal{P}(\omega)/\mathcal{I}_g$.

23. Remark. Frankiewicz has shown in [F] that $MA \Rightarrow \mathcal{P}(\omega)/\mathcal{I}_f \simeq \mathcal{P}(\omega)/\mathcal{I}_g$. His proof contains a substantial gap. In view of the results of this note it seems rather unlikely that this gap can be bridged.

Acknowledgment. This research was supported by NSF grant DMS-8505550 during the author's stay at the Mathematical Sciences Research Institute in 1989–90.

References

[E] Erdős, P., *My Scottish Book "Problems"*, The Scottish Book (D. Mauldin, ed.), Birkhäuser, Boston, Mass., 1981.

[F] Frankiewicz, R., *Some remarks on embeddings of Boolean algebras*, Measure Theory, Oberwolfach 1983 (D. Kölzow and D. Maharam-Stone, eds.), SLNM 1089, pp. 64-68.

[J] Just, W., *Ph.D. Thesis (in Polish)*, University of Warsaw (1987).

[J1] Just, W., *A modification of Shelah's oracle-c.c. with applications*, Trans. AMS (to appear).

[J2] Just, W., *The space* $(\omega^*)^{n+1}$ *is not always a continuous image of the space* $(\omega^*)^n$, Fund. Math. **132** (1989), 59-72.

[J3] Just, W., *Nowhere dense P-subsets of* ω^*, Proc. AMS **104** (1989), 1145-1146.

[J4] Just, W., *Repercussions of a problem of Erdős and Ulam*, Canad. J. of Math. **XLII**, No. 5 (1990), 902-914.

[JK] Just, W., Krawczyk, A., *On certain Boolean Algebras* $\mathcal{P}(\omega)/I$, Trans. AMS **285**, No. 1 (1984), 411-429.

[K] Kechris, A.S., *Measure and category in effective descriptive set theory*, Ann. of Math. Logic **5** (1973).

[M] Mazur, K., F_σ*-ideals and* $\omega_1\omega_1^*$*-gaps in the Boolean algebras* $\mathcal{P}(\omega)/\mathcal{I}$, Preprint, University of Warsaw.

[Mo] Moschovakis, Y., *Descriptive Set Theory*, North Holland, 1980.

[S] Shelah, S., *Proper forcing*, SLNM 940.

[T] Todorčević, S., *Partition Problems in Topology*, Contemporary Mathematics **84** (1989).

[V] Veličković, B., *OCA and automorphisms of* $\mathcal{P}(\omega)/fin$, Submitted for publication.

Department of Mathematics, Ohio University, Athens OH 46701

ONE CANNOT SHOW FROM ZFC THAT THERE IS AN ULM-TYPE CLASSIFICATION OF THE COUNTABLE TORSION-FREE ABELIAN GROUPS

Garvin Melles

Abstract. Using generalized recursive set functions, we define some notions of classification and prove that if the universe has a Cohen real over L, then there is no Ulm-type classification of the countable torsion-free abelian groups.

1. Introduction

We consider a class of models $\mathcal{C}$ to be classifiable (i.e., has a structure theory) if there is an effective construction within set theory from the models in $\mathcal{C}$ to invariants depending only on the isomorphism types of $\mathcal{C}$, and if there is a way to effectively construct from each invariant, an example with that invariant. We call a classification canonical if the construction from invariants back to models with those invariants can be made canonical.

The divisible abelian groups, rank one abelian torsion-free groups, models of the language $< E_i : i < \alpha >$ with α an ordinal and each E_i an equivalence relation such that for every $i < \alpha$ E_i refines E_{i+1}, models of a language with countably many unary relations, the countable homogeneous models of a finite relational language and the models of the theory of $({}^{<\omega}\omega, f)$ where f is a unary function s.t. $f(\eta \frown i) = \eta$ for $\eta \neq < >$ and $f(< >) =< >$, are all examples of classes classifiable in the canonical sense. The last example is interesting because although it is ω-stable and has NDOP, it is deep and therefore has 2^λ many models for every uncountable λ. Ulm's classification of the countable torsion groups is not an example of a canonical classification, but it is close to one. From a given admissible Ulm sequence one can effectively construct a group with that Ulm sequence, but the construction is canonical only up to the choice of a bijection from ω to the Ulm sequence's length. There is a natural strengthening of the notion of canonical classification which we call Ulm-type classification. The countable torsion abelian groups have such a classification by

Ulm's theorem, whereas if the universe has a Cohen real over L, there is no Ulm-type classification of the countable torsion-free abelian groups.

In order to prove non-classification we need an exact definition for the intuitive notion of effective construction within set theory. We define the class of recursive set functions and define a set B as effectively constructible from a set A if there is a set function F recursive in the cardinality function such that $F(A) = B$. It should be noted that one needs a hypothesis stronger than ZFC to prove the non-classifibility of the countable torsion-free abelian groups because the canonical well ordering of L is recursive in our sense, so that within L any reasonable class $\mathcal{C}$ of countable models (reasonable in a sense to be defined later) is canonically classifiable by using Scott sentences as invariants and by taking as the canonical example of a model with a given Scott sentence φ as the least countable model M in the sense of the well ordering of L such that $M \models \varphi$.

What is the connection between Shelah's classification theory and the type of classification defined above? It follows from Theorem 2.4 and the proof of Lemma 2.9 in [Sh] that for a countable first order theory T which is not superstable or has OTOP or DOP, isomorphism type is not absolute under extensions of the universe preserving cardinals. As a result, such theories cannot have a classification via generalized recursive functions. How about for countable superstable T without OTOP or DOP? This is still an open problem, but at the MSRI conference, Shelah and Hrushovski found a finitary structure theorem for the superstable $\aleph_\epsilon$-saturated models with NDOP which is of the type defined here [Sh401].

For another approach to the classifiability of countable structures see [F & S] . Also, Hodges & Shelah ask some questions of a similar flavor, cf. [H], [H & S] . The author would like to thank his thesis advisor Paul Eklof for many helpful discussions.

Definition 1. A set function F is recursive in $F_1, \ldots, F_k$ if it is a member of the smallest class containing the initial functions and closed under substitution, definition by recursion, the μ-operator, and random well ordering by ordinals. The class of primitive recursive set functions in $F_1, \ldots, F_k$ is the smallest class containing the initial functions, closed under substitution, and definition by recursion.

Initial functions:

(1) $F(x) = F_i(x),\ 1 \leq i \leq k$

(2) $P_{n,i}(x_1, \ldots, x_n) = x_i$

(3) $F(x) = 0$

(4) $F(x, y) = x \cup \{y\}$

(5) $C(x, y, u, v) = x$ if $u \in v, y$ otherwise

Substitution:

(1) $F(x_1, \ldots, x_n, y_1, \ldots, y_m) = G(x_1, \ldots, x_n, H(x_1, \ldots, x_n), y_1, \ldots, y_m)$
(2) $F(x_1, \ldots, x_n, y_1, \ldots, y_m) = G(H(x_1, \ldots, x_n), y_1, \ldots, y_m)$

Recursion:

$$F(x_1, \ldots, x_n, z) = \\ G(\cup\{(x_1, \ldots, x_n, u, F(x_1, \ldots, x_n, u)) : u \in z\}, x_1, \ldots, x_n, z)$$

The μ-operator:

If $\forall x_1, \ldots, x_n \exists \alpha \in ORD(G(x_1, \ldots, x_n, \alpha) = 0)$ where G is a $n+1$-ary set function then μG is the n-ary set function such that for every $x_1, \ldots, x_n \in V, \mu G(x_1, \ldots, x_n) =$ the least ordinal α such that $G(x_1, \ldots, x_n, \alpha) = 0$.

Random Well Ordering by Ordinals

If for every $x_1, \ldots, x_n \in V$

$$\forall x[(x = (\alpha, f, x_1) \wedge \alpha \text{ is an ordinal} \wedge \\ f \text{ is a bijection from } \alpha \text{ to } x_1) \rightarrow G(x, x_1, \ldots, x_n) = \phi]$$

$$\leftrightarrow \exists x[(x = (\alpha, f, x_1) \wedge \alpha \text{ is an ordinal} \wedge \\ f \text{ is a bijection from } \alpha \text{ to } x_1) \wedge G(x, x_1, \ldots, x_n) = \phi]$$

Then the function $F(x_1, \ldots, x_n)$ defined from the recursive functions $G(x, x_1, \ldots, x_n)$ by letting $F(x_1, \ldots, x_n) = \phi$ if

$$\exists x[x = (\alpha, f, x_1) \wedge \alpha \text{ is an ordinal} \wedge \\ f \text{ is a bijection from } \alpha \text{ to } x_1) \wedge G(x, x_1, \ldots, x_n) = \phi]$$

and by letting $F(x_1, \ldots, x_n) = 1$ otherwise is recursive.

Definition 2. # is the set function which takes every set to its cardinality.

All recursive set functions are Δ_1 definable and if $V = L$, then the class of set recursive functions is the class of Δ_1 definable functions. We are now ready to give a formal definition of classification.

Definition 3. A Class $\mathcal{C}$ of models is canonically classifiable if there are set functions F_1, F_2, F_3 recursive in # such that:

(1) $\forall M_1, M_2 \in \mathcal{C}, F_1(M_1) = F_2(M_2) \leftrightarrow M_1 \cong M_2$. If $M \notin \mathcal{C}$, then $F_1(M) = \phi$.
(2) $\forall x \in V$ $F_2(x) = 1$ if $x \in \{F_1(M) : M \in \mathcal{C}\}$ and $F_2(x) = 0$ otherwise
(3) If $F_1(M) \in Inv(\mathcal{C}, F_1) = \{F_1(M) : M \in \mathcal{C}\}$, then $F_3(F_1(M)) \cong M$.

In the formal definition of canonical classification F_1 is an set algorithm which calculates invariants from models in $\mathcal{C}$, F_2 determines whether a given set is an invariant, and F_3 constructs from a given invariant a canonical model with that invariant.

Definition 4. If $\mathcal{C}$ is a class of models, a choice function F on the isomorphism types of $\mathcal{C}$ is a set function such that for every M_1 and M_2 in $\mathcal{C}, F(M_1) = F(M_2) \leftrightarrow M_1 \cong M_2$ and $F(M_1) \cong M_1$, and $\forall x \in V$, if $x \notin \mathcal{C}$, then $F(x) = \phi$.

Theorem 1. *If $\mathcal{C}$ is a class of models, then $\mathcal{C}$ is canonically classifiable if and only if there is a choice function F on the isomorphism types of $\mathcal{C}$ which is recursive in #.*

Proof. Suppose $\mathcal{C}$ is canonically classifiable, and let F_1, F_2, and F_3 be functions recursive in # that witness the classifibility of $\mathcal{C}$. Then let $F = F_3(F_1)$. Now suppose that F is a choice function on the isomorphism types of $\mathcal{C}$ which is recursive in #. Define $F_1(x) = F(x)$. Let $F_2(x) = 1$ if $F(F(x)) = F(x)$ and ϕ otherwise. Let $F_3 =$ the identity function.

If $\mathcal{C}$ is a class of countable models, we change the definition of classification slightly. We use recursive functions defined on HC and we can drop the random well ordering clause in the definition of recursive; otherwise the definition is the same.

Definition 3′. A bijection witnessing function is a function from the countable infinite ordinals which takes each countable ordinal to a bijection from ω to the given ordinal.

Definition 3″. A class $\mathcal{C}$ of countable models is canonically classifiable if there are recursive functions of HC such that (1), (2), and (3) from Definition 3 hold for $\mathcal{C} \cap HC$. $\mathcal{C}$ has an Ulm-type classification if there are functions (on HC) F_1, F_2, and F_3 recursive in oracle G such that for every substitution of G by a bijection witnessing function H, (1), (2), and (3) from Definition 3 hold for $\mathcal{C} \cap HC$.

The Main Lemma. *If M is a transitive model of ZFC and G is a Cohen real over M, then there is no Σ_1 formula of set theory which in $M[G]$ defines a choice function on the isomorphism types of the hereditarily countable torsion-free abelian groups.*

Corollary 1. *If $M \models ZFC$ and G is a Cohen real over M, then there is no canonical classification of the countable torsion-free abelian groups within $M[G]$.*

Theorem 2. *(Levy-Schoenfield Absoluteness) If $L \subseteq M \subseteq V, M \models ZFC$, $m \in HC^M$ and $\varphi(x, y)$ is a Δ_0-formula, then*

$$V \models \exists x \varphi(x, m) \quad \rightarrow \quad M \models \exists x \varphi(x, m).$$

Corollary 2. *If V has a Cohen real over L, then there is no Ulm type classification of the countable torsion-free abelian groups.*

Proof. Assume there is a function F defined on HC such that F is recursive in oracle G and for every substitution of a bijection witnessing function H (which we denote $F(H)$) for $G, F(H)$ defines a choice function on the isomorphism types of the countable torsion-free abelian groups within HC. Let H' be any bijection witnessing function such that for α an ordinal countable in $L, H'(\alpha) =$ the least bijection from ω to α in the sense of the canonical well ordering of L. Using Levy-Shoenfield Absoluteness one can show that for every $m \in HC^{L[G]}, F(H)(m) = F^{[G]}(H' \upharpoonright L)(m)$. This a contradiction since $F(H' \upharpoonright L)$ in $L[G]$ would be a Σ_1 definable choice function on the isomorphism types in HC of the countable torsion-free abelian groups.

The author would like to thank Menachem Magidor and Tomek Bartoszynski for pointing out the following stronger version of the Main Lemma and the resulting corollary.

Extended Main Lemma. *If M is a transitive model of ZFC and G is a Cohen real over M, then there is no Σ_1 formula of set theory with parameters in HC^M which defines in $M[G]$ a choice function on the isomorphism types of the hereditarily countable torsion-free abelian groups.*

Proof. Exactly the same as the proof of the Main Lemma.

Corollary 3. *There is a model M of ZFC such that for no Σ_1 formula $\varphi(x, y, y_1, \ldots, y_n)$ and $p_1, \cdots, p_n$ hereditarily countable parameters does $\varphi(x, y, p_1, \ldots, p_n)$ define a choice function on the isomorphism types of the torsion-free abelian groups.*

Proof. Iterate Cohen forcing ω_1 times.

2. Outline of the Proof of the Main Lemma

Here we give an informal idea of the direction of the proof. First we let $M \models ZFC$ and define a set of fractions $S \subseteq \mathbb{Q}$ and a countable partial ordering $(P, \leq)$ such that the generic set G has closely associated with it a group $A \subseteq \mathbb{Q}^{(\omega)}$. Next we prove that the automorphism group of A is isomorphic to S (under multiplication) and suppose in contradiction to the Main Lemma that there is a Σ_1 formula $\varphi(x,y)$ of set theory and a canonical copy C_A of A such that

$$M[G] \models \forall B \forall X[B \cong A \rightarrow (\varphi(B, C_A) \wedge \varphi(B, X) \rightarrow X = C_A)]$$

so there is a $p \in G$ such that

$$p \Vdash \forall B \forall X[B \cong \dot{A} \rightarrow (\varphi(B, C_A) \wedge \varphi(B, X) \rightarrow X = \dot{C}_A))]$$

so for every $\pi \in Aut(P, \leq)$

$$\pi p \Vdash \forall B \forall X[B \cong \pi\dot{A} \rightarrow (\varphi(B, \pi\dot{C}_A) \wedge (\varphi(B, X) \rightarrow X = \pi\dot{C}_A))]$$

By the construction of $\dot{A}$, $\forall \pi \in \mathrm{Aut}(P, \leq)$ $i_G(\pi\dot{A}) \cong A$ by π^{-1}. Therefore, for all $\pi \in Aut(P, \leq)$ such that $\pi p \in G$ we have $C_A = i_G(\pi\dot{C}_A)$. Now by applying the Stable Names Lemma and using the rigidity of A (and thereby of C_A) we get a contradiction by exploiting the richness of $\mathrm{Aut}(P, \leq)$.

3. Preliminaries to the Proof of the Main Lemma

Theorem 3. *(cf. Jech 19.14) If $\varphi(x_1, \ldots, x_n)$ is a formula and π is an automorphism of B, then for all $\dot{x}_1, \ldots \dot{x}_n \in M^B$*

$$\|\varphi(\pi\dot{x}_1, \ldots \pi\dot{x}_n)\| = \pi(\|\varphi(\dot{x}_1, \ldots, \dot{x}_n)\|)$$

Lemma 2. *(cf. Jech 19.16) If P is a separative partially ordered set, $p \in P, \pi \in \mathrm{Aut}\, P$, and $\dot{x}_1, \ldots, \dot{x}_n \in M^B$ then*

$$p \Vdash \varphi(\dot{x}_1, \ldots, \dot{x}_n) \quad \textit{iff} \quad \pi(p) \Vdash \varphi(\pi\dot{x}_1, \ldots, \pi\dot{x}_n)$$

Definition 5. We define a set of prime numbers $N_1 = \{p_i : i < \omega\}$ and a set of integers $N_2 = \{m_i : i < \omega\}$ which are defined by induction on $i < \omega$ as follows: Let $p_0 = 2$. If p_i has been defined let $m_i = p_0^{i+1} \cdot \ldots \cdot p_i^{i+1} + 1$. If m_i has been defined, let p_{i+1} be a prime greater than m_i. Let $N_2' = \{m \in \omega : m$ is relatively prime to every element in $N_1\}$. Note that $N_2 \subseteq N_2'$. Let $S =$ the set of fractions with denominators and numerators relatively prime to N_1.

Definition 6. Let $P'' = \{f : f$ is a function such that $\operatorname{dom} f$ is a finite subset of $\mathbb{Q}^{(\omega)}$ and $\operatorname{ran} f \subseteq \{0,1\}\}$. If $p \in P''$, let $s_p = \{\bar{q} \in \mathbb{Q}^{(\omega)} : \bar{q} \in \operatorname{dom} p$ and $p(\bar{q}) = 1\}$. Let $t_p = \{\bar{q} \in \mathbb{Q}^{(\omega)} : \bar{q} \in \operatorname{dom} p$ and $p(\bar{q}) = 0\}$. Let $H_p =$ the universe of the subgroup of $\mathbb{Q}^{(\omega)}$ generated by s_p. Let $H_p^* =$ the set of fractions that are equal to elements of H_p divided by integers in N_2'. Let $P' = \{p \in P'' : H_p^* \cap t_p = \emptyset$ and for every $\bar{q} \in t_p$ there exists a $\bar{q}' \in s_p$ such that $\bar{q}' = n\bar{q}$ where n is some integer whose prime factors are elements of $N_1\}$. If $p \in P'$ then define $O_p = \bigcap\{\mathbb{Q}^{(\omega)} - H : H \cap t_p = \emptyset$ and H is the universe a subgroup of $\mathbb{Q}^{(\omega)}$ such the $H_p^* \subseteq H$ and every element of H is infinitely divisible by every prime in $N_2'\}$. Define an equivalence relation $\sim$ on P' by defining $p_1 \sim p_2$ if and only if $H_{p_1}^* = H_{p_2}^*$ and $O_{p_1} = O_{p_2}$. Let $P = P'/\sim$. Define a partial ordering $\leq$ on P by defining $\tilde{p}_2 \leq \tilde{p}_1$ if and only if $H_{p_1}^* \leq H_{p_2}^*$ and $O_{p_1} \leq O_{p_2}$.

Definition 7. If p is an element of P' then $l(p) = \sup\{n : \bar{q}(n) \neq 0$ and $\bar{q} \in \operatorname{dom} p\}$. If $\tilde{p}$ is an element of P then $l(\tilde{p}) = l(p)$.

We use elements of $(\mathbb{Q} - \{0\})^\omega$ and automorphisms of ω and compositions of them to induce automorphisms of P. We define $\operatorname{Aut}(P, \leq)$ as the set of all such automorphisms. If $p \in P'$ and $\pi \in (\mathbb{Q} - \{0\})^\omega$ then πp is the function with domain $= \{\pi\bar{q} : \bar{q} \in \operatorname{dom} p\}$ and such that $\pi p(\pi\bar{q}) = 0$ iff $p(\bar{q}) = 0$ where $\pi(q_1, \cdots, q_n) = \big(\pi(1) \cdot (q_1), \ldots, \pi(n) \cdot (q_n)\big)$. Let $\pi\tilde{p} = \widetilde{\pi p}$. If $p \in P'$ and π is an automorphism of ω then πp is the function with domain $= \{\pi\bar{q} : \bar{q} \in \operatorname{dom} p\}$ and such that $\pi p(\pi\bar{q}) = 0$ iff $p(\bar{q}) = 0$ where $(\pi\bar{q})(\pi n) = \bar{q}(n)$. Let $\pi\tilde{p} = \widetilde{\pi p}$.

Lemma 3. *If $p \in P'$ and $i < \omega$, then there is an $m \in N_2$ such that if $\pi \in (\mathbb{Q} - \{0\})^\omega$ such that $\pi(j) = 1$ for $j < i$ and $\pi(j) = m$ for $j \geq i$ then $p^\wedge \pi p \in P'$.*

Proof. Let $s_p = \{\bar{q}_0, \ldots, \bar{q}_s\}$. For each $j < \omega$ and each $r \leq s$, let $\bar{q}_r'(j) = 0 \quad j < i$ and let $\bar{q}_r'(j) = \bar{q}_r(j)$ if $j \geq i$. We can assume that for each $r \leq s$ there is an integer n_r such that $n_r\bar{q}_r' \in H_p$. If not, extend p to the smallest $p' \in P'$ extending p such that $\bar{q}_r' \in s_{p'}$ and work with p' in place of p. Let $n^* = n_0 \cdot \ldots \cdot n_s$. Let $k_1 =$ the smallest integer such that for no $i \geq k_1$, does p_i divides $n^* \cdot n$ where n is an integer such that for some $r \leq s, \bar{q}_r/n \in t_p$. Let $k_2 =$ the smallest integer such that for no $i \geq k_2$, is there a $p \in N$ such that p^i divides $n^* \cdot n$ where n is an integer such that for some $r \leq s, \bar{q}_r/n \in t_p$. Let $k = \max\{k_1, k_2\}$. Let $m = m_k$. $H_p^* = H_{p^\wedge\pi p}^*$ since for every $r \leq s \;\; \pi\bar{q}_r = \bar{q}_r + p_0^{k+1} \cdot \ldots \cdot p_k^{k+1}\bar{q}_r'$. Since $H_p^* = H_{p^\wedge\pi p}^*$, we have $H_{p^\wedge\pi p}^* \cap t_p = \emptyset$. Suppose $H_{p^\wedge\pi p}^* \cap t_{\pi p} \neq \emptyset$. Suppose $\bar{q} \in t_p$ such that $\pi\bar{q} \in H_{p^\wedge\pi p}^*$. For each $r \leq s$ we have $\bar{q}_r + (m-1)\bar{q}_r' = \pi\bar{q}_r$ where $m - 1 =$

$p_0^{k+1} \cdot \ldots \cdot p_k^{k+1}$ and $\bar{q} = \bar{q}_r/n$ where n divides $p_0^{k+1} \cdot \ldots \cdot p_k^{k+1}$. Therefore, if $a = p_0^{k+1} \cdot \ldots \cdot p_k^{k+1}/n, a\bar{q}'_r \in H_p^*$ and $\bar{q} = \bar{q}_r/n = \pi\bar{q}_r/n - a\bar{q}'_r$ which implies that $\bar{q} \in H_p^* = H_{p \wedge \pi p}$ which is a contradiction of $H_{p\wedge \pi p} \cap t_p = \emptyset$. □

Lemma 4. *Let $p \in P'$ and let $\bar{q}_1$ and $\bar{q}_2$ be independent elements of $\mathbb{Q}^{(\omega)}$ in H_p. Then there is a $p_i \in N_1, t$ a positive integer and $p'' \in P'$ such that $\operatorname{dom} p'' = \operatorname{dom} p \cup \{\bar{q}_1/p_i^t, \bar{q}_2, \bar{q}_2/p_i^t\}$ and $p''(\bar{q}_1/p_i^t) = 1, p''(\bar{q}_2) = 1$, and $p''(\bar{q}_2/p_i^t) = 0$.*

Proof. Let $s_p = \{\bar{q}'_0, \ldots, \bar{q}'_s\}$. Without loss of generality, $\bar{q}_2 \in s_p$. Without loss of generality, for each $i, j < l(\bar{q}_1)$ $\bar{q}_1(i) = \bar{q}_1(j)$ or $\bar{q}_1(i) \cdot \bar{q}_1(j) = 0$. (If not, multiply s_p by a $\pi \in (\mathbb{Q} - \{0\})^\omega$ so that the assumption is true about πs_p. Lemma 4 holds for s_p if and only if it holds for πs_p since π is an automorphism of $\mathbb{Q}^{(\omega)}$). Without loss of generality, if $\bar{q} \in H_p$ then for each $i < l(\bar{q})$ $\bar{q}(i)$ is an integer. If not, multiply s_p by a $\pi \in (\mathbb{Q}-\{0\})^\omega$ so that the assumption is true about π_p. Lemma 4 holds for s_p if and only if it holds for πs_p since π is an automorphism of $\mathbb{Q}^{(\omega)}$. Let $p_i \in N_1$ such that p_i does not divide n for any n such that $\bar{q}'_r/n \in t_p$. Extend p to p' by letting $\operatorname{dom} p' = \operatorname{dom} p \cup \{\bar{q}_1/p_i^t\}$ where t is a positive integer such that $p_i^t > 2|\bar{q}_2(j)|$ for every $j < l(\bar{q}_2)$ and let $p'(\bar{q}_1/p_i^t) = 1$. Let $\bar{q}'_r/n \in t_p$. To see that $p' \in P$, suppose $a_0\bar{q}'_0 + \ldots + a_s\bar{q}'_s + a\bar{q}_1/p_i^t = m\bar{q}'_r/n$ where m is an integer whose prime factors are not in N_1. Then $p_i^t\bar{q}'_r/n \in H_p$, and since n and p_i^t are relatively prime and $n(m\bar{q}'_r/n) \in H_p, m\bar{q}'_r/n \in H_p$ so $\bar{q}'_r/n \in H_p^*$, a contradiction of $\bar{q}'_r/n \in t_p$. Extend p' to $p'' \in P'$ by letting $\operatorname{dom} p'' = \operatorname{dom} p' \cup \{\bar{q}_2/p_i^t\}$ and let $p''(\bar{q}_2/p_i^t) = 0$. Suppose $a_0\bar{q}'_0 + \ldots + a_s\bar{q}'_s + \bar{q}_1/p_i^t = \bar{q}_2/p_i^t$. Then $a\bar{q}_1 = \bar{q}_2 \mod p_i^t$ which implies that for every $i < \omega$ if $\bar{q}_1(i) = 0$ then $\bar{q}_2(i) = 0$ and that for every $i, j < \omega$ if $\bar{q}_1(i) = \bar{q}_1(j)$ then $\bar{q}_2(i) = \bar{q}_2(j)$. This is a contradiction of our assumption that $\bar{q}_1$ and $\bar{q}_2$ are independent. Suppose $a_0\bar{q}'_0 + \ldots + a_s\bar{q}'_s + a\bar{q}_1/p_i^t = m\bar{q}_2/p_i^t$ where m is an integer whose prime factors are not in N_1. Since m and p_i^t are relatively prime, $\bar{q}_2/p_i^t \in H_p$ which is a contradiction of what we have just shown.

Lemma 5. *$(P, \leq)$ is separative.*

Let M be a countable transitive model of ZFC. Let G be a generic subset of P.

Definition 8. If $x \in M[G]^B$ define by induction on $\rho(x), i_G(x)$ as follows:

i) $i_G(\emptyset) = \emptyset$
ii) $i_G(x) = \{i_G(y) : y \in \operatorname{dom} x \text{ and } x(y) \bigcap G \neq \emptyset\}$

Definition 9. If $x \in M[G]^B$ define by induction on $n \in \omega, D_n(x)$ as follows:

i) $D_0(x) = \operatorname{dom} x$
ii) $D_{n+1}(x) = \bigcup\{\operatorname{dom} y : y \in D_n(x)n\}$

Let $D(x) = \bigcup\{D_n(x) : n \in \omega\}$.

Definition 10. If $\dot{x}_1, \ldots, \dot{x}_n \in M^B$ then $\pi \in G\operatorname{Aut}(\dot{x}_1, \ldots, \dot{x}_n)$ if and only if $\pi \in [(\mathbb{Q} - \{0\})^w]^{M[G]}$ and for every Δ_0-formula $\varphi(v_1, \ldots, v_m)$ if $\{\dot{y}_1, \ldots, \dot{y}_m\} \subseteq D(\dot{x}_1) \cup \ldots \cup D(\dot{x}_n) \cup \{\dot{x}_1, \ldots, \dot{x}_n\}$ then there is a $p_{\pi,\varphi(\dot{y}_1,\ldots,\dot{y}_m)}$ $\in P$ such that $\pi(p_{\pi,\varphi(\dot{y}_1,\ldots,\dot{y}_m)}) \in G$ and $p_\pi, \varphi(\dot{y}_1, \ldots, \dot{y}_m) \Vdash \varphi(\dot{y}_1, \ldots, \dot{y}_m)$ or $p_{\pi,\varphi(\dot{y}_1,\ldots,\dot{y}_m)} \Vdash \neg\varphi(\dot{y}_1, \ldots, \dot{y}_m)$ and if $\varphi(v_1, \ldots v_m)$ if of the form $\exists w \in v_1 \Psi(w, v_1 \ldots, v_m)$ then $\exists p \in P$ such that $\pi p \in G$ and for some $\dot{z} \in \operatorname{dom} \dot{x}_1 \quad p \Vdash \Psi(\dot{z}, \dot{x}_1, \ldots, \dot{x}_n)$. We often drop the π on the subscript of $p_{\pi,\varphi(\dot{y}_1,\ldots,\dot{y}_m)}$.

Remark. If $x_1, \ldots, x_n \in M^B$ then every $\pi \in [(\mathbb{Q} - \{0\}^\omega]^M$ is an element of $G\operatorname{Aut}(x_1, \ldots, x_n)$.

Theorem 4. *If $\varphi(v_1, \ldots, v_m)$ is a Δ_0 formula and $\{y_1, \ldots y_m\} \subseteq D(\dot{x}_1) \cup \ldots \cup D(\dot{x}_n) \cup \{\dot{x}_1, \ldots, \dot{x}_n\}$, and $\pi \in G\operatorname{Aut}(\dot{x}_1, \ldots, x_n)$ then $p_{\pi,\varphi(\dot{y}_1,\ldots,\dot{y}_m)} \Vdash \varphi(\dot{y}_1, \ldots, \dot{y}_m)$ if and only if $M[G] \models \varphi(i_G(\pi\dot{y}_1), \ldots, i_G(\pi\dot{y}_m))$.*

Proof. First we prove the theorem for $x \in y, x \subseteq y, y \subseteq x$, and $x = y$ by induction on $\Gamma(\rho(x), \rho(y))$ where Γ is the canonical well ordering of Ord $\times$ Ord. Then we prove the theorem by induction on the complexity of $\varphi(v_1, \ldots, v_m)$.

Definition 11. If $\dot{x} \in M^B, \{\bar{q}\} \in (\mathbb{Q} - \{0\})^{(\omega)}$ and $\dot{a} \in D(\dot{x})$ then $\bar{q}$ <u>insures</u> $\dot{a}$ (relative to $\dot{x}$) if there exists $\{\dot{x}_0, \ldots, \dot{x}_n\} \subseteq D(\dot{x}) \cup \{\dot{x}\}$ such that $\dot{a} = \dot{x}_0, \dot{x}_i \in \operatorname{dom} \dot{x}_{i+1}$ for each $i < n, \dot{x}_n = \dot{x}$, and there is a $p \in P$ such that $p \in \dot{x}_n(\dot{x}_{n-1}) \cdot \ldots \cdot \dot{x}_1(\dot{x}_0)$ such that $l(p) \le l(\bar{q})$ and such that $\bar{q}(p) \in G$.

Remark. If $\bar{q}$ insures $\dot{a}$ (relative to $\dot{x}$) and $\pi \in G\operatorname{Aut}(\dot{x})$ is such that $\pi|l(\bar{q}) = \bar{q}$ then $i_G(\pi\dot{a}) = i_G(\pi\dot{x}_0) \in trcl\ i_G(\pi\dot{x})$ since $\pi p \in \pi\dot{x}_n(\pi\dot{x}_{n-1}) \cdot \ldots \cdot \pi\dot{x}_1(\pi\dot{x}_0)$ and $\pi p \in G$. If $i_G(\pi\dot{a}) \in trcli_G(\pi\dot{x})$ then there is a $k \in \omega$ such that $\pi|k$ insures $\dot{a}$. (Let $k \ge l(p)$ for some $p \in \dot{x}_n(\dot{x}_{n-1}) \cdot \ldots \cdot \dot{x}_1(\dot{x}_0)$).

Definition 12. If $\dot{x} \in M^B, \dot{a} \in D(\dot{x})$ and $\bar{q} \in (\mathbb{Q} - \{0\})^\omega$ then $\bar{q}$ <u>fixes</u> $\dot{a}$ if for every $\pi_1, \pi_2 \in [(\mathbb{Q} - \{0\}^w]^M \ \pi_1|l(\bar{q}) = \bar{q} = \pi_1|l(\bar{q})$ implies $i_G(\pi_1\dot{a}) = i_G(\pi_2\dot{a})$. $\bar{q}$ <u>fixes $\dot{a}$ at z</u> if for every $\pi_1, \pi_2 \in [(\mathbb{Q} - \{0\})^w]^M \quad \pi_1|l(\bar{q}) = \bar{q} = \pi_2|l(\bar{q})$ implies $i_G(\pi_1\dot{a}) = i_G(\pi_2\dot{a}) = z$.

Definition 13. If $\dot{x} \in M^B$, and $\pi \in G\operatorname{Aut}(\dot{x})$ then $\pi \in UGAut(\dot{x})$ if for every $\dot{a} \in D(\dot{x})$ and for every $k \in \omega$ if $\pi|k$ insures $\dot{a}$ and for every

$\bar{q} \in (\mathbb{Q}-\{0\})^{(\omega)}$ extending $\pi \restriction k$ there is a $\bar{q}'$ extending $\bar{q}$ which fixes $\dot{a}$ then there exists $k' \geq k$ such that $\pi \restriction k'$ fixes $\dot{a}$.

Definition 14. If $\dot{x} \in M^B$, and $\pi \in UG\,\mathrm{Aut}(\dot{x})$ then $\pi \in VUG\,\mathrm{Aut}(\dot{x})$ if for every $\dot{y} \in D(\dot{x}) - \{\emptyset\}, z \in M[G]$ and $k \in \omega$ such that $\pi \restriction k$ insures and fixes $\dot{y}$, if for every extension $\bar{q}$ of $\pi \restriction k$ there exists a $\bar{q}'$ extending $\bar{q}$ and $\dot{b} \in \mathrm{dom}\,\dot{y}$ such that $\bar{q}'$ insures and fixes $\dot{b}$ at z then there exists $\dot{a} \in \mathrm{dom}\,\dot{y}$ and $k' \geq k$ such that $\pi \restriction k'$ insures and fixes $\dot{a}$ at z.

Stable Names Lemma. *If $\dot{x}, \dot{x}_1, \cdots, \dot{x}_n \in M^B$ and $D(\dot{x}) \cup D(\dot{x}_1) \cup \cdots \cup D(\dot{x}_n)$ is countable in M and if $k \in \omega$ is such that $i_G(\dot{x}) = i_G(\pi\dot{x})$ for every $\pi \in G\,\mathrm{Aut}(\dot{x}, \dot{x}_1, \cdots, \dot{x}_n)$ with $\pi(i) = 1$ for $i < k$, then for all $\dot{y} \in D(\dot{x})$*

I. Whenever $\pi \in [(\mathbb{Q}-\{0\})^{\omega}]^M$ such that $\pi(i) = 1$ for $i < k$ and there is a $k' \geq k$ such that $\pi \restriction k'$ insures $\dot{y}$ (relative to $\dot{x}$) then there is a $k'' \geq k'$ such that $\pi \restriction k''$ fixes $\dot{y}$.

II. If $\pi \in VUG\,\mathrm{Aut}(\dot{x}) \cap G\,\mathrm{Aut}(\dot{x}, \dot{x}_1, \cdots, \dot{x}_n)$ such that $\pi(i) = 1$ for $i < k$ and there is a $k'' \geq k$ such that $\pi \restriction k''$ insures and fixes $\dot{y}$ (insures $\dot{y}$ relative to $\dot{x}$) then $i_G(\pi\dot{y}) = i_G(\pi'\dot{y})$ for every $\pi' \in [(\mathbb{Q}-\{0\})^{\omega}]^M$ such that $\pi' \restriction k'' = \pi \restriction k''$.

Proof. By induction on $\rho(\dot{y})$. Without loss of generality $k = 0$. Let $\pi \in [(\mathbb{Q}-\{0\})^{\omega}]^M$ and let $k' \in \omega$ such that $\pi \restriction k'$ insures $\dot{y}$.

Definition 15. $s_{\dot{y}}$ is the name with $\mathrm{dom}\,s_{\dot{y}} = \{(n, s_{n\dot{y}})^{\vee} : n \in \omega\}$ and if $\dot{x} \in \mathrm{dom}\,s_{\dot{y}}, s_{\dot{y}}(\dot{x}) = 1$. $s_{n\dot{y}}$ is the name with $\mathrm{dom}\,s_{n\dot{y}} = \{\pi\dot{y} : \pi \in [(\mathbb{Q}-\{0\})^{\omega}]^M$ and $\pi(i) = 1$ for $i < n$ and if $\dot{x} \in \mathrm{dom}\,s_{n\dot{y}}, s_{n\dot{y}}(\dot{x}) = 1$.

Let $\Theta(x)$ be the formula which says x is a function with domain ω and such that $\forall n \in \omega, x(n)$ has at least two elements. Suppose that I. of the Stable Names Lemma does not hold. Then $M[G] \models \Theta(i_G(\pi s_{\dot{y}}))$ and by Jech 19.14, 19.16, and the Forcing Theorem there is a $k'' \geq k'$ such that $\forall \pi' \in [(\mathbb{Q}-\{0\})^{\omega}]^M \pi' \restriction k'' = \pi \restriction k'' \Longrightarrow M[G] \models \Theta(i_G(\pi' s_{\dot{y}}))$.

Let $S = \{i_G(\pi'\dot{y}) : \pi' \in G\,\mathrm{Aut}(\dot{x}, \dot{x}_1, \cdots \dot{x}_n)$ and $\pi' \restriction k' = \pi \restriction k'\}$. S is countable in $M[G]$ since $i_G(\pi'\dot{y}) \in \mathit{trcl}\ i_G(\pi'\dot{x}) = \mathit{trcl}\ i_G(\dot{x})$ for π' such that $\pi' \restriction k' = \pi \restriction k'$ and $i_G(\dot{x})$ is hereditarily countable in $M[G]$. Well order S as $\{s_i : i < \omega\}$. We will define $\{\pi_i : i < \omega\} \in {}^{\omega}((\mathbb{Q}-\{0\})^{(\omega)})$ such that if we let $\pi^* = \bigcup\{\pi_i : i < \omega\}$, then $\pi^* \restriction k' = \pi \restriction k'$, $\pi^* \in VUG\,\mathrm{Aut}(\dot{x}) \bigcap G\,\mathrm{Aut}(\dot{x}, \dot{x}_1, \cdots, \dot{x}_n)$ and $i_G(\pi^*\dot{y}) \notin S$. This will contradict the definition of S, so I. of the Stable Names Lemma must hold.

Definition 16. If $\bar{q} \in (\mathbb{Q}-\{0\})^{(\omega)}$ extends $\pi \restriction k'$ then label $\bar{q}$ with (yes, z) if there is an $\dot{a} \in \mathrm{dom}\,\dot{y}$ such that $\bar{q}$ insures (relative to $\dot{y}$ thru $(\dot{a}, \dot{y})$) and fixes $\dot{a}$ at z. Label $\bar{q}$ with (no, z) if and only if for all $\pi' \in [(\mathbb{Q}-\{0\})^{\omega}]^M$ such that $\pi' \restriction l(\bar{q}) = \bar{q}, z \notin i_G(\pi'\dot{y})$.

Claim. *(Assuming that I. of Stable Names Lemma does not hold for $\dot{y}$). For every $\bar{q}$ extending $\pi\restriction k'$ there exist $\bar{q}_1$ and $\bar{q}_2$ extending $\bar{q}$ such that for some $z, \bar{q}_1$ is labeled (yes, z) and $\bar{q}_2$ is labeled (no, z).*

Proof. Let $\bar{q}$ extend $\pi\restriction k'$. Let π_1 and π_2 be elements of $[(\mathbb{Q}-\{0\})^\omega]^M$ such that $\pi_1\restriction l(\bar{q}) = \bar{q} = \pi_2\restriction l(\bar{q})$ and $i_G(\pi_1\dot{y}) \neq i_G(\pi_2\dot{y})$ and let $z \in i_G(\pi_1\dot{y})$ and $\dot{z} \notin i_G(\pi_2\dot{y})$. Let $\dot{a} \in \operatorname{dom}\dot{y}$ such that $i_G(\pi_1\dot{a}) = z$. Since there is an extension of the form $\pi_1\restriction k''$ which insures $\dot{a}$ (relative to $\dot{y}$ thru $(\dot{a},\dot{y})$), by the induction hypothesis there is a k''' such that $\pi_1\restriction k'''$ insures and fixes $\dot{a}$. Let $\bar{q}_1$ be $\pi_1\restriction k''$. If there is a $\dot{b} \in \operatorname{dom}\dot{y}$ such that $i_G(\pi_2\dot{b}) = z$ then $\|\pi_2\dot{b} \notin \pi_2\dot{y}\| \bigcap G \neq \emptyset$, and $\|\pi_1\dot{a} = \pi_2\dot{b}\| \bigcap G \neq \emptyset$ so there is a $k_2 \geq k''' \in \omega$ such that for every $\pi' \in [(\mathbb{Q}-\{0\})^\omega]^M$ such that $\pi'\restriction k_2 = \pi_2\restriction k_2$, $\|\pi'\dot{b} \notin \pi'\dot{y}\| \bigcap G \neq \emptyset, \|\pi'\pi_2^{-1}\pi_1\dot{a} = \pi'\dot{b}\| \bigcap G \neq \emptyset$ and $i_G(\pi'\pi_2^{-1}\pi_1\dot{a}) = z$. Let $\bar{q}_2 = \pi_2\restriction k_2$. If there is no $\dot{b} \in \operatorname{dom}\dot{y}$ such that $i_G(\pi_2\dot{b}) = z$ then let $d_{\dot{y}}$ be the name with $\operatorname{dom} d_{\dot{y}} = \operatorname{dom}\dot{y}$ and for $\dot{c} \in \operatorname{dom} d_{dy}$ let $d_{\dot{y}}(\dot{c}) = 1$. $M[G] \models z \notin i_G(\pi_2 d_{dy}) \wedge z = i_G(\pi_1\dot{a})$ so since $\pi_1\restriction k'''$ fixes $\dot{a}$ at z, for all $\pi' \in [(\mathbb{Q}-\{0\})^\omega]^M$ with $\pi'\restriction k''' = \pi_2\restriction k''', M[G] \models z = i_G(\pi'\pi_2^{-1}\pi_1\dot{a})$ and therefore there is a $k_2' \geq k'''$ such that for all $\pi' \in [(\mathbb{Q}-\{0\})^\omega]^M$ such that $\pi'\restriction k_2' = \pi_2\restriction k_2', M[G] \models z \notin i_G(\pi' d_{\dot{y}})$. Therefore $M[G] \models z \notin i_G(\pi'\dot{y})$. Let $\bar{q}_2 = \pi_2\restriction k_2'$. $\square$

Construction of π^*. Let $\pi_0 = \pi\restriction k'$. If $F = \{\varphi(w_1,\cdots,w_n) : \varphi$ is a Δ_0 formula and $\{w_1,\cdots w_n\} \subseteq D(\dot{x})\bigcup\{\dot{x}\}\bigcup D(\dot{x}_1)\bigcup\cdots\bigcup D(\dot{x}_n)\bigcup\{\dot{x}_1,\cdots,\dot{x}_n\}\}$, then well order F as $\{\varphi_i(\omega_{i_1},\cdots,w_{i_{n_i}}) : i < \omega\}$. Order $D(\dot{x})$ as $\{\dot{a}_i : i < \omega\}$ so that for every $i \in \omega$ there exists an $j \in \omega$ such that $j > i$ and $\dot{a}_i = \dot{a}_j$. For each $\dot{a} \in D(\dot{x})$ and each $\bar{q}$ such that $\bar{q}$ insures (relative to $\dot{x}$) and fixes $\dot{a}$, let $Z_{\dot{a},\bar{q}} = \{z : z \in i_G(\pi\dot{a})$ for some $\pi \in [(\mathbb{Q}-\{0\})^\omega]^M$ such that $\pi\restriction l(\bar{q}) = \bar{q}\}$. Let $Z_{\dot{a}} = \{z : z \in Z_{\dot{a},\bar{q}}$ for some $\bar{q} \in (\mathbb{Q}-\{0\})^{(\omega)}\}$. Order $Z_{\dot{a}}$ as $\{z_{\dot{a},i} : i \in \omega\}$ so that for every $i \in \omega$ there exists an $j \in \omega$ such that $j > i$ and such that $z_{\dot{a},i} = z_{\dot{a},j}$. If π_i has been defined, define π_{i+1} as follows: By the claim, there are extensions of π_i, π_{i_1} and π_{i_2} such that for some z, π_{i_1} is marked (yes,z) and π_{i_2} is marked (no,z). If $z \in s_i$ then let $\pi_i' = \pi_{i_2}$ else let $\pi_i' = \pi_{i_1}$. Now if π_i' insures $\dot{a}_i$ let π_i'' be an extension of π_i' which fixes $\dot{a}_i$ if such an extension exists, else let $\pi_i'' = \pi_i'$. If π_i'' insures (relative to $\dot{x}$) and fixes $\dot{a}_i$ and if for every extension $\bar{q}$ of π_i'' there exists a $\bar{q}'$ extending $\bar{q}$ and $\dot{b} \in \operatorname{dom}\dot{a}_i$ such that $\bar{q}'$ insures (thru $(\dot{b},\dot{a}_i)$) and fixes $\dot{b}$ at $z_{\dot{a}_i,k}$ where k equals the number of indices j such that $j < i$ and $\dot{a}_i = \dot{a}_j$ then let $\pi_i''' =$ one of the $\bar{q}'$ else let $\pi_i''' = \pi_i''$. Let $\pi' \in [(\mathbb{Q}-\{0\})^\omega]^M$ be any extension of π_i'''. If $M[G] \models \varphi_i(i_G(\pi'\dot{w}_{i_1}),\cdots,i_G(\pi'\dot{w}_{i_{n_i}}))$ then let $p_{\varphi_i(\dot{w}_{i_1},\cdot\dot{w}_{i_{n_i}})} \in \|\varphi_i(\dot{w}_{i_1},\cdots,\dot{w}_{i_{n_i}})\|$ such that $\pi' p_{\varphi_i(\dot{w}_{i_1},\cdots,\dot{w}_{i_{n_i}})} \in G$ and if $\varphi_i(v_{i_1},\cdots,v_{i_{n_i}})$ is of the form $\exists w \in v_{i_1}\Psi_i(w,v_{i_1},\cdots,v_{i_{n_i}})$ find $\dot{w} \in \operatorname{dom}\dot{w}_{i_1}$ such that $M[G] \models \Psi_i(i_G(\pi'\dot{w}), i_G(\pi'\dot{w}_{i_1}),\cdots,i_G(\pi'\dot{w}_{i_{n_i}}))$ and let $p' \in \|\Psi_i(\dot{w}, \dot{w}_{i_1}, \cdots, \dot{w}_{i_{n_i}})\|$ such that $\pi' p' \in G$. Let $\pi_{i+1} = \pi' \restriction$ (maximum of $\{l(\pi_i'' y'), l(p'), l(p_{\varphi i}(\dot{w}_{i_1}, \cdots, \dot{w}_{i_{n_i}})\})$. If $M[G] \models \neg\varphi_i(i_G(\pi' \dot{w}_{i_1}), \cdots, i_G(\pi' \dot{w}_{i_{n_i}}))$ then let $p_{\varphi_i}(\dot{w}_{i_1}, \cdots, \dot{w}_{i_{n_i}}) \in \|\neg\varphi_i(i_G(\pi' \dot{w}_{i_1}), \cdots, i_G(\pi' \dot{w}_{i_{n_i}}))\|$ such that $\pi' p_{\varphi_i(\dot{w}_{i_1}, \cdots, \dot{w}_{i_{n_i}})} \in G$. Let $\pi_{i+1} = \pi' \restriction$(maximum of$\{l(\pi_i'''),$ $l(p_{\varphi_i(\dot{w}_{i_1}, \cdots, \dot{w}_{i_{n_i}})})\})$. Let $\pi^* = \bigcup\{\pi_i : i < \omega\}$.

Claim. *If π_i' is labeled (yes, z) then $z \in i_G(\pi^* \dot{y})$. If π_i' is labeled (no,z) then $z \notin i_G(\pi^* \dot{y})$.*

Proof. If π_i' is labeled (yes, z) then for some $\dot{a} \in \operatorname{dom}(\dot{y})$ for every $\pi' \in [(\mathbb{Q} - \{0\})^\omega]^M$ such that $\pi' \restriction l(\pi_i') = \pi_i'$, $z = i_G(\pi' \dot{a})$ and π_i' insures $\dot{a}$ (relative to $\dot{y}$ thru $(\dot{a}, \dot{y})$ and relative to $\dot{x}$). We have by the induction hypothesis (II) and by the fact that $\pi^* \in VUG\operatorname{Aut}(\dot{x})$ that $z = i_G(\pi^* \dot{a}) \in i_G(\pi^* \dot{y})$. If π_i' is labeled (no,z) and if $z \in i_G(\pi^* \dot{y})$ then for some $\dot{a} \in \operatorname{dom}(\dot{y})$, $z = i_G(\pi^* \dot{a}) \in i_G(\pi^* \dot{y})$ but then there is a $k \in \omega$ such that $k > l(\pi_i')$ and $\pi^* \restriction k$ insures $\dot{a}$. By the induction hypothesis (I) and since $\pi^* \in UG\operatorname{Aut}(\dot{x})$ there is a $k' > k$ such that $\pi^* \restriction k'$ fixes $\dot{a}$. We have by the induction hypothesis (II) and by the fact that $\pi^* \in VUG\operatorname{Aut}(\dot{x})$ and $z = i_G(\pi^* \dot{a}) \in i_G(\pi^* \dot{y})$ that $\pi^* \restriction k'$ is labeled (yes,z), but this contradicts π_i' having label (no,z). □(of claim and of part I. of the Stable Names Lemma).

Proof of II. Let $\pi^* \in VUG\operatorname{Aut}(\dot{x}) \bigcap G\operatorname{Aut}(\dot{x}, \dot{x}_1, \ldots, \dot{x}_n)$ such that $\pi^*(i) = 1$ for $i < k$ and let k'' be as in the hypothesis of II. Let $\bar{q} = \pi^* \restriction k''$. Let $\pi' \in [(\mathbb{Q} - \{0\})^\omega]^M$ such that $\pi' \restriction k'' = \pi^* \restriction k''$. We must show that $z \in i_G(\pi^* \dot{y})$ if and only if $\dot{z} \in i_G(\pi' \dot{y})$.

If $z \in i_G(\pi^* \dot{y})$ then for some $\dot{a} \in \operatorname{dom}(\dot{y})$, $z = i_G(\pi^* \dot{y})$ but then there is a $k_1 \geq k''$ such that $\pi^* \restriction k_1$ insures $\dot{a}$ thru $(\dot{a}, \dot{y})$. By the induction hypothesis (I) for every $\dot{a} \in D(\dot{y})$ and for every $k \in \omega$ if $\pi \restriction k$ insures $\dot{a}$ then for every $\bar{q} \in (\mathbb{Q} - \{0\})^{(\omega)}$ extending $\pi \restriction k$ there is a $\bar{q}'$ extending $\bar{q}$ which fixes $\dot{a}$. Therefore, since $\pi^* \in UG\operatorname{Aut}(\dot{x})$ there is a $k' > k_1$ such that $\pi^* \restriction k'$ fixes $\dot{a}$. By the induction hypothesis (II), if $\pi'' \in [(\mathbb{Q} - \{0\})^\omega]^M$ such that $\pi'' \restriction k' = \pi^* \restriction k'$, $i_G(\pi^* \dot{a}) = i_G(\pi'' \dot{a})$ since $\pi^* \restriction k'$ insures and fixes $\dot{a}$ (relative to $\dot{x}$). Thus $z = i_G(\pi^* \dot{a})$. Now $\pi^* \restriction k_1$ insures $\dot{a}$ thru $(\dot{a}, \dot{y})$, so $z = i_G(\pi'' \dot{a}) \in i_G(\pi'' \dot{y})$. Since $i_G(\pi'' \dot{y}) = i_G(\pi' \dot{y})$, $z \in i_G(\pi' \dot{y})$.

Now let $z \in i_G(\pi' \dot{y})$. We claim that for every extension $\bar{q}'$ of $\pi^* \restriction k''$ there exists $\bar{q}''$ an extension of $\bar{q}'$ and $\dot{b} \in \operatorname{dom} \dot{y}$ such that $\bar{q}''$ insures and fixes $\dot{b}$ at z. Indeed, given $\bar{q}'$, let $\pi \in [(\mathbb{Q} - \{0\})^\omega]^M$ such that $\pi' \restriction l(\bar{q}') = \bar{q}'$; since $\pi^* \restriction k'' = \pi \restriction k''$ and $\pi^* \restriction k''$ fixes $\dot{y}$, $z \in i_G(\pi \dot{y})$. So for some $\dot{b} \in \operatorname{dom} \dot{y}$, $z = i_G(\pi \dot{b}) \in i_G(\pi \dot{y})$. Then for some $k_3 \geq l(\bar{q}')$, $\pi \restriction k_3$ insures $\dot{b}$ thru $(\dot{b}, \dot{y})$, so $\pi \restriction k_3$ insures $\dot{b}$ relative to $\dot{x}$ (because $\pi^* \restriction k''$ insures $\dot{y}$ relative to $\dot{x}$). Then by induction hypothesis (I), there exists $k_4 \geq k_3$ such that $\pi \restriction k_4$ fixes $\dot{b}$. Let $\bar{q}'' = \pi \restriction k_4$ and the claim is proved. Since $\pi^* \in VUG\operatorname{Aut}(\dot{x})$ there exists

$\dot{a} \in \operatorname{dom} \dot{y}$ and $k^* \geq k''$ such that $\pi^* \restriction k^*$ insures and fixes $\dot{a}$ (relative to $\dot{x}$) at z. By inductive hypotheses (II), $i_G(\pi^* \dot{a}) = z$, so $z \in i_G(\pi^* \dot{y})$. $\square$

4. Proof of the Main Lemma

Notation. If $\bar{q} \in \mathbb{Q}^{(\omega)}$ and $\dot{a}$ is a name, then $\{\bar{q}, \dot{a}\}^{\vee}$ is the name with domain $\{\bar{q}^{\vee}, \dot{a}\}$ and such that $\{\bar{q}, \dot{a}\}^{\vee}$ and $(\bar{q}, \dot{a})^{\vee}(\dot{a}) = 1$ and $\{\bar{q}, \dot{a}\}^{\vee}, (\dot{a}) = 1$. $(\bar{q}, \dot{a})^{\vee}$ is the name with domain $\{\{\bar{q}\}^{\vee}, \{\bar{q}, \dot{a}^{\vee}\}$ and such that $(\bar{q}, \dot{a})^{\vee}(\{\bar{q}\}^{\vee}) = 1$ and $(\bar{q}, \dot{a})^{\vee}(\{\bar{q}, \dot{a}\}^{\vee}) = 1$.

Let A be the group whose universe is $\bigcup\{H_p^* : \tilde{p} \in G\}$ and let addition on A be the restriction of addition on $\mathbb{Q}^{(\omega)}$ to the universe of A. Suppose there is a Σ_1 formula $\varphi(x, y)$ such that in M[G], $\varphi(x, y)$ defines a function such that for each hereditarily countable torsion-free group H there is a unique isomorphic copy C_H of H such that $\varphi(H, C_H)$ and if H_1 and H_2 are hereditarily countable torsion-free groups then $C_{H_1} = C_{H_2}$ if and only if $H_1 \cong H_2$. Then for some sets C_A, f and U_{C_A}

M[G] $\models \varphi(A, C_A) \wedge$ (f is an isomorphism from A to C_A)$\wedge$ (The universe of C_A is U_{C_A})

$\varphi(x, y)$ is of the form $\exists z \psi(z, x, y)$ where $\psi(z, x, y)$ is a Δ_0 formula. Let x be a hereditarily countable set such that

M[G] $\models \psi(x, A, C_A) \wedge$ (f is an isomorphism from A to C_A)$\wedge$ (The universe of C_A is U_{C_A}).

One can show there exists hereditarily countable x which satisfies $\psi(x, A, C_A)$ by using the fact that A is hereditarily countable, the Mostowski Collapse, and the Reflection Principle in order to find a countable transitive model N containing A such that $\exists z \psi(z, x, y)$ defines a function in N. Then C_A and a witness for $\exists z \psi(z, A, C_A)$ must be in N by the absoluteness of $\psi(z, x, y)$, and by the absoluteness of $\psi(z, x, y), C_A$ and the witness for $\exists z \psi(z, A, C_A)$ in N satisfy $\psi(z, A, C_A)$ in M[G].

Let $\dot{C}_A$ and $\dot{U}_{C_A}$ be names for C_A and U_{C_A} such that $D(\dot{C}_A)$ and $D(\dot{U}_{C_A})$ are hereditarily countable. Let $\dot{f}$ be a name for f such that $\operatorname{dom} \dot{f} = \{(\bar{q}, \pi\dot{a})^{\vee} : \bar{q} \in \mathbb{Q}^{(\omega)}, \pi \in \operatorname{Aut}(P, \leq)$ and $\dot{a} \in \operatorname{dom} \dot{U}_{C_A}\}$. Let $\dot{A}$ be the name for A such that $\operatorname{dom} \dot{A} = \{(\bar{q})^{\vee} : \bar{q} \in \mathbb{Q}^{(\omega)}\}$ and $\dot{A}((\bar{q})^{\vee}) = u(\bar{q} \to 1)^{\sim}$ where $\bar{q} \to 1$ denotes the element of P' whose domain is $\{\bar{q}\}$ and whose value at $\bar{q}$ is 1, and $u(\bar{q} \to 1)^{\sim}$ is the set of extensions of $(\bar{q} \to 1)^{\sim}$ in P. Let $\dot{x}$ be a name of x such that $D(\dot{x})$ is hereditarily countable. By the Forcing Theorem there is a $p^* \in G$ such that

$p^* \Vdash \psi(\dot{x}, \dot{A}, \dot{C}_A) \wedge (\dot{f}$ is an isomorphism from A to $\dot{C}_A)\wedge$ (The universe of $\dot{C}_A$ is $\dot{U}_{C_A}$).

By Jech 19.16 for every $\pi \in \mathrm{Aut}(P, \leq)$

$\pi p^* \Vdash \psi(\pi\dot{x}, \pi\dot{A}, \pi\dot{C}_A) \wedge (\pi\dot{f}$ is an isomorphism from $\pi\dot{A}$ to $\pi\dot{C}_A)\wedge$ (The universe of $\pi\dot{C}_A$ is $\pi\dot{U}_{C_A}$)

So for every $\pi \in \mathrm{Aut}(P, \leq)$ such that $\pi p^* \in G$,

M[G] $\models \psi(i_G(\pi\dot{x}), i_G(\pi\dot{A}), i_G(\pi(\dot{C}_A)) \wedge (i_G(\pi\dot{f})$ is an isomorphism from $i_G(\pi\dot{A})$ to $i_G(\pi\dot{C}_A))\wedge$ (The universe of $i_G(\pi\dot{C}_A)$ is $i_G(\pi\dot{U}_{C_A})$)

By Theorem 4 for every $\pi \in G\,\mathrm{Aut}(\dot{x}, \dot{f}, \dot{A}, \dot{C}_A, \dot{U}_{C_A})$ such that π is the identity on the length of p^* and for every $\pi \in \mathrm{Aut}(P, \leq)$ such that $\pi p^* \in G$,

M[G] $\models \psi(i_G(\pi\dot{x}), i_G(\pi\dot{A}), i_G(\pi\dot{C}_A)) \wedge (i_G(\pi\dot{f})$ is an isomorphism from $i_G(\pi\dot{A})$ to $i_G(\pi\dot{C}_A))\wedge$ (The universe of $i_G(\pi\dot{C}_A)$ is $i_G(\pi\dot{U}_{C_A})$)

So, for every $\pi \in G\,\mathrm{Aut}(\dot{x}, \dot{f}, \dot{A}, \dot{C}_A, \dot{U}_{C_A})$ such that π is the identity on the length of p^* and for every $\pi \in \mathrm{Aut}(P, \leq)$ such that $\pi p^* \in G$,

$M[G] \models \varphi(i_G(\pi\dot{A}), i_G(\pi\dot{C}_A)) \wedge (i_G(\pi\dot{f})$ is an isomorphism from $i_G(\pi\dot{A})$ to $i_G(\pi\dot{C}_A)\wedge$ (The universe of $i_G(\pi\dot{C}_A)$ is $i_G(\pi\dot{U}_{C_A})$).

Note that for every $\pi \in G\,\mathrm{Aut}(\dot{C}_{\dot{A}}, \dot{U}_{C_A}\dot{A}, \dot{f})$ and for every $\pi \in \mathrm{Aut}(P\ \leq)$, π^{-1} is an isomorphism from A to $i_G(\pi\dot{A})$ since $\bar{q} \in A$ iff $\pi^{-1}\bar{q} \in i_G(\pi\dot{A})$. So, for every $\pi \in G\,\mathrm{Aut}(\dot{x}, \dot{f}, \dot{A}, \dot{C}_A, \dot{U}_{C_A})$ such that π is the identity on the length of p^* and for every $\pi \in \mathrm{Aut}(P, \leq)$ such that $\pi p^* \in G$,

$$M[G] \models A \cong i_G(\pi\dot{A}) \wedge \varphi(A, C_A) \wedge \varphi(i_G(\pi\dot{A}), i_G(\pi\dot{C}_A))$$

By our assumption about $\varphi(x, y)$ and the preceding equation, we have that for every $\pi \in G\,\mathrm{Aut}(\dot{x}, \dot{f}, \dot{A}, \dot{C}\dot{A}, \dot{U}_{C_A})$ such that π is the identity on the length of p^* and for every $\pi \in \mathrm{Aut}(P, \leq)$ such that $\pi p^* \in G$

$$(\spadesuit)\ M[G] \models C_A = i_G(\pi\dot{C}_A)$$

So, for every $\pi \in G\,\mathrm{Aut}(\dot{x}, \dot{f}, \dot{A}, \dot{C}_A, \dot{U}_{C_A})$ such that π is the identity on the length of p^*, and for every $\pi \in \mathrm{Aut})$ such that $\pi p^* \in G$, $i_G(\pi\dot{f})\pi^{-1}f^{-1}$ is an automorphism of C_A.

$$\begin{array}{ccc} A & \xrightarrow[\cong]{f} & C_A \\ {\scriptstyle \pi^{-1}}\Big\downarrow{\scriptstyle \cong} & & \Big\| \Big\downarrow {\scriptstyle \mathrm{id}} \\ i_G(\pi\dot{A}) & \xrightarrow[\cong]{i_G(\pi\dot{f})} & i_G(\pi\dot{C}_A) \end{array}$$

For every $\pi \in G\,\mathrm{Aut}(\dot{x}, \dot{f}, \dot{A}, \dot{C}_{\dot{A}}, \dot{U}_{\dot{C}_{\dot{A}}})$ such that π is the identity on the length of p^* and for every $\pi \in \mathrm{Aut}(P, \leq)$ such that $\pi p^* \in G$, the above picture holds (the diagram is not commutative).

Claim. *The only automorphisms of A are multiplication by fractions in S.*

Proof. If $\bar{q}_1$ and $\bar{q}_2$ are elements of A which are independent in $\mathbb{Q}^{(\omega)}$ then there can be no automorphism of A which sends $\bar{q}_1$ to $\bar{q}_2$ because by Lemma 4 the set $D\bar{q}_1\bar{q}_2 = \{p \in P : p \leq \{\bar{q}_1 \to 1, \bar{q}_2 \to 1\}^{\sim}$ and such that for some $p_i \in N_1$ and some positive integer $t, p \leq \{\bar{q}_1/p_i^t \to 1\}^{\sim}$ and $p \leq \{\bar{q}_2/p_i^t \to 0\}^{\sim}\}$ is dense below $\{\bar{q}_1 \to 1, \bar{q}_2 \to 1\}^{\sim}$; thus since $G \bigcap D\bar{q}_1\bar{q}_2 \neq \emptyset$, for some positive integer t we have $\bar{q}_1/p_i^t \in A$ and $\bar{q}_2/p_i^t \neq A$. Since the set $D\bar{q}p_i = \{p \in P : p \leq \{\bar{q} \to 1, \bar{q}/p_i^t \to 0\}^{\sim}$ for some positive integer $t\}$ is dense below $\{\bar{q} \to 1\}^{\sim}$ for all $p_i \in N_1$, no element of A but 0 is infinitely divisible by any element of N_1, so the only possible automorphisms of A are multiplication by fractions of the form m_1/m_2 where m_1 and m_2 are elements of N_2'.

So, by the above claim, if $\pi \in G\,\mathrm{Aut}(\dot{x}, \dot{f}, \dot{A}, \dot{C}_{\dot{A}}, \dot{U}_{\dot{C}_{\dot{A}}})$ such that π is the identity on the length of p^* or if $\pi \in \mathrm{Aut}(P, \leq)$ such that $\pi p^* \in G$, then for some m_1 and m_2 elements of N_2'

$$(I) \qquad M[G] \models i_G(\pi\dot{f}) = m_1/m_2 f\pi$$

Let $\bar{r} = (r_0, \cdots, r_n)$ be an element of A. Let $\dot{a} \in D_0(\dot{U}_{\dot{C}_{\dot{A}}})$ such that $M[G] \models (\bar{r}, i_G(\dot{a})) \in f$. We have $\dot{f}((\bar{r}, \dot{a})^{\vee}) \bigcap G \neq \emptyset$. By (♠) and the Stable Names Lemma, there is $n' \in \omega$ such that the identity sequence in $\mathbb{Q}^{(n')}$ insures and fixes $\dot{a}$ (In applying the Stable Names Lemma, $\dot{U}_{\dot{C}_{\dot{A}}}$ takes the place of $\dot{x}$ and $\dot{a}$ takes the place of $\dot{y}$). Let $p \in P$ such that $p \in \dot{f}((\bar{r}, \dot{a})^{\vee}) \bigcap G$. Let $s\dot{a}$ the name with $dom s\dot{a} = \{\pi\dot{a} : \pi \in (\mathbb{Q} - \{0\})^{\omega}$ and π is the identity on $n'\}$ and for every name in the domain let its value be 1. Let $p'' \in G$ such that $p'' \Vdash s\dot{a}$ has only one element (there is one since $M[G] \models i_G(s\dot{a})$ has only one element).

Definition 17. If $k \in \omega$, let π_{sk} be the permutation of ω such that for $i < k$ $\pi_{sk}(i) = k + i$, for $k \leq i < 2k$ $\pi_{sk}(i) = i - k$, and for $i \geq 2k$ $\pi_{sk}(i) = i$.

Let $n \in \omega$ such that the identity sequence in $\mathbb{Q}^{(n)}$ insures and fixes $\dot{a}$, such that $n > l(p), n > 1(\bar{r})$, and such that $\pi_{sn}p \in G, \pi_{sn}p'' \in G$ and $\pi_{sn}p^* \in G$ (By a simple denseness argument if $p \in G$ then there are infinitely many $n \in \omega$ such that $\pi_{sn}p \in G$). Since $\pi_{sn}(\bar{r}, \dot{a})^\vee = (\bar{r}, \pi_{sn}\dot{a})^\vee$ and $\pi_{sn}[\dot{f}((\bar{r}, \dot{a})^\vee)] \bigcap G = \pi_{sn}\dot{f}((\bar{r}, \pi_{sn}\dot{a})^\vee) \bigcap G \neq \emptyset$,

$$M[G] \models (\bar{r}, i_G(\pi_{sn}\dot{a})) \in i_G(\pi_{sn}\dot{f})$$

By I. there are m_1 and m_2 elements of N_2' so that

$$(II) \qquad M[G] \models (m_1/m_2\pi_{sn}\bar{r}, i_G(\pi_{sn}\dot{a})) \in i_G(\dot{f}) = f$$

So there is a $p' \in P$ such that $p' \in \dot{f}((m_1/m_2\pi_{sn}\bar{r}, \pi_{sn}\dot{a})^\vee) \bigcap G$. If $m \in \omega$ let $\pi_{1,m}$ be the element of $(\mathbb{Q} - \{0\})^\omega$ such that $\pi(j) = 1$ for $j < n$ and $\pi(j) = m$, for $j \geq n$. By Lemma 3 and a simple denseness argument there is an $m \in N_2'$ such that $m \neq 1$ and such that $\pi_{1,m}p' \in G$. Let $\pi_{m,1}$ be the element of $(\mathbb{Q} - \{0\})^\omega$ such that $\pi(j) = 1$ for $n \leq j < 2n$ and $\pi(j) = m$ for $j < n$ or $j \geq 2n$. Note that $\pi_{1,m}p \in G$ since $l(p) < n$. Note also that $i_G(\pi_{1,m}\dot{a}) = i_G(\dot{a})$ since the map π with domain n such that $\pi(i) = 1$ for $i < n$ fixes $\dot{a}$. Therefore

$$M[G] \models (\bar{r}, i_G(\pi_{i,m}\dot{a})) = (\bar{r}, i_G(\dot{a})) \in i_G(\pi_{1,m}\dot{f})$$

which implies by I and $\pi_{1,m}\bar{r} = \bar{r}$ that

$$(III) \qquad M[G] \models i_G(\pi_{1,m}\dot{f}) = f\pi_{1,m}$$

By II and since $\pi_{1,m}p' \in G$ we must have

$$(IV) \quad M[G] \models (m_1/m_2\pi_{sn}\bar{r}, i_G(\pi_{1,m}\pi_{sn}\dot{a})) = (m_1/m_2\pi_{sn}\bar{r}, i_G(\pi_{sn}\pi_{m,1}\dot{a})) \in i_G(\pi_{1,m}\dot{f})$$

Since $(1/m\pi_{m,1})\dot{a} = \pi_{m,1}\dot{a}$ (by construction of P, multiplying or dividing any equivalence class of P' by an element of N_2' leaves the equivalence class unchanged, and therefore multiplying or dividing any element of M^B by an element of N_2' leaves the name unchanged) and $p'' \Vdash \dot{a} = (1/m\pi_{m,1})\dot{a}$ we have $p'' \Vdash \dot{a} = \pi_{m,1}\dot{a}$ and $\pi_{sn}p'' \Vdash \pi_{sn}\dot{a} = \pi_{sn}\pi_{m,1}\dot{a}$. Since $\pi_{sn}p'' \in G$, by IV,

$$(V) \quad M[G] \models (m_1/m_2\pi_{sn}\bar{r}, i_G(\pi_{sn}\dot{a})) = (m_1/m_2\pi_{sn}\bar{r}, i_G(\pi_{sn}\pi_{m,1}\dot{a})) \in i_G(\pi_{1,m}\dot{f})$$

This is a contradiction since by II. and V. we have that

$$M[G] \models i_G(\pi_{1,m}\dot{f})(m_1/m_2\pi_{sn}\bar{r}) = f(m_1/m_2\pi_{sn}\bar{r})$$

but by III. we know that $i_G(\pi_{1,m}\dot{f}) = f\pi_{1,m}$, so

$$M[G] \models i_G(\pi_{1,m}\dot{f})(m_1/m_2\pi_{sn}\bar{r}) = m \cdot f(m_1/m_2\pi_{sn}\bar{r}) \quad (m \neq 1).$$

REFERENCES

[B] Jon Barwise, *Back and Forth through Infinitary Logic*, Studies in Model Theory (1973), The Mathematical Association of America, 5–34.

[B1] ———, *Admissible Sets and Structures* (1975), Springer-Verlag, New York.

[B&E] Jon Barwise and Paul Eklof, *Infinitary properties of Abelian torsion groups*, Ann. Math. Logic **2** (1970), 25–68.

[F] Laszlo Fuchs, *Infinite Abelian Groups II*, Academic Press, New York, 1970.

[FS] Harvey Friedman & Lee Stanley, *A Borel reducibility theory for classes of countable structures*, J. Sym. Logic **54**(3) (1989).

[H] Wilfred Hodges, *On the Effectivity of some Field Constructions*, Proceedings of the London Mathematical Society **32** (1976), 133–162.

[H1] ———, *Constructing Pure Injective Hulls*, The Journal of Symbolic Logic **45**(3) (Sept. 1980).

[H2] ———, *What is a structure theory?*, preprint.

[HS1] Wildred Hodges & Saharon Shelah, *Naturality and definability I*, J. London Math. Soc. (2) **33** (1986), 1–12.

[HS2] ———, *Naturality and definability II*, preprint..

[J] Thomas Jech, *Set Theory*, Academic Press, New York, 1978.

[J & K] Ronald B. Jenson and Carol Karp, *Primitive Recursive Set Functions*, Proceedings of Symposia in Pure Mathematics, vol. 13 part I (1971), American Mathematical Society, 143–176.

[K] Irving Kaplansky, *Infinite Abelian Groups*, The University of Michigan Press, Ann Arbor, 1962.

[M] Garvin Melles, *Classification Theory and Generalized Recursive Functions*, University of California at Irvine, 1989.

[S] Saharon Shelah, *Classification of First Order Theories which have a Structure Theorem*, Bulletin of the American Mathematical Society **12**(2) (April 1985).

[Sh] ———, *Existence of Many $L_{\infty\lambda}$-equivalent non isomorphic models of T of power λ*, Annals of Pure and Applied Logic **34** (1987), 291-310.

[Sh401] ———, *Finitary Structure Theorem*, Preprint.

INSTITUTE OF MATHEMATICS, THE HEBREW UNIVERSITY, JERUSALEM, ISRAEL

$\mathbf{\Sigma}^1_3$-ABSOLUTENESS FOR SEQUENCES OF MEASURES

WILLIAM J. MITCHELL

ABSTRACT. We extend Jensen's $\mathbf{\Sigma}^1_3$-absoluteness result to apply to the core model for sequences of measures, provided that sharps exist and there is no inner model of $\exists\kappa\, o(\kappa) = \kappa^{++}$. The proof includes a result on the patterns of indiscernibles analogous to the one which arises in Jensen's proof.

1. INTRODUCTION

We say that a transitive model M of set theory is *correct* for a formula ϕ (or, equivalently, that ϕ is absolute for M) if for all reals $x \in M$ we have $M \models \phi(x)$ iff $V \models \phi(x)$. Shoenfield proved in [Sho61] that

Theorem 1.1. *Any transitive model M of set theory containing ω_1^V is correct for $\mathbf{\Sigma}^1_2$ formulas.*

Jensen [Jen81] later extended theorem 1.1 by showing that if the sharp of every real exists and there is no inner model with a measurable cardinal, then the core model K for mice with a single measure is correct for $\mathbf{\Sigma}^1_3$ formulas. In this note we extend this result to the core model for sequences of measures.

We write $K = L(\mathcal{E})$ for the maximal core model for sequences of measures originally constructed in [Mi84] and [Mi]. The preferred construction of K has changed since those papers were originally written; we will follow the new style but will try to make the exposition accessible to a reader without prior knowledge of these developments. See [MiS90] for an exposition of this new construction of the core model. In particular the model which we call K is the model which was called $K(\mathcal{F})$ in [Mi84] and [Mi].

Theorem 1.2. *Suppose that there is no model of $\exists\kappa\, o(\kappa) = \kappa^{++}$, that $a^\sharp$ exists for every real a, and that $\mathcal{M}$ is a transitive model of set theory such that $K^{\mathcal{M}}$ is an iterated ultrapower of K. Then $\mathcal{M}$ is correct for $\mathbf{\Sigma}^1_3$ formulas.*

If $0^\dagger$ does not exist then K is the Dodd-Jensen core model and theorem 1.2 reduces to Jensen's result. Our statement is actually slightly stronger than that given in [Jen81] even this case, since it allows for parameters from $\mathcal{M}$, but Jensen's proof does give this stronger version. Theorem 1.2 follows immediately from theorem 1.3 below (we write $(a,b)^\sharp$ for the sharp of a real coding the pair (a,b) of reals).

Theorem 1.3. *Suppose that there is no model of $\exists\kappa\, o(\kappa) = \kappa^{++}$, that a and b are reals such that $(a,b)^\sharp$ exists, and that ϕ is a Π^1_2 formula such that $V \models \phi(a,b)$. Suppose further that $K^{\mathcal{M}}$ is an iterated ultrapower of K. Then $K^{\mathcal{M}}[b]$, the smallest model of ZF containing $K^{\mathcal{M}}$ and b, satisfies $\exists x\, \phi(x,b)$.*

Theorem 1.2 can be given a slightly stronger statement which is somewhat analogous to the statement of Shoenfield's theorem: it is sufficient that $\mathcal{M}$ contain an iterated ultrapower of any countable premouse. The following conjecture probably gives something close to the correct analog of Shoenfield Absoluteness for $\mathbf{\Sigma}^1_3$ formulas.

Conjecture 1.4. *Suppose that the class of ordinals is measurable and that K is the core model for cardinals up through a Woodin cardinal together with a sharp for the Woodin cardinal, and let $\mathcal{M}$ be a model which contains an iterated ultrapower of any countable iterable premouse $\mathfrak{m}$ which does not iterate out to be longer than K. Then $\mathcal{M}$ is correct for $\mathbf{\Sigma}^1_3$ formulas.*

This conjecture, if true, would be the "correct analog" to theorem 1.1 in the sense that the negative large cardinal hypothesis $\neg\exists\kappa\, o(\kappa) = \kappa^{++}$ of theorem 1.2 has been entirely eliminated. The assumption that the class of ordinals is measurable is a strengthening of the assumption of theorem 1.2 that sharps exist. The assumption that sharps exist is necessary both for the conjecture and for theorem 1.2: for example if $0^\sharp$ does not exist then K is equal to L, which need not be correct for the $\mathbf{\Sigma}^1_3$ formula "there is a nonconstructible real". Steel has pointed out that the core model for a Woodin cardinal alone would not be enough and that some stronger assumption than the existence of sharps is necessary: If M is any fully iterable minimal model for "there exists a Woodin cardinal" then the set of reals of M is Q_3, which is Π^1_3 definable. The assumption that the class of ordinals is measurable can probably be reduced to a strong form of ineffability, but some such assumption appears to be necessary to define the core models (see [Ste90]).

Magidor (private communication) has given an alternate proof of Jensen's absoluteness result. This proof is somewhat easier than Jensen's and does not require the covering lemma but it does not give the 'pattern of indiscernibles' result which Jensen's does—that for any real a there is a mouse

$\mathfrak{m}$ and an ordinal $\gamma < \omega_1^V$ such that beyond some point the indiscernibles for $L(a)$ are obtained by taking every γth indiscernible for $\mathfrak{m}$. Our proof, which is a direct extension of that of Jensen, uses the covering lemma even more heavily that his does and gives a 'patterns of indiscernibles' result, lemma 5.6, which is almost as strong as that of Jensen though much more difficult to state. Steel has recently extended Magidor's method to give an alternate proof of theorem 1.2 which works so long as there is no model with a strong cardinal. This proof is substantially easier than ours, but does not include the "pattern of indiscernibles" results and does not prove the stronger theorem 1.3.

Prerequisites. It is assumed throughout this paper that the reader has a good understanding of the basic theory [Mi74] of models $L(\mathcal{U})$ of sets constructed from a coherent sequence of measures, including iterated ultrapowers of such models, the use of iterated ultrapowers to compare two such models, and indiscernibles generated by iterated ultrapowers. This paper also depends heavily on the theory of the core model and in particular on the covering lemma. The following paragraphs summarize the fine structure and other core model theory used in the paper, but it is recommended that the reader be previously acquainted with this theory, as described in [Mi84], [Mi] and [MiS89].

We recommend one of two strategies for reading this paper. A full understanding of the proof will require a good understanding of the fine structure, and for this the reader should be familiar with [Mi84], [Mi] and [MiS89]. An understanding of the basic ideas of this paper, however, should be possible with a considerably more superficial understanding of the fine structure. This section was originally written with the aim that the paper should be accessible at this level to a reader with only an understanding of [Mi74] and some acquaintance with the fine structure of L. This aim is probably impossible, but the next paragraphs should make it possible to read this paper with a somewhat unsophisticated understanding of the details of fine structure. These paragraphs should also be read by more sophisticated readers, if only as an explanation of the notation and conventions used in this paper.

The core model K which we are considering is exactly the same model as the core model $K(\mathcal{F})$ described in [Mi] but we use a more modern presentation based on an observation of S. Baldwin. A detailed exposition of fine structure theory using this presentation can be found in [MiS90]. It should be noted the paper [MiS90] is primarily concerned with cardinals larger than a strong cardinal and thus involves some complexity, notably the use of iteration trees, which is not needed here. The first four chapters of [MiS90] are particularly recommended.

There are three basic differences between the the presentation of the core model given in [Mi84] and that used in this paper. A minor difference is a change in the indexing of the sequence of measures: A coherent sequence $\mathcal{E}$ is a function $\mathcal{E}(\gamma)$ of a single variable, rather than a function $\mathcal{F}(\kappa,\beta)$ of two variables as in [Mi74] and [Mi]. A measure U on κ on the sequence $\mathcal{E}$ will be written $\mathcal{E}(\gamma)$, where γ is an ordinal in the interval $\kappa^+ < \gamma < \kappa^{++}$, instead of $\mathcal{F}(\kappa,\beta)$ as in [Mi84]. A more significant difference is that the the members $\mathcal{E}$ of the sequence are extenders rather than measures but in this paper all of the extenders are equivalent to measures so the reader can safely ignore this difference. We will in fact simply refer to and deal with the extenders on $\mathcal{E}$ as measures. The third, and most basic, difference is that the sequence $\mathcal{E}$ now contains partial as well as total measures. A member $\mathcal{E}(\gamma)$ of the sequence $\mathcal{E}$ is a total measure on the sets in $L(\mathcal{E}\restriction\gamma)$, and only on those sets, so that $\mathcal{E}(\gamma)$ is a total measure on $L(\mathcal{E})$ only if every subset of $\operatorname{crit}(\mathcal{E}(\gamma))$ in $L(\mathcal{E})$ is already a member of $L(\mathcal{E}\restriction\gamma)$. The effect of this gambit is to code the mice into the sequence $\mathcal{E}$. In [Mi84] it was necessary to to define the core model $K[\mathcal{F}]$ to be the class of sets constructible using both the sequence $\mathcal{F}$ and a class coding all of the $\mathcal{F}$ mice, but in the new presentation the core model K is a model of the form $L(\mathcal{E})$ and (as explained more fully in the next paragraph) the mice of K are simply the initial segments of K. The partial measures are important to the fine structure and as such will come up in section 2, but for the most part they can be ignored in reading this paper. Except when we specify otherwise, the word "measure" will always refer to a total measure.

The models $M = L(\mathcal{E})$ or $M = J_\gamma(\mathcal{E})$ used in this paper will all be *iterable premise*, which means that M satisfies three conditions: (1) $\mathcal{E}$ is good, (2) M is iterable, and (3) every initial segment of M is a mouse. The first condition essentially says that $\mathcal{E}$ is a coherent sequence of extenders; see [MiS90] for the details. The second condition says that every iterated ultrapower of M is well founded (although in [MiS90] this is stated in terms of iteration trees). The third condition asserts that $J_\alpha(\mathcal{E})$ is a mouse for all ordinals α if $M = L(\mathcal{E})$ or all ordinals $\alpha < \gamma$ if $M = J_\gamma(\mathcal{E})$. Because of this requirement the definitions of a *premouse* and a *mouse* use a simultaneous recursion on γ. A *mouse* is a premouse $J_\alpha(\mathcal{E})$ which satisfies an additional condition which makes it look like what Dodd and Jensen call a *core mouse* in [Dod82]. To explain this more fully we will have to go a bit more into the fine structure of the models of $J_\alpha(\mathcal{E})$. We will, of course, only give a brief outline of the fine structure. The approach to fine structure which we will use is different from, though basically equivalent to, that of [MiS90]. One advantage of this approach is that the definition is almost identical to Jensen's definition in [Jen72] of the fine structure of L, thereby highlighting the superiority of the new presentation of the core model (although the proofs for K are, of course, still more difficult than those for L). The only basic difference from the definition in L is at the beginning: if α is an ordinal such that $\mathcal{E}(\alpha)$ is defined then it is necessary to define a Σ_0 code $\mathcal{A}_0 = (J_{\rho_0}, A_0)$ for $J_\alpha(\mathcal{E})$ in order to get a amenable structure with a predicate for $\mathcal{E}(\alpha)$. This definition uses the same idea as the Σ_0 codes used in [Mi]. Once this is done we define the Σ_n projectum ρ_n, the Σ_n code $\mathcal{A}_n = (J_{\rho_n}(\mathcal{E}), A_n)$, and the Σ_n standard parameter p_n by recursion on n using exactly the same definition that Jensen used in L: If ρ_n, p_n and A_n have been defined then

- ρ_{n+1} is the least ordinal ρ such that there is a subset of ρ which is Σ_1 definable from parameters in $\mathcal{A}_n = (J_{\rho_n}(\mathcal{E}), A_n)$, but is not a member of $J_\alpha(\mathcal{E})$.
- p_{n+1} is the least finite set of ordinals such that there is such a subset definable from parameters from $\rho_{n+1} \cup p_{n+1}$.
- $A_{n+1} \subset \rho_{n+1}$ codes the set of Σ_1-sentences with parameters from $\rho_{n+1} \cup \{p_{n+1}\}$.

The *projectum*, $\rho = \operatorname{proj}(J_{\alpha(\mathcal{E})}$, of a premouse $J_\alpha(\mathcal{E})$ is defined to be $\min\{\, \rho_n : n \in \omega \,\}$.

A fundamental theorem of the fine structure of L states that for each $n \in \omega$ the nth projectum ρ_n is contained in the Σ_1-hull in $\mathcal{A}_n$ of $\rho_{n+1} \cup p_{n+1}$. This need not be true for an arbitrary premouse $J_\alpha(\mathcal{E})$. A premouse $J_\alpha(\mathcal{E})$ is said to be *m-sound* for $m \le \omega$ if for all $n < m$ the Σ_n projectum ρ_n is contained in the Σ_1-hull in $\mathcal{A}_n$ of $\rho_{n+1} \cup p_{n+1}$, and the premouse $J_\alpha(\mathcal{E})$ is a *mouse* if $J_\alpha(\mathcal{E})$ is ω-sound.

An embedding $i\colon J_\alpha(\mathcal{E}) \to J_{\alpha'}(\mathcal{E}')$ of premise is said to be *fine structure preserving* if for each n the restriction of i to $J_{\rho_n}(\mathcal{E})$ is a Σ_1-elementary embedding between the Σ_n codes. We will define ultrapowers of mice in such a way that the canonical embeddings do preserve fine structure. Two observations about the iterated ultrapowers used in this paper will be useful for this definition. The first observation is that every iterated ultrapower will have increasing critical points, that is, if the iterated ultrapower is $(\, M_\nu : \nu < \phi \,)$ with $M_{\nu+1} = \operatorname{ult}(M_\nu, E_\nu)$ then $\operatorname{crit}(E_\nu) < \operatorname{crit}(E'_\nu)$ whenever $\nu < \nu'$. The second observation is that in every ultrapower of a mouse, and in fact in every embedding $k\colon \mathfrak{m} \to \mathfrak{m}'$ of a mouse $\mathfrak{m}$, the critical point of i will be at least as large as the projectum of $\mathfrak{m}$, so that $i \restriction \operatorname{proj}(\mathfrak{m}) = \mathrm{id}$. There is actually one exception to this second observation, which will be discussed below: we will use nontrivial embeddings of $K = L(\mathcal{E})$ even though K is, according to the definition, a mouse with $\operatorname{proj}(K) = \Omega$, the order type of the class of all ordinals.

Now suppose that $M = J_\alpha(\mathcal{E})$ is a mouse, or more generally an $n+1$-sound premouse, and that E is a measure on M with critical point κ such that $\rho_{n+1} \leq \kappa < \rho_n$. Then the mouse ultrapower of M by E is defined as follows: start by taking the ordinary ultrapower $i^E\colon \mathcal{A}_n \to \mathcal{A}'_n = ult(\mathcal{A}_n, E)$ of the Σ_n code $\mathcal{A}_n$ of M. Then $\mathcal{A}'_n$ will be the Σ_n code of a n-sound premouse $M' = J_{\alpha'}(\mathcal{E}')$, and i will extend to a fine structure preserving embedding $i^*\colon M \to M'$. This embedding i^* will be what we call the *mouse ultrapower* of M by E. This definition of the mouse ultrapower can be readily extended to an iterated ultrapower provided that the critical points of the iterated ultrapower are increasing, as is always the case in this paper. Every ultrapower or iterated ultrapower of a mouse in this paper will be a mouse ultrapower.

The following facts will cover most of our use of fine structure in this paper.

Theorem 1.5. *(1) Suppose that $x \in \mathcal{P}(\kappa) \cap J_{\alpha+1}(\mathcal{E})$ but $x \notin J_\alpha(\mathcal{E})$, and let $i\colon J_\alpha(\mathcal{E}) \to J_{\alpha'}(\mathcal{E}')$ be any iterated ultrapower of $J_\alpha(\mathcal{E})$ such that $i{\restriction}\kappa$ is the identity. Then x is definable in $M = J_{\alpha'}(\mathcal{E}')$ in the same way that it is definable in $J_\alpha(\mathcal{E})$.*

(2) If $i\colon J_\alpha(\mathcal{E}) \to J_{\alpha'}(\mathcal{E}')$ is any cofinal fine structure preserving embedding such that $i{\restriction}\rho$ is the identity, where ρ is the projectum of $J_\alpha(\mathcal{E})$, then i is an iterated ultrapower of $J_\alpha(\mathcal{E})$.

This theorem has two major consequences. The first follows from clause (1) above:

Corollary 1.6. *(Comparability of mice) If $J_\alpha(\mathcal{E})$ and $J_{\alpha'}(\mathcal{E}')$ are mice and $\mathcal{E}{\restriction}\rho = \mathcal{E}'{\restriction}\rho$ then either $\mathcal{P}(\rho) \cap J_\alpha(\mathcal{E}) \subseteq J_{\alpha'}(\mathcal{E}')$ or $\mathcal{P}(\rho) \cap J_{\alpha'}(\mathcal{E}') \subseteq J_\alpha(\mathcal{E})$.*

The second follows from clause (2):

Corollary 1.7. *(The maximality principle) If $J_\alpha(\mathcal{E})$ is a mouse, $\kappa \geq \operatorname{proj}(J_\alpha(\mathcal{E}))$, and E is a measure on $\mathcal{P}(\kappa) \cap J_\alpha(\mathcal{E})$ which is coherent for adding to $J_\alpha(\mathcal{E})$ such that $\operatorname{ult}(J_\alpha(\mathcal{E}), E)$ is well founded then E is already on the sequence $\mathcal{E}$, that is, $E = \mathcal{E}(\gamma)$ for some index γ. Furthermore if $i\colon J_\alpha(\mathcal{E}) \to J'_\alpha(\mathcal{E}')$ is an iterated ultrapower then the same is true of $J_{\alpha'}(\mathcal{E}')$, provided that $i{\restriction}\operatorname{proj}(J_\alpha(\mathcal{E})) = \mathrm{id}$ and E was not used in the iterated ultrapower i.*

The phrase "is coherent for adding" means that there is an ordinal γ such that the sequence $\mathcal{E}' = \mathcal{E}{\restriction}\gamma \cup (\gamma, E)$, which includes E as $\mathcal{E}'(\gamma)$, is *good*, which it will be recalled is the first criterion for $J_{\gamma+1}(\mathcal{E}')$ to be a premouse.

The reader might wish to consider reading this paper under the (false) assumption that only Σ_1 codes are necessary. This means that Σ_0 codes do not arise and that a mouse is simply a structure $\mathfrak{m} = J_\alpha(\mathcal{E})$ with $\rho = \text{proj}(\mathfrak{m})$ equal to $\rho_1^{\mathfrak{m}}$, so that there is a new Σ_1-subset of ρ definable in $\mathfrak{m}$, every subset of an ordinal $\nu < \rho$ definable over $\mathfrak{m}$ is in $\mathfrak{m}$, and $\mathfrak{m}$ is equal to the Σ_1-hull of $\rho \cup p_1$ in $\mathfrak{m}$. Then mouse ultrapowers are ordinary ultrapowers, and for any embedding which is the identity on ρ the term "fine structure preserving" means Σ_1-elementary. As in most fine structure arguments, all of the arguments of this paper apply to this special case exactly as they apply to the general case—in fact the reduction of the Σ_n case to the Σ_1 case, repeated n times, is the basic idea behind the use of fine structure.

The covering lemma is used in this paper in two different ways. First, chapter 2 uses the fact that the core model K satisfies theorem 1.5(2), and hence also the maximality principle 1.7 (with the mention of the projectum ρ omitted). Thus the covering lemma allows us to make the core model K an exception to the rule asserted above that $i \restriction \text{proj}(\mathfrak{m})$ is always equal to the identity whenever $\mathfrak{m}$ is a mouse. See [Mi] for a proof of these facts.

The second, and more basic, application of the covering lemma comes in the proof of lemma 3.7. The versions of the covering lemma which are used are stated there as lemmas 3.8 and 3.9 and should cause no problems to the reader.

Notation. As is usual in descriptive set theory, we identify the real numbers with ${}^\omega\omega$. If X is a set of ordinals then we write $[X]^n$ for the set of size n subsets of X, which we identify with the set of increasing sequences of members of X of length n. If $\mathbf{c}, \mathbf{c}' \in [X]^{<\omega}$ then we write $\mathbf{c} \equiv \mathbf{c}'$ if $\mathbf{c}$ and $\mathbf{c}'$ have the same length. If in addition $\mathbf{d}, \mathbf{d}' \in [X]^{<\omega}$ then we write $\mathbf{c}, \mathbf{c}' \equiv \mathbf{d}, \mathbf{d}'$ if $\mathbf{c} \equiv \mathbf{d}$, $\mathbf{c}' \equiv \mathbf{d}'$, and for all i and j we have $c_i \leq c'_j$ iff $d_i \leq d'_j$. We use Ω to stand for the class of all ordinals.

Martin-Solovay trees. Most of the proof of theorem 1.3 is concerned with indiscernibles rather than with the Π^1_2 formula ϕ. We make the connection between the two via the Martin-Solovay analysis of $\mathbf{\Sigma}^1_3$ sentences, which we use instead of the infinitary logic used by Jensen in [Jen81]. This analysis was introduced in [MaS69] in a much more delicate form than we will require here. In order to provide a visible destination for the main body of the proof we will describe this analysis here rather than at the end where it logically belongs.

Suppose that $\phi(x,b)$ is a Π^1_2 formula with parameter b. We will define a tree $T \in K[b]$ such that any branch through T constructs a real r such that $\phi(r,b)$. The construction of the tree T depends on the real b and on a pair $(J,\mathcal{T}) \in K$. The rest of the paper after this section will be concerned with constructing the pair $(J,\mathcal{T})$. The definition in K of the pair $(J,\mathcal{T})$ depends in turn on a finite set of parameters, and much of the work of this paper will take place in the universe V, with a knowledge of a real $(a,b)^\sharp$ such that $V \models \phi(a,b)$, in order to show that this finite set of parameters can be chosen in such a way that the real $(a,b)^\sharp$ determines a branch in V through the tree T. By the absoluteness of well foundedness it follows that there is also a branch in $K[b]$, and hence there is a real $(r,b)^\sharp \in K[b]$ such that $\phi(r,b)$.

Definition 1.8. A pair $(J,\mathcal{T})$ is *suitable* provided that J is an uncountable set of ordinals and for each $\tau \in \mathcal{T}$, there is $n \in \omega$ such that τ is a function from $[J]^n$ into the ordinals.

We write $\mathcal{T}_n$ for the set of n-ary functions in $\mathcal{T}$. Assume that $(J,\mathcal{T})$ is suitable. By theorem 1.1, $V \models \phi(r,b)$ if and only if $L(r,b) \models \phi(r,b)$, so we only need to consider the truth of $\phi(r,b)$ in $L(r,b)$. The tree T will be defined so that any infinite branch of T constructs a pair (e,σ), where $e = (r,b)^\sharp$ for some real r such that $L(r,b) \models \phi(r,b)$ and σ is a function which will ensure that the alleged sharp e is well founded.

Definition 1.9. A branch through the tree T will be a pair (e,σ) which satisfies the 6 clauses listed below. Each of these clauses specifies a closed set, that is, if any of these clauses fail for a pair (e,σ) then there is $n \in \omega$ such that the failure is evident in $(e{\restriction}n, \sigma{\restriction}n)$. Thus the tree T can be defined to be the set of pairs $(\hat{e},\hat{\sigma})$ of finite sequences for which none of the clauses have yet failed.

(1) e is the set of Gödel numbers of the first order theory of a structure

$$\mathcal{B} = (B, \dot{r}, \dot{b}, \dot{c}_0, \dot{c}_1, \dots).$$

(2) The Gödel numbers of "$ZFC + V = L(\dot{r},\dot{b})$" and of the sentences asserting that the ordinals c_i form a remarkable set of indiscernibles (see [Sil71]) are in e.
(3) For $n \in \omega$ the Gödel number of "$n \in \dot{b}$" is in e if and only if $n \in b$.
(4) The Gödel number of "$\phi(\dot{r},\dot{b})$" is in e.
(5) If s is a term with n free variables in the language of $\mathcal{M}$ which does not use any of the constants $\dot{c}_i$ and the Gödel number of "$s(\dot{c}_0,\dots,\dot{c}_{n-1})$ is an ordinal" is in e then $\sigma(s) \in \mathcal{T}_n$.

(6) Suppose that s and s' are in the domain of σ, that $\mathbf{c}$ and $\mathbf{c}'$ are finite sequences from $\{\, \dot{c}_i : i \in \omega \,\}$ so that $s(\mathbf{c})$ and $s'(\mathbf{c}')$ make sense, and that $\mathbf{j}$ and $\mathbf{j}'$ are arbitrary sequences in $[J]^{<\omega}$ such that $\mathbf{c}, \mathbf{c}' \equiv \mathbf{j}, \mathbf{j}'$. Then the Gödel number of "$s(\mathbf{c}) \le s'(\mathbf{c}')$" is in e if and only if $\sigma(s)(\mathbf{j}) \le \sigma(s')(\mathbf{j}')$.

Clauses (1) to (4) ensure that the first order theory coded by the alleged sharp e is correct, while clauses (5) and (6) ensure that e is the theory of a well founded model $L(r, b)$ with a class of indiscernibles.

The rest of this paper will be concerned with defining the set J of ordinals and the set $\mathcal{T}$ of functions, with the ultimate definition taking place inside K although most of the actual work will take place outside it.

Summary of the proof. We conclude this introduction with an outline of a proof of Jensen's result, followed by a discussion of what is necessary to extend this proof to a proof of theorem 1.3. I believe that the proof of Jensen's result outlined here is essentially the same as that given in [Jen81], though it is not easy to make the comparison.

The proof of Jensen's result proceeds under the assumption that $L(\mu)$ does not exist, so all mice contain at most one total measure and K contains no total measures. Let K^a be the core model as defined in $L[a, b]$. Then $K^a \neq K$, since $(a, b)^\sharp$ gives an ultrafilter on $L(a, b)$ and hence on K^a, so let M_0 be the least mouse not in K^a and let $j\colon M_0 \longrightarrow M = \operatorname{ult}_\Omega(M_0, U_0)$, the Ω-fold iteration of M_0 by its measure U_0. We now have two classes of indiscernibles: the class I of Silver indiscernibles for $L[a, b]$ given by $(a, b)^\sharp$ and the class C of indiscernibles in M generated by the iterated ultrapower j. It is not hard to prove that $I \subset C$. Now there are two major lemmas to be proved: the first lemma is that any map $\pi : L(a, b) \longrightarrow L(a, b)$ yields a map $\pi^*\colon M \longrightarrow M$ such that $\pi^* \restriction \Omega = \pi \restriction \Omega$ and $\pi^* \restriction \operatorname{range}(j) = \mathrm{id}$, and the second lemma is that if c and c' are adjacent members of I then there are at most countably many members of C in the interval (c, c'). By the first lemma, together with the fact that $\pi^* ``C \subset C$, it follows that the order type of $C \cap (c, c')$ is nondecreasing as c increases, and it follows by the second lemma that for c sufficiently large the order type ξ of $C \cap (c, c')$ is constant. Then there is $\delta \in \Omega$ such that $I \setminus \delta$ consists of every ξth member of $C \setminus \delta$ and hence $I \setminus \delta$ is definable in K. The set J of ordinals for the Martin-Solovay tree is taken to be $I \setminus \delta$. To get the set $\mathcal{T}$ of functions, define the function s_ι for $\iota < \xi$ by letting $s_\iota(\nu)$ be the ιth member of C above ν. Now let λ be least such that $j(\lambda) \ge \Omega$ and let $\mathcal{T}$ be the closure under composition of the set containing (i) the constant functions, $c_\nu(x) = \nu$ for $\nu < \delta$, (ii) the functions $j(g)$ where $g \in M_0$ and $g\colon [\lambda]^n \to \lambda$ for some $n \in \omega$, and (iii) the functions s_ι for $\iota < \xi$.

This choice of J and $\mathcal{T}$ is suitable, so any branch through the Martin-Solovay tree T generated by $(J, \mathcal{T})$ will give a solution to $\exists x\, \phi(x, b)$. Furthermore any embedding π of $L(a, b)$ preserves the functions in $\mathcal{T}$, that is, $\pi^*(\tau(\mathbf{c})) = \tau(\pi(\mathbf{c}))$ for any $\tau \in \mathcal{T}$ and $\mathbf{c} \in [I]^n$. Now set $e = (a, b)^\sharp$, and if s is any n-term in the language of $\mathcal{B}$ let $\sigma(s)$ be any member of $\mathcal{T}_n$ such that $s^{L(a,b)}(\mathbf{c}) = \sigma(s)(\mathbf{c})$ for some sequence $\mathbf{c} \in [I \setminus \xi]^n$. This definition doesn't depend on the choice of $\mathbf{c}$: to see this, let $\mathbf{c}' \equiv \mathbf{c}$ be the first n members of $I \setminus \delta$ and suppose that $\sigma(s)(\mathbf{c}') = s(\mathbf{c}')$. Then there is an embedding $\pi\colon L(a, b) \to L(a, b)$ such that $\pi(\mathbf{c}') = \mathbf{c}$. Then $\sigma(s)(\mathbf{c}) = \sigma(s)(\pi(\mathbf{c}')) = \pi^*(\sigma(s)(\mathbf{c}')) = \pi(\sigma(s)(\mathbf{c}')) = \pi(s(\mathbf{c}')) = s(\pi(\mathbf{c}')) = s(\mathbf{c})$. Clause (6) of definition 1.9 may be verified similarly, completing the proof that (e, σ) is a branch through T. It follows by absoluteness that there is a branch through T which is a member of $K[b]$ and hence there is a real $r \in K[b]$ such that $\phi(r, b)$. This completes the proof of the theorem.

In our proof, K^a does contain measures. The first problem which this gives rise to is that it is not obvious what is meant by "the first mouse not in K^a", since a mouse $\mathfrak{m}$ in K will be a mouse for $\mathcal{E}^K{\restriction}\rho$ for some ordinal ρ, and $\mathcal{E}^K{\restriction}\rho$ will not in general be in K^a. We solve this problem in section 2 by using iterated ultrapowers to carry out a modified version of the standard comparison of the corelike models K and K^a. During the construction K may be replaced by a mouse which is in the current iteration of K, but not in the current iteration of K^a, and then by successively smaller mice until "the least mouse not in K^a" is reached.

In practice we construct the iterated ultrapower N of K^a first so that we can then iterate K against N without moving N. The reason for this procedure is that we need to have N definable in K^a.

Since we were dealing with sequences of measures, instead of a single measure, Jensen's class C of indiscernibles is now replaced by a function $\mathcal{C}$. If $\mathcal{E}(\gamma)$ is one of the measures in M, then $\mathcal{C}(\gamma)$ is the set of indiscernibles for $\mathcal{E}(\gamma)$. The second of the lemmas of the proof of Jensen's result becomes our main lemma 3.1, which says essentially that the same measure is never used more than $\rho^+ + \omega_1^V$ times in the iteration j, where ρ is an ordinal coming out of the construction of the iterated ultrapower M of K. More precisely, if the order type of $\mathcal{C}(\gamma) \cap (\beta \setminus \alpha)$ is greater than $\rho^+ + \omega_1^V$ then either $I \cap \beta \not\subset \alpha$ or there is $\mathcal{E}(\gamma')$ with $\gamma' > \gamma$ such that $\mathcal{C}(\gamma') \cap \beta \not\subset \alpha$.

At this point we have to deal with a complication which does not come up in Jensen's proof. In his argument M is taken to be $\mathrm{ult}_\Omega(M_0, U_0)$, where both M_0 and U_0 are in K, and it follows that M is definable in K. In our construction the iterated ultrapower $j\colon K \longrightarrow M$ involves a choice of which measure to use at each stage and hence need not be definable in K. In addition, we not yet mentioned the model $\mathcal{M}$ from theorem 1.3, but have worked exclusively with the true core model K. Both of these points are dealt with in section 4, where we give a definition inside $K^{\mathcal{M}}$ of a complete iteration $s\colon \overline{M}_0 \longrightarrow \overline{M}$ which mimics the construction of $j\colon M_0 \longrightarrow M$. The basic idea is to start with mice in $K^{\mathcal{M}}$ which are iterates of the mice used in j and then use every measure at least as many times in the iteration s as the main lemma would permit it to have been used in the iteration j, thus ensuring that j can be embedded into s. The definition of s depends on a finite sequence of parameters, and we show that for the proper choice of these parameters we can define an embedding $t\colon M \longrightarrow \overline{M}$ mapping indiscernibles from $\mathcal{C}$ into the corresponding indiscernibles in the system $\overline{\mathcal{C}}$ generated by $\bar{\jmath}$. Then the pair $(J, \mathcal{T})$ is suitable, where $J = t``I$ and $\mathcal{T}$ is the set of terms arising from the system $\overline{\mathcal{C}}$ of indiscernibles.

Finally, in section 5 we define terms in M, and map these terms from M to the terms in $\overline{M}$ in such a way as to use $L(a,b)$ to construct a branch in V through the Martin-Solovay tree T obtained from $(J, \mathcal{T})$. It follows that there is a branch through T in $K^{\mathcal{M}}[b]$ and this completes the proof of theorem 1.3.

2. The Comparison

Notation. Before starting the actual construction we will define some general notions. As was pointed out in the last section we will follow the new presentation of the core model (see [MiS90]) rather than that of [Mi], but since we will avoid use of fine details of the core model, and because for sequences of measures the model is in fact identical to that in [Mi84] and [Mi], the difference should not cause major problems. The core model has the form $L(\mathcal{E})$, where $\mathcal{E}$ is a sequence of extenders. Some of the extenders in $\mathcal{E}$ are only partial extenders, but except for a brief mention of this in the proof of the main lemma we will deal only with the total extenders in $\mathcal{E}$. In addition the assumption that there is no model of $\exists\kappa\, o(\kappa) = \kappa^{++}$ implies that all of the extenders in $\mathcal{E}$ are isomorphic to measures, and from this point on we will forget we ever knew that in some sense they really are extenders. Thus, unless specified otherwise, the term "measure" always refers to a measure which is total in the model M in which is occurs, and when we refer to a measure $\mathcal{E}(\gamma)$ we will, unless specified otherwise, assume not only that γ is in the domain of $\mathcal{E}$ but that $\mathcal{E}(\gamma)$ is a total measure in the premouse being considered.

If $\mathcal{E}(\gamma)$ is a measure on κ then we say that κ is the critical point of $\mathcal{E}(\gamma)$, written $\kappa = \operatorname{crit}(\mathcal{E}(\gamma))$. We use $O(\kappa)$ for $\sup\{\,\gamma : \kappa = \operatorname{crit}(\mathcal{E}(\gamma))\,\}$ and $o(\kappa)$ for the order type of $\{\,\gamma : \mathcal{E}(\gamma)$ is a total measure and $\kappa = \operatorname{crit}(\mathcal{E}(\gamma))\,\}$.

If $\gamma' < \gamma$ and $\mathcal{E}(\gamma)$ is a measure on κ then we write $\mathfrak{C}^{\mathcal{E}}(\gamma',\gamma)$ for the *coherence function,* the least function $g\colon \kappa \longrightarrow \kappa$ in the order of construction of $L(\mathcal{E})$ such that $[g]_{\mathcal{E}(\gamma)} = \gamma'$. Thus if $\mathcal{E}(\gamma')$ exists then for all $f \in L(\mathcal{E})$

$$\{\,\nu < \kappa : \forall x \in f``\nu\,(x \cap \nu \in \mathcal{E}(\mathfrak{C}^{\mathcal{E}}(\gamma',\gamma)(\nu)) \iff x \in \mathcal{E}(\gamma'))\,\} \in \mathcal{E}(\gamma).$$

While the definition of the coherence function depends on the the sequence $\mathcal{E}$, the function only depends on $\mathcal{E}{\restriction}(\gamma+1)$. For this reason we will normally omit the superscript.

Definition 2.1. Suppose that $j_{0,\nu}\colon M_0 \longrightarrow M_\nu$ is an arbitrary iterated ultrapower with strictly increasing critical points, $M_{\nu+1} = \operatorname{ult}(M_\nu, \mathcal{E}_\nu(\gamma_\nu))$ where $\mathcal{E}_\nu = j_{0,\nu}(\mathcal{E})$. We define the *sequence $\mathcal{C}_\nu$ of indiscernibles generated by the iteration $j_{0,\nu}$* by induction on ν: $\mathcal{C}_0(\gamma) = \varnothing$ for all γ; at successor ordinals we have

$$\mathcal{C}_{\nu+1}(\gamma) = \begin{cases} \mathcal{C}_\nu(\gamma) & \text{if } \gamma < \gamma_\nu, \\ \mathcal{C}_\nu(\gamma_\nu) \cup \{\operatorname{crit}(\mathcal{E}_\nu(\gamma_\nu))\} & \text{if } \gamma = j_{\nu,\nu+1}(\gamma_\nu), \\ \mathcal{C}_\nu(\gamma') & \text{if } \gamma = j_{\nu,\nu+1}(\gamma') \text{ for } \gamma' \neq \gamma_\nu, \\ \varnothing & \text{otherwise;} \end{cases}$$

and if ν is a limit ordinal then $c \in \mathcal{C}_\nu(\gamma)$ if and only if there are ordinals $\nu' < \nu$ and γ' such that $c < \kappa_{\nu'} = \operatorname{crit}(j_{\nu',\nu})$, $\gamma = j_{\nu',\nu}(\gamma')$ and $c \in \mathcal{C}_{\nu'}(\gamma')$.

Definition 2.2. If the sequence $\mathcal{D}_\nu$ of indiscernibles is generated by the iterated ultrapower $j_{0,\nu}$ then α is an *accumulation point* for γ in $\mathcal{D}_\nu$ if one of the following clauses holds:

(1) α is measurable in $\mathcal{M}_\nu$, $\alpha < \gamma \le O^{\mathcal{E}_\nu}(\alpha)$, and for every $\delta < \alpha$ and $\beta < \gamma$ there is λ such that $\beta \le \lambda < O^{\mathcal{E}_\nu}(\alpha)$ and $\mathcal{D}_\nu(\lambda) \not\subset \delta$.

(2) There is $\nu' < \nu$ such that if $\kappa = j_{\nu',\nu}(\alpha)$ then $\kappa < \gamma \le O^{\mathcal{E}_\nu}(\kappa)$ and for all $\delta < \alpha$ and $\beta < \gamma$ in the range of $j_{\nu',\nu}$ there is λ such that $\beta \le \lambda < O^{\mathcal{E}_\nu}(\kappa)$ and $\mathcal{D}_\nu(\lambda) \cap \alpha \not\subset \delta$.

We say that α is a *strict accumulation point* for γ in $\mathcal{D}_\nu$ if α is an accumulation point for γ and $\alpha \in \mathcal{D}_\nu(\beta)$ for some $\beta < \gamma$.

There are several observations to be made on this definition. First, note that if α is an accumulation point for γ and $\alpha < \gamma' < \gamma$ for clause (1) or $\kappa < \gamma' < \gamma$ for clause (2) then α is also an accumulation point for γ'. However we will be interested primarily in the largest ordinal γ such that α is an accumulation point for γ, so that either $\gamma = O^{\mathcal{E}_\nu}(\kappa)$ or $\mathcal{D}_\nu(\gamma)$ is bounded in α. If α is an accumulation point for this γ by clause (2) but α is not a strict accumulation point for γ then $\gamma' = \sup\{\eta : j_{\nu',\nu}(\eta) < \gamma\} \leq O^{\mathcal{E}_\nu}(\alpha)$ and α is an accumulation point for γ' by clause (1). Finally notice that if κ is a limit point of $\mathcal{C}_\nu(\gamma)$, then κ is an accumulation point for $\gamma + 1$ in $\mathcal{C}_\nu$. A cardinal κ is an accumulation point for $O^{\mathcal{E}_\nu}(\kappa)$ if every measure on κ in $\mathcal{E}_\nu$ is generated by indiscernibles. For the construction of N in this section this will be true if "generated" means simply that each such measure $\mathcal{E}_\nu(\lambda)$ is the filter of subsets of κ which contain all but a bounded subset of $\mathcal{C}_\nu(\lambda)$. The general case is given by definition 3.5.

The construction of M and N. We are now ready to begin the actual construction. Recall that ϕ is a Π^1_2 formula and a and b are reals such that $V \models \phi(a, b)$, and hence $L(a, b) \models \phi(a, b)$; and we want to prove that $K[b] \models \exists r\, \phi(r, b)$. Let K^a be the core model K as defined in $L(a, b)$. We will work inside $L(a, b)$ to define an iterated ultrapower N of K^a and then we will define the model M to be a modified iterated ultrapower of K in such a way that M is an iterated ultrapower of the least mouse not in N.

Set $N_0 = K^a = L(\mathcal{F}_0)$, and suppose that $i_{0,\nu}\colon N_0 \longrightarrow N_\nu$ has already been defined. We will write $\mathcal{F}_\nu$ for $i_{0,\nu}(\mathcal{F}_0)$ and $\mathcal{D}_\nu$ for the system of indiscernibles generated by $i_{0,\nu}$.

Definition 2.3. Let κ be the least cardinal in $N_\nu = L(\mathcal{F}_\nu)$ such that one of the following two conditions fails:

(1) If κ is measurable in N_ν then $\mathrm{cf}^{L(a,b)}(\kappa) = \omega_1^V$.
(2) $\mathcal{D}_\nu(\gamma)$ is unbounded in κ for all γ such that $\mathcal{F}_\nu(\gamma)$ is a total measure on κ.

If clause (1) fails then $N_{\nu+1} = \mathrm{ult}(N_\nu, \mathcal{F}_\nu(\gamma))$ where $\mathcal{F}_\nu(\gamma)$ is the order 0 measure on κ, while if clause (2) fails then $N_{\nu+1} = \mathrm{ult}(N_\nu, \mathcal{F}_\nu(\gamma))$ where $\gamma < O^{\mathcal{F}_\nu}(\kappa)$ is the least ordinal such that $\mathcal{F}_\nu(\gamma)$ exists and $\mathcal{D}_\nu(\gamma)$ is bounded in κ.

If there is an ordinal ν such that both conditions are true in N_ν for all $\kappa \in \Omega$ then we set $N_{\nu'} = N_\nu$ for all $\nu' > \nu$. Set $i = i_{0,\Omega}$, $N = N_\Omega = L(\mathcal{F})$, and $\mathcal{D} = \mathcal{D}_\Omega$, where Ω denotes the order type of the ordinals.

Proposition 2.4.

(1) *The iterated ultrapower $i\colon K^a \longrightarrow N$ is definable in $L(a, b)$ from the parameter ω_1^V.*
(2) *For all κ such that $o^N(\kappa) > 0$ we have $\mathrm{cf}^{L(a,b)}(\kappa) = \omega_1^V$.*
(3) *Every ordinal κ is an accumulation point for $O^N(\kappa)$ in $\mathcal{D}$.*
(4) *Each measure $\mathcal{F}(\gamma)$ in N is countably complete.*
(5) *The ordinals of N have order type Ω.*

Proof. The proof is easy □

The definition of M is more complicated, since we are effectively looking for a minimal mouse which is not in N. We will define a series of iterated ultrapowers,

$$j^k_{0,\nu}\colon M^k_0 \to M^k_\nu$$

by recursion on k and ν, using the standard procedure for comparing K with N, but using proposition 2.4 to show that N is not be moved in the comparison. For $k = 0$, the model M^0_ν will be $L(\mathcal{E}^0_\nu)$. For $k > 0$ the model M^k_ν will be a mouse iteration of the mouse $\mathfrak{m} = M^k_0$.

Definition 2.5. We define a sequence $(M^k_0 : k < \hat{k})$, by recursion on $k < \hat{k} < \omega$, using a secondary recursion on ν to define iterated ultrapowers $j^k_{0,\nu}\colon M^k_0 \to M^k_\nu$ for $\nu < \nu_k$. First set $M^0_0 = K$. Now suppose that M^k_0 has been defined, together with an iterated ultrapower $j^k_{0,\nu}\colon M^k_0 \to M^k_\nu$. Let κ be the least cardinal in M^k_ν, if there is one, such that one of the following statements fails:

(1) $\mathcal{P}^N(\kappa) = \mathcal{P}^{M^k_\nu}(\kappa)$,

(2) $\mathcal{E}^k_\nu{\restriction}\delta = \mathcal{F}{\restriction}\delta$, where $\delta = O^{M^k_\nu}(\kappa) = O^N(\kappa)$, *ie*, M^k_ν and N have the same measures on κ.

If clause (1) fails, with $\mathcal{P}^N(\kappa) \subsetneqq \mathcal{P}^{M^k_\nu}(\kappa)$, then *drop to a mouse*: set $\nu_k = \nu$ and if $M^k_{\nu_k} = J_\alpha(\mathcal{E}^k_{\nu_k})$ then set $M^{k+1}_0 = J_\beta(\mathcal{E}^k_{\nu_k})$ where β is the least ordinal such that $\mathcal{P}(\kappa) \cap J_{\beta+1}(\mathcal{E}^k_{\nu_k})$ is not contained in N. Thus M^{k+1}_0 is the least mouse in $M^k_{\nu_k}$ with projectum at most κ which is not a member of N. If clause (2) fails in such a way that there is a $\gamma \in \operatorname{domain}(\mathcal{E}^k_\nu)$ such that $\gamma \geq O^{\mathcal{F}}(\kappa)$ and $\mathcal{F}{\restriction}\gamma = \mathcal{E}^k_\nu{\restriction}\gamma$ then set $M^k_{\nu+1} = \operatorname{ult}(M^k_\nu, \mathcal{E}_\nu(\gamma))$.

This definition is justified by the following proposition, which implies that the clauses above will always fail in the way described in the construction.

Proposition 2.6. *If κ is least such that one of the clauses of definition 2.6 fails then $\mathcal{P}^N(\kappa) \subseteq \mathcal{P}^{M^k_\nu}(\kappa)$, and if $\mathcal{P}^N(\kappa) = \mathcal{P}^{M^k_\nu}(\kappa)$ then $O^N(\kappa) < O^{M^k_\nu}(\kappa)$ and $\mathcal{F}{\restriction}O^N(\kappa) = \mathcal{E}^k_\nu{\restriction}O^N(\kappa)$.*

Proof. For $k = 0$, the model M^0_ν is an iterated ultrapower of K. By [Mi, theorem 5.2] it follows that M^0_ν contains all mice over its sequence of measures, and in particular all mice in N over the initial segment of $\mathcal{F}$ on which they agree. Thus M^0_ν contains every subset of κ which is in N. By [Mi, lemma 7.8] it contains all countably complete measures except those which were used in the iteration, and in particular all of those in N. This implies the second clause of the proposition.

If $k > 0$ then M^k_ν is an iterate of a mouse $\mathfrak{m} = M^k_0$ in which there is a subset of $\text{proj}(M^k_\nu) < \kappa$ which is definable in M^k_ν but is not a member of N. Since the mice are linearly ordered it follows that the κ-mice in N are an initial segment of those in M^k_ν. Thus M^k_ν contains all subsets of κ which are in N. Furthermore, because M^k_ν is an iterated ultrapower of a mouse it contains all measures E which fit on the sequence $\mathcal{E}_\nu$ such that $\text{crit}(E)$ is at least as large as the projectum of M^k_ν and $\text{ult}(M^k_\nu, E)$ is iterable, except those which have been used in the iteration. By proposition 2.4, the measures in N are all countably complete, and hence preserve the iterability of M, and the measures on κ can't have been used in the iteration since the critical points are increasing. □

Since the sequence of models M^k_ν can only drop to a mouse finitely often, clause (1) can only fail finitely often, so that the sequence of mice M^k_0 has a last member, $M^{\hat{k}}_0$. We do not know whether it is consistent with ZFC that this construction never drops to a mouse, that is, that $\hat{k} = 0$.

The construction will stop at some $\nu \leq \Omega$, and we will take M to be the final model M_ν. The next lemma implies that in fact the construction does not stop before Ω, so that $M = M_\Omega$. The model M is well founded and iterable, but

Proposition 2.7. *The order type of the ordinals of M is longer than Ω; hence $\Omega \in M$.*

Proof. If the proposition is false then either $M = N$ or M is an initial segment of N. Now if $\hat{k} > 0$ then M^k_0 is a mouse and there is a subset x of $\kappa = \text{proj}(M^k_0)$ which is definable in M^k_0 but is not a member of N. Then x is definable in M but since N satisfies ZF x is not definable in N, so M must not be equal to N or an initial segment of N. Thus we can assume that $\hat{k} = 0$. In that case M cannot be a proper initial segment of N, since M is an iterated ultrapower of K which contains all the ordinals. Thus we must have $M = N$.

The model N was defined in $L(a, b)$ from the parameter ω^V_1, so if π is any embedding of $L(a, b)$ into itself such that $\pi(\omega^V_1) = \omega^V_1$ then $\pi{\restriction}N \colon N \to N$. Now let π be an embedding which is not the identity such that $\pi(\omega^V_1) = \omega^V_1$ Since $M = N$, the map $\pi{\restriction}N$ takes M into M. Now consider the embedding $\pi \cdot j^0_{0,\Omega} \colon K = M^0_0 \to M \to M$. As stated following Corollary 1.7 the covering lemma implies that any embedding of K into a well founded model is an iterated ultrapower of K. In particular $\pi \cdot j^0_{0,\Omega}$ is an iterated ultrapower, but there can only be the one iterated ultrapower $j^0_{0,\Omega}$ from K to M with increasing critical points, since any such iterated ultrapower is determined by the sequence of measures in the model M. Thus $j^0_{0,\Omega} = \pi \cdot j^0_{0,\Omega}$. Now the existence of $(a,b)^\sharp$ implies that we can choose the embedding π so that for some $c \in I$ we have $\pi(c) > j^0_{0,\Omega}(c)$. Then $\pi\big(j^0_{0,\Omega}(c)\big) \geq \pi(c) > j^0_{0,\Omega}(c)$, so that $\pi \cdot j^0_{0,\Omega}$ cannot be equal to $j^0_{0,\Omega}$. □

We are interested almost exclusively in j^k, and it will be useful later to restrict our attention to a tail of that map. Let n^k_ν be the least integer n such that the critical point of $j^{\hat{k}}_{\nu,\nu+1}$ is at least as large as the Σ_n projectum of M^k_ν. Then n^k_ν is nonincreasing with ν, and hence is eventually constant. Let $\hat{\nu}$ be least ordinal such that $\Omega \in j^{\hat{k}}_{\hat{\nu},\Omega}$"$M^{\hat{k}}_{\hat{\nu}}$, and if $\hat{k} > 0$ then n^k_ν is constant for $\nu \geq \hat{\nu}$. Then we write

$$M_0 = M^{\hat{k}}_{\hat{\nu}}, \qquad M_\nu = M^{\hat{k}}_{\hat{\nu}+\nu}, \qquad M = M_\Omega = M^{\hat{k}}_\Omega$$
$$j_{0,\nu} = j^{\hat{k}}_{\hat{\nu},\hat{\nu}+\nu} \colon M_0 \to M_\nu$$
$$j = j_{0,\Omega} \colon M_0 \to M$$

Embeddings of M. As we have already observed in the proof of proposition 2.7, the existence of $(a,b)^\sharp$ implies that the elementary embeddings from $L(a,b)$ into $L(a,b)$ are the same as the order preserving maps $\pi\colon I \to I$ on the indiscernibles I of $L(a,b)$. We now want to extend such maps π to maps $\pi^*\colon M \to M$. Again, we observe that N is defined in $L(a,b)$ from the parameter ω^V_1 and hence if $\pi(\omega^V_1) = \omega^V_1$ then $\pi{\restriction}N\colon N \to N$.

Lemma 2.8. *There is an ordinal ρ such that for any embedding π: $L(a,b) \to L(a,b)$ with $\pi{\restriction}\rho = \mathrm{id}$ there is an embedding $\pi^*\colon M \to M$ such that $\pi^*{\restriction}(j$"$M_0) = \mathrm{id}$ and $\pi^*{\restriction}\Omega = \pi{\restriction}\Omega$.*

Proof. If such an embedding π^* exists, then it is clear what it must be. Every member z of M can be written in the form $z = j(f)(\alpha)$ where $f \in M_0$ and $\alpha \in \Omega$. Since $\pi^*(j(f)) = j(f)$ and $\pi^*(\alpha) = \pi(\alpha)$ we must have $\pi^*(z) = \pi^*(j(f)(\alpha)) = j(f)(\pi^*(\alpha)) = j(f)(\pi(\alpha))$. Now we must prove that this definition works. Note that if $x \in \mathcal{P}^M(\Omega)$ then, regardless of whether the general definition works we can write $\pi^*(x) = \bigcup\{\,\pi(x \cap \xi) : \xi \in \Omega\,\}$.

Claim. *For each $x \in \mathcal{P}^M(\Omega)$ there is $\rho_x \in \Omega$ such that $\pi^*(x) = x$ whenever $\pi{\restriction}\rho_x = \mathrm{id}$.*

Proof. Since $M \cap V_\Omega = N$, the initial segments $x \cap \xi$ of x are members of $L(a,b)$ and thus can be written in the form $x \cap \xi = g_\xi(\mathbf{c}_\xi) \cap \xi$ where $g_\xi\colon [\Omega]^{m_\xi} \to N$ is definable in $L(a,b)$ without parameters and $\mathbf{c}_\xi \in [I]^{m_\xi}$ for some $m_\xi \in \omega$. Let $n_\xi \leq m_\xi$ be the largest integer n such that $\mathbf{c}_\xi{\restriction}n \subset \xi$. By Fodor's theorem we can find a stationary subclass I' of I such that $g_\xi = g$, $n_\xi = n$, and $\mathbf{c}{\restriction}n_\xi = \mathbf{d}$ are constant on I'.

Set $\rho_x = \max(\mathbf{d}) + 1$ and suppose that $\pi{\restriction}\rho_x = \mathrm{id}$. We will show that $\pi^*(x) = x$. If not then pick $\xi \in I'$ such that $\pi(x \cap \xi) \neq x \cap \pi(\xi)$. First suppose that $\pi(\xi) = \xi' \in I'$. Then $\pi(x \cap \xi) = \pi(g(\mathbf{c}_\xi) \cap \xi) = g(\pi(\mathbf{c}_\xi)) \cap \xi'$, but since $\pi(\mathbf{c}_\xi){\restriction}n = \mathbf{c}_\xi{\restriction}n = \mathbf{d} = \mathbf{c}_{\xi'}{\restriction}n$ this is equal to $g(\mathbf{c}_{\xi'}) \cap \xi' = x \cap \xi'$, contrary to assumption.

Now if $\xi' \notin I'$ then pick π' so that $\pi'{\restriction}\xi' = \mathrm{id}$ and $\xi'' = \pi'(\xi') \in I'$. Then $(\pi' \cdot \pi)(x) \cap \xi'' = x \cap \xi''$ by the last paragraph, so $\pi(x) \cap \xi' = (\pi' \cdot \pi)(x) \cap \xi' = (x \cap \xi'') \cap \xi' = x \cap \xi'$. □

Now let $\rho = \sup\{\, \rho_x : x \in \mathcal{P}(\Omega) \cap j``M_0 \,\}$. If $\chi = j^{-1}(\Omega) \in M_0$ then $|\mathcal{P}(\Omega) \cap j``M_0| = |\mathcal{P}(\chi) \cap M_0| < \Omega$, so $\rho < \Omega$. Now we will show that if $\pi{\restriction}\rho = \mathrm{id}$ then the map π^* defined by the equation $\pi^*(j(f)(a)) = j(f)(\pi(a))$ is well defined and one to one and preserves fine structure. If $z_0 = j(f_0)(\xi_0)$ and $z_1 = j(f_1)(\xi_1)$ then we have

$$\begin{aligned}
\pi^*(z_0) = \pi^*(z_1) &\iff \pi^*(j(f_0)(\xi_0)) = \pi^*(j(f_1)(\xi_1)) \\
&\iff j(f_0)(\pi(\xi_0)) = j(f_1)(\pi(\xi_1)) \\
&\iff \pi(\xi_0, \xi_1) \in x = \{\, (\nu, \nu') : j(f_0)(\nu) = j(f_1)(\nu') \,\}. \qquad (1)
\end{aligned}$$

Now $x = j(\{\, (\nu, \nu') : f_0(\nu) = f_1(\nu') \,\})$ so $\rho \geq \rho_x$ and hence $\pi(x) = x$ and (1) is equivalent to $\pi(\xi_0, \xi_1) \in \pi(x)$ and hence to $(\xi_0, \xi_1) \in x$, that is, to $z_0 = j(f_0)(\xi_0) = j(f_1)(\xi_1) = z_1$. Thus $\pi^*(z_0) = \pi^*(z_1)$ if and only if $z_0 = z_1$, and a similar argument shows that $\pi^*(z_0) \in \pi^*(z_1)$ if and only if $z_0 \in z_1$.

We now give an equivalent alternate definition of π^* as a generalized ultrapower $\hat{\pi}$. In order to simplify the description we will give the detailed construction without considering fine structure. If $\hat{k} = 0$, so that M is an iterated ultrapower of K, then the construction given in the next paragraph is accurate. Otherwise the construction is properly treated as a mouse ultrapower: Let $n = n^{\hat{k}}_{\hat{\nu}}$, the least integer such that the Σ_n projectum of M is smaller than Ω. Then the construction described is applied to the Σ_{n-1} code of M and since M is an $n-1$ sound premouse the resulting embedding can be extended to a fine structure preserving embedding of all of M.

The embedding $\hat{\pi}\colon M \longrightarrow M'$ is defined by treating π as an extender, setting $M' = \{\, [(f, a)]_\sim : f \in M \text{ and } a \in \Omega \,\}$, where $(f, a) \sim (f', a')$ iff $(a, a') \in \pi(\{\, (\nu, \nu') : f(\nu) = f'(\nu') \,\})$, and setting $\hat{\pi}(z) = [(\mathfrak{c}_z, 0)]_\sim$ where $\mathfrak{c}_z$ is the constant function, $\mathfrak{c}_z(\xi) = z$ for all ξ.

Now we will show that $\hat{\pi}$ is the same as π^*. We can define $k\colon M \longrightarrow M'$ by $k\big(j(f)(a)\big) = [(j(f), a)]_\sim$. It is easy to see that this embedding is well defined. But k is onto, since if $f \in M$ then there is $g \in M_0$ and

$\xi \in \Omega$ such that $f = j(g)(\xi)$ so that if we define the function $\bar{g}$ on ordinals $<\alpha, \beta>$ coding pairs (α, β) of ordinals by setting $\bar{g}(\langle \nu, \nu' \rangle) = g(\nu)(\nu')$ then $[(f, a)]_{\sim} = [(j(\bar{g}), \langle \pi(\xi), a \rangle)]_{\sim} = k\big((j(\bar{g})(\langle \pi(\xi), a \rangle)\big)$. Thus $k \colon M \cong M'$. In order to show that $\hat{\pi} = k \cdot \pi^*$ it is enough to show that $\hat{\pi} \restriction \Omega = \pi \restriction \Omega$ and that $\hat{\pi} \restriction j``M_0 = k \restriction j``M_0$. The first is immediate, since $\hat{\pi} \restriction N = \pi \restriction N$. For the second, if we again write $\mathfrak{c}_z$ for the constant function then we have $\hat{\pi}(j(z)) = [(\mathfrak{c}_{j(z)}, 0)]_{\sim} = [(j(\mathfrak{c}_z), 0)]_{\sim} = k\big(j(\mathfrak{c}_z)(0)\big) = k(j(z))$. □

Now fix, for the remainder of the paper, an ordinal ρ which satisfies proposition 2.8 and in addition is at least as large as $|{}^{O(\chi)}\chi \cap M_0|$, where $\chi = j^{-1}(\Omega) \in M_0$.

We will write $\mathcal{C}_\nu$ for the indiscernibles for M generated by $j_{0,\nu}$ and $\mathcal{D}_\nu$ for the indiscernibles for N generated by $i_{0,\nu}$. Let $\mathcal{C}$ be $\mathcal{C}_\Omega$ and let $\mathcal{D}$ be $\mathcal{D}_\Omega$. The following is a general fact about iterated ultrapowers.

Fact 2.9. *For each $\alpha \in M_\nu$ there is $f \in j``M_0$ such that either $\alpha \in f``\alpha$ or there is $\gamma \in f``\alpha$ such that $\alpha \in \mathcal{C}(\gamma)$. In the latter case γ is definable from α, using j, as follows:*

(1) *$\kappa = \operatorname{crit}(\mathcal{E}(\gamma))$ is the smallest ordinal such that $\kappa \geq \alpha$ and there is $g \in j``M_0$ such that $\kappa \in g``\alpha$, and*
(2) *γ is the unique ordinal such that $\mathcal{E}(\gamma)$ is a measure on κ, there is $g \in j``M_0$ such that $\gamma \in g``\alpha$, and for all $h \in j``M_0$ and $x \in h``\alpha$ we have $\alpha \in x$ iff $x \in \mathcal{E}(\gamma)$.*

Proposition 2.10. *$I \setminus \rho$ is a subset of $\mathcal{C}(\gamma_0)$, where $\mathcal{E}(\gamma_0)$ is the order 0 measure on Ω.*

Proof. First, notice that if $c \in I \setminus \rho$ then for all $f \in j``M_0$ we have $f``c \cap \Omega \subset c$. Suppose to the contrary that $\xi < c \leq f(\xi) < \Omega$ and pick $\pi : L(a, b) \to L(a, b)$ so that $\pi \restriction c = \mathrm{id}$ and $\pi(c) > f(\xi)$. Then $\pi^*(f(\xi)) \geq \pi^*(c) = \pi(c)$, which contradicts the fact that $\pi^*(f(\xi)) = f(\pi(\xi)) = f(\xi) < \pi(c)$.

In particular there is no $f \in j``M_0$ such that $c \in f``c$, so $c \in \mathcal{C}(\gamma)$ for some ordinal γ by fact 2.9. Furthermore $\operatorname{crit}(\mathcal{E}(\gamma)) \geq \Omega$, but $\operatorname{crit}(\mathcal{E}(\gamma)) \leq \Omega$ since $\Omega = j(\mathfrak{c}_\chi)(0)$ where $\mathfrak{c}_\chi$ is the constant function. Thus $\operatorname{crit}(\mathcal{C}(\gamma)) = \Omega$. Since c is an $L(a, b)$ indiscernible c is regular in $L(a, b)$ and hence $o^N(c) = 0$ by clause (1) of definition 2.3. Thus $\mathcal{E}(\gamma)$ must be the order 0 measure on Ω. □

3. The Main Lemma

The main lemma below corresponds to the second of the two lemmas we referred to in our summary of the proof of Jensen's result. The connection between lemma 3.1 and Jensen's lemma, which stated that there are at most countably many members of the class C of indiscernibles for M in the interval between any two adjacent members of the class I of Silver indiscernibles for $L(a,b)$, is made by corollary 3.10 at the end of this section, though unfortunately the statement of that corollary is a good deal messier than that of Jensen's lemma. Corollary 3.10 will be used in the next section to show that it is possible to work in $K^{\mathcal{M}}$ and still define an iterated ultrapower which is rich enough that j can be embedded into it.

Main Lemma 3.1. *Let $\nu \in \Omega$ be an arbitrary ordinal such that the critical point κ of $j_{\nu,\Omega}$ is not in I and $\mathrm{cf}(\kappa) \geq \rho^+$, write ν' for the least ordinal such that $\mathrm{crit}(i_{\nu',\Omega}) \geq \kappa$, and suppose that κ is an accumulation point for γ in $\mathcal{C}_\nu$, the sequence of indiscernibles generated by $j_{0,\nu}$. Then $\mathcal{E}_\nu{\restriction}\gamma = \mathcal{F}_{\nu'}{\restriction}\gamma$.*

Proof. Fix an arbitrary ordinal $\lambda < \gamma$ such that $\mathcal{E}_\nu(\lambda)$ is a measure on κ. Since κ is an accumulation point for $\gamma > \lambda$ in $\mathcal{C}_\nu$, the set $C_0 = \bigcup\{\,\mathcal{C}_\nu(\gamma') : \gamma' \geq \lambda\}$ is unbounded in κ and can be used to generate $\mathcal{E}_\nu(\lambda)$. In the lemma 3.7 below we will find a set D in $L(a,b)$ such that $D \cap C_0$ is unbounded in κ and D generates one of the measures $\mathcal{F}_{\nu'}(\eta)$ in exactly the same way that C_0 generated $\mathcal{E}_\nu(\lambda)$. Then $\mathcal{E}_\nu(\lambda) = \mathcal{F}_{\nu'}(\eta)$ since $D \cap C_0$ is unbounded, and it follows that $\lambda = \eta$ by coherence. Since λ was arbitrary this will complete the proof of the main lemma.

In the course of the proof we will use the symbol C to denote an unbounded subset of C_0. At various places we will put conditions on members of C which have the effect of making C smaller, so that at the end we will have $C \subset D$.

We will be using some ideas from [Mi91b], beginning with the following definition.

Definition 3.2. A set $C \subset \kappa$ is a *set of indiscernibles* in κ over a model M with a sequence $\mathcal{F}$ of measures if there is an *assignment* for C, that is, a function $\beta\colon \kappa \to O(\kappa)$ such that for all functions $f \in M$ there is a $\delta < \kappa$ such that for all $\alpha \in C \setminus \delta$ and all $x \in \mathcal{P}(\kappa) \cap f``\alpha$ we have $\alpha \in x \iff x \in \mathcal{F}(\beta(\alpha))$.

Proposition 3.3. *The set C_0 is a set of indiscernibles in κ for M_ν, with the assignment defined by $\beta(\alpha) = \eta$ if and only if $\mathcal{E}_\nu(\eta)$ is a measure on κ and $\alpha \in \mathcal{C}_\nu(\eta)$.*

It is proved in [Mi91b] that if C is a set of indiscernibles in κ for the core model K then there is a function $h \in K$ so that for all $\alpha \in C$ we have $\beta(\alpha) \in h``\alpha$. This clearly need not be true for C_0, but the next proposition says that it is true for a cofinal subset of C_0. It is the only place in which we use the assumption that $\mathrm{cf}(\kappa) \geq \rho^+$.

Proposition 3.4. *There is a function $h \in j_{0,\nu}$“M_0 such that $C = \{\alpha \in C_0 : \beta(\alpha) \in h“\alpha\}$ is cofinal in κ.*

Proof. By Fact 2.9, for each $\alpha \in C_0$ there is $h_\alpha \in j_{0,\nu}$“M_0 such that $\beta(\alpha) \in h_\alpha“\alpha$. Since $\mathrm{cf}(\kappa) \geq \rho^+ > |{}^{O(x)}\chi \cap M_0|$ it follows that there must be a single function $h \in j$“M_0 such that $\{\alpha : h_\alpha = h\}$, and hence $\{\alpha \in C_0 : \beta(\alpha) \in h“\alpha\}$, is cofinal in C_0. □

Let h and C be as given by this proposition, so that κ is still an accumulation point for $\lambda + 1$ in C although it may not be an accumulation point for γ in C. Now if β is an assignment for C then there is a function β^C in M_ν such that $\beta(\alpha) = \beta^C(\alpha)$ for all sufficiently large $\alpha \in C$, and in particular there is essentially only one assignment having such a function h. To see this, define $x_h(\gamma)$ for $\gamma < \kappa$ to be the least set x in the ordering of M_ν such that

$$\forall\eta \in h“(\gamma+1)\,\big(x \in \mathcal{E}_\nu(\eta) \iff \eta = h(\gamma)\big).$$

Then for each sufficiently large $\alpha \in C$ the ordinal $\gamma = \beta(\alpha)$ satisfies

$$\forall\xi < \alpha\,\big(\alpha \in x_h(\xi) \iff x_h(\xi) \in \mathcal{E}_\nu(\gamma)\big). \tag{1}$$

Furthermore there can be only one ordinal $\gamma \in h“\alpha$ satisfying formula (1), since if $\xi < \xi' < \alpha$ and $h(\xi) \neq h(\xi')$ then $x_h(\xi') \in \mathcal{E}_\nu(\xi') \setminus \mathcal{E}_\nu(\xi)$, so that the right hand side of formula (1) differs for $\gamma = h(\xi)$ and $\gamma' = h(\xi')$ at $x_h(\xi')$ while the left hand side does not involve γ. Hence formula (1) can be used to define, in M_ν, a function β^C such that $\beta^C(\alpha) = \beta(\alpha)$ for all sufficiently large $\alpha \in C$.

Definition 3.5. We say that a set D *generates a measure* U in M via a function g if $\alpha < g(\alpha) \leq O(\alpha)$ for all $\alpha \in D$ and U is the filter of sets $x \subset \kappa$ such that for all sufficiently large ordinals $\alpha \in D$.

$$\begin{aligned} \alpha &\in x && \text{if } g(\alpha) = O(\alpha) \\ x \cap \alpha &\in \mathcal{E}(g(\alpha)) && \text{if } g(\alpha) < O(\alpha). \end{aligned}$$

In particular, the set C generates $\mathcal{E}_\nu(\lambda)$ in M_ν via the function g defined by

$$g(\alpha) = \begin{cases} O(\alpha) & \text{if } \lambda = \beta^C(\alpha) \\ \mathfrak{C}(\lambda, \beta^C(\alpha))(\alpha) & \text{if } \lambda < \beta^C(\alpha). \end{cases} \tag{$*$}$$

To see this, note that if $\alpha \in C$ and $\beta^C(\alpha) = \lambda$ then $\alpha \in \mathcal{C}(\lambda)$ and hence for sufficiently large α we have $\alpha \in x$ if and only if $x \in \mathcal{E}_\nu(\lambda)$; while if $\alpha \in C$ and $\beta^C(\alpha) > \lambda$ then $x \in \mathcal{E}_\nu(\lambda)$ if and only if $\{\, \xi \in \kappa : x \cap \xi \in \mathcal{E}_\nu(\mathfrak{C}(\lambda, \beta^C(\xi))(\xi) \,\} \in \mathcal{E}_\nu(\beta^C(\alpha))$, and for sufficiently large $\alpha \in C$ this is equivalent to $x \cap \alpha \in \mathcal{E}(g(\alpha))$ since $\mathcal{E}{\restriction}\lambda = \mathcal{E}_\nu{\restriction}\lambda$. This function g is in M_ν because β^C is in M_ν, and since the models $N_{\nu'}$ and M_ν have the same subsets of κ it follows that the function g is a member of $N_{\nu'}$. We now switch to working in $N_{\nu'}$ with the aim of using g there. Let c be the largest member of I below κ. This exists since by the hypothesis κ is not a member of I.

Proposition 3.6. *There is a set $X \in L(a,b)$ such that $|X|^{L(a,b)} = c$ and $X \cap C$ is cofinal in κ.*

Proof. Every member of C may be written in the form $g_n(\alpha, c, c_1, \dots, c_n)$ where $n \in \omega$, g_n is the universal Σ_1 function on $n+2$ variables in $L(a,b)$, $\alpha < c$, and $(c_1, \dots, c_n)$ are the first n members of I above c. Since $\mathrm{cf}(\kappa) > \omega$ there is an unbounded subset of C on which n is constant, so that if we set $X = \{\, g_n(\alpha, c, c_1, \dots, c_n) : \alpha \in c \,\}$ then $X \cap C$ is cofinal in κ. □

We can assume *wlog* that $C \subset X$. Now we have to find a set D of indiscernibles in $L(a,b)$ which contains a cofinal subset of C. The next lemma abstracts the properties we need for this set.

Lemma 3.7. *The following is true in $L(a,b)$, where κ and ν' are as in the main lemma, X is given by proposition 3.6, and g is defined by $(*)$ above. There is a set D of indiscernibles in κ for $N_{\nu'}$ such that the assignment β^D for D is in $N_{\nu'}$ and there is a function $h \in N_{\nu'}$ such that for sufficiently large α in X*

$$\begin{aligned} \beta^D(\alpha) &\in h``\alpha && \text{if } \alpha \in D \\ h``\alpha \cap (\kappa \setminus \alpha) &\neq \varnothing && \text{if } \alpha \notin D. \end{aligned}$$

We will defer the proof of lemma 3.7 until after we have finished the proof of the main lemma. The first part of the proof works with any function h satisfying the conditions of lemma 3.7, but in the course of the proof we will choose h to also satisfy a further closure property.

Claim. *$C \setminus D$ is bounded in κ.*

Proof. Let $\eta < \nu$ be such that h is in the range of $j_{\eta,\nu}$ and suppose $\alpha = \mathrm{crit}(j_{\eta,\nu})$ is in $C \setminus D$ and is large enough that lemma 3.7 applies. Since $C \subset X$ it follows that there is $\beta < \alpha$ such that $h(\beta) \in \kappa \setminus \alpha$, but this is impossible because $h(\beta) \in \mathrm{range}(j_{\eta,\nu})$ while $j_{\eta,\nu}(\alpha) = \kappa$. □

Thus we can assume *wlog* that $C \subset D$, and in fact since any subset of D also satisfies lemma 3.7 we can assume that $C = D$. We will continue to use β^C and β^D for the assignment functions for C and D, respectively, since these functions also depend on the sequence $\mathcal{E}_\nu$ and $\mathcal{F}_{\nu'}$ and hence need not be the same.

This proof breaks down into two cases, depending on whether the set of ordinals α such that $\beta^C(\alpha) = \lambda$ is unbounded in C. We will begin with the easier case, that in which $\beta^C(\alpha)$ is equal to λ on an unbounded subset of C. In this case we can assume *wlog* that $\beta^C(\alpha) = \lambda$ for all $\alpha \in C$. Define a function q by setting $q(\alpha)$ equal to the least ordinal η, if there is one, such that $\beta^D(\alpha) = h(\eta)$. Then q is a member of $N_{\nu'}$ since h and β^D are members of $N_{\nu'}$, and $q(\alpha)$ is defined and $q(\alpha) < \alpha$ for all members of $D = C$. Now q is in M_ν since it is in $N_{\nu'}$, and since $C = C(\lambda)$ it follows that $\{\alpha : q(\alpha) < \alpha\} \in \mathcal{E}_{\nu'}(\lambda)$ and hence there is an ordinal $\xi < \kappa$ such that $\{\alpha < \kappa : q(\alpha) = \xi\} \in \mathcal{E}_{\nu'}(\lambda)$. Hence $q(\alpha) = \xi$ for all sufficiently large $\alpha \in C = D$, so $\beta^D(\alpha) = h(\xi)$ for all sufficiently large $\alpha \in D$. Then both $\mathcal{F}_{\nu'}(h(\xi))$ and $\mathcal{E}_\nu(\lambda)$ are generated by C and hence $\mathcal{F}_{\nu'}(h(\xi)) = \mathcal{E}_\nu(\lambda)$. This implies that $h(\xi) = \lambda$, so that $\mathcal{F}_{\nu'}(\lambda) = \mathcal{E}_\nu(\lambda)$ as was to be shown, completing the proof of the first case.

Thus we can assume *wlog* that $\beta^C(\alpha) > \lambda$ for all $\alpha \in C$, so that $\mathcal{E}_\nu(\lambda)$ is the set of subsets x of κ such that $x \cap \alpha \in \mathcal{E}_\nu(g(\alpha))$ for all sufficiently large $\alpha \in C$. Then $g(\alpha) < O^{\mathcal{E}_\nu}(\alpha) = O^{\mathcal{F}_{\nu'}}(\alpha)$ for all sufficiently large $\alpha \in C$. We will begin this second case by enhancing the function h from lemma 3.7.

Claim. *There is a function $h \in N_{\nu'}$ which satisfies clauses* (1) *and* (2) *of lemma 3.7 such that for all $\alpha \in C$ there is an ordinal $\eta \in h``\alpha$ such that $g(\alpha) = \mathfrak{C}^{\mathcal{F}_{\nu'}}(\eta, \beta^D(\alpha))(\alpha)$.*

Proof. Let h_1 be any function satisfying clauses (1) and (2) of lemma 3.7 and set $\eta_\alpha = [g]_{\mathcal{F}_{\nu'}(\beta^D(\alpha))}$. Since $\beta^D(\alpha)$ is in $h_1``\alpha$ and $\eta_\alpha$ is definable from $\beta^D(\alpha)$ together with finitely many parameters which do not depend on $\alpha$ there is a function $h_2 \in N_{\nu'}$ such that $\eta_\alpha \in h_2``\alpha$ for all $\alpha \in D$. Now $g(\alpha) < O(\alpha)$, that is, $\alpha \in \{\nu : g(\nu) < O(\nu)\}$, for all $\alpha \in D$. It follows that for all sufficiently large $\alpha \in D$ the set $\{\nu : g(\nu) < O(\nu)\}$ is in $\mathcal{F}_{\nu'}(\beta^D(\alpha))$, so $\eta_\alpha < \beta^D(\alpha)$ and thus the coherence function $\mathfrak{C}^{\mathcal{F}_{\nu'}}(\eta_\alpha, \beta^D(\alpha))$ exists. By the definition of the coherence function we have $[g]_{\mathcal{F}_{\nu'}(\beta^D(\alpha))} = \eta_\alpha = [\mathfrak{C}^{\mathcal{F}_{\nu'}}(\eta_\alpha, \beta^D(\alpha))]_{\mathcal{F}_{\nu'}(\beta^D(\alpha))}$. Thus $B_\alpha = \{\xi < \kappa : g(\xi) = \mathfrak{C}^{\mathcal{F}_{\nu'}}(\eta_\alpha, \beta^D(\alpha))(\xi)\}$ is in $\mathcal{F}_{\nu'}(\beta^D(\alpha))$. Now B_α is definable from η_α and $\beta^D(\alpha)$ together with parameters which do not depend on α, so there is a function $h_3 \in N_{\nu'}$ such that $B_\alpha \in h_3``\alpha$ for all sufficiently large $\alpha \in D$. It follows that $\alpha \in B_\alpha$, that is, that $g(\alpha) = \eta_\alpha$, for all sufficiently large

$\alpha \in D$. Then any function h in $N_{\nu'}$ such that $h``\nu = h_1``\nu \cup h_2``\nu$ for limit ordinals ν will satisfy the conditions of the claim. □

Now define the function q in $N_{\nu'}$ by setting $q(\alpha)$ equal to the least ordinal η (if one exists) such that $g(\alpha) = \mathfrak{C}^{\mathcal{F}_{\nu'}}(h(\eta), \gamma^D(\alpha))$. By the claim, $q(\alpha)$ does exist and $q(\alpha) < \alpha$ for all ordinals $\alpha \in D = C$. We will show that $q(\alpha)$ is constant for all sufficiently large $\alpha \in C$. Define x_η, for $\eta < \kappa$, to be the least set $x \subset \kappa$ in the order of construction of $L(\mathcal{F}_{\nu'})$ such that $x \in \mathcal{F}_{\nu'}(h(\eta))$ but $x \notin \mathcal{F}_{\nu'}(h(\eta'))$ for any $\eta' \in \kappa$ such that $h(\eta') \neq h(\eta)$. Now $x_{q(\alpha)}$ is defined in $N_{\nu'}$ from parameters h and $q(\alpha) < \alpha$, so there is a function $k \in N_{\nu'}$ such that $x_{q(\alpha)} \in k``\alpha$. Since $x_{q(\alpha)} \in \mathcal{F}_{\nu'}(h(q(\alpha)))$ it follows that $x_{q(\alpha)} \cap \alpha \in \mathcal{F}_{\nu'}\big(\mathfrak{C}^{\mathcal{F}_{\nu'}}\big(h(q(\alpha)), \beta^D(\alpha)\big)(\alpha)\big) = \mathcal{F}_{\nu'}(g(\alpha)) = \mathcal{E}(g(\alpha))$. Now k is in M_ν since it is in $N_{\nu'}$, so $x_{q(\alpha)} \in \mathcal{E}(\lambda)$ for all sufficiently large α in C and (again using $x_{q(\alpha)} \in k``\alpha$) it follows that

$$x_{q(\alpha)} \cap \alpha' \in \mathcal{E}(g(\alpha')) = \mathcal{F}_{\nu'}(\mathfrak{C}^{\mathcal{F}_{\nu'}}(h(q(\alpha')), \beta^D(\alpha'))(\alpha'))$$

for all $\alpha' > \alpha$ in $C = D$. This implies that $x_{q(\alpha)} \in \mathcal{F}_{\nu'}(h(q(\alpha')))$ and hence $h(q(\alpha')) = h(q(\alpha))$, so $q(\alpha') = q(\alpha)$. Thus there is an ordinal ξ such that $q(\alpha) = \xi$ for every sufficiently large $\alpha \in D$. Then, as in the first case, we get that $\mathcal{F}_{\nu'}(h(\xi))$ is equal to the set of subsets x of κ such that $x \cap \alpha \in \mathcal{E}(g(\alpha))$ for every sufficiently large $\alpha \in C$, so that $\mathcal{F}_{\nu'}(h(\xi)) = \mathcal{E}_\nu(\lambda)$. Thus $h(\xi) = \lambda$ and $\mathcal{F}_{\nu'}(\lambda) = \mathcal{E}(\lambda)$, as was to be shown. □

This completes the proof of the lemma 3.1 assuming lemma 3.7.

Proof of lemma 3.7. With the exception of one step the proof of this lemma takes place entirely inside $L(a, b)$, and all calculations are carried out in $L(a, b)$ unless otherwise noted. We will have two cases, depending on whether range$(i_{0,\nu'})$ is cofinal in κ. In the simpler of the two cases, that in which range$(i_{0,\nu'})$ is not cofinal in κ, the required set D is taken from $\mathcal{D}_{\nu'}$ just as C was taken from $\mathcal{C}_\nu$ and the required version of the covering lemma is a simple modification of the basic lemma from [Mi]. In the case in which range$(i_{0,\nu'})$ is cofinal in κ the set D of indiscernibles comes out of the covering lemma itself and hence a stronger form of the covering lemma will be necessary. In neither case is it necessary to know anything of the proof of the covering lemma, or anything of core model techniques beyond those which have already been used in this paper.

Suppose first that range$(i_{0,\nu'}) \cap \kappa \subset \kappa' < \kappa$. The set of indiscernibles in this case is the set $D = \bigcup\{\, \mathcal{D}_{\nu'}(\eta) : \mathrm{crit}(\mathcal{F}(\eta)) = \kappa \,\}$. Then D is a set of indiscernibles, with $\beta^D(\alpha)$ equal to the ordinal η such that $\alpha \in \mathcal{D}_{\nu'}(\eta)$. Now for each $\alpha \in X$ there is $h_\alpha \in N_0$ such that either $\alpha \in i_{0,\nu'}(h_\alpha)``\alpha$

or $\alpha \in \mathcal{D}_{\nu'}(\eta)$ for some $\eta \in i_{0,\nu'}(h_\alpha)``\alpha$. We will use the covering lemma to show that there is an ordinal $\delta < \kappa$ and a function $\sigma \in N_{\nu'}$ such that $\{ i_{0,\nu'}(h_\alpha) : \alpha \in X \} \subset \sigma``\delta$. Then the function $h(<\alpha,\xi>) = \sigma(\alpha)(\xi)$ satisfies the requirements of lemma 3.7.

Recall that c is the largest member of I below κ and set $c^* = c^{+K^a}$. We will first show that $c^* = c^{+L(a,b)}$. Suppose the contrary, so that $|c^*| = c$ in $L(a,b)$. This is the point at which we have to move out of $L(a,b)$. Since $c \in I$ there is an elementary embedding $\pi\colon L(a,b) \to L(a,b)$ such that $\pi\restriction c = \text{id}$ and $\pi(c) > c$. Let $U = \{ x \subset c : c \in \pi(x) \}$. Then $U \cap K^a \in L(a,b)$ since $|\mathcal{P}(c) \cap K^a|^{L(a,b)} = |(c^+)^{K^a}|^{L(a,b)} = c$, and by the maximality of the core model it follows that there is an ordinal γ such that $U \cap K^a = \mathcal{F}_0(\gamma)$, where $K^a = L(\mathcal{F}_0)$. Now $\pi\restriction K^a\colon K^a \to K^a$, and we can define an elementary embedding $k\colon \text{ult}(K^a, U) \to K^a$ by setting $k([f]_U) = \pi(f)(c)$, so that $k \cdot i^U = \pi$. Then $k\restriction(c+1)$ is the identity, but if $\xi = O^{\text{ult}(K^a,U)}(c)$ then $k(\xi) > \xi$ and hence $\xi \geq (c^{++})^{\text{ult}(K^a,U)}$. This contradicts the fact that $O^{K^a}(c) = k(\xi) < (c^{++})^{K^a}$.

Let η be the least ordinal such that $i_{0,\nu'}(\eta) \geq \kappa$, so that $c \leq \eta < \kappa$ since $c \in I$ implies that $i_{0,\Omega}``c \subset c$. Then each h_α is a function from η into $O(\eta)$, and since $K^a = N_0 \models O(\eta) < \eta^{++}$ it follows that $K^a \models O(\eta)^\eta = \eta^+$. We will show that $\eta^* = (\eta^+)^{K^a}$ has cofinality greater than c. Since X has cardinality c it follows that there is a function $\sigma' \in N_0$ such that $h_\alpha \in \sigma'``(\eta)$ for all $\alpha \in X$ and so we can take $\delta = \sup(i_{0,\nu'}``\eta) < \kappa$ and $\sigma = i_{0,\nu'}(\sigma')\restriction\delta$.

Suppose that $\text{cf}(\eta^*) < c$ in $L(a,b)$. If $\eta = c$ then $\eta^* = c^* = c^+$, so we must have $\eta \geq c^+$. Now we use

Covering Lemma I 3.8. ([Mi]) *If α is any successor cardinal of K then $(\text{cf}(\alpha))^\omega \geq |\alpha|$.*

In [Mi] this was stated for the special case in which α is the successor in K of a singular strong limit cardinal μ, and was used to show that in this case α is still the successor of μ in V. In our case we apply the lemma inside $L(a,b)$, so that the core model is K^a, and we take $\alpha = \eta^*$. Suppose that $\text{cf}(\eta^*) \leq c$. Then the lemma implies that $|\eta^*| = |\,\text{cf}^{K^a}(\eta^*)| \leq (\text{cf}(\eta^*))^\omega \leq c^\omega = c < c^+ \leq |\eta^*|$ This contradiction shows that $\text{cf}(\eta^*) > c$ and this completes the proof of the first case.

Now suppose that $\text{range}(i_{0,\nu'})$ is cofinal in κ. Then there are no indiscernibles $\mathcal{D}_\nu(\eta)$ for measures η on κ, so we take the indiscernibles from the following version of the covering lemma instead.

Covering Lemma II 3.9. ([Mi], [Mi91a], [Mi91b]) *Suppose κ is an ordinal and y is a set such that $|y|^\omega < |\kappa|$. Then there is a set D of indiscernibles*

and a function $h \in K$ such that

(1) *If $\alpha \in (y \cap \kappa) \setminus D$ then $h``\alpha \cap (\kappa \setminus \alpha) \neq \varnothing$, and if $\alpha \in D$ then $\beta^D(\alpha) \in h``\alpha$.*

(2) *If $w \in y$ and $w \subset \xi \leq \kappa$ then there is $w' \in h``\xi$ such that $w = w' \cap \xi$.*

(3) *If $g \in y$ and $g\colon \alpha \to O(\alpha)$ for some $\alpha \in D$ then there is a function $g' \in h``\alpha$ such that for all $\xi < \alpha$ we have $g(\xi) = \mathfrak{C}(g'(\xi), \beta(\alpha))(\alpha)$.*

In applying lemma 3.9 we have the difficulty that it only gives indiscernibles for the sequence of measures in the core model. Since we will be working in $L(a,b)$ the core model will be K^a and the sequence of measures will be $\mathcal{F}_0$. What we need instead are indiscernibles for the sequence $\mathcal{F}_{\nu'}$ of measures in the iterated ultrapower $N_{\nu'}$ of K^a. We will begin by using the set X to define a set X' of the same cardinality which will be used as the set y of the lemma. Applying the lemma for $K^a = N_0$ will give a set D' of indiscernibles and a function h' in N_0, and we will then use these to define the required set D of indiscernibles and function $h \in N_{\nu'}$.

First note that $\kappa \in \operatorname{range} i_{0,\nu'}$ since otherwise the facts that $\operatorname{range} i_{0,\nu'}$ is confinal in κ and κ is regular in $N_{\nu'}$ would imply that there is an ordinal $\nu'' < \nu'$ such that $\kappa = \operatorname{crit}(i_{\nu'',\nu'})$, contradicting the minimality of ν'. Set $\kappa' = i^{-1}_{0,\nu'}(\kappa)$. Now X' is defined as follows:

- For each ordinal α in X the least ordinal α' such that $i_{0,\nu'}(\alpha') \geq \alpha$ is a member of X'.
- If $\alpha < i_{0,\nu'}(\alpha')$ and $\alpha \in \mathcal{D}_{\nu'}(\eta)$ for some measure $\mathcal{F}_{\nu'}(\eta)$ on $i_{0,\nu'}(\alpha')$ then there is a function $f\colon \alpha' \to O(\alpha')$ in X' such that $\eta \in i_{0,\nu'}(f)``\alpha$.
- If $\alpha < i_{0,\nu'}(\alpha')$ and there is no such η as in the last clause then there is $f\colon \alpha' \to \alpha'$ in X' such that $i_{0,\nu'}(f)``\alpha \not\subset \alpha$.

To see that the function f required in the third clause always exists, note that if $\alpha \notin \mathcal{D}_{\nu'}(\eta)$ for any ordinal η then f can be chosen so that $\alpha \in i_{0,\nu'}(f)``\alpha$, while if $\alpha \in \mathcal{D}_{\nu'}(\eta)$ for some ordinal $\eta$ then there is a function $f$ and an $\eta \in i_{0,\nu'}(f)``\alpha$ such that $\alpha \in \mathcal{D}_{\nu'}(\eta)$. If the critical point of $\mathcal{F}_{\nu'}(\eta)$ is equal to $i_{0,\nu'}(\alpha')$ then the second clause holds for α. If it is not equal to $i_{0,\nu'}(\alpha')$ then it must be smaller than $i_{0,\nu'}(\alpha')$ so that we can take $f' : \alpha' \to \alpha'$ and $\eta \in i_{0,\nu'}(f)``\alpha \setminus \alpha$, satisfying the third clause.

Now we can apply lemma 3.9 in $L(a,b)$, using X' and κ' for y and κ, since $i_{0,\nu'}``c \subset c$ implies $\kappa' > c$ and $|X'|^\omega = c < c^+ \leq \kappa'$. This yields a set D' of indiscernibles for κ' over K^a and a function $h' \in K^a$ satisfying the conditions of the lemma. Now define $D \subset \kappa$ to be

$$D = i_{0,\nu'}``(D') \cup \bigcup\{\, \mathcal{D}_{\nu'}(\eta) : \operatorname{crit}(\mathcal{F}_{\nu'}(\eta)) \in i_{0,\nu'}``(D')\,\}.$$

Now we will show that D satisfies the lemma. As before, let α be any ordinal less then κ and let α' be least such that $i_{0,\nu'}(\alpha') \geq \alpha$. First we will deal with the ordinals α in $X \setminus D$. For these we need only show that there is a function $h \in N_{\nu'}$ such that $h``\alpha \cap (\kappa \setminus \alpha) \neq \varnothing$ for all $\alpha \in X \setminus D$. First suppose that $\alpha' \notin D'$. Then $h'``(\alpha') \cap (\kappa' \setminus \alpha') \neq \varnothing$ and since $i_{0,\nu'}``\alpha' \subset \alpha$ it follows that if we set $h_1 = i_{0,\nu'}(h')$ then $h_1``\alpha \cap (\kappa \setminus \alpha) \neq \varnothing$. Thus we can assume for the rest of the proof that $\alpha' \in D'$.

Since $\alpha \notin D$ it follows that $\alpha \notin \mathcal{D}_{\nu'}(\eta)$ for any measure $\mathcal{F}_{\nu'}(\eta)$ on $i_{0,\nu'}(\alpha')$, so that there is $f\colon \alpha' \to \alpha'$ in X' such that $i_{0,\nu'}(f)``\alpha \not\subset \alpha$. By lemma 3.9 there is $f' \in h'``(\alpha')$ such that $f = f' \restriction \alpha'$, so that $i_{0,\nu'}(f')``\alpha \cap (\kappa \setminus \alpha) \neq \varnothing$. Then $i_{0,\nu'}(f') \in h_1``\alpha$, so if we set $h_2(\xi, \xi') = h_1(\xi)(\xi')$ then $h_2``(\alpha \times \alpha) \cap (\kappa \setminus \alpha) \neq \varnothing$.

Thus for the rest of the proof we will assume that $\alpha \in D$. We will have to define $\beta^D(\alpha)$, show that it works as an assignment, and prove that $\beta^D(\alpha) \in h``\alpha$. First suppose that $\alpha = i_{0,\nu'}(\alpha')$, in which case set $\beta^D(\alpha) = i_{0,\nu'}(\beta^{D'}(\alpha')) \in h_1``\alpha$. We must show that this function β^D is an assignment on $i_{0,\nu'}``(D')$. If $f\colon \kappa \to \mathcal{P}(\kappa)$ is a function in $N_{\nu'}$ then there is a function $f' \in N_0$ and an ordinal $\delta_0 < \kappa$ such that $f = i_{0,\nu'}(f')(\delta_0)$. There is $\delta_1 < \kappa'$ such that for all $\varepsilon \in D' \setminus \delta_1$

$$\forall \xi, \xi' < \varepsilon \left(\varepsilon \in f'(\xi)(\xi') \iff f'(\xi)(\xi') \in \mathcal{F}_0(\beta^{D'}(\varepsilon))\right).$$

Then whenever $\alpha = i_{0,\nu'}(\alpha') > i_{0,\nu'}(\delta_1)$ we have

$$\forall \xi, \xi' < \alpha \big(\alpha \in i_{0,\nu'}(f')(\xi)(\xi') \iff i_{0,\nu'}(f')(\xi)(\xi') \in \mathcal{F}_{\nu'}(\beta^D(\alpha))\big),$$

and it follows that if $\delta = \max(\delta_0, i_{0,\nu'}(\delta_1))$ then for all α in $i_{0,\nu'}``(D') \setminus \delta$

$$\forall \xi' < \alpha \left(\alpha \in f(\xi') \iff f(\xi') \in \mathcal{F}_{\nu'}(\beta^D(\alpha))\right)$$

and hence β^D is an assignment for $i_{0,\nu'}``(D')$.

Thus we are left with the final case, in which $\alpha' \in D'$ and $\alpha \in \mathcal{D}_{\nu'}(\eta)$ for some ordinal η_α such that $\operatorname{crit}(\mathcal{F}_{\nu'}(\eta_\alpha)) = i_{0,\nu'}(\alpha')$. By the choice of X' there is $f\colon \alpha' \to O(\alpha')$ in X' such that $\eta_\alpha = i_{0,\nu'}(f)(\delta_0)$ for some $\delta_0 \in \alpha$, and by lemma 3.9(3) there is a function $f'\colon \kappa' \to O(\kappa')$ in $h'``(\alpha')$ such that $f(\xi') = \mathfrak{C}(f'(\xi'), \beta^{D'}(\alpha'))(\alpha')$ for all $\xi' < \alpha'$. Then $\eta(\alpha) = \mathfrak{C}(i_{0,\nu'}(f')(\delta_0), \beta^D(\alpha))(\alpha)$. Set $\beta^D(\alpha) = i_{0,\nu'}(f')(\delta_0) \in h_2``(\alpha \times \alpha)$.

Now suppose that $f\colon \kappa \to \mathcal{P}(\kappa)$ in $N_{\nu'}$. For all sufficiently large indiscernibles $\alpha \in D \setminus i_{0,\nu'}``D'$ and for all $\xi < \alpha$

$$\begin{aligned} \alpha \in f(\xi) &\iff f(\xi) \cap i_{0,\nu'}(\alpha') \in \mathcal{F}_{\nu'}(\beta^D(\alpha)) \\ &\iff f(\xi) \cap i_{0,\nu'}(\alpha') \in \mathcal{F}_{\nu'}\big(\mathfrak{C}(\beta^D(\alpha), \beta^{D'}(i_{0,\nu'}(\alpha')))(i_{0,\nu'}(\alpha'))\big). \end{aligned}$$

The last expression is equivalent to

$$(1)\quad \{\chi : f(\xi)\cap\chi \in \mathcal{F}_{\nu'}\big(\mathfrak{C}(\beta^D(\alpha),\beta^D(i_{0,\nu'}(\alpha')))(\chi)\}\in\mathcal{F}(\beta^D(i_{0,\nu'}(\alpha'))$$

since $\beta^D(\alpha)$ and $\beta^D(i_{0,\nu'}(\alpha')) = i_{0,\nu'}(\beta^{D'}(\alpha')$ are each in $h``\alpha'$, and (1) is equivalent to $f(\xi)\in\mathcal{F}_{\nu'}(\beta^D(\alpha))$. This completes the proof that D is a set of indiscernibles and hence the proof of lemma 3.7. □

This completes the proof of lemma 3.7 and hence of the main result of this section, lemma 3.1.

Corollary 3.10. *For any ordinal d*

(1) *If* $\delta = \sup\big(d\cap(I\cup\bigcup_{\gamma'>\gamma}\mathcal{C}(\gamma'))\big)$ *then the order type of* $\mathcal{C}(\gamma)\cap(d\setminus\delta)$ *is at most* $\rho^+ + \omega_1^V$.

(2) *If* $\delta = \sup\big(d\cap(I\cup\bigcup_{\gamma'\geq\gamma}\mathcal{C}(\gamma'))\big)$ *then the set of ordinals* $\alpha\in d\setminus\delta$ *such that* α *is an accumulation point for* γ *in* $\mathcal{C}$ *has order type at most* $\rho^+ + \omega_1^V$.

Proof. Suppose first that the hypothesis to clause (1) holds. If the order type of $\mathcal{C}(\gamma)\cap(d\setminus\delta)$ is not greater than ρ^+ then we are done, so we can assume that the limit κ of the first ρ^+ members of $\mathcal{C}(\gamma)$ above δ is less than d. Then κ is an indiscernible in $\mathcal{C}$, so there is an ordinal ν such that $\kappa = \operatorname{crit}(j_{\nu,\Omega})$. We may assume that there is also an ordinal λ such that $\gamma = j_{\nu,\Omega}(\lambda)$, since otherwise $\mathcal{C}(\gamma)\cap\kappa$ would be empty. The definition of κ implies that $\kappa\notin I$ and κ is an accumulation point for $\lambda+1$ in $\mathcal{C}_\nu$, so if ν' is the least ordinal such that $\operatorname{crit}(i_{\nu',\Omega})>\kappa$ then the main lemma implies that $\mathcal{F}_{\nu'}\restriction(\lambda+1) = \mathcal{E}_\nu\restriction(\lambda+1)$.

Set $\kappa_\alpha = i_{\nu',\nu'+\alpha}(\kappa)$ and let ξ be the least ordinal such that either $\kappa_\xi = i_{\nu',\Omega}(\kappa)$, so that κ_ξ is not an indiscernible in $\mathcal{D}$, or κ_ξ is in $\mathcal{D}(\eta)$ for some $\eta>\gamma' = i_{\nu',\Omega}(\lambda)$. Set $\kappa' = \kappa_\xi$. We will show that

i) $\mathcal{D}(\gamma')\cap(\kappa'\setminus\kappa)$ has order type at most ω_1^V,

ii) $\kappa' = j_{\nu,\nu+\xi}(\kappa)$, and either $\kappa' = j_{\nu,\Omega}(\kappa)$, so that κ' is not an indiscernible in $\mathcal{C}$, or $\kappa'\in\mathcal{C}(\eta)$ for some $\eta>\gamma$, and

iii) $\mathcal{D}(\gamma')\cap(\kappa'\setminus\kappa) = \mathcal{C}(\gamma)\cap(\kappa'\setminus\kappa)$.

The corollary follows easily from this: (ii) implies that either $\kappa'\geq d$ or $\mathcal{E}(\gamma)\subset\kappa'$, and then (i) and (iii) imply that the order type of $\mathcal{C}(\gamma)\cap(d\setminus\delta)$ is at most the sum of the order types of $\mathcal{C}(\gamma)\cap(\kappa\setminus\delta)$ and $\mathcal{D}(\gamma')\cap(\kappa'\setminus\kappa)$, which is at most $\rho^+ + \omega_1^V$.

First we prove clause (i). For $\alpha\leq\xi$ set $\lambda_\alpha = i_{\nu',\nu'+\alpha}(\lambda)$. Take α so that κ_α is the supremum of the first ω_1^V members of $\mathcal{D}(\gamma')$ above κ. If no such α exists or $\kappa_\alpha\geq d$ then we are done, so we can assume that

$\kappa_\alpha < d$. It follows that $\mathcal{D}_{\nu'+\alpha}(\lambda_\alpha)$ is unbounded in κ_α, and by the definition 2.3 of N this implies that $\mathcal{D}_{\nu'+\alpha}(\varepsilon)$ is unbounded in κ_α for every measure $\mathcal{F}_{\nu'+\alpha}(\varepsilon)$ on κ_α with $\varepsilon < \lambda_\alpha$. Furthermore $\mathrm{cf}^{L(a,b)}(\kappa_\alpha) = \omega_1^V$ since $\mathcal{D}_{\nu'+\alpha}(\lambda_\alpha) = \mathcal{D}(\lambda') \cap \kappa_\alpha$ is in $L(a,b)$. By definition 2.3 this implies that $N_{\nu'+\alpha+1} = \mathrm{ult}(N_{\nu'+\alpha}, \mathcal{F}_{\nu'+\alpha}(\eta))$ where $\eta > \lambda_\alpha$, so that either $\kappa_\alpha = i_{\nu',\Omega}(\kappa)$ (if $\mathrm{crit}(\mathcal{F}_{\nu'+\alpha}(\eta)) > \kappa_\alpha$) or $\kappa_\alpha \in \mathcal{D}(i_{\nu',\Omega}(\eta))$ where $i_{\nu',\Omega}(\eta) > \gamma'$. In either case this implies that $\alpha = \xi$, and this completes the proof of clause (i).

Now we show by induction on $\alpha \leq \xi$ that

$$\kappa_\alpha = i_{\nu',\nu'+\alpha}(\kappa) = j_{\nu,\nu+\alpha}(\kappa)$$
$$\mathcal{P}^{M_{\nu+\alpha}}(\kappa_\alpha) = \mathcal{P}^{N_{\nu'+\alpha}}(\kappa_\alpha)$$
$$\mathcal{F}_{\nu'+\alpha}\restriction(\lambda_\alpha + 1) = \mathcal{E}_{\nu_\alpha}\restriction(\lambda_\alpha + 1).$$

For $\alpha = 0$ this has already been shown to be a consequence of the main lemma. If it is true for α then since $\alpha < \xi$ there is $\eta_\alpha < \lambda$ such that $N_{\nu'+\alpha+1} = \mathrm{ult}(N_{\nu'+\alpha}, E_\alpha)$ where $E_\alpha = \mathcal{F}_{\nu'+\alpha}(\eta_\alpha)$. But then $\mathcal{E}_{\nu+\alpha}(\eta_\alpha)$ is also equal to E_α and since $\mathcal{E}_{\nu+\alpha}\restriction\eta_\alpha = \mathcal{F}_{\nu'+\alpha}\restriction\eta_\alpha = \mathcal{F}\restriction\eta_\alpha$ and $\eta_\alpha \notin$ domain $\mathcal{F}$ we also have $M_{\nu+\alpha+1} = \mathrm{ult}(M_{\nu+\alpha}, E_\alpha)$. Thus $M_{\nu+\alpha+1}$ and $N_{\nu'+\alpha+1}$ also match as required.

This implies clause (iii), and that $\kappa' = j_{\nu,\nu+\xi}(\kappa)$. Now we note that $\mathcal{E}_{\nu+\xi}\restriction(\lambda_\xi + 1) = \mathcal{F}_{\nu'+\xi}\restriction(\lambda_\xi + 1) = \mathcal{F}\restriction(\lambda_\xi + 1)$. Thus $M_{\nu+\xi+1}$ must be an ultrapower of $M_{\nu+\xi}$ by some measure $\mathcal{E}_{\nu+\xi}(\eta)$ with $\eta > \lambda_\alpha$. Then $\kappa' = j_{\nu+\Omega}$ if $\mathrm{crit}(\mathcal{E}_{\nu+\xi}(\eta)) > \kappa'$, and otherwise $\kappa' \in \mathcal{C}(j_{\nu+\xi,\Omega}(\eta))$ with $j_{\nu+\xi,\Omega}(\eta) > \gamma$. This concludes the proof of clause (ii) and hence of clause (1) of the lemma.

The proof of clause (2) of the lemma is similar. There is a slight complication in this case since γ need not be a member of range$(j_{\nu',\Omega})$, but if we take $\bar{\gamma}$ to be the least member of range$(j_{\nu',\Omega}) \setminus \gamma$ and do the first part of the construction with $\bar{\gamma}$ instead of γ then all of the accumulation points for γ below κ are also accumulation points for $\bar{\gamma}$, so that if $\bar{\gamma} = j_{\nu',\Omega}(\bar{\lambda})$ then as before $\mathcal{F}_{\nu'}\restriction\bar{\lambda} = \mathcal{E}_\nu\restriction\bar{\lambda}$. Since $\gamma \leq \bar{\gamma}$ the second part of the argument still shows that there cannot be more than ω_1^V accumulation points for γ between κ' and the first member of $\bigcup_{\gamma' > \gamma} \mathcal{C}(\gamma')$, and this completes the proof of clause (2). □

4. Terms in $K^{\mathcal{M}}$

This and the next section will complete the proof of theorem 1.3. In this section we will work inside $K^{\mathcal{M}}$ to construct a suitable pair $(J^*, \mathcal{T}^*)$ and in the next section we will show that $L(a,b)$ yields a branch through the Martin-Solovay tree T associated with $(J^*, \mathcal{T}^*)$ in V. Since the tree T

is in $K^{\mathcal{M}}[b]$ it follows that T has a branch in $K^{\mathcal{M}}[b]$ and hence $K^{\mathcal{M}}[b] \models \exists x\, \phi(x,b)$.

Notice that although the work takes place inside $K^{\mathcal{M}}$ it does use a finite sequence of parameters which are chosen with knowledge from V. The section can be divided into two parts. The first part constructs in $K^{\mathcal{M}}$ a model $\overline{M}_0$ which mimics the construction of M_0 and defines in V a fine structure preserving embedding $t_0 \colon M_0 \longrightarrow \overline{M}_0$. The second part constructs an iterated ultrapower $s \colon \overline{M}_0 \longrightarrow \overline{M}$ of $\overline{M}_0$ which mimics the construction of $j \colon M_0 \longrightarrow M$. The iterated ultrapower s gives us the class J of indiscernibles and the set $\mathcal{T}$ of terms. The connection between $\overline{M}$ and M will be made in section 5 where we define a map t so that the following diagram commutes:

$$\begin{array}{ccc} M & \xrightarrow{\ t\ } & \overline{M} \\ {\scriptstyle j}\uparrow & & \uparrow{\scriptstyle s} \\ M_0 & \xrightarrow{\ t_0\ } & \overline{M}_0 \end{array}$$

Recall that I is the set of Silver indiscernibles for $L(a,b)$ which are larger than ρ^+. The map t will be defined by first letting $t{\restriction}I$ map I isomorphically onto J, and then observing that the set $\mathcal{T}$ of terms for $\overline{M}$ can be used to define a set $\mathcal{T}^M$ of terms for M. Every member of M can be written in the form $\tau^M(\mathbf{i})$ for some $\tau^M \in \mathcal{T}^M$ and $\mathbf{i} \in [I]^{<\omega}$ (although not all of these expressions will denote any member of M), and thus we can define $t(\tau^M(\mathbf{i}))$ to be $\tau^{\overline{M}}(t(\mathbf{i}))$.

In order to show that $L(a,b)$ induces a branch through the Martin-Solovay tree associated with $(J,\mathcal{T})$ we would like to show that if $\pi \colon J \longrightarrow J$ is any order preserving map then there is a map $\pi^* \colon M \longrightarrow M$ such that the diagram

$$\begin{array}{ccc} M & \xrightarrow{\ t\ } & \overline{M} \\ {\scriptstyle \pi^*}\uparrow & & {\scriptstyle \pi}\uparrow \\ M & \xrightarrow{\ t\ } & \overline{M} \end{array}$$

commutes. We don't know if this is true in general, but we are able to prove it for maps π which preserve successors and ωth successors in J. For $\xi < \omega^2$ let $s^J_\xi(\nu)$ be the ξth member of J larger than ν. We modify the pair $(J,\mathcal{T})$ by letting J^* be the set of members of J which are not of the form $s^J_\xi(\nu)$ for any $\xi < \omega^2$, and letting $\mathcal{T}^*$ be the terms in $\mathcal{T}$, augmented by the functions s^J_ξ for $\xi < \omega^2$. The suitability of $(J,\mathcal{T})$ implies that of $(J^*,\mathcal{T}^*)$, and the existence of the maps π^* implies that there is a branch through the Martin-Solovay tree associated with $(J^*,\mathcal{T}^*)$, and this completes the proof.

The construction of $\overline{M}_0$ and of $t_0\colon M_0 \to \overline{M}_0$. The construction of $\overline{M}_0$ will depend on a finite set $\{\hat{k}, (\nu_k : k < \hat{k}), \hat{\nu}, (\overline{M}_0^k : 0 < k \le \hat{k}), \lambda\}$ of parameters in K. The first three have already been defined, and the last two are defined below. The choice of these parameters comes from our previous work in V and depends on a knowledge of the real number a, but given this choice of parameters the construction of $\overline{M}_0$ takes place inside K.

We define models $\overline{M}_0^k$ by recursion on $k \le \hat{k}$, along with maps $t_0^k\colon M_0^k \to \overline{M}_0^k$. Set $\overline{M}_0^0 = M_0^0 = K$ and let $t_0^0\colon M_0^0 = K \to K^{\mathcal{M}}$ be the iterated ultrapower asserted to exist by the hypothesis of theorem 1.3. Suppose that $t_0^k\colon M_0^k \to \overline{M}_0^k$ has been defined. We first define an iterated ultrapower $s_{0,\nu}^k\colon \overline{M}_0^k \to \overline{M}_\nu^k$ by a subsidiary recursion on ν. Recall that ν_k was the length of the iteration of M_0^k, which stopped with the definition of M_0^{k+1} as a mouse in $M_{\nu_k}^k$. We write $\overline{\mathcal{E}}_\nu^k$ for the sequence of measures in $\overline{M}_\nu^k$ and $\overline{\mathcal{C}}_\nu^k$ for the system of indiscernibles for $\overline{M}_\nu^k$ generated by $s_{0,\nu}^k$.

If $k = 0$ then let λ be the least ordinal such that $\operatorname{crit}(j_{\nu,\nu+1}^0) < j_{0,\nu}^0(\lambda)$ for all $\nu < \nu_0$. Let $\overline{\mathcal{C}}_\nu^k$ be the system of indiscernibles generated by $s_{0,\nu}^k$, and let κ be least the least measurable cardinal in $\overline{M}_\nu^k$ such that one of the two following conditions fails, and such that if $k = 0$ then $\kappa < s_{0,\nu}^0(\lambda)$:

(1) If κ is measurable in $\mathcal{F}(\kappa)$ then $\operatorname{cf}(\kappa) = \nu_k^+$ in K.

(2) κ is an accumulation point for $O^{\overline{M}_\nu^k}(\kappa)$ in $\overline{\mathcal{C}}_\nu^k$.

If such an ordinal κ exists then we set $\overline{M}_{\nu+1}^k = \operatorname{ult}(\overline{M}_\nu^k, \overline{\mathcal{E}}_\nu^k(\gamma))$ where if case (1) failed at κ then $\overline{\mathcal{E}}_\nu^k(\gamma)$ is the order 0 measure on κ in $\overline{M}_\nu^k$ and if case (2) failed then γ is the least ordinal such that $\overline{\mathcal{E}}_\nu^k(\gamma)$ is a measure on κ and $\overline{\mathcal{C}}_\nu^k(\gamma)$ is bounded in κ. This construction will stop at some ordinal $\bar{\nu}_k$.

Now we use recursion on ν to define an increasing function $\sigma^k\colon \nu_k \to \bar{\nu}_k$, together with fine structure preserving embeddings $t_\nu^k\colon M_\nu^k \to \overline{M}_{\sigma^k(\nu)}^k$ so that the following diagram commutes for $\nu' < \nu \le \nu_k$:

$$
(1)\qquad
\begin{array}{ccc}
M_{\nu'}^k & \xrightarrow{\;j_{\nu',\nu}^k\;} & M_\nu^k \\
{\scriptstyle t_{\nu'}^k}\big\downarrow & & \big\downarrow{\scriptstyle t_\nu^k} \\
\overline{M}_{\sigma^k(\nu')}^k & \xrightarrow{\;s_{\sigma(\nu'),\sigma(\nu)}^k\;} & \overline{M}_{\sigma^k(\nu)}^k
\end{array}
\;.
$$

We set $\sigma^k(0) = 0$, and the map t_0^k is given by the induction hypothesis. If ν is a limit ordinal then $\sigma^k(\nu) = \sup\{\sigma^k(\nu') : \nu' < \nu\}$ and t_ν^k is defined so that the rectangles (1) commute for $\nu' < \nu$. Now suppose that t_ν^k and

$\sigma^k(\nu)$ are given, and $M^k_{\nu+1} = \text{ult}(M^k_\nu, E)$. Then let $\sigma^k(\nu+1)$ be $\nu'+1$ where ν' is least such that $\overline{M}^k_{\nu'+1} = \text{ult}(\overline{M}^k_{\nu'}, E')$ for $E' = s^k_{\sigma^k(\nu),\nu'}(t^k_\nu(E))$. This ν' always exists by the construction of $s^k_{0,\bar\nu_k}$. Define $t^k_{\nu+1}$ by $t^k_{\nu+1}([f]_E) = \left[s^k_{\sigma(\nu),\nu'}(t^k_\nu(f))\right]_{E'}$.

Finally, set

$$t^*_k : M^k_{\nu_k} \xrightarrow{t^k_\nu} M^k_{\sigma^k(\nu_k)} \xrightarrow{s^k_{\sigma^k(\nu_k),\bar\nu_k}} \overline{M}^k_{\bar\nu_k}$$

and define $\overline{M}^{k+1}_0 = t^*_k(M^{k+1}_0)$ and $t^{k+1}_0 = t^*_k \restriction M^{k+1}_0$.

For $k = \hat k$ recall that $M_0 = M^{\hat k}_{\hat\nu}$, where $\hat\nu$ is least such that $\Omega \in j_{\hat\nu,\Omega}$"$M^{\hat k}_{\hat\nu}$. We use exactly the same procedure as for $k < \hat k$ to define

$$s^{\hat k}_{0,\hat\nu} : \overline{M}^{\hat k}_0 \longrightarrow \overline{M}^{\hat k}_{0,\sigma^{\hat k}(\hat\nu)} = \overline{M}_0.$$

together with an embedding $t_0 = t^{\hat k}_{\hat\nu}$ so that $t_0 : M_0 \longrightarrow \overline{M}_0$.

The construction of $\overline{M}$ and of the pair $(J, \mathcal{T})$. We are now just about ready to define the iterated ultrapower $s_{0,\nu} : \overline{M}_0 \longrightarrow \overline{M}_\nu$ by recursion on ν, together with the set $\mathcal{T}$ of terms and J of ordinals. Because the definition of $s_{0,\nu}$ is determined by the definition of the terms and the desired properties of the terms, we will work backwards. First we will state a proposition which gives the properties which we expect of the embedding and the terms, and then we give the definition of the terms and of J assuming that the embedding $s = s_{0,\Omega} : \overline{M}_0 \longrightarrow \overline{M}_\Omega = \overline{M}$ has been defined. This definition will then dictate the definition of the iteration s, since at each stage we will take an ultrapower to generate an indiscernible that is needed as the denotation of some term in $\mathcal{T}$.

Proposition 4.1.

(1) *For each $n \in \omega$, $\tau \in \mathcal{T}_n$ and $\mathbf{c} \in [J]^n$ there is a member x of $\overline{M}$ such that $x = \tau^{\overline{M}}(\mathbf{c})$.*

(2) *For each $x \in \overline{M}$ there are $n \in \omega$, $\tau \in \mathcal{T}_n$ and $\mathbf{c} \in [J]^n$ such that $x = \tau^{\overline{M}}(\mathbf{c})$.*

(3) *$J = \{\, c \in \overline{\mathcal{C}}(\gamma) : \forall\tau \in \mathcal{T}\, \forall\mathbf{c} \in [J \cap c]^n\ \left(c \neq \tau^{\overline{M}}(\mathbf{c})\right) \}$, where $\overline{\mathcal{E}}(\gamma)$ is the order 0 measure on Ω in $\overline{M}$.*

(4) *If π is any order preserving map from J into J then π extends to a map $\pi^* : \overline{M} \longrightarrow \overline{M}$ defined by $\pi^*(\tau(\mathbf{c})) = \tau(\pi(\mathbf{c}))$.*

We now define the set $\mathcal{T}$ and class J, assuming that $\overline{M}$ and $s = s_{0,\Omega}$ have been defined. Recall that $\chi = j^{-1}(\Omega)$. We write $\bar\chi$ for $t_0(\chi)$, which will be equal to $s^{-1}(\Omega)$. The definition given below has been simplified by ignoring fine structure. The functions f in clause (2) of definition 4.2 are actually members of the Σ_{n-1} code of $\overline{M}$, where $n = n^{\hat{k}}_{\hat{\nu}}$ is the least integer such that the Σ_n projectum of M is smaller that than Ω. The Σ_n projectum ρ_n of $\overline{M}$ is equal to the Σ_n projectum of $\overline{M}_0$ and $s{\restriction}\rho$ is the identity, and M_0 is $(n-1)$-sound so that every member of $\overline{M}$ is definable from members of the Σ_{n-1} code of $\overline{M}$. Furthermore the last sentence is also true with s, $\overline{M}_0$ and $\overline{M}$ replaced by j, M_0 and M.

Definition 4.2. The set $\mathcal{T} \subset \overline{M}_0$ is obtained by starting with the following four classes of basic terms and closing under composition.

(1) If x is any variable then $\dot{x}$ is a unary term.
(2) If f is any function in $\overline{M}_0$ with domain in $[\bar{\chi}]^n$ for some $n < \omega$ then $\dot{f}$ is an n-ary term.
(3) If ξ is any ordinal smaller than $\rho^+ + \omega_1^V$ then $\dot{\ell}_\xi$ is a binary term.
(4) If ξ is any ordinal smaller than $\rho^+ + \omega_1^V$ then $\dot{a}_\xi$ is a ternary term.

The following definition gives the meaning of the basic terms from definition 4.2. The meaning of a term obtained from these terms by composition is then given by recursion on the length of the term.

Definition 4.3.

(1) $\dot{x}^{\overline{M}}(c) = c$.
(2) $\dot{f}^{\overline{M}}(\xi_0, \dots, \xi_{n-1}) = s(f)(\xi_0, \dots, \xi_{n-1})$.
(3) $\dot{\ell}_\xi{}^{\overline{M}}(\gamma, \delta)$ is equal to 0 unless $\gamma \in \operatorname{domain}(\overline{\mathcal{E}})$ and $\delta < \operatorname{crit}(\overline{\mathcal{E}}(\gamma))$, in which case $\dot{\ell}_\xi{}^{\overline{M}}(\gamma, \delta)$ is the ξth member of $\overline{\mathcal{C}}(\gamma)$ above δ.
(4) $\dot{a}_\xi{}^{\overline{M}}(\eta, \gamma, \delta)$ is equal to 0 unless $\gamma \in \operatorname{domain}(\overline{\mathcal{E}})$ and $\delta < \kappa = \operatorname{crit}(\overline{\mathcal{E}}(\gamma)) < \gamma < \eta \le O^{\overline{\mathcal{E}}}(\kappa)$, in which case $\dot{a}_\xi{}^{\overline{M}}(\eta, \gamma, \delta)$ is the ξth member ν of $\overline{\mathcal{C}}(\gamma)$ above ξ such that ν is an accumulation point for $\overline{\mathcal{E}}(\eta)$.

Now we can take clause 4.1(3) as a definition of J. Note that if τ is a term, $\mathbf{c} \in [J]^n$, and $\tau^{\overline{M}}(\mathbf{c}) = \nu \in \Omega$ then there is a term τ' and sequence $\mathbf{c}' \in [J \cap (\nu+1)]^{n'}$ such that ${\tau'}^{\overline{M}}(\mathbf{c}') = \nu$.

Now we define the iterated ultrapower $s\colon \overline{M}_0 \to \overline{M}_\Omega = \overline{M}$, defining $s_{0,\nu}\colon \overline{M}_0 \to \overline{M}_\nu$ by recursion on ν. The strategy in deciding which ultrapower to use at each stage is to check whether there is an indiscernible which is needed as a denotation of a term from clause (3) or (4), but which does not yet exist. If there is such a missing indiscernible then the next ultrapower is chosen so as to add it; otherwise the next ultrapower is chosen to add a new member of J. The first clause of definition 4.4 will add an instance of clause 4.2(4), the second will add an instance of clause 4.2(3), and the default case will add a member of J.

Definition 4.4. Suppose that $s_{0,\nu}\colon \overline{M}_0 \to \overline{M}_\nu$ has been defined, giving the system $\overline{\mathcal{C}}_\nu$ of indiscernibles. Then $\overline{M}_{\nu+1} = \operatorname{ult}(\overline{M}_\nu, \overline{\mathcal{E}}_\nu(\gamma_\nu))$, where γ_ν is chosen as follows: Let (κ, η, λ) be the lexicographically least triple (if there is one) such that $\kappa \le s_{0,\nu}(\bar{\chi})$ and one of the two following clauses is true:

(1) $\kappa < \lambda < \eta \le O^{\overline{M}_\nu}(\kappa)$, $\overline{\mathcal{E}}_\nu(\lambda)$ is a measure on κ, and there is $\delta < \kappa$ such that the order type of the set of ordinals $d \in \overline{\mathcal{C}}_\nu(\lambda) \cap (\delta, \kappa)$ such that d is an accumulation point for η in $\overline{\mathcal{C}}_\nu$ is less then $\rho^+ + \omega_1^V$.
(2) $\overline{\mathcal{E}}_\nu(\eta)$ is a measure on κ but there is $\delta < \kappa$ such that the order type of $\overline{\mathcal{C}}_\nu(\eta) \cap (\kappa \setminus \delta)$ is less than $\rho^+ + \omega_1^V$.

If clause (1) holds then set $\gamma_\nu = \lambda$, if clause (2) holds then set $\gamma_\nu = \eta$, and if neither of the clauses holds for any triple (κ, η, λ) then set $\gamma_\nu = s_{0,\nu}(\gamma')$ where $\overline{\mathcal{E}}_0(\gamma')$ is the order 0 measure on $\bar{\chi}$ in $\overline{M}_0$.

Proof of proposition 4.1. Clause 4.1(1), which asserts that every term denotes an ordinal, is proved by induction on the complexity of the term τ. Clause 4.1(2) is also proved by an induction: Every member of $\overline{M}$ is of the form $s(f)(\mathbf{c})$ where $f \in \overline{M}_0$ and $\mathbf{c}$ is a sequence of indiscernibles arising from the iterated ultrapower s. Thus it is enough to show that each of these indiscernibles can be denoted by a term, but this follows easily from the fact that indiscernibles from clauses (1) and (2) were only added because they were required as the denotation of some term and all of the other indiscernibles are in J. Clause 4.1(3) follows easily from the definition of J.

The proof of clause 4.1(4) uses a normal form for the terms $\mathcal{T}$. The next definition is a start towards the definition of this normal form.

Definition 4.5. A *support sequence* is a finite sequence $\mathfrak{s} = (\,(\varepsilon_i, \xi_i, f_i, q_i) : i < k)\,)$ of quadruples such that

(1) $\varepsilon_i \in \{\,1, 2, 3\,\}$ for each $i < k$.
(2) $\xi_i < \rho^+ + \omega_1^V$ for each $i < k$.
(3) f_i and g_i are in $s``\overline{M}_0$, and are each functions such that $\operatorname{domain}(f_i) = \operatorname{domain}(g_i) = [\Omega]^i$.

Definition 4.6. A support sequence $\mathfrak{s}$ is a *support sequence for* $\mathbf{d}$ if $\mathbf{d}$ is a finite sequence of ordinals, and for all $i < \operatorname{len}(\mathbf{d})$

(1) If $\varepsilon_i = 1$ then $\mathcal{E}(f_i(\mathbf{d}{\restriction}i))$ is a measure on some ordinal κ, $f_i(\mathbf{d}{\restriction}i) < g_i(\mathbf{d}{\restriction}i) \le O^{\overline{M}}(\kappa)$, and $d_i = a_{\xi_i}(f_i(\mathbf{d}{\restriction}i), g_i(\mathbf{d}{\restriction}i), d_{i-1})$.
(2) If $\varepsilon_i = 2$ then $g_i(\mathbf{d}{\restriction}i) = 0$, $\mathcal{E}_{f_i}(\mathbf{d}{\restriction}i)$ is a measure, and $d_i = s_{\xi_i}(f_i(\mathbf{d}{\restriction}i), d_{i-1})$.

(3) If $\varepsilon_i = 3$ then $\xi_i = g_i(\mathbf{d}\restriction i) = f_i(\mathbf{d}\restriction i) = 0$ and d_i is in J with $d_i > d_{i-1}$.

A sequence $\mathbf{d}$ is a *support* if it has a support sequence.

Notice that any support sequence corresponds, in a natural way, to a term. If $\mathbf{c}$ is an increasing sequence from J of length equal to the number of $i < k$ such that $\varepsilon_i = 3$ then we write $\mathfrak{s}(\mathbf{c})$ for the unique sequence $\mathbf{d}$, if there is one, such that $\mathfrak{s}$ is a support sequence for $\mathbf{d}$ and $\mathbf{c} = \{ d_i : \varepsilon_i = 3 \}$. We say that a term $\tau(\mathbf{c})$ is in *standard form* if it has the form $\tau(\mathbf{c}) = f(\mathfrak{s}(\mathbf{c}))$ where $f \in \text{range}(s)$ and where there is no function $h \in \text{range}(s)$ such that $\tau(\mathbf{c}) \in h``[\sup \mathfrak{s}(\mathbf{c})]^{<\omega}$.

Proposition 4.7. *If $\mathfrak{s}$ is a support sequence of length n and $\mathbf{c}$ and $\mathbf{c}'$ are in $[J]^n$ then $\mathfrak{s}(\mathbf{c})$ exists if and only if $\mathfrak{s}(\mathbf{c}')$ does. Furthermore, if ϕ is any Σ_1-formula with parameters from $s``\overline{M}_0$ then the $\Sigma_{n^{\hat{k}}_{\bar{\nu}}}$ code of $\overline{M}$ satisfies $\phi(\mathfrak{s}(\mathbf{c})) \iff \phi(\mathfrak{s}(\mathbf{c}'))$. If $\hat{k} = 0$ and ϕ is any formula then $\overline{M} \models \phi(\mathfrak{s}(\mathbf{c})) \iff \phi(\mathfrak{s}(\mathbf{c}'))$.*

Proof. This is a straightforward induction on n. It depends on two facts: one is the completeness of the iterated ultrapower j, which ensures that any desired indiscernible or accumulation point exists, and the other is the fact that C is a sequence of indiscernibles for $\overline{M}$ over $s``M_0$. □

We will show that for every term τ there is a term τ' in standard form such that $\tau(\mathbf{c}) = \tau'(\mathbf{c})$ for all $\mathbf{c} \in [J]^{<\omega}$. Thus proposition 4.7 implies that if $\pi : J \to J$ is any order preserving embedding then the extension $\pi' : \overline{M} \to \overline{M}$ of π defined by setting $\pi'(\tau(\mathbf{c})) = \tau(\pi(\mathbf{c}))$ preserves fine structure.

Lemma 4.8. *Suppose that $\mathbf{d}$ and $\mathbf{e}$ are supports, with support sequences $\mathfrak{s}$ and $\mathfrak{t}$ respectively. Then $\mathbf{d} \cup \mathbf{e}$ is also a support, and the support sequence for $\mathbf{d} \cup \mathbf{e}$ depends only on $\mathfrak{s}$ and $\mathfrak{t}$.*

Proof. Let j be the least integer such that $d_j \neq e_j$ or $\mathfrak{s}_j \neq \mathfrak{t}_j$, so that $\mathbf{d}\restriction j = \mathbf{e}\restriction j$ and $\mathfrak{s}\restriction j = \mathfrak{t}\restriction j$. We can also assume without loss of generality that $\mathbf{d} = \mathfrak{s}(\mathbf{c})$ and $\mathbf{e} = \mathfrak{t}(\mathbf{c})$ for the same sequence $\mathbf{c} \in [J]^{<\omega}$. If either of the sequences $\mathbf{d}$ or $\mathbf{e}$ has length j then $\mathbf{d} \cup \mathbf{c}$ is equal to the longer of the two sequences and hence has a support sequence, so we can assume that e_j and d_j both exist. We can assume without loss of generality that $d = d_j \leq e_j$. We will construct a support sequence $\mathfrak{t}'$ for $\mathbf{e} \cup \{d\}$ such that $\mathfrak{t}'\restriction j+1 = \mathfrak{s}\restriction j+1$. The required support sequence for $\mathbf{d} \cup \mathbf{e}$ is obtained by recursion on j.

Write (ε, ξ, f, g) for $\mathfrak{s}_j$, and write $\mathfrak{t}_i = (\varepsilon_i, \xi_i, f_i, g_i)$ for $i < \text{len}(\mathbf{e})$. If $d = e_j$ then the sequence $\mathfrak{t}'$ defined by $\mathfrak{t}'_j = \mathfrak{s}_j$ and $\mathfrak{t}'_i = \mathfrak{t}_i$ for $i \neq j$ is a support sequence for $\mathbf{d}$ such that $\mathfrak{t}'\restriction j+1 = \mathfrak{s}\restriction j+1$, as required. Thus we can assume that $d < e_j$. In this case we define $\mathfrak{t}'\restriction j+1 = \mathfrak{s}\restriction j+1$, as required, and $\mathfrak{t}'_{i+1} = \mathfrak{t}_i$ for $i > j+1$. We can also set $\mathfrak{t}'_{j+2} = \mathfrak{t}_{j+1}$ except when $\varepsilon_{j+1} \neq 3$ and $\varepsilon_{j+1} = \varepsilon$, $f_{j+1} = f$ and $g_{j+1} = g$, that is when d and e_j are indiscernibles or accumulation points of the same type, and are respectively the ξth and ξ_jth such indiscernibles or accumulation points above $e_{j-1} = d_{j-1}$. In this case we set $\mathfrak{t}'_{j+2} = (\,\varepsilon_{j+1}, \xi', f_{j+1}, g_{j+1}\,)$ where $\xi + \xi' = \xi_{j+1}$. Then $\mathfrak{t}'$ is the required support sequence for $\mathbf{e} \cup \{d\}$.

To complete the proof of the lemma we need to show that $\mathfrak{t}'$ depends only on the support sequences $\mathfrak{s}$ and $\mathfrak{t}$. The construction described above also depends on whether $d_j < e_j$, $d_j = e_j$, or $d_j > e_j$, but the completeness of the iterated ultrapower s implies that the order of d_j and e_j is the same as the lexicographic ordering of the quadruples

$$(*) \qquad (\,\varepsilon, f(\mathbf{d}\restriction j), g(\mathbf{d}\restriction j), \xi\,) \text{ and } (\,\varepsilon_j, f_j(\mathbf{e}\restriction j), g_j(\mathbf{e}\restriction j), \xi_j\,).$$

Since $\mathbf{e}\restriction j = \mathbf{d}\restriction j$, proposition 4.7 implies that this order depends only on $\mathfrak{s}_j$ and $\mathfrak{t}_j$. □

In order to complete the proof of proposition 4.1(4) we need to show that for any term τ there is a term τ' in standard form such that $\tau(\mathbf{c}) = \tau'(\mathbf{c})$ for every sequence $\mathbf{c}$ from J. Since any term will have the form $\tau = t(\tau_0, \dots, \tau_{n-1})$ where t is one of the terms given by the four clauses of the definition, the proof breaks down into four cases:

Case 1. If t comes from clause (1) then τ is just a variable x_i and $\tau(\mathbf{c}) = c_i$, which is a term in standard form.

Case 2. If t comes from clause (2) then we can assume that each of the terms τ_i is in standard form, $\tau_i = f_i(\mathfrak{s}_i(\mathbf{c}))$. Furthermore we can assume that f_i is the identity, so that $\tau_i(\mathbf{c}) = f_i(\mathfrak{s}_i(\mathbf{c})) = \sup(\mathfrak{s}_i(\mathbf{c}))$, since otherwise f could be replaced by the function f' defined by $f'(\vec{\nu}) = f(f_0(\vec{\nu}), \dots, f_{n-1}(\vec{\nu}))$, which is also in range$(s)$. By the lemma, the union of the sequences $\mathfrak{s}_i(\mathbf{c})$ is a support $\mathbf{d}$, with support sequence $\mathfrak{s}$ depending only on the support sequences $\mathfrak{s}_i$, so that we can write $\tau(\mathbf{c}) = f(\mathfrak{s}(\mathbf{c})) = f(\mathbf{d})$. This is a term in standard form unless there is a sequence $\vec{\alpha}$ and a function $h \in \text{range}(s)$ such that $f(\mathbf{d}) = h(\vec{\alpha})$ and $\sup(\vec{\alpha}) < \sup(\mathbf{d})$. In this case define a function f' by setting $f'(\vec{\nu})$ equal to the least sequence $\vec{\alpha}$ such that $h(\vec{\alpha}) = f(\vec{\nu})$, where the sequences $\vec{\alpha}$ of ordinals are ordered lexicographically as decreasing sequences. Then $f(\mathbf{d}) = h(f'(\mathbf{d}))$. If $f'(\mathbf{d}) < \sup(\mathbf{d})$ then by using the normality of the measures in the sequence $\overline{\mathcal{E}}$ we can define a function $f'' \in \text{range}(s)$ such that $f'(\mathbf{d}) = f''(\mathbf{d} \cap \sup(f'(\mathbf{d}))+1)$. If j is least such that $d_j > \sup(f'(\mathbf{d}))$ then $\tau(\mathbf{c}) = f(\mathbf{d}) = (hf'')(\mathbf{d}\restriction j) = (hf'')((\mathfrak{s}\restriction j)(\mathbf{c}))$, and this last is a term in standard form.

Cases 3 and 4. In these cases $\tau(\mathbf{c})$ is either of the form $\dot{\ell}_\xi(\tau_0(\mathcal{C}), \tau_2(\mathcal{C}))$ or of the form $\dot{a}_\xi(\tau_0(\mathbf{c}), \tau_1(\mathbf{c}), \tau_2(\mathbf{c}))$, where each of the terms τ_i can be taken to be in standard form, $\tau_i = f_i(\mathfrak{s}_i(\mathbf{c}))$. By merging the three support sequences we can assume that all the sequences $\mathfrak{s}_i$ are the same (actually, initial parts of the same sequence). We can always take f_2 equal to the identity, so that $\tau_2(\mathbf{c}) = \sup(\mathfrak{s}_2(\mathbf{c}))$. Since $\tau_i(\mathbf{c}) \in h``(\tau(\mathbf{c}))$ for some $h \in \text{range}(s)$ we have that $\sup(\mathfrak{s}_i(\mathbf{c})) < \tau(\mathbf{c})$ for $i = 1, 2$. Thus $\mathfrak{s}(\mathbf{c}) \cup \{\tau(\mathbf{c})\}$ is a support with support sequence obtained by adding either $(1, f_1, 0, \xi)$ or $(2, f_2, f_1, \xi)$ to $\mathfrak{s}$.

This concludes the definition, in $K^{\mathcal{M}}$, of J and $\mathcal{T}$. Proposition 4.1 implies that J and $\mathcal{T}$ are suitable. As stated at the beginning of the section, we will actually use a modified pair $(J^*, \mathcal{T}^*)$. The set $\mathcal{T}^*$ is obtained from $\mathcal{T}$ by adding the functions $s^J_\xi(\nu)$ for $\xi < \omega^2$ and closing under composition, where $s^J_\xi(\nu)$ is the ξth member of J above ν, and $J^* = \{\nu \in J : \forall \xi < \omega^2\ \nu \notin s^J_\xi``\nu\}$. It is easy to see that the suitability of $(J, \mathcal{T})$ implies that of $(J^*, \mathcal{T}^*)$, so if T is the Martin-Solovay tree defined in $K^{\mathcal{M}}[b]$ using J^* and $\mathcal{T}^*$ then any branch through T will give a solution to the formula $\exists x\, \phi(x, b)$. All that remains is to prove that there exists a branch through T in V, and hence in $K^{\mathcal{M}}[b]$.

5. Terms in M

Since $\mathcal{T} \subset \overline{M}_0$, we can define $\mathcal{T}^M$ to be $\{x \in M_0 : t_0(x) \in \mathcal{T}\}$. The meaning of an expression $\tau^M(\mathbf{c})$ for $\tau \in \mathcal{T}^M$ and $\mathbf{c} \in [I]^n$, is given by definition 5.1 below.

Definition 5.1.

(1) $\dot{x}^M(c) = c$.

(2) $\dot{f}^M(\xi_0, \dots, \xi_{n-1}) = j(f)(\xi_0, \dots, \xi_{n-1})$.

(3) If $\gamma \in \text{domain}(\mathcal{E})$ and $\delta < \text{crit}(\mathcal{E}(\gamma))$ then $\dot{\ell}_\xi{}^M(\gamma, \delta)$ is the ξth member ν of $\mathcal{C}(\gamma)$ above δ, provided that such an ordinal ν exists and that $(\nu + 1) \cap I \subset \delta$ and $\mathcal{C}(\gamma') \cap \nu \subset \delta$ for all $\gamma' > \gamma$. Otherwise $\dot{\ell}_\xi{}^M(\gamma, \delta)$ is undefined.

(4) If $\gamma \in \text{domain}(\mathcal{E})$ and $\delta < \kappa = \text{crit}(\mathcal{E}(\gamma)) < \gamma < \eta \le O^{\mathcal{E}}(\kappa)$ then $\dot{a}_\xi{}^M(\eta, \gamma, \delta)$ is the ξth member ν of $\mathcal{C}(\gamma)$ which is an accumulation point in $\mathcal{C}$ for η, provided that such an ordinal ν exists and that $(\nu{+}1) \cap I \subset \delta$ and $\mathcal{C}(\gamma') \cap \nu \subset \delta$ for all $\gamma' \ge \eta$. Otherwise $\dot{a}_\xi{}^M(\eta, \gamma, \delta)$ is undefined.

Notice that this definition is similar to definition 4.3 except for the phrases beginning with the words "provided that" in clauses (3) and (4).

The corresponding clauses would have been redundant in definition 4.3, since they are always true in $\overline{M}$ because of the completeness of the iterated ultrapower s. In the present context it is quite possible that, for example, $\mathcal{E}(\gamma)$ and $\mathcal{E}(\gamma')$ might be measures on the same cardinal κ, with $\gamma < \gamma'$, and that nevertheless the least member c of $\mathcal{C}(\gamma)$ is larger than the least member c' of $\mathcal{C}(\gamma')$. Then $\dot{\ell}_1{}^M(\gamma', 0) = c'$ by definition 5.1, but $\dot{\ell}_1{}^M(\gamma, 0)$ does not exist. To allow $\dot{\ell}_1{}^M(\gamma, 0) = c$ would be wrong, as it would be violate the natural order of the terms as given by the lexicographic order of the quadruples $(*)$ in the proof of lemma 4.8.

The analogue to clause (2) of proposition 4.1 is true in this context:

Proposition 5.2. *Every set in M has the form $\tau^M(\mathbf{c})$ for some $\tau \in \mathcal{T}^M_n$ and $\mathbf{c} \in [I]^n$.*

Proof. As in the proof of clause (2) of proposition 4.1, it is enough to show that every ordinal in $\bigcup_\lambda \mathcal{C}(\lambda) \setminus I$ has this form. We do this by induction on ν. Suppose $\nu \in \mathcal{C}(\lambda)$ where $\lambda \in f``\nu$ for some $f \in \mathrm{range}(j)$. First suppose that ν is not an accumulation point for any $\gamma > \lambda$. Then by corollary 3.10 there is $\delta < \nu$ and $\xi < \rho^+ + \omega_1^V$ such that ν is the ξth member of $\mathcal{C}(\lambda)$ above δ. Then $\nu = \dot{\ell}_\xi{}^M(\lambda, \delta)$ and the induction hypothesis implies that δ and λ are denoted by terms. Hence ν is also denoted by a term.

If, on the other hand, ν is an accumulation point for some $\gamma > \lambda$, then the largest such γ is in $f``\nu$ for some $f \in \mathrm{range}(j)$. Then by corollary 3.10 there is $\delta < \nu$ and $\xi < \rho^+ + \omega_1^V$ such that ν is the ξth member of $\mathcal{C}(\lambda)$ above δ which is an accumulation point for γ. Then $\nu = \dot{a}_\xi{}^M(\gamma, \lambda, \delta)$, and hence ν is denoted by a term since by the induction hypothesis δ, λ and γ are denoted by terms. □

We also have

Proposition 5.3. *For every term $\tau \in \mathcal{T}^M$ there is a term τ' in standard form such that whenever $\mathbf{c} \in [I]^n$ is a sequence such that $\tau(\mathbf{c})$ is defined then $\tau'(\mathbf{c})$ is also defined and $\tau(\mathbf{c}) = \tau'(\mathbf{c})$.*

Proof. This is identical to the proof of the same proposition for $\overline{M}$, which is the last part of the proof of clause (4) of proposition 4.1. The proof shows, in fact, that $t_0(\tau')$ is the term of $\mathcal{T}^{\overline{M}}$ which is in standard form and is equal to $t_0(\tau)$. □

Lemma 5.4. *Let $t{\restriction}I$ be the map taking I isomorphically onto J. Then t can be extended to a fine structure preserving embedding from M to $\overline{M}$ such that $t(\tau^M(\mathbf{c})) = \tau^{\overline{M}}(t(\mathbf{c}))$ for all terms τ and all $\mathbf{c} \in [I]^{<\omega}$ such that $\tau^m(\mathbf{c})$ is defined.*

Proof. By proposition 5.2, every member x of M is represented by an expression $x = \tau^M(\mathbf{c})$, with $\tau \in \mathcal{T}^M$ and $\mathbf{c} \in [I]^{<\omega}$, and by proposition 5.3 we can assume that τ is in standard form. Now define $t(x) = \tau^{\overline{M}}(t(\mathbf{c}))$. We need to verify that $\tau^{\overline{M}}(t(\mathbf{c}))$ always exists and that this map preserves fine structure, but this follows by the same proof as proposition 4.7. □ □

We already know, by lemma 2.8, that if $\pi : L(a,b) \longrightarrow L(a,b)$ with $\pi{\restriction}\rho = \text{id}$ then $\pi{\restriction}N$ extends to $\pi^*\colon M \longrightarrow M$. We would now like to show that π^* preserves terms, that is, that if $\tau \in \mathcal{T}^M$ then $\pi^*(\tau^M(\mathbf{c})) = \tau^M(\pi(\mathbf{c}))$. The problem is that $\tau^M(\pi(\mathbf{c}))$ might not exists, even if $\tau^M(\mathbf{c})$ does. We do not know whether this in fact can happen, but we will show that it does not happen for sufficiently well behaved embeddings π. The main result of this section is lemma 5.6 below. Immediately after the statement of lemma 5.6 we will use it to complete the proof of the main theorem, theorem 1.3, and after that the proof of lemma 5.6 will take up the rest of the paper.

Through the rest of this section we will always use $\mathbf{c}$ and $\mathbf{d}$ to denote members of $[I]^{<\omega}$.

Definition 5.5.

(1) $s^I_\xi(\delta)$ is the ξth member of I larger than δ.

(2) $\mathbf{d}$ is a η-conservative extension of $\mathbf{c}$ if for every member d of $\mathbf{d} \setminus \mathbf{c}$ there is a $c \in \mathbf{c} \cup \{0\}$ and $\xi < \eta$ such that either $d = s^I_\xi(c)$ or $c = s^I_\xi(d)$

(3) $\mathbf{c} \equiv_\eta \mathbf{d}$ iff $\mathbf{c} \equiv \mathbf{d}$ and for each $\xi < \eta$ we have (i) $c_0 = s^I_\xi(0)$ if and only if $d_0 = s^I_\xi(0)$, (ii) for each i, $c_{i+1} = s^I_\xi(c_i)$ if and only if $d_{i+1} = s^I_\xi(d_i)$, and (iii) for each i there an ordinal $\delta < c_i$ such that $c_i = s^I_\xi(\delta)$ if and only if there is an ordinal $\delta < d_i$ such that $d_i = s^I_\xi(\delta)$.

(4) $\mathbf{c}, \mathbf{c}' \equiv_\eta \mathbf{d}, \mathbf{d}'$ if and only if $\mathbf{c}, \mathbf{c}' \equiv \mathbf{d}, \mathbf{d}'$ and $\mathbf{c} \cup \mathbf{c}' \equiv_\eta \mathbf{d} \cup \mathbf{d}'$.

In the rest of this paper, the letters f and g will always be used in accordance with the following convention: the letter g is always used to denote functions of the form

$$g(\mathbf{c}) = \bar{g}(\mathbf{c}, d_0, \ldots, d_{n-1}),$$

where $\bar{g}$ is definable in $L(a,b)$ from parameters in ρ and $d_0, \ldots, d_{n-1}$ are arbitrary members of I such that $d_i > g(\mathbf{c})$, while the letter f always denotes functions in $j``M_0$. Thus every ordinal $\nu \in \Omega$ can be written in the form $g(\mathbf{c})$ for some function g and $\mathbf{c} \in [I \cap (\nu+1)]^{<\omega}$, and every set in M can be written in the form $f \cdot g(\mathbf{c})$ where $\mathbf{c} \in [I \cap (g(\mathbf{c})+1)]^{<\omega}$.

Lemma 5.6.

(1) *If* $\mathbf{c} \equiv_{\omega^2} \mathbf{d}$ *and* $\tau \in \mathcal{T}^M$ *then* $\tau(\mathbf{c})$ *exists iff* $\tau(\mathbf{d})$ *exists.*
(2) *If* $\mathbf{d} \equiv_{\omega^2} \mathbf{c}$ *and* $f(g(\mathbf{c})) = \tau^M(\mathbf{c})$ *then* $f(g(\mathbf{d})) = \tau^M(\mathbf{d})$.
(3) *For any* f, g *and* $\mathbf{c}$, *if* $x = f(g(\mathbf{c}))$ *then there is a term* $\tau \in \mathcal{T}^M$ *and an* ω^2*-conservative extension* $\mathbf{c}'$ *of* $\mathbf{c}$ *such that* $x = \tau(\mathbf{c}')$

Proof of theorem 1.3, assuming lemma 5.6. We will use lemma 5.6 to show that there is a branch through the Martin-Solovay tree T associated with $(J^*, \mathcal{T}^*)$. Let I^* be the set of ω^2-limit points of I and define $\mathcal{T}^{*M}$ like $\mathcal{T}^*$, by augmenting $\mathcal{T}^M$ with the functions s^I_ξ for $i < \omega^2$ and closing under composition. Thus lemma 5.6 is true for $\mathcal{T}^{*M}$, and in fact it remains true for $\mathcal{T}^{*M}$ even if $\equiv_{\omega^2}$ is replaced by $\equiv$ and instead of allowing an ω^2-conservative extension in clause (3) x is required to be equal to $\tau(\mathbf{c})$. Recall that a branch through the Martin-Solovay tree T is a pair (e, σ) where e is the sharp of a pair (r, b) of reals and σ is a function which takes terms of the structure $\mathcal{B}$ from definition 1.9 into the set J^* of ordinals. The real e will be $(a, b)^\sharp$, and we need to define the map σ. By lemma 5.6(3) we can define a map σ^M taking terms s of $\mathcal{B}$ with n free variables into terms in $\mathcal{T}^{*M}_n$ by letting $\sigma^M(s)$ be some term $\tau \in \mathcal{T}^{*M}$ such that $\tau^M(\mathbf{c}) = s^{\mathcal{B}}(\mathbf{c})$ for an arbitrary sequence $\mathbf{c} \in [I]^{<\omega}$. Clauses (1) and (2) of lemma 5.6 imply that $\sigma^M(s)$ does not depend on the choice of $\mathbf{c}$. The final clause of the definition of the Martin-Solovay tree clearly holds, since $s(\mathbf{c})$ and $\sigma^M(s)(\mathbf{c})$ are the same ordinal. Now if we set $\sigma = t_0 \cdot \sigma^M$, so that σ maps the terms of $\mathcal{B}$ to members of $\mathcal{T}^*$, then the pair $((a, b)^\sharp, \sigma)$ is the desired branch through T in V. It follows by absoluteness that there is such a branch in $K[b]$, and hence there is a solution to the Π^1_2 formula $\phi(x, b)$. □

The proof of lemma 5.6 will involve a series of observations:

Proposition 5.7. *If* $f(g(\mathbf{c})) = f(g(\mathbf{c}'))$ *then there is* g' *such that* $f(g(\mathbf{c})) = f(g'(\mathbf{c} \cap \mathbf{c}'))$.

Proof. Let $\mathbf{d} = \mathbf{c} \cap \mathbf{c}'$. Order $[I]^{<\omega}$, regarded as a class of descending sequences of ordinals, lexicographically, and suppose that $\mathbf{e}$ is the least sequence such that $\mathbf{e} \supset \mathbf{d}$ and there is g' such that $f(g'(\mathbf{e})) = f(g(\mathbf{c}))$. Let e_i be the least member of $\mathbf{e}$ such that $e_i \notin \mathbf{d}$. It will be enough to show that

$$\exists \nu < e_i \left(fg'(e_0, \ldots, e_{i-1}, \nu, e_{i+1}, \ldots, e_{n-1}) = fg'(\mathbf{e}) \right), \tag{1}$$

for then we will have $\nu = g''(\mathbf{e}')$ for some $\mathbf{e}' \in [I \cap (\nu + 1)]^{<\omega}$, so that

$$fg'(\mathbf{e}) = fg'(e_0, \ldots, e_{i-1}, g''(\mathbf{e}'), e_{i+1}, \ldots, e_{n-1}),$$

and since $(\mathbf{e} \setminus \{e_i\}) \cup \mathbf{e}' <_{\text{lex}} \mathbf{e}$ this contradicts the minimality of $\mathbf{e}$.

Now $e_i \notin \mathbf{c} \cap \mathbf{c}'$ so assume *wlog* that e_i is not in $\mathbf{c}$ and let π_0 and π_1 be embeddings such that $\pi_0(\mathbf{c}) = \pi_1(\mathbf{c})$, $\pi_0(\mathbf{e} \setminus \{e_i\}) = \pi_1(\mathbf{e} \setminus \{e_i\})$, and $\pi_0(e_i) < \pi_1(e_i)$. Then we have

$$\pi_i^*(fg(\mathbf{c})) = f(\pi_i(g(\mathbf{c}))) = fg(\pi_i(\mathbf{c}))$$

for $i = 0, 1$ and since $\pi_0(\mathbf{c}) = \pi_1(\mathbf{c})$ it follows that

$$(2) \quad fg'(\pi_0(\mathbf{e})) = \pi_0^*(fg'(\mathbf{e})) = \pi_0^*(fg(\mathbf{c})) \\ = \pi_1^*(fg(\mathbf{c})) = \pi_1^*(fg'(\mathbf{e})) = fg'(\pi_1(\mathbf{e}))$$

Now if $x = \{ (\xi, \xi') : f(\xi) = f(\xi') \}$ then (2) says that $(g'(\pi_0(\mathbf{e})), g'(\pi_1(\mathbf{e})) \in x$. But $x \in \text{range}(j)$, so if $\nu > \max(g'(\pi_0(\mathbf{e})), g'(\pi_1(\mathbf{e}))$ then $x \cap \nu \in N$ and $\pi_1(x \cap \nu) = x \cap \pi_1(\nu)$. Thus $\big(g(\pi_0(\mathbf{e})), g(\pi_1(\mathbf{e}))\big) \in \pi_1(x)$. Then since $\pi_0(\mathbf{e})$ and $\pi_1(\mathbf{e})$ agree except at e_i, if we set $\mathbf{e}' = \pi_1(\mathbf{e})$ then we have

$$\exists \nu < e'_i \quad (g'((e'_0, \dots, e'_{i-1}, \nu, e'_{i+1}, \dots, e'_{n-1}), g'(\mathbf{e}')) \in \pi_1(x).$$

Since $\pi_1 \colon L(a,b) \longrightarrow L(a,b)$ is elementary this implies that

$$\exists \nu < e_i \quad (g'(e_0, \dots, e_{i-1}, \nu, e_{i+1}, \dots, e_{n-1}), g'(\mathbf{e})) \in x,$$

which is equivalent to (1). □

Proposition 5.8. *For all ordinals α and γ and all embeddings π from I into I we have $\alpha \in \mathcal{C}(\gamma)$ iff $\pi^*(\alpha) \in \mathcal{C}(\pi^*(\gamma))$.*

Proof. Recall that by fact 2.9 we have $\alpha \in \mathcal{C}(\gamma)$ if and only if (1) $\gamma \in f``\alpha$ for some $f \in j``M_0$ and (2) if f is any function in $j``M_0$ then $\forall x \in f``\alpha\, (\alpha \in x \iff x \in \mathcal{E}(\gamma))$. Since π^* is a fine structure preserving embedding of M into itself such that $\pi^*(f) = f$ for all $f \in j``M_0$ the conditions (1) and (2) are both preserved by π^*. □

Corollary 5.9. *If $g_0(\mathbf{c}) \in \mathcal{C}(f_1(g_1(\mathbf{c})))$ and $\mathbf{d} \equiv \mathbf{c}$ then*

$$g_0(\mathbf{d}) \in \mathcal{C}(f_1(g_1(\mathbf{d}))).$$

Proof. Pick π_0 and π_1 so that $\pi_0(\mathbf{c}) = \pi_1(\mathbf{d})$. Then by proposition 5.8

$$(1) \qquad g_0(\mathbf{c}) \in \mathcal{C}(f_1(g_1(\mathbf{c}))) \quad \text{iff} \quad \pi_0^*(g_0(\mathbf{c})) \in \mathcal{C}(\pi_0^*(f_1 g_1(\mathbf{c}))).$$

But

$$\pi_0^*(g_0(\mathbf{c})) = g_0(\pi_0(\mathbf{c})) = g_0(\pi_1(\mathbf{d})) = \pi_1^*(g_0(\mathbf{d}))$$

and

$$\pi_0^*(f_1 g_1(\mathbf{c})) = f_1 g_1(\pi_0(\mathbf{c})) = f_1 g_1(\pi_1(\mathbf{d})) = \pi_1^*(f_1 g_1(\mathbf{d}))$$

so the right hand side of (1) is also equivalent to $g_1(\mathbf{d}) \in \mathcal{C}(f_1(g_1(\mathbf{d})))$. □

Lemma 5.10. *Consider formulas ϕ in the language of M with added predicates for $\{(\nu,\lambda) : \nu \in \mathcal{C}(\lambda)\}$ and $\{f : f \in j``M_0\}$ and constants for members of $j``M_0$, and say that a formula ϕ is Δ_0 if it has no quantifiers (not even bounded quantifiers).*

(1) *Suppose that ϕ is Σ_1 and that $\mathbf{c} \equiv_\omega \mathbf{c}'$. Then $M \models \phi(\mathbf{c}) \iff \phi(\mathbf{c}')$.*
(2) *Suppose that ϕ is Σ_2 and $\mathbf{c} \equiv_{\omega^2} \mathbf{c}'$. Then $M \models \phi(\mathbf{c}) \iff \phi(\mathbf{c}')$.*

Proof. By corollary 5.9 we see that if ϕ is Δ_0 and $\mathbf{c} \equiv \mathbf{c}'$ then $M \models \phi(\mathbf{c}) \iff \phi(\mathbf{c}')$. Now let $\phi(\mathbf{c})$ be the Σ_1 formula $\exists x\, \psi(x, \mathbf{c})$, and suppose that $M \models \phi(\mathbf{c})$. Let $x = fg(\mathbf{d})$ be such that $M \models \psi(x, \mathbf{c})$. If $\mathbf{c}' \equiv_\omega \mathbf{c}$ then we can find $\mathbf{d}'$ extending $\mathbf{c}'$ so that $\mathbf{c}', \mathbf{d}' \equiv \mathbf{c}, \mathbf{d}$, but this implies that $M \models \psi(fg(\mathbf{d}'), \mathbf{c}')$ and hence $M \models \phi(\mathbf{c}')$. This proves clause (1).

Now suppose that $\mathbf{c} \equiv_{\omega^2} \mathbf{c}'$ and that $M \models \phi(\mathbf{c})$, where $\phi(\mathbf{c})$ is the Σ_2 formula $\exists x\, \psi(x, \mathbf{c})$. Pick $x = fg(\mathbf{d})$ so that $M \models \psi(x, \mathbf{c})$. Then there is an extension $\mathbf{d}'$ of $\mathbf{c}'$ such that $\mathbf{c}', \mathbf{d}' \equiv_\omega \mathbf{c}, \mathbf{d}$. Since ψ is a Π_1 formula, clause (1) implies that $M \models \psi(fg(\mathbf{d}), \mathbf{c}) \iff \psi(fg(\mathbf{d}'), \mathbf{c}')$. Thus we have $M \models \psi(fg(\mathbf{d}'), \mathbf{c}')$ and hence $M \models \phi(\mathbf{c}')$. □

Suppose that $\nu \in \bigcup_\lambda \mathcal{C}(\lambda) \setminus I$. Then there is a unique λ such that $\nu \in \mathcal{C}(\lambda)$ and $\lambda \in f``\nu$ for some $f \in j``M_0$. We write $\lambda(\nu)$ for this unique λ. If ν is a strict accumulation point, ie, ν is an accumulation point for some ordinal $\gamma > \lambda$, then there is a largest such γ, which will either be $O^M(\text{crit}(\mathcal{E}(\lambda(\nu))))$ or the least ordinal such that $\gamma \in f``\nu$ for some function $f \in j``M_0$ and $\nu \cap \bigcup_{\gamma' \geq \gamma} \mathcal{C}(\gamma')$ is bounded in ν. In either case $\gamma \in f``\nu$ for some f. We write $\gamma(\nu)$ for this γ. If ν is not a strict accumulation point then we set $\gamma(\nu) = \lambda(\nu)+1$. Finally, let $\delta(\nu) < \nu$ be the larger of $\sup(I \cap \nu)$ and $\sup \bigcup \{\mathcal{C}(\lambda') \cap \nu : \lambda' \geq \gamma(\nu)\}$. Thus $\delta(\nu) < \nu$, and in this notation the ordinal δ which appeared in either case of corollary 3.10 is written $\delta(d)$.

Lemma 5.11. *Suppose that $\nu = g(\mathbf{c}) \in \bigcup_\lambda \mathcal{C}(\lambda)$ and $\nu \notin I$. Then there is an ω^2-conservative extension $\mathbf{d}$ of $\mathbf{c}$ and functions g_0, f_1, g_1, f_2 and g_2 such that $g_i(\mathbf{d}) < \nu$ for $i = 1, 2, 3$ and*

$$\delta(\nu) = g_0(\mathbf{d}) \qquad \lambda(\nu) = f_1 g_1(\mathbf{d}) \qquad \gamma(\nu) = f_2 g_2(\mathbf{d}).$$

Furthermore, if $\mathbf{c}', \mathbf{d}' \equiv_{\omega^2} \mathbf{c}, \mathbf{d}$ and $\nu' = g(\mathbf{c}')$ then $\delta(\nu') = g_0(\mathbf{d}')$, $\lambda(\nu') = f_1 g_1(\mathbf{d}')$, and $\gamma(\nu') = f_2 g_2(\mathbf{d}')$.

Proof. We will first find the required functions f_i and g_i and the ω^2-conservative extension $\mathbf{d}$. For $\lambda(\nu)$ no extension is required. Let f_1 be such that $\lambda(\nu) \in f_1``\nu$, and take $\mathbf{d} \in [\nu \cap I]^{<\omega}$ so that $\lambda(\nu) = f_1 g_1(\mathbf{d})$ for some function g_1, with $\mathbf{d} \supset \mathbf{c}$ and $\mathbf{d}$ as small as possible. We claim that $\mathbf{d} = \mathbf{c}$.

If it is not then we can find sequences $\mathbf{c}'$, $\mathbf{d}'_0$, and $\mathbf{d}'_1$ so that $\mathbf{c},\mathbf{d} \equiv \mathbf{c}',\mathbf{d}'_0$ and $\mathbf{c},\mathbf{d} \equiv \mathbf{c}',\mathbf{d}'_1$ but $\mathbf{d}'_0 \cap \mathbf{d}'_1 = \mathbf{c}'$. Then $f_1 g_1(\mathbf{d}'_0) \neq f_1 g_1(\mathbf{d}'_1)$, since otherwise proposition 5.7 implies that there is g' so that $f_1 g_1(\mathbf{d}'_0) = f_1 g'(\mathbf{c}')$ and hence $f_1 g'(\mathbf{c}) = f_1 g_1(\mathbf{d})$, contrary to assumption. By indiscernibility and corollary 5.9 $g(\mathbf{c}')$ is in both $\mathcal{C}(f_1(g_1(\mathbf{d}'_0)))$ and $\mathcal{C}(f_1(g_1(\mathbf{d}'_1)))$, and since both $g_1(\mathbf{d}'_0)$ and $g_1(\mathbf{d}'_1)$ are less than $g(\mathbf{c}')$ this implies that $f_1 g_1(\mathbf{d}'_0) = \lambda(g(\mathbf{c}')) = f_1 g_1(\mathbf{d}'_1)$, contradicting the claim.

Pick g_0, f_2, g_2 and $\mathbf{d}'$ so that $\gamma(\nu) = f_2(g_2(\mathbf{d}'))$ and $\delta(\nu) = g_0(\mathbf{d}')$, with $\mathbf{d}' \in [I \cap \nu]^{<\omega}$. Let $\mathbf{d}$ be the sequence such that $\mathbf{d}',\mathbf{c} \equiv_\omega \mathbf{d},\mathbf{c}$ and each member of $\mathbf{d} \setminus \mathbf{c}$ is as small as possible, so that $\mathbf{d}$ is a ω^2-conservative extension of $\mathbf{c}$. We will show that $\gamma(\nu) = f_2 g_2(\mathbf{d})$ and $\delta(\nu) = g_0(\mathbf{d})$.

First we consider $\gamma(\nu)$. We can assume that ν is a strict accumulation point, since otherwise $\gamma(\nu)$ equals $\lambda(\nu)+1$. Certainly $f_2 g_2(\mathbf{d}) \leq f_2 g_2(\mathbf{d}') = \gamma(\nu)$ and it follows immediately that ν is an accumulation point for $f_2 g_2(\mathbf{d})$. Now consider the formula $\phi(\delta,\gamma,\nu)$:

$$\forall\beta\forall\alpha\forall f\ \ (\delta < \alpha < \nu \wedge f(\beta) \geq \gamma \implies \alpha \notin \mathcal{C}(f(\beta)))\,.$$

The formula $\phi(\delta,\nu,\gamma)$ implies that ν is not an accumulation point for any ordinal larger than γ. It is a Π_1 formula and $\phi(g_0(\mathbf{d}'), f_2 g_2(\mathbf{d}'), g(\mathbf{c}))$ is true, so lemma 5.10 implies that $\phi(g_0(\mathbf{d}), f_2 g_2(\mathbf{d}), g(\mathbf{c}))$ is also true. Thus $f_2 g_2(\mathbf{d}) \geq f_2 g_2(\mathbf{d}')$ so $f_2 g_2(\mathbf{d}) = f_2 g_2(\mathbf{d}') = \gamma(\nu)$

Now we show that $g_0(\mathbf{d}) = g_0(\mathbf{d}') = \delta(\nu)$. If $\delta(\nu) = \max(I \cap \nu)$, then $\delta(\nu) = \max(\mathbf{c} \cap \nu)$ since $g(\mathbf{c}) < \alpha$ for any $\alpha > \max(c \cap \nu)$ in I. In that case $g_0(\mathbf{d}') = \max(\mathbf{c}) = g_0(\mathbf{d})$. Now assume $\delta(\nu) \notin I$. Again we have $g_0(\mathbf{d}) \leq g_0(\mathbf{d}') = \delta(\nu)$. As before, the formula $\phi(g_0(\mathbf{d}), f_2 g_2(\mathbf{d}), g(\mathbf{c}))$ is true. Since $f_2 g_2(\mathbf{d}) = \gamma(\nu)$ and $g(\mathbf{c}) = \nu$ this says $\phi(g_0(\mathbf{d}), \gamma(\nu), \nu)$ which asserts that $\mathcal{C}(\gamma') \cap \nu \subset \delta$ for all $\gamma' \geq \gamma(\nu)$. But this implies that $g_0(\mathbf{d}) \geq \delta(\nu)$, so $g_0(\mathbf{d}) = \delta(\nu)$.

Now we prove the last clause of the lemma. Suppose that $\mathbf{c}',\mathbf{d}' \equiv_{\omega^2} \mathbf{c},\mathbf{d}$ where $\mathbf{d}$ and the functions g_i and f_i are as above. Corollary 5.9 implies that $g(\mathbf{c}') \in \mathcal{C}(f_1 g_1(\mathbf{d}'))$ and hence $\lambda(g(\mathbf{c}')) = f_1 g_1(\mathbf{d}')$. The formula $\phi(g_0(\mathbf{d}'), f_2 g_2(\mathbf{d}'), g(\mathbf{c}'))$ is true because $\phi(g_0(\mathbf{d}), f_2 g_2(\mathbf{d}), g(\mathbf{c}))$ is true, and thus $f_2 g_2(\mathbf{d}') \geq \gamma(g(\mathbf{c}'))$, so it is enough to show that $f_2 g_2(\mathbf{d}')$ is an accumulation point for $g(\mathbf{c}')$. The statement that ν is an accumulation point for γ is

$$\begin{aligned}&\forall\alpha\forall f'\forall\beta\exists f''\,\exists\xi\exists\mu\\ &\qquad\qquad(\alpha < \nu \wedge f'(\beta) < \gamma \wedge \beta < \nu \implies\\ &\qquad\qquad\qquad \alpha < \xi < \nu\ \wedge\ f'(\beta) \leq f''(\mu)\ \wedge\ \xi \in \mathcal{C}(f''(\mu)))\,.\end{aligned}$$

Since this is a Π_2 formula which is true of $(g(\mathbf{c}), f_2g_2(\mathbf{d}))$, lemma 5.10 implies that it holds of $(g(\mathbf{c}'), f_2g_2(\mathbf{d}'))$, so that $f_2g_2(\mathbf{d}')$ is an accumulation point for $g(\mathbf{c}')$. Thus $f_2g_2(\mathbf{d}') = \gamma(g(\mathbf{c}'))$.

If $\delta(g(\mathbf{c})) = \max(I \cap (\mathbf{c} \setminus g(\mathbf{c})))$ then $g_0(\mathbf{d}') = \max(I \cap (\mathbf{c}' \setminus g(\mathbf{c}'))) = \delta(g(\mathbf{c}'))$, so we can assume that $\delta(g(\mathbf{c}')) \in \mathcal{C}(\gamma)$ for some $\gamma \geq \gamma(g(\mathbf{c}))$. Then since we know that $f_2g_2(\mathbf{d}') = \gamma(g(\mathbf{c}'))$, the truth of $\phi(g_0(\mathbf{d}'), f_2g_2(\mathbf{d}'), g(\mathbf{c}'))$ implies that $g_0(\mathbf{d}') \geq \delta(g(\mathbf{c}'))$. The statement that $\delta \leq \delta(\nu)$ is made by the Π_2 formula

$$\forall \alpha < \delta \exists \xi \exists f' \exists \beta \big(\alpha < \xi < \nu \wedge f'(\beta) \geq \gamma(\nu) \wedge \xi \in \mathcal{C}(f(\beta))\big).$$

This is true for $(g_0(\mathbf{d}), g(\mathbf{c}))$, so it is true of $(g_0(\mathbf{d}'), g(\mathbf{c}'))$ and hence $g_0(\mathbf{d}') = \delta(g(\mathbf{c}'))$. $\square$

Proof of clause (3) *of lemma 5.6.* It will be sufficient to prove that if $g(\mathbf{c})$ is an ordinal then there is a ω^2-conservative extension $\mathbf{d}$ of $\mathbf{c}$ and a term such that $g(\mathbf{c}) = \tau(\mathbf{d})$. We prove this by induction on the size of the ordinal $g(\mathbf{c})$. Assume that it is true of every ordinal less than $\nu = g(\mathbf{c})$. If $\nu = f'(\alpha)$ for some function f' and $\alpha < \nu$ then using proposition 5.7 we can assume that $\alpha = g'(\mathbf{c})$ for some g'. By the induction hypothesis $\alpha = \tau'(\mathbf{d})$ for some term τ' and ω^2-conservative extension $\mathbf{d}$ of $\mathbf{c}$, so $\nu = \dot{f}(\tau(\mathbf{d}))$ and $\tau = \dot{f} \cdot \tau'$ is the required term.

Otherwise ν is an indiscernible, and one of the clauses of corollary 3.10 holds for $d = \nu$. Then by lemma 5.11 we can write $\delta(\nu) = g_0(\mathbf{c}')$, $\lambda(\nu) = f_1(g_1(\mathbf{c}'))$, and in the case of clause (2) $\gamma(\nu) = f_2(g_2(\mathbf{c}'))$ where $\mathbf{c}'$ is a ω^2-conservative extension of $\mathbf{c}$ and $g_i(\mathbf{c}') < \nu$ for $i = 0, 1, 2$. It follows by the induction hypothesis that there are terms τ_i and a ω^2-conservative extension $\mathbf{d}$ of $\mathbf{c}'$ such that $g_i(\mathbf{c}') = \tau_i(\mathbf{d})$ for $i = 1, 2, 3$. Then ν can be written in the form $\tau(\mathbf{c}')$, where τ is a term given by clause (4) of definition 4.2 if ν is a strict accumulation point, and clause (3) otherwise, taking $\xi < \rho^+ + \omega_1^V$ as given by corollary 3.10. $\square$

We need one more fact before we can prove the rest of lemma 5.6. By the results so far we can use induction on ξ to show, for example, that for each λ, δ and ξ there is $\xi' \geq \xi$ such that $\pi^*\big(\dot{\ell}^M_\xi(\lambda, \delta)\big) = \dot{\ell}^M_{\xi'}\big(\pi^*(\lambda), \pi^*(\delta)\big)$. The next lemma shows that the no new indiscernibles appear, and hence $\xi = \xi'$.

Lemma 5.12. *Suppose that $\nu = g(\mathbf{c})$, that $\delta(\nu) < \nu' < \nu$, and that $\nu' \in \mathcal{C}(\lambda(\nu))$ and (if ν is a strict accumulation point) ν' is an accumulation point for $\gamma(\nu)$ in $\mathcal{C}$. Then $\nu' = g'(\mathbf{c}')$ for some function g' and some ω^2-conservative extension $\mathbf{c}'$ of $\mathbf{c}$.*

Proof. If the lemma is false then $\nu' = g'(\mathbf{d})$, where $\mathbf{d}$ is not a ω^2-conservative extension of $\mathbf{c}$ and if Q is the set of $i < \text{len}(\mathbf{d})$ such that d_i is not in any ω^2-conservative extension of $\mathbf{c}$ then $g'(\mathbf{d})$ actually depends on the members d_i of $\mathbf{d}$ such that $i \in Q$. Let g_i for $i = 0, 1, 2$ and f_i for $i = 1, 2$ be as in the last lemma. Then there is a sequence $\mathbf{c}'$ such that $\mathbf{c}' \equiv_{\omega^2} \mathbf{c}$ and the members of $\mathbf{c}'$ which are not related by the functions s^I_ξ are spread out enough that there is room for $\rho^+ + \omega_1^V + 1$ many disjoint sequences, $(\mathbf{d}^\alpha : \alpha \leq \rho^+ + \omega_1^V)$, such that

$$\begin{aligned} \mathbf{d}^\alpha, \mathbf{c}' \equiv_{\omega^2} \mathbf{d}, \mathbf{c} \qquad & \text{for } \alpha \leq \rho^+ + \omega_1^V \\ d_i^\alpha < d_i^{\alpha'} \qquad & \text{for } \alpha < \alpha' \leq \rho^+ + \omega_1^V \text{ and } i \in Q. \end{aligned}$$

Then if we set $\nu_\alpha = g'(\mathbf{d}^\alpha)$ we have $\nu_\alpha < \nu_{\alpha'}$ for $\alpha < \alpha' < \rho^+ + \omega_1^V + 1$, and for all $\alpha \leq \rho^+ + \omega_1^V$ we have

$$\delta(\nu_\alpha) = g_0(\mathbf{c}') = \delta(\nu'), \qquad \lambda(\nu_\alpha) = f_1 g_1(\mathbf{c}') = \lambda(\nu'),$$

and if ν is a strict accumulation point

$$\gamma(\nu_\alpha) = f_2 g_2(\mathbf{c}') = \gamma(\nu').$$

This contradicts corollary 3.10 which says that the order type of the set of such ordinals cannot be greater than $\rho^+ + \omega_1^V$. $\square$

The remaining clauses, (1) and (2), of lemma 5.6 follow easily by an induction on the complexity of the term τ, using lemmas 5.11 and 5.12. This completes the proof of lemma 5.6 and hence of the main theorem.

Acknowledgments

The author would like to thank Tom Jech and Jean Larson, who listened to the proof and insisted on understanding it, and John Steel who read the original draft and made many useful suggestions.

This work was partially completed while the author was visiting the University of California, Los Angeles and the California Institute of Technology, and was partially supported by grant number DMS-8614447 from the National Science Foundation.

REFERENCES

[Dod82] A. Dodd, *The Core Model*, London Math. Soc. Lecture Notes 61, Cambridge University Press, Cambridge, 1982.

[Jen72] R. Jensen, *The Fine Structure of the Constructible Hierarchy*, Ann. Math. Logic **4** (1972), 229–309.

[Jen81] ———, *Some Applications of the Core Model*, Set Theory and Model Theory, Lecture Notes in Mathematics **872** (B. Jensen and A. Prestel, eds.), Springer-Verlag, New York, 1981, pp. 55–97.

[MaS69] D. A. Martin and R. Solovay, *A Basis Theorem for Π_3^1 Sets of Reals*, Annals of Mathemtics **89** (1969), 138–160.

[Mi74] W. J. Mitchell, *Sets Constructible from Sequences of Ultrafilters*, Journal of Symbolic Logic **39** (1974), 57–66.

[Mi84] ———, *The Core Model for Sequences of Measures, I*, Math. Proc. of the Cambridge Philosophical Society **95** (1984), 41–58.

[Mi87] ———, *Applications of the Core Model for sequences of measures*, trans. of the American Mathematics Society **299** (1987), 41–58.

[Mi] ———, *The Core Model for Sequences of Measures, II*, submitted to *Math. Proc. of the Cambridge Phil. Soc.*.

[Mi91a] ———, *On the Singular Cardinal Hypothesis*, Transactions of the American Mathematical Society (to appear).

[Mi91b] ———, *Definable Singularity*, Transactions of the American Mathimatical Society (to appear).

[MiS89] W. J. Mitchell and J. Steel, *Fine Structure and Iteration Trees*, (in preparation).

[Sho61] J. R. Shoenfield, *The Problem of Predicativity*, Essays on the Foundations of Mathematics (Y. Bar-Hillel, E. I. J Poznanski, M. O. Rabin and A. Robinson, eds.), The Magnes Press, Jerusalem, 1961, pp. 132–142.

[Sil71] J. Silver, *Some Applications of Set Theory in Model Theory*, Annals of Mathematical Logic **3** (1971), 45-110.

[Ste90] J. Steel, *The Core Model Iterability Problem*, Handwritten notes (June 1990).

DEPARTMENT OF MATHEMATICS, UNIVERSITY OF FLORIDA, GAINESVILLE FL 32611
E-mail address: *mitchell@math.ufl.edu*

VIVE LA DIFFÉRENCE I: NONISOMORPHISM OF ULTRAPOWERS OF COUNTABLE MODELS

SAHARON SHELAH

ABSTRACT. We show that it is not provable in ZFC that any two countable elementarily equivalent structures have isomorphic ultrapowers relative to some ultrafilter on ω.

SUMMARY

§2. Elementarily equivalent structures do not have isomorphic ultrapowers.

If V is a model of CH then in a generic extension we make $2^{\aleph_0} = \aleph_2$ and we find countable elementarily equivalent graphs Γ, Δ such that for *every* ultrafilter $\mathcal{F}$ on ω, $\Gamma^\omega/\mathcal{F} \not\simeq \Delta^\omega/\mathcal{F}$. In this model there is an ultrafilter $\mathcal{F}$ such that any ultraproduct with respect to $\mathcal{F}$ of finite structures is saturated.

§3. The case of finite graphs.

By a variant of the construction in §2 we show that there is a generic extension of V in which for some explicitly defined sequences of finite graphs Γ_n, Δ_n, all nonprincipal ultraproducts $\prod_n \Gamma_n/\mathcal{F}_1$ or $\prod_n \Delta_n/\mathcal{F}_2$, are elementarily equivalent, but no countable ultraproduct of the Γ_n is isomorphic to a countable ultraproduct of the Δ_n.

§4. The effect of $\aleph_3$ Cohen reals.

We prove that if we simply add $\aleph_3$ Cohen reals to a model of GCH, then there is at least one ultrafilter $\mathcal{F}$ such that for certain pseudorandom finite graphs Γ_n, Δ_n, the ultraproducts $\prod_n \Gamma_n/\mathcal{F}$, $\prod_n \Delta_n/\mathcal{F}$ are elementarily equivalent but not isomorphic. This implies that there are also countably infinite graphs Γ, Δ such that for the same ultrafilter $\mathcal{F}$, the ultrapowers $\Gamma^\omega/\mathcal{F}$, $\Delta^\omega/\mathcal{F}$ are elementarily equivalent and not isomorphic.

§A. Appendix.

We discuss proper forcing, iteration theorems, and the use of $(Dl)_{\aleph_2}$ in §4.

1. Introduction

Any two elementarily equivalent structures of cardinality λ have isomorphic ultrapowers (by [Sh 13], in 1971) with respect to an ultrafilter on 2^λ. Earlier, as the culmination of work in the sixties, Keisler showed, assuming $2^\lambda = \lambda^+$, that the ultrafilter may be taken to be on λ [Keisler]. In particular, assuming the continuum hypothesis, for countable structures any nonprincipal ultrafilter on ω will do. As a special case, the continuum hypothesis implies that an ultraproduct of power series rings over prime fields F_p is isomorphic to the ultrapower of the corresponding rings of p-adic integers ; this has number-theoretic consequences [AxKo]. Kim has conjectured that the isomorphism $\prod_p F_p[[t]]/\mathcal{F} \simeq \prod_p \mathbb{Z}_p/\mathcal{F}$ is valid for any nonprincipal ultrafilter over ω, regardless of the status of the continuum hypothesis. In fact it has not previously been clear what could be said about isomorphism of nonprincipal ultrapowers or ultraproducts over ω *in general*, in the absence of the continuum hypothesis; it has long been suspected that such questions do involve set theoretic issues going beyond ZFC, but there have been no concrete results in this area. For the case of two different ultrafilters and on higher cardinals, see [Sh a VI]. In particular, ([Sh a VI, 3.13]) if $M = (\omega, <)^\omega/D$ (D an ultrafilter on ω), the cofinality of $(\{a \in M : a > n$ for every natural number $n\}, >)$ can be any regular $\kappa \in (\aleph_0, 2^{\aleph_0}]$.

It does follow from the results of [Sh 13] that there is always an ultrafilter $\mathcal{F}$ on λ such that for any two elementarily equivalent models M, N of cardinality λ, $M^\omega/\mathcal{F}$ embeds elementarily into $N^\omega/\mathcal{F}$. On the other hand, we show here that it is easy to find some countable elementarily equivalent structures with nonisomorphic ultrapowers relative to a certain nonprincipal ultrafilter on ω: given enough Cohen reals, some ultrafilter will do the trick (§4), and with more complicated forcing any ultrafilter will do the trick (§2, refined in §3). The (first order theories of the) models involved have the independence property but do not have the strict order property. Every unstable theory either has the independence property or the strict order property (or both) (in nontechnical terms, in the theory we can interprate in a way the theory of the random graph or the theory of a linear order), and our work here clearly makes use of the independence property. The rings occurring in the Ax-Kochen isomorphism are unstable, but do not have the independence property, so the results given here certainly do not apply directly to Kim's problem. However it does appear that the methods used in §4 can be modified to refute Kim's conjecture, and we intend to return to this elsewhere [Sh 405].

A final technical remark: the forcing notions used here are $<\omega_1$-proper, strongly proper, and Borel. Because of improvements made in the iteration

ERRATUM

SET THEORY OF THE CONTINUUM

HAIM JUDAH, WINFRIED JUST, AND HUGH WOODIN, EDS.

In the paper "Vive la Différence I: Nonisomorphism of Ultrapowers of Countable Models" by Saharon Shelah, the following corrections should be made:

Lines 6–12 from the bottom of page 359 should read:

2.3 Notation. We work with the language of bipartite graphs (with a specified bipartition P, Q). $\Gamma_{k,l}$ is a bipartite graph with bipartition $U = U_{k,\ell}, V = V_{k,\ell}$, $|U| = k$ and $V = \dot{\bigcup}_{m<l} \binom{U}{m}$, where $\binom{U}{m}$ denotes the set of all subsets of U of cardinality m. The edge relation is membership. The theory of $\Gamma_{k,l}$ converges as $l, k/l \longrightarrow \infty$ to a complete theory which we call T_∞. Let Γ_∞ be a model of T_∞ of power $\aleph_0$ such that $\omega \subset U$ and for every $b \in V$ the set $b \cap \omega$ is finite.

theorems for proper forcing [Sh 177, Sh f], we just need the properness; in earlier versions ω-properness was somehow used.

In the appendix we give a full presentation of a less general variant of the preservation theorem of [Sh f] VI §1.

The forcing notions introduced in §2, §3 here (see 2.15, 2.16) are of interest per se. Subsequently specific cases have found more applications; see Bartoszynski, Judah and Shelah [BJSh 368], Shelah and Fremlin [ShFr 406].

2. All Ultrafilters on ω Can Be Inadequate

Starting with a model V of CH, in a generic extension we will make $2^{\aleph_0} = \aleph_2$ and find countable elementarily equivalent graphs Γ, Δ such that for any pair of ultrafilters $\mathcal{F}, \mathcal{F}'$ on ω, $\Gamma^\omega/\mathcal{F} \not\simeq \Delta^\omega/\mathcal{F}'$. More precisely:

2.1 Theorem. *Suppose $V \models$CH. Then there is a proper forcing notion $\mathcal{P}$ with the $\aleph_2$-chain condition, of cardinality $\aleph_2$ (and hence $\mathcal{P}$ collapses no cardinal and changes no cofinality) which makes $2^{\aleph_0} = \aleph_2$ and has the following effects on ultraproducts:*

(i) *There are countable elementarily equivalent graphs Γ, Δ such that no ultrapowers $\Gamma^\omega/\mathcal{F}_1$, $\Delta^\omega/\mathcal{F}_2$ are isomorphic.*

(ii) *There is a nonprincipal ultrafilter $\mathcal{F}$ on ω such that for any two sequences Γ_n, Δ_n of finite models for a countable language, if their ultrapowers with respect to $\mathcal{F}$ are elementarily equivalent, then these ultrapowers are isomorphic, and in fact saturated.*

2.2 Remark. The two properties *(i,ii)* are handled quite independently by the forcing, and in particular *(ii)* can be obtained just by adding random reals.

2.3 Notation. We work with the language of bipartite graphs (with a specified bipartition P, Q). $\Gamma_{k,l}$ is a bipartite graph with bipartition $U = U_{k,\ell}, V = V_{k,\ell}$, $|U| = k$ and $V = \dot{\bigcup}_{m<l} \binom{U}{m}$, where $\binom{U}{m}$ denotes the set of all subsets of U of cardinality m. The edge relation is membership. We also let Γ_∞ be the bipartite graph with $|U| = \aleph_0$ specifically $U = \omega$ and V the set of all finite subsets of U. The theory of the $\Gamma_{k,l}$ converges to that of Γ_∞ as $l, k/l \longrightarrow \infty$.

2.4 Remark. Our construction will ensure that for any sequence (k_n, l_n) with $l_n, k_n/l_n \longrightarrow \infty$ and any ultrafilters $\mathcal{F}_1, \mathcal{F}_2$ the ultraproducts $\prod_i \Gamma_{k_{n_i}, l_{n_i}}/\mathcal{F}_1$ and $\Gamma_\infty^\omega/\mathcal{F}_2$ are nonisomorphic. In particular, if Γ_{fin} is the disjoint union of the graphs $\Gamma_{2^n,n}$, and Γ is the disjoint union of Γ_{fin} and Γ_∞, then Γ_{fin} and Γ are elementarily equivalent, but any isomorphism of

$\Gamma^\omega/\mathcal{F}$ and $\Gamma^\omega_{\text{fin}}/\mathcal{F}$ would induce an isomorphism of an ultrapower of Γ_∞ with some ultraproduct $\prod_i \Gamma_{2^{n_i},n_i}/\mathcal{F}$. (Note that the graphs under consideration have connected components of diameter at most 4.)

2.5 The model. We will build a model N of ZFC by iterating suitable proper forcing notions with countable support [Sh b], see also [Jech]. The model N will have the following combinatorial properties:

- P1. If $(A_n)_{n<\omega}$ is a collection of finite sets with $|A_n| \longrightarrow \infty$, and $g : \omega \longrightarrow \omega$ with $g(n) \longrightarrow \infty$, and f_i $(i < \omega_1)$ are functions from ω to ω with $f_i \in \prod_n A_n$ for all $i < \omega_1$, then there is a function H from ω to finite subsets of ω such that: $H(n)$ has size at most $g(n)$; $H(n) \subseteq A_n$; and for each i, $H(n)$ contains $f_i(n)$ if n is sufficiently large (depending on i).
- P2. ${}^\omega\omega$ has true cofinality ω_1, that is: there is a sequence $(f_i)_{i<\omega_1}$ which is cofinal in ${}^\omega\omega$ with respect to the partial ordering of eventual domination (given by "$f(n) < g(n)$ for sufficiently large n").
- P3. For every sequence $(A_k : k < \omega)$ of finite sets, for any collection $B_i (i < \omega_1)$ of infinite subsets of ω, and for any collection $(g_i)_{i<\omega_1}$ of functions in $\prod_k A_k$, there is a function $f \in \prod_k A_k$ such that for all $i, j < \omega_1$, the set $\{n \in B_i : f(n) = g_j(n)\}$ is infinite.
- P4. $2^{\aleph_1} = \aleph_2$.

Note that (P3,P4) imply $2^{\aleph_0} = \aleph_2$.

2.6 Proposition. *Any model N of ZFC with properties (P1-P2) satisfies part (i) of Theorem 2.1. More precisely, the following weak saturation property holds for any ultraproduct $\Gamma^* = \prod_n \Gamma_{k_n,l_n}/\mathcal{F}$ for which $l_n \longrightarrow \infty$, $(\ell_n < k_n)$ and fails in any countably indexed ultrapower of Γ_∞ : (†) Given ω_1 elements of U^{Γ^*}, some element of V^{Γ^*} is linked to each of them.*

Proof. Our discussion in Remark 2.4 shows that it suffices to check the claim regarding (†). First consider an ultraproduct $\Gamma^* = \prod_n \Gamma_{k_n,l_n}/\mathcal{F}$ for which $l_n \longrightarrow \infty$, $l_n < k_n$.

Given $\aleph_1$ elements $a_i = f_i/\mathcal{F} \in \Gamma^*$ we apply (P1) with $g(n) = l_n - 1, A_n = U_{k_n,l_n}$. H picks out a sequence of small subsets of U_{k_n,l_n}, and if $b \in V^{\Gamma^*}$ is chosen so that its n-th coordinate is linked to all the elements of $H(n)$, then this does the trick.

Now let Γ^* be of the form $\Gamma^\omega_\infty/\mathcal{F}$. We will show that (†) fails in this model. Let $(f_i : i < \omega_1)$ be a cofinal increasing sequence in ${}^\omega\omega$, under the partial ordering given by eventual domination. Remember $U^{\Gamma_\infty} = \omega$. Let $a_i = f_i/\mathcal{F}$ for $i < \omega_1$. Let $b \in V^{\Gamma^*}$ be represented by the sequence b_n of elements of V in Γ_∞. Let B_n be the subset of U^{Γ_∞} coded by b_n; we may suppose it is never empty. Define $g(n) = \sup B_n$ and let i be chosen so that f_i dominates g eventually. Then off a finite set we have $f_i(n) \notin B_n$, and hence in Γ^*, a_i and b are unlinked. □

2.7 Proposition. *Any model N of ZFC with the properties (P3,P4) satisfies part (ii) of Theorem 2.1.*

Proof. We must construct an ultrafilter $\mathcal{F}$ on ω such that any ultraproduct of finite structures with respect to $\mathcal{F}$ is saturated. The construction takes place in $\aleph_2$ steps; at stage $\alpha < \aleph_2$ we have a filter $\mathcal{F}_\alpha$ generated by a subfilter of at most $\aleph_1$ sets $(B_i)_{i<\omega_1}$ containing the cobounded subsets of ω, and we have a type $p = (\varphi_i)_{i<\omega_1}$ over some ultraproduct $\prod_k A_k/\mathcal{F}$ of finite structures to realize. (More precisely, since the filter $\mathcal{F}$ has not yet been constructed, the "type" p is given as a set of pairs $(\varphi_i(x;\bar{y}), \bar{f}^{(i)})$ where $\bar{f}^{(i)} = \langle f_1^{(i)}, \ldots\rangle$ with $f_j^{(i)} \in \prod_k A_k$, p is closed under conjunction, and p is consistent in a strong sense: for each φ_i there is a function g_i such that $\varphi_i(g_i(n); f_1^{(i)}(n), \ldots)$ holds for all n in some set which has already been put into $\mathcal{F}$.) By (P4) we can arrange the construction so that at a given stage α we only have to deal with one such type.

By (P3) there is a function $f \in \prod_k A_k/\mathcal{F}$ such that for all $i, j < \omega_1$, the set $\{n \in B_j : f(n) = g_i(n)\}$ is infinite, where g_i witnesses the consistency of φ_i. We adjoin to $\mathcal{F}$ all of the sets $X_i = \{n \in \omega : \varphi_i(f(n); f_1^{(i)}(n), \ldots)\}$. The resulting filter is nontrivial, and is again generated by at most $\aleph_1$ sets. Furthermore our construction ensures that $f/\mathcal{F}$ will realize the type $p = \{\varphi_i(x; f_1^{(i)}/\mathcal{F}, \ldots)\}$ in the ultraproduct.

One may also take care as one proceeds to ensure that the filter which is being constructed will be an ultrafilter. □

2.8 Outline of the construction

In the remainder of this section we will manufacture a model N of ZFC with the properties P1-P4 specified in 2.5. We will use a countable support iteration of length ω_2 of ${}^\omega\omega$-bounding proper forcing notions of cardinality at most $\aleph_1$, starting from a model M of GCH. (See the Appendix for definitions and an outline of relevant results.) By [Sh 177] or [Sh f] VI§2 or A2.3 here, improving the iteration theorem of [Sh b, Theorem V.4.3], countable support iteration preserves the property:

"${}^\omega\omega$-bounding and proper".

Thus every function $f : \omega \longrightarrow \omega$ in N is eventually dominated by one in M, and property P2 follows: ${}^\omega\omega$ has true cofinality ω_1 in N. Our construction also yields P4: $2^{\aleph_1} = \aleph_2$. The other two properties are more specifically combinatorial, and will be ensured by the particular choice of forcing notions in the iteration. The next two propositions state explicitly that suitable forcing notions exist to ensure each of these two properties; it will then remain only to prove these two propositions.

2.9 Proposition. *Suppose that $(A_n)_{n<\omega}$ is a collection of finite sets with $|A_n| \longrightarrow \infty$, and $g : \omega \longrightarrow \omega$ with $g(n) \longrightarrow \infty$. Then there is a proper ${}^\omega\omega$-bounding forcing notion $\mathcal{P}$ such that for some $\mathcal{P}$-name $\underset{\sim}{H}$ the following holds in the corresponding generic extension:*

> *$\underset{\sim}{H}$ is a function with domain ω with $\underset{\sim}{H}(n) \subseteq A_n$ and $|\underset{\sim}{H}(n)| \leq g(n)$ for all relevant n, and for every $f \in \prod_n A_n$ in the ground model, we have $f(n) \in \underset{\sim}{H}(n)$ if n is sufficiently large (depending on f).*

2.10 Proposition. *Suppose M is a model of ZFC, and $(A_k : k < \omega)$ is a sequence of finite sets in M. Then there is an ${}^\omega\omega$-bounding proper forcing notion such that in the corresponding generic extension we have a function $\underset{\sim}{\eta} \in \prod_k A_k$ satisfying: for all $f \in \prod_k A_k$ and infinite $B \subseteq \omega$, both in M, $\underset{\sim}{\eta}$ agrees with f on an infinite subset of B.*

We give the proof of Proposition 2.10 first.

2.11 Definition. For $\mathcal{A} = (A_k : k < \omega)$ a sequence of finite sets of natural numbers, for simplicity $|A_k| \geq 2$ for every k, let $\mathcal{Q}(\mathcal{A})$ be the set of pairs (T, K) where $T \subseteq {}^{\omega >}\omega$ is a tree and $K : T \longrightarrow \omega$, such that for all η in T we have:

1. $\eta(l) \in A_l$ for $l < \text{len}(\eta)$.
2. For any $k \geq K(\eta)$ and $x \in A_k$ there is ρ in T extending η with $\rho(k) = x$.

We take $(T', K') \geq (T, K)$ iff T' is a subtree of T. By abuse of notation, we may write "T" for "(T, K)" with $K(\eta)$ the minimal possible value, and we may ignore the presence of K in other ways.

We use $\mathcal{Q}(\mathcal{A})$ as a forcing notion: the intersection of a generic set of conditions defines a function $\underset{\sim}{\eta} \in \prod_k A_k$, called the *generic branch.*

We also define partial orders $\leq_m$ on $\mathcal{Q}(\mathcal{A})$ as follows. $T \leq_m T'$ iff $T \leq T'$ and:

1. $T \cap {}^{m \geq}\omega = T' \cap {}^{m \geq}\omega$;
2. $K(\eta) = K'(\eta)$ for $\eta \in T \cap {}^{m \geq}\omega$.

Note the *fusion* property: if (T_n) is a sequence of conditions with $T_n \leq_n T_{n+1}$ for all n, then sup T_n exists (and is a condition). We pay attention to K in this context.

2.12 Remark. With the notation of 2.11, $\mathcal{Q}(\mathcal{A})$ forces:

For any $f \in \prod_k A_k$ and infinite $B \subseteq \omega$, both in the ground model, the generic branch $\underset{\sim}{\eta}$ agrees with f on an infinite subset of B.

2.13 Proof of Proposition 1.10. It suffices to check that $\mathcal{Q}(\mathcal{A})$ is an ${}^{\omega}\omega$-bounding proper forcing notion. We claim in fact:

(*) Let $(T, K) \in \mathcal{Q}(\mathcal{A})$, $m < \omega$, and let $\underset{\sim}{\alpha}$ be a $\mathcal{Q}(\mathcal{A})$-name for an ordinal. Then there is T', $T \leq_m T'$ such that for some finite set w of ordinals, $T' \Vdash$ "$\underset{\sim}{\alpha} \in w$".

This condition implies that $\mathcal{Q}(\mathcal{A})$ is ${}^{\omega}\omega$-bounding, since given a name $\underset{\sim}{f}$ of a function in ${}^{\omega}\omega$, we can find a sequence of conditions T_n and finite sets w_n of integers such that (T_n) is a fusion sequence (i.e. $T_n \leq_n T_{n+1}$ for all n) and $T_n \Vdash$ "$\underset{\sim}{f}(n) \in w_n$"; then $T = \sup\ T_n$ forces "$\underset{\sim}{f}(n) \leq \max\ w_n$ for all n".

At the same time, the condition $(*)$ is stronger than Baumgartner's Axiom A, which implies α-properness for all countable α.

It remains to check $(*)$. We fix T (and the corresponding function $K : T \longrightarrow \omega$), $\underset{\sim}{\alpha}$, m as in $(*)$. For $\nu \in T$ let T^{ν} be the restriction of T to the set of nodes comparable with ν. For ν in T, pick a condition (T_ν, K_ν) by induction on $\text{len}(\nu)$ such that $T_\nu \geq T^{\nu}$ and $\eta \lhd \nu \& \nu \in T_\eta \Longrightarrow T_\nu \geq T_\eta$ and $T_\nu \Vdash$ "$\underset{\sim}{\alpha} = \alpha_\nu$" for some α_ν. We may suppose $K_\nu \geq K$ on T_ν. Set $k_0 = m$, and define k_l inductively by

$$k_{l+1} =: \max(k_l + 1, \max\{K_\eta(\eta) + 1 : \eta \in T \cap {}^{k_l}\omega\}).$$

Let $(\eta_j)_{j=2,\ldots,N}$ be an enumeration of $T \cap {}^{\leq k_1}\omega$. (It is convenient to begin counting with 2 here.) For $\nu \in T$ with $\nu \restriction k_1 = \eta_j$, we will write $j = j(\nu)$.

Let T' be:

$$\{\eta : \exists \nu \in T \text{ extending } \eta,\ \text{len}(\nu) \geq k_N,\ \text{and } \nu \in T_{\nu \restriction k_{j(\nu)}}\}$$

Observe that for η of length at least k_N, the only relevant ν in the definition of T' is η itself. That is, $\eta \in T'$ if and only if $\eta \in T_{\eta \restriction k_{j(\eta)}}$. In particular T' is a condition (with $K'(\eta) \leq K_{\eta \restriction k_{j(\eta)}}(\eta)$ for $\text{len}(\eta) \geq k_N$). Also, since $T' \cap {}^{k_N \geq}\omega \subseteq \bigcup\{T_\nu : \nu \in T \cap {}^{k_N \geq}\omega\}$, we find $T' \Vdash$ "$\underset{\sim}{\alpha} \in \{\alpha_\nu : \nu \in T \cap {}^{k_N \geq}\omega\}$". Notice also that $T' \restriction k_1 = T \restriction k_1$.

The main point, finally, is to check that we can take $K' = K$ on $T' \cap {}^{m \geq}\omega$. Fix $\eta_j \in T' \cap {}^{m \geq}\omega$, $k \geq K(\eta_j)$, and $x \in A_k$; we have to produce an extension ν of η_j in T', with $\nu(k) = x$. Let η_h be an extension of η_j of length k_1, such that η_h has an extension $\nu \in T$ with $\nu(k) = x$. If $k < k_h$, then $\nu{\restriction}(k+1) \in T'$, as required.

Now suppose $k \geq k_{h+1}$, and let η be an extension of η_h of length k_h. Then $T_\eta \subseteq T'$, and $k \geq K_\eta(\eta)$. Thus a suitable ν extending η exists.

We are left only with the case: $k \in [k_h, k_{h+1})$. In particular $k \geq k_2$, so $k > K(\eta_h)$ for all η_h in $T \cap {}^{k_1\geq}\omega$. This means that any extension of $\eta_{h'}$ of η of length k_1 could be used in place of our original choice of η_h. Easily there is such $h' \neq h$ (remember $|A_k| \geq 2$ and demand on K). But k cannot lie in two intervals of the form $[k_h, k_{h+1})$, so we must succeed on the second try. $\square$

2.14 Logarithmic measures

We will define the forcing used to prove Proposition 2.9 in 2.16 below. Conditions will be perfect trees carrying extra information in the form of a (very weak) "measure" associated with each node. These measures may be defined as follows.

For a a set, we write $P^+(a)$ for $P(a) \setminus \{\emptyset\}$. A *logarithmic measure* on a is a function $\| \, \| : P^+(a) \longrightarrow \mathbb{N}$ such that:

1. $x \subseteq y \Longrightarrow \|x\| \leq \|y\|$;
2. If $x = x_1 \cup x_2$ then for some $i = 1$ or 2, $\|x_i\| \geq \|x\| - 1$.

By (1), $\| \, \|$ has finite range. If a is finite (as will generally be the case in the present context), one such logarithmic measure is $\|x\| = \lfloor \ln_2 |x| \rfloor$.

2.15 The forcing notion $\mathcal{L}T$

We will force with trees such that the set of successors of any node carries a specified logarithmic measure; the measures will be used to prevent the tree from being pruned too rapidly. The formal definition is as follows.

1. $\mathcal{L}T$ is the set of pairs (T, t) where:
   1.1. T is a subtree of ${}^{\omega>}\omega$ with finite stem; this is the longest branch in T before ramification occurs. We call the set of nodes of T which contain the whole stem the *essential part* of T; so T will consist of its essential part together with the proper initial segments of its stem. We denote the essential part of T by $\mathrm{ess}(T)$.
   1.2. t is a function defined on the essential part of T, with $t(\eta)$ a logarithmic measure on the set $\mathrm{succ}_T(\eta)$ of all successors of η in T; we often write $\| \, \|_\eta$ (or possibly $\| \, \|_\eta^T$) for $t(\eta)$. For η a proper initial segment of the stem of T, we stipulate $t(\eta)[\mathrm{succ}(\eta)] = 0$.
2. The partial order on $\mathcal{L}T$ is defined by: $(T_2, t_2) \geq (T_1, t_1)$ iff $T_2 \subseteq T_1$, and for $\eta \in T_2$ $t_2(\eta)$ is the restriction of $t_1(\eta)$ to $P^+(\mathrm{succ}_{T_2}(\eta))$.

3. We define $\mathcal{LT}^{[(T,t)]}$ to be $\{(T',t') \in \mathcal{LT} : (T',t') \geq (T,t)\}$ with the induced order. Similarly for $\mathcal{LT}^f$, $\mathcal{LT}_d$, and $\mathcal{LT}_d^f$ (see below).

2.16 The forcing notion $\mathcal{LT}_d^f$

$\mathcal{LT}^f$ is the set of pairs $(T,t) \in \mathcal{LT}$ in which T has only *finite* ramification at each node.

$\mathcal{LT}_d$ is the set of pairs $(T,t) \in \mathcal{LT}$ such that for any m, every branch of T is almost contained in the set $\{\eta \in T : \forall \nu \geq \eta \; \|\operatorname{succ}_T(\nu)\|_\nu \geq m\}$ (i.e. the set difference is finite).

$\mathcal{LT}_d^f$ is $\mathcal{LT}^f \cap \mathcal{LT}_d$. For $T \in \mathcal{LT}^f$, an equivalent condition for being in $\mathcal{LT}_d^f$ is: $\lim_k \inf\{\|\operatorname{succ}_T(\eta)\|_\eta : \operatorname{len}(\eta) = k\} = \infty$. Note: $\mathcal{LT}_d^f$ is an upward closed subset of $\mathcal{LT}_d$.

We make an observation concerning fusion in this connection. Define:

1. $(T_1,t_1) \leq^* (T_2,t_2)$ if $(T_1,t_1) \leq (T_2,t_2)$ and in addition for all $\eta \in \operatorname{ess} T_2$, $\|\operatorname{succ}_{T_2}(\eta)\|_\eta^{T_2} \geq \|\operatorname{succ}_{T_1}(\eta)\|_\eta^{T_1} - 1$.
2. $(T_1,t_1) \leq_m (T_2,t_2)$ if $(T_1,t_1) \leq (T_2,t_2)$ and for all $\eta \in T_2$ with $\|\operatorname{succ}_{T_1}(\eta)\|_\eta \geq m$, (so $\eta \in \operatorname{ess}(T_1)$) we have $\|\operatorname{succ}_{T_2}(\eta)\|_\eta \geq m$ (hence $\eta \in \operatorname{ess}(T_2)$ when $m > 0$).
3. $(T_1,t_1) \leq_m^* (T_2,t_2)$ if $(T_1,t_1) \leq_m (T_2,t_2)$ and for all $\eta \in T_2$ with $\|\operatorname{succ}(\eta)\|_\eta^{T_1} \leq m$, we have $\operatorname{succ}_{T_1}(\eta) \subseteq T_2$.

If (T_n,t_n) is a sequence of conditions in $\mathcal{LT}_d^f$ with $(T_n,t_n) \leq_n^* (T_{n+1},t_{n+1})$ for all n, then $\sup(T_n,t_n)$ exists in $\mathcal{LT}_d^f$.

We also mention in passing that a similar statement holds for $\mathcal{LT}_d$, with a more complicated notation. Using arguments like those given here one can show that $\mathcal{LT}_d$ is also proper. This will not be done here.

For $\eta \in T, (T,t) \in \mathcal{LT}$ we let T^η be the set of $\nu \in T$ comparable with η, $t^\eta = t \restriction \operatorname{ess}(T^\eta)$: so $(T,t) \leq (T^\eta, t^\eta)$; we may write $(T,t)^\eta$ or (T^η, t) instead of (T^η, t^η).

We will now restate Proposition 2.9 more explicitly, in two parts.

2.17 Proposition. *Suppose that $(A_n)_{n<\omega}$ is a collection of finite sets with $|A_n| \longrightarrow \infty$, and that $g : \omega \longrightarrow \omega$ with $g \longrightarrow \infty$. Then there is a condition (T_0,t_0) in $\mathcal{LT}_d^f$ such that (T_0,t_0) forces:*

There is a function $\underset{\sim}{H}$ such that $|\underset{\sim}{H}(n)| < g(n)$ for all n [more exactly, $|\underset{\sim}{H}(n)| < \max\{g(n),1\}$] , and for every f in the ground model,

$f(n) \in \underset{\sim}{H}(n)$ for n sufficiently large.

Proof. Without loss of generality $g(n) > 1$ and A_n is nonempty for every n. Let $a_n = \{A \subseteq A_n : |A| = g(n) - 1\}$, $T_0 = \bigcup_N \prod_{n<N} a_n$, and define

a logarithmic measure $\| \|_n$ on a_n by $\|x\|_n = \max\{l :$ if $A' \subseteq A_n$ has cardinality 2^l, then there is $A \in x$ containing $A'\}$. Set $t_0(\eta) = \| \|_{\text{len}\,\eta}$.

Obviously $(T_0, t_0) \in \mathcal{LT}_d^f$, (a pedantic reader will note $T_0 \not\subseteq {}^{\omega>}\omega$ and rename) For a generic branch $\underset{\sim}{\eta}$ of T_0:

$(T_0, t_0) \Vdash_{\mathcal{LT}_d^f}$ "$|\underset{\sim}{\eta}(n)| < g(n)$ for all n;"
$(T_0, t_0) \Vdash_{\mathcal{LT}_d^f}$ "For f in the ground model, $f(n) \in \underset{\sim}{\eta}(n)$ for all large n."

□

2.18 Proposition. *The forcing notion $\mathcal{LT}_d^f$ is ${}^\omega\omega$-bounding and proper.*

It remains only to prove this proposition.

2.19 Lemma. *If $(T,t) \in \mathcal{LT}_d$ and W is a subset of T, then there is some $(T',t') \in \mathcal{LT}_d$ with $(T,t) \leq^* (T',t')$ such that either:*

(+) *every branch of T' meets W; or else*
(−) *T' is disjoint from W.*

Proof. Let T^W be the set of all $\eta \in T$ for which there is a condition (T',t') such that T' has stem η, $(T^\eta, t) \leq^* (T',t')$, and every infinite branch of T' meets W. (T^η is the set of $\nu \in T$ comparable to η; so it is a tree whose stem contains η.)

If the stem of T is in T^W we get (+). Otherwise we will construct $(T',t') \in \mathcal{LT}_d$ such that (−) holds, $(T,t) \leq^* (T',t')$, and $T' \cap T^W = \emptyset$. For this we define $T' \cap {}^n\omega$ (and $t' = t \restriction \text{ess}(T')$) inductively.

If $n \leq \text{len}(\text{stem}(T))$ then we let $T' \cap {}^n\omega$ be $\{\text{stem}(T)\restriction n\}$.

So suppose that $n \geq \text{len}(\text{stem}T)$ and that we have defined everything for $n' \leq n$. Let $\nu \in T' \cap {}^n\omega$, and in particular, $\nu \notin T^W$. Let $a = \text{succ}_T(\nu)$, $a_1 = a \cap T^W$, $a_2 = a \backslash a_1$. Then for some $i = 1$ or 2, $\|a_i\|_\nu \geq \|a\|_\nu - 1$.

Since $\nu \notin T^W$, it follows easily that $\|a_1\|_\nu < \|a\|_\nu - 1$; otherwise one pastes together the conditions $(T_{\nu'}, t_{\nu'})$ associated with $\nu' \in a_1$ to show $\nu \in T^W$. Thus $\|a_2\|_\nu \geq \|a\|_\nu - 1$. Let $T' \cap (\text{succ}_T(\nu))$ be a_2. As we can do this for all $\nu \in T' \cap {}^n\omega$, this completes the induction step. □

2.20 Lemma. *If $\underset{\sim}{\alpha}$ is an $\mathcal{LT}_d^f$-name of an ordinal, $(T,t) \in \mathcal{LT}_d^f$, $m < \omega$, and $\|\text{succ}_T\,\eta\|_\eta > m$ for $\eta \in \text{ess}(T)$, then there is $(T',t') \in \mathcal{LT}_d^f$ with $(T,t) \leq_m (T',t')$, and a finite set w of ordinals, such that $(T',t') \Vdash_{\mathcal{LT}_d^f}$ "$\underset{\sim}{\alpha} \in w$".*

Proof. Let W be the set of nodes ν of T for which there is a condition (T_ν, t_ν) with $(T_\nu, t_\nu)\,{}_m\!\geq (T^\nu, t^\nu)$ such that (T_ν, t_ν) forces a value on $\underset{\sim}{\alpha}$. We claim that for any $(T_1, t_1)\;{}^*\!\geq (T,t)$, T_1 must meet W. Indeed, fix $(T_2, t_2) \geq (T_1, t_1)$ forcing "$\underset{\sim}{\alpha} = \beta$" for some β. Then for some $\nu \in T_2$, all extensions η of ν in T_2 will satisfy $\|\operatorname{succ}_{T_2}(\eta)\|_\eta \geq m$, and $(T_2, t_2)^\nu$ witnesses the fact that $\nu \in W$. Thus if we apply Lemma 2.19, the alternative $(-)$ is not possible.

Accordingly we have some $(T_1, t_1) \,{}^*\!\geq (T, t)$ such that every branch of (T_1, t_1) meets W. Let W_0 be the set of minimal elements of W in T_1. Then W_0 is finite. For $\nu \in W_0$ select (T_ν, t_ν) with $(T_\nu, t_\nu) \,{}_m\!\geq (T, t)^\nu$ and $(T_\nu, t_\nu) \Vdash$ "$\underset{\sim}{\alpha} = \alpha_\nu$" for some α_ν. Form $T' = \bigcup\{T^\nu : \nu \in W_0\}$. □

2.21 Lemma. *If $(T, t) \in \mathcal{LT}_d^f$, $\underset{\sim}{\alpha}$ is an $\mathcal{LT}_d^f$-name of an ordinal, $m < \omega$, then there is $(T', t') \in \mathcal{LT}_d^f$ with $(T, t) \leq_m^* (T', t')$, and a finite set of ordinals w, such that $(T', t') \Vdash$ "$\underset{\sim}{\alpha} \in w$".*

Proof. Fix k so that $\|\operatorname{succ}(\eta)\|_\eta > m$ for $\operatorname{len}(\eta) \geq k$. Apply 2.20 to each T^ν for $\nu \in T$ of length $k + 1$. □

2.22 Proof of 2.18. As in 2.13, using 2.21. □

This completes the verification that the desired model N can be constructed by iterating forcing.

3. Nonisomorphic Ultraproducts of Finite Models

We continue to use the bipartite graphs $\Gamma_{k,l}$ introduced in 2.3. Varying the forcing used in §2, we will get:

3.1 Theorem. *Suppose that V satisfies CH, and that $(k_n, l_n), (k'_n, l'_n)$ are monotonically increasing sequences of pairs (and $2 < l'_n < k'_n < l_n < k_n < l'_{n+1}$) such that:*

(1) $k'_n/l'_n \longrightarrow \infty$;
(2) $(k_n/l_n) > (k'_n)^{ndl'_n}$, *for each $d > 0$, for n large enough;*
(3) $\ln l'_n > k^n_{n-1}$.

Then there is a proper forcing $\mathcal{P}$ satisfying the $\aleph_2$-cc, of size $\aleph_2$, such that in $V^{\mathcal{P}}$ no two ultraproducts $\prod \Gamma_{k_i,l_i}/\mathcal{F}_1$, $\prod \Gamma_{k'_i,l'_i}/\mathcal{F}_2$ are isomorphic.

More precisely, we will call a bipartite graph with bipartition (U, V) $\aleph_1$-*complete* if every set of ω_1 elements of U is linked to a single common element of V (property (†) of Proposition 2.6), and then our claim is that in $V^{\mathcal{P}}$, no nonprincipal ultraproduct of the first sequence Γ_{k_n,l_n} is $\aleph_1$-complete, and every nonprincipal ultraproduct of the second sequence $\Gamma_{k'_n,l'_n}$ is; furthermore, as indicated, this phenomenon can be controlled by the rates of growth of k and of l/k.

3.2 Definition. Let f, g be functions in ${}^\omega\omega$. A model N of ZFC is (f,g)-*bounded* if for any sequence $(A_n)_{n<\omega}$ of finite sets with $|A_n| = f(n)$, there are $\aleph_1$ sequences $\mathcal{B}_i = (B_{i,n} : n < \omega)$, indexed by $i < \omega_1$, with:

(1) $B_{i,n} \subseteq A_n$ for all n
(2) For all $i < \omega_1$, $|B_{i,n}| < g(n)$ eventually
(3) $\bigcup_i \prod_n B_{i,n} = \prod_n A_n$ in N

3.3 Lemma. *Let $(k_n), (l_n)$ be sequences with $l_n, k_n/l_n \longrightarrow \infty$, and let $f(n) = \binom{k_n}{l_n}$, $g(n) = k_n/l_n$. Suppose that N is a model of ZFC which is (f,g)-bounded. Then no ultraproduct $\prod_n \Gamma_{k_n,l_n}/\mathcal{F}$ can be $\aleph_1$-complete.*

Proof. Let $\mathcal{B}_i$ have properties (1-3) of 3.2 with respect to $A_n = V_{k_n,l_n}$. For each i, choose $a_i \in \prod_n U_{k_n,l_n}$ so that $a_i(n)$ is not linked to any $b \in B_{i,n}$, as long as $|B_{i,n}| < g(n)$ (so $l_n|B_{i,n}| < k_n$). Then $a_i/\mathcal{F}(i < \omega_1)$ cannot all be linked to any single b in $\prod_n \Gamma_{k_n,l_n}/\mathcal{F}$, for any ultrafilter $\mathcal{F}$. □

3.4 Definition. For functions $f, g \in {}^\omega\omega$ we say that a forcing notion $\mathcal{P}$ has the (f,g)-*bounding property* provided that:

(∗) For any sequence $(A_k : k < \omega)$ in the ground model, with $|A_k| = f(k)$, and any $\underset{\sim}{\eta} \in \prod_k A_k$ in the generic extension, there is a "cover" $\mathcal{B} = (B_k : k < \omega)$ in the ground model with $B_k \subseteq A_k$, $|B_k| < g(k)$ (more exactly, $< \text{Max}\{g(k), 2\}$), and $\underset{\sim}{\eta}(k) \in B_k$ for each k.

Similarly a forcing notion has the $(\boldsymbol{F}, g)$-bounding property, for $\boldsymbol{F}$ a collection of functions, if it has the (f, g^ε)-bounding property for each $f \in \boldsymbol{F}$ and each $\varepsilon > 0$. In this terminology, notice that $(\{f\}, g)$-bounding is a stronger condition than (f,g)-bounding.

3.4A Definition. Call a family $\boldsymbol{F}$ *g-closed* if it satisfies the following two closure conditions:

1. For $f \in \boldsymbol{F}$, the function $F(n) = \prod_{m<n}(f(m)+1)$ lies in $\boldsymbol{F}$;
2. For $f \in \boldsymbol{F}$, f^g is in $\boldsymbol{F}$.

Proof of 3.1. We build a model N of ZFC by an iteration of length ω_2 with countable support of proper forcing notions with the $(\boldsymbol{F}, g)$ bounding property for a suitable family $\boldsymbol{F}$, all of which are of the form $(\mathcal{L}\mathcal{T}_d^f)^{[(T,t)]}$; and we arrange that all of the forcing notions of this form which are actually $(\boldsymbol{F}, g)$-bounding will occur cofinally often. (In order to carry this out one actually makes use of auxiliary functions (f_1, g_1) with f_1 eventually dominating $\boldsymbol{F}$ and g_1 eventually dominated by any positive power of g, but these details are best left to the discussion after 3.5.)

One can show that a countable support iteration of proper $(\boldsymbol{F}, g)$-bounding forcing notions is again $(\boldsymbol{F}, g)$-bounding. This is an instance of a general iteration theorem of [Sh f, VI] but we make our presentation self-contained by giving a proof in the appendix-A2.5. If we force over a ground model with CH (so that CH holds at intermediate points in the iteration) then our final model is $(\boldsymbol{F}, g)$-bounded, and by 3.3 no ultraproduct of the Γ_{k_n,l_n} can be $\aleph_1$-complete.

One very important point still remains to be checked. It may be formulated as follows.

3.5 Proposition. *Let $f_0, g_0, h : \omega \longrightarrow \omega \backslash \{0,1\}$ and suppose that $(A_n)_{n<\omega}$ is a sequence of finite nonempty sets with $|A_n| \longrightarrow \infty$. Assume:*

$$\text{(1)} \qquad \prod_{m \le n} |A_m|^{h(m)} < g_0(n) \text{ for every } n \text{ large enough };$$

$$\text{(2)} \qquad \frac{\ln h(n)}{\ln \prod_{i<n} f_0(i)} \longrightarrow \infty.$$

Then there is a condition $(T, t) \in \mathcal{L}T_d^f$ such that $(\mathcal{L}T_d^f)^{[(T,t)]}$ is (f_0, g_0)-bounding and (T, t) forces:

There is a function $\underset{\sim}{H}$ such that $\underset{\sim}{H}(n) \subseteq A_n$, $|\underset{\sim}{H}(n)| < h(n)$ for all n, and for every f in the ground model, $f(n) \in \underset{\sim}{H}(n)$ for n sufficiently large.

Continuation of the proof of 3.1. We will now check that the proof of theorem 3.1 can be completed using this proposition.

We set $f^*(n) = \binom{k_n}{l_n}$, $g(n) = k_n/l_n$, $h(n) = l'_n$, and $A_n = U_{k'_n,l'_n}$. (So $|A_n| = k'_n$.) Let $\boldsymbol{F}_0$ be the set of increasing functions f satisfying

$$\lim_{n \longrightarrow \infty} \ln h(n)/(g^d(n-1) \ln f(n-1)) \longrightarrow \infty \qquad \text{for all } d > 0.$$

If $f_0 \in \boldsymbol{F}_0$ and g_0 is a positive power of g, then conditions (1,2) of 3.5 hold by condition (2) of 3.1 (for (2) of 3.5 note for $d = 2$ that $g^d(n-1) > n$). Furthermore $\boldsymbol{F}_0$ is g-closed (this uses the fact that $g(n) \ge n$ eventually by (2) of 3.1), and $f^* \in \boldsymbol{F}_0$. By diagonalization find f_1, g_1 satisfying $(1, 2)$ of 3.5 so that f_1 eventually dominates any function in the g-closure of f^*, and g_1 is eventually dominated by any positive power of g. Apply the proposition to (f_1, g_1, h) and observe that an (f_1, g_1)-bounding forcing notion is (g-closure of f^*, g)-bounding. We let $\boldsymbol{F} = g$-closure of $\{f^*\}$.

Forcing with the corresponding $(\mathcal{LT}_d^f)^{[(T,t)]}$ produces a branch $\underset{\sim}{H}$ so that if $\underset{\sim}{H}(n)$ is thought of as an element $b_n \in V_{k'_n,l'_n}$, then for all $f \in \prod_n A_n$ in the ground model, and any ultrafilter $\mathcal{F}$ on ω, $f/\mathcal{F}$ is linked to $\underset{\sim}{H}(n)/\mathcal{F}$ in $\prod_n \Gamma_{k'_n,l'_n}/\mathcal{F}$. $\square$

3.6 Terminology

A logarithmic measure $\| \ \|$ on a is called *m-additive* if for every choice of $(a_i)_{i<m}$ with $\bigcup_i a_i = a$, there is $i < m$ with $\|a_i\| \geq \|a\| - 1$.

3.7 Lemma. *Suppose $f, g : \omega \longrightarrow \omega \setminus \{0,1\}$, $(T,t) \in \mathcal{LT}_d^f$, and:*

i. *for every $\eta \in \mathrm{ess}(T)$, $t(\eta)$ is $\prod_{i<\mathrm{len}\,\eta} f(i)$-additive;*
ii. *for every n we have $|T \cap {}^{(n+1)}\omega| < g(n)$.*

Then $(\mathcal{LT}_d^f)^{[(T,t)]}$ is (f,g)-bounding.

Proof. Let $F(n) = \prod_{i<n} f(i)$. Suppose that $(A_n)_{n<\omega} \in V$, $|A_n| \leq f(n)$, and $(T,t) \Vdash$ "$\underset{\sim}{\eta} \in \prod_n A_n$". By fusion as in 2.19–2.22 there is $(T',t') \in \mathcal{LT}_d^f$ with $(T',t') \geq (T,t)$ such that for every n the set

$$W =: \{\nu \in T' : (T'^{\nu}, t') \text{ forces a value on } \underset{\sim}{\eta}(n)\}$$

meets every branch of (T',t').

For each n, choose $N(n)$ large enough that (T'^{ν}, t') forces a value η_ν^n on $\underset{\sim}{\eta}\restriction n$ for each $\nu \in T' \cap {}^{N(n)}\omega$. Thus $\eta_\nu^n \in \prod_{i<n} A_i$. By downward induction on $k < N(n)$, for $\nu \in T' \cap {}^k\omega$ choose $\eta_\nu^n \in {}^n\omega$ and $s(\nu,n) \subseteq \mathrm{succ}_{T'}(\nu)$ so that:

$$\|s(\nu,n)\|_\nu \geq \|\mathrm{succ}_{T'}(\nu)\|_\nu - 1;$$
$$\eta_\nu^n \restriction \min\{k,n\} = \eta_{\nu'}^n \restriction \min\{k,n\} \text{ for } \nu' \in s(\nu,n).$$

Since $|\{\eta_{\nu'}\restriction \min(k,n) : \nu' \in \mathrm{succ}_{T'}(\nu)\}| \leq F(k)$ and $\| \ \|_\nu$ is $F(k)$-additive, this is easily done. Let $T'_n = \{\nu \in T' : (\forall l < \mathrm{len}(\nu) \cap N(n))\ \nu\restriction(l+1) \in s(\nu\restriction l, n)\}$.

We now define $T'' \subseteq T'$ so that for all k the set X_k of n for which $T'' \cap {}^{k\geq}\omega = T'_n \cap {}^{k\geq}\omega$ is infinite. For this we proceed by induction on k. If $T'' \cap {}^{k\geq}\omega$ has been defined, then we can select $X \subseteq X_k$ infinite such that for $n \in X$ and $\nu \in T'' \cap {}^k\omega$, $s(\nu,n) = s(\nu)$ is independent of n. We then define

$$T'' \cap {}^{(k+1)}\omega = \{\nu \in T' \cap {}^{k+1}\omega : \nu\restriction k \in T'' \cap {}^k\omega \text{ and } \nu \in s(\nu\restriction k)\}$$

Observe that $(T'', t\restriction T'') \mathrel{{}^*{\geq}} (T',t')$, and $(T'', t\restriction T'')$ forces:

"For any k, if $n \in X_{N(k)}$ and $n \geq k$, then

$$\underset{\sim}{\eta}\restriction k = \eta_\nu^n \restriction k \text{ for some } \nu \in T'' \cap {}^k\omega\text{''}.$$

Indeed, for any ν' of length $N(k)$ in T'', if $\nu' \in T'_n$ then $\eta^k_{\nu'} = \eta^n_{\nu'}\restriction k = \eta^n_{\nu'\restriction k}\restriction k$. Since $|T'' \cap {}^{k+1}\omega| \le |T \cap {}^{k+1}\omega| < g(k)$, this yields the stated bounding principle. □

3.8 Proof of 3.5. Let $F_0(n) = \prod_{i<n} f_0(i)$. Let $a_n = \{A \subseteq A_n : |A| = h(n) - 1\}$, $T_0 = \bigcup_N \prod_{n<N} a_n$, and define a logarithmic measure $\| \; \|_n$ on a_n by: for $a \subseteq a_n$

$$\|a\|_n = \max\{l : \text{ for all } A' \subseteq A_n \text{ of cardinality} \le F_0(n)^l, \\ \text{there is } A \in a \text{ containing } A'\}.$$

Set $t_0(\eta) = \| \; \|_{\mathrm{len}\,\eta}$.

Obviously $\| \; \|_n$ is $F_0(n)$-additive and $|T \cap {}^{(n+1)}\omega| = \prod_{m \le n}(|A_m|)^{(h(m)-1)}$ which is (by condition (1) of 3.5) $< g_0(n)$, so $(\mathcal{L}T^f_d)^{[(T_0,t_0)]}$ is (f_0, g_0)-bounding by lemma 3.7.

We need to check that $\|a_n\|_n \longrightarrow \infty$:

$$\|a_n\|_n = \max\{l : F_0(n)^l < h(n)\} \sim \frac{\ln h(n)}{\ln F_0(n)}.$$

So (2) from 3.5 guarantees it. □

4. Adding Cohen Reals Creates a Bad Ultrafilter

In this section we show that a weaker form of the results in §§2, 3 is obtained just by adding $\aleph_3$ Cohen reals to a suitable ground model. This result was actually the first one obtained in this direction. This construction is also used in [Sh 345] and again in [Sh 405].

4.1 Theorem. *If we add $\aleph_3$ Cohen reals to a model of $[2^{\aleph_i} = \aleph_{i+1}$ $(i = 1, 2)$ & $\diamondsuit_{\{\delta<\aleph_3:\mathrm{cof}\,\delta=\aleph_2\}}]$, then there will be a nonprincipal ultrafilter $\mathcal{F}$ on ω and two sequences of pseudorandom finite graphs (Γ^1_n), (Γ^2_n) such that $\prod_n \Gamma^1_n/\mathcal{F} \not\simeq \prod_n \Gamma^2_n/\mathcal{F}$. In fact the same result will apply if the sequences Γ^1_n, Γ^2_n are replaced by any subsequences.*

Here we call a sequence (Γ^1_n) of finite graphs *pseudorandom* if the theory of Γ^1_n converges fairly rapidly to the theory of the random infinite graph; cf. 4.4 below. The only condition needed on the two sequences in Theorem 4.1 is that the Γ^1_m and Γ^2_n are of radically different sizes (4.5 below). As a variant (with very much the same proof) we can take all Γ^2_n equal to the random infinite graph, keeping (Γ^1_n) a sequence of pseudorandom finite graphs, and obtain the same result for a suitable ultrafilter.

4.2 Corollary. *Under the hypotheses of Theorem 4.1 there are elementarily equivalent countable graphs* Γ^1_ω, Γ^2_ω *and a nonprincipal ultrafilter* $\mathcal{F}$ *on* ω *with* $(\Gamma^1_\omega)^\omega/\mathcal{F} \not\simeq (\Gamma^2_\omega)^\omega/\mathcal{F}$.

This is proved much as in Remark 2.4, noting that large pseudorandom graphs are connected of diameter 2.

4.3 Remark. With more effort we can replace the hypotheses on the ground model in Theorem 4.1 by:

$$2^{\aleph_i} = \aleph_{i+1}\ (i = 0, 1)\ \&\ \diamondsuit_{\{\delta<\aleph_2:\mathrm{cof}\,\delta=\aleph_1\}},$$

adding only $\aleph_2$ Cohen reals. In the definition of $\mathcal{AP}$ below, $\underset{\sim}{\mathcal{F}}$ would then not be an arbitrary name of an ultrafilter; instead $\mathcal{AP}$ would be replaced by a family of $\aleph_1$ isomorphism types of members of $\mathcal{AP}$, (using $\aleph_0$ in place of $\aleph_1$ in clause 4.8 (i) below) which is closed under the operations used in the proof.

The same approach allows us to eliminate the $\diamondsuit$ from Theorem 4.1. With the modified version of $\mathcal{AP}$ and $\aleph_3$ Cohen reals, we can replace $\diamondsuit_{\{\delta<\aleph_3:\mathrm{cof}\,\delta=\aleph_2\}}$ by $\diamondsuit_{\{\delta<\aleph_3:\mathrm{cof}\,\delta=\aleph_1\}}$, which in fact follows from the other hypotheses [Gregory, Sh 82].

We will not enlarge on these remarks any further here.

4.4 Definition. A finite graph Γ on n vertices is *sufficiently random* if:

i. For any two disjoint sets of vertices V_1, V_2 with $|V_1 \cup V_2| \leq (\log n)/3$, there is a vertex v linked to all vertices of V_1, and none in V_2;
ii. For any sets of vertices V_1, V_2 with $|V_i| > 3\log n$ there are adjacent and nonadjacent pairs of vertices in $V_1 \times V_2$.
iii. If V_1, V_2, V are three disjoint sets of vertices and $P \subseteq V_1 \times V_2$, with $|P|, |V| > 5\log n$, and if all pairs in P have distinct first entries, then some $v \in V$ *separates* some pair $(v^1, v^2) \in P$ in the sense that: $[R(v^1, v) \iff \neg R(v^2, v)]$. Here R is the edge relation (in the appropriate graph).

For sufficiently large n most graphs of size n are sufficiently random. We call any sequence of sufficiently random graphs of size tending to infinity a sequence of pseudorandom graphs.

(See [Bollobas] for background on random graphs.)

4.5 Notation

i. (Γ^1_n), (Γ^2_n) are two sequences of sufficiently random graphs such that for any m, n we have $\|\Gamma^1_m\| > \|\Gamma^2_n\|^5$ or $\|\Gamma^2_n\| > \|\Gamma^1_m\|^5$. ($\|\Gamma\|$ is the number of vertices of Γ.) These sequences are kept fixed. Γ

is the infinite random (homogeneous) graph. If we replace $\prod \Gamma_n^2/\mathcal{F}$ by $\Gamma^\omega/\mathcal{F}$ throughout, the argument is much the same, with slight simplifications.

ii. $\mathbb{P}$ is the forcing notion that adds $\aleph_3$ Cohen reals to V. $\underset{\sim}{x}_\alpha$ is the name of the α-th Cohen real as an element of ${}^\omega\omega$. For $\mathcal{A} \subseteq \aleph_3$, $\mathbb{P}\restriction\mathcal{A}$ denotes $\{\mathtt{p} \in \mathbb{P} : \operatorname{dom}\mathtt{p} \subseteq \mathcal{A}\}$.

4.6 Discussion

Working in the ground model we will build a $\mathbb{P}$-name for a suitable nonprincipal ultrafilter $\underset{\sim}{\mathcal{F}}$. We will view the reals $\underset{\sim}{x}_\alpha$ as (for example) potential members of the ultraproduct $\prod \Gamma_n^1$. We will consider candidates $\underset{\sim}{y}_\alpha$ for (representatives of) their images under a putative isomorphism, and defeat them by arranging (for example) that the set of n for which

$$R(\underset{\sim}{x}_\alpha(n), \underset{\sim}{x}_\beta(n)) \text{ iff } \neg R(\underset{\sim}{y}_\alpha(n), \underset{\sim}{y}_\beta(n))$$

gets into $\underset{\sim}{\mathcal{F}}$.

Note however that this must be done for every two potential sequences $(\underset{\sim}{k}^1(n))$ and $(\underset{\sim}{k}^2(n))$ indexing the ultraproducts $\prod_n \Gamma^1_{\underset{\sim}{k}^1(n)}/\underset{\sim}{\mathcal{F}}$, $\prod_n \Gamma^2_{\underset{\sim}{k}^2(n)}/\underset{\sim}{\mathcal{F}}$ to be formed. At stage α we deal with sequences $\underset{\sim}{k}^1_\alpha(n), \underset{\sim}{k}^2_\alpha(n) \in V^{\mathbb{P}\restriction\alpha}$ (which are guessed by the diamond). We require $\{n : x_\alpha(n) \in \Gamma^{\varepsilon_\alpha}_{\underset{\sim}{k}^{\varepsilon_\alpha}_\alpha(n)}\} \in \underset{\sim}{\mathcal{F}}$ where $\varepsilon_\alpha \in \{1,2\}$ is a label, and another very important requirement is that for any sequence $(\underset{\sim}{A}_n : n < \omega) \in V^{\mathbb{P}\restriction\alpha}$ with $\underset{\sim}{A}_n \subseteq \Gamma^{\varepsilon_\alpha}_{\underset{\sim}{k}^{\varepsilon_\alpha}_\alpha}(n)$ and $|\underset{\sim}{A}_n|/\|\Gamma^{\varepsilon_\alpha}_{\underset{\sim}{k}^{\varepsilon_\alpha}_\alpha(n)}\|$ small enough, the set $\{n : \underset{\sim}{x}_\alpha(n) \notin \underset{\sim}{A}_n\} \in \underset{\sim}{\mathcal{F}}$. (This sort of condition is an analog of the notion of a Γ-*big* type in [Sh 107].) It will be used in combination with clause (ii) in the definition of sufficient randomness.

The name $\underset{\sim}{\mathcal{F}}$ is built by carefully amalgamating a large set of approximations to the final object, using the combinatorial principle $\diamondsuit_{\aleph_2}$, which follows from the cardinal arithmetic [Gregory]; this method, which was illustrated in [Sh 107], is based on the theorem from [ShHL 162]. (The comparatively elaborate tree construction of [ShHL 162] can be simplified in the presence of $\diamondsuit$; it is designed to work when $\aleph_2$ is replaced by a limit cardinal and $\diamondsuit$ is weakened to the principle Dl_λ.) In what follows, the connection with [ShHL 162] is left somewhat vague; the details will be found in §A3 of the Appendix. In particular, in §A3.5 we show how the present $\mathcal{AP}$ fits the framework of §A3.1-3.

4.7 A notion of smallness

If $\mathcal{F}$ is a filter on ω, $k \in {}^\omega\omega$, $\varepsilon \in \{1,2\}$, then a sequence $(A_n : n < \omega)$ of subsets of the $\Gamma^\varepsilon_{k(n)}$ (i.e. $A_n \subseteq \Gamma^\varepsilon_{k(n)}$) is $(\mathcal{F}, k, \varepsilon)$-*slow* if there is some d

such that $\mathcal{F}\text{-lim}\left[|A_n|/\left(\sqrt{\|\Gamma^\varepsilon_{k(n)}\|}\cdot(\log\|\Gamma^\varepsilon_{k(n)}\|)^d\right)\right]=0$. Later on we will deal primarily with the case $\varepsilon=1$, to lighten the notation, and we will then write "$(\mathcal{F},k)$-slow" in place of "$(\mathcal{F},k,1)$-slow".

It should perhaps be emphasized that here (as opposed to §3) ε is merely a label.

4.8 Definition. We define the partially ordered set $\mathcal{AP}$ of approximations as follows. The intent is that the approximations should build the name of a suitable ultrafilter $\underset{\sim}{\mathcal{F}}$. Recall that the sequences (Γ^ε_n) (with $\varepsilon\in\{1,2\}$) are fixed (4.5(i)). Also bear in mind that the ultrafilter must eventually "defeat" a potential isomorphism between two ultraproducts $\prod_n\Gamma^\varepsilon_{\underset{\sim}{k}^\varepsilon(n)}/\underset{\sim}{\mathcal{F}}$.

1. An element $q\in\mathcal{AP}$ is a quadruple $(\mathcal{A},\underset{\sim}{\mathcal{F}},\boldsymbol{\varepsilon},\underset{\sim}{\boldsymbol{k}})=(\mathcal{A}^q,\underset{\sim}{\mathcal{F}}^q,\boldsymbol{\varepsilon}^q,\underset{\sim}{\boldsymbol{k}}^q)$ where
   i. $\mathcal{A}\subseteq\aleph_3$ has cardinality $\aleph_1$; $\boldsymbol{\varepsilon}=(\varepsilon_\alpha:\alpha\in\mathcal{A})$ with each ε_α an element of $\{1,2\}$;
   ii. $\underset{\sim}{\mathcal{F}}$ is a $\mathbb{P}{\restriction}\mathcal{A}$-name of a nonprincipal ultrafilter on ω, and if we set $\underset{\sim}{\mathcal{F}}{\restriction}(\mathcal{A}\cap\alpha)=:\quad\underset{\sim}{\mathcal{F}}{\restriction}\{\underset{\sim}{X}:\underset{\sim}{X}$ is a $\mathbb{P}{\restriction}(\mathcal{A}\cap\alpha)$-name for a subset of $\omega\}$, then $\underset{\sim}{\mathcal{F}}{\restriction}(\mathcal{A}\cap\alpha)$ is a $\mathbb{P}{\restriction}(\mathcal{A}\cap\alpha)$-name for all α;
   iii. $\underset{\sim}{\boldsymbol{k}}=(\underset{\sim}{k}_\alpha:\alpha\in\mathcal{A})$ with $\underset{\sim}{k}_\alpha$ a $\mathbb{P}{\restriction}(\mathcal{A}\cap\alpha)$-name of a function from ω to ω;
   iv. For each $\alpha\in\mathcal{A}$, and each $\mathbb{P}{\restriction}(\mathcal{A}\cap\alpha)$-name $(\underset{\sim}{A}_n:n<\omega)$;
   if $\Vdash_{\mathbb{P}\restriction(\mathcal{A}\cap\alpha)}$ "$(\underset{\sim}{A}_n)_{n<\omega}$ is $(\underset{\sim}{\mathcal{F}}{\restriction}\alpha,\,\underset{\sim}{k}_\alpha,\,\varepsilon_\alpha)$-slow" then
   $\Vdash_{\mathbb{P}}$ "$\{n:\underset{\sim}{x}_\alpha(n)\in\Gamma^{\varepsilon_\alpha}_{\underset{\sim}{k}_\alpha(n)}\setminus A_n\}\in\underset{\sim}{\mathcal{F}}$". We write $\mathcal{A}=\mathcal{A}^q$, $\mathcal{F}=\mathcal{F}^q$, and so on, when necessary.

2. We take $q\le q'$ if $\mathcal{A}^q\subseteq\mathcal{A}^{q'}$ and $q'{\restriction}\mathcal{A}^q=q$.

Some further comment is in order here. When we begin to check that $\underset{\sim}{\mathcal{F}}$ is indeed the name of an ultrafilter such that for any pair of sequences $\underset{\sim}{k}^1(n),\underset{\sim}{k}^2(n)$, the ultraproducts $\prod\Gamma^\varepsilon_{\underset{\sim}{k}^\varepsilon(n)}/\underset{\sim}{\mathcal{F}}$ are nonisomorphic, we will notice that there is an automatic asymmetry because the sequences (Γ^1_n) and (Γ^2_n) are so different: on some set in $\underset{\sim}{\mathcal{F}}$ we will have $|\Gamma^\varepsilon_{\underset{\sim}{k}^\varepsilon(n)}|>|\Gamma^{\varepsilon^*}_{\underset{\sim}{k}^{\varepsilon^*}(n)}|^5$ holding with $\{\varepsilon,\varepsilon^*\}=\{1,2\}$ in some order. The parameter ε_α in an approximation can be viewed as a guess as to the direction in which this asymmetry goes (after adding Cohen reals); the notion of an approximation includes a clause (iv) designed to be useful when $\underset{\sim}{k}_\alpha$ coincides with a particular $\underset{\sim}{k}^\varepsilon$ in the context just described.

On the other hand, we could first use $\diamondsuit$ to guess ε_α, $\underset{\sim}{k}^{\varepsilon_\alpha}_\alpha$, and many other things; in this case we do not actually need to include these kinds of data in the approximations themselves, though it would still be necessary to mention them in clause (iv). Alternatively, the set $\mathcal{AP}$ could also be used as a forcing notion, without $\diamondsuit$, and in this case the ε and $\underset{\sim}{k}$ would have to be included. So the version given here is the most flexible one.

4.9 Claim (Amalgamation)

1. *Suppose that $q_0, q_1, q_2 \in \mathcal{AP}$, $\mathcal{A}^{q_1} \subseteq \delta$, $\mathcal{A}^{q_2} = \mathcal{A}^{q_0} \cup \{\delta\}$, and $q_0 \leq q_1, q_2$. Then we can find $r \geq q_1, q_2$ in $\mathcal{AP}$.*

2. *If $q_1, q_2 \in \mathcal{AP}$, $\alpha < \aleph_3$, $\operatorname{dom} q_1 \subseteq \alpha$, and $q_2 \restriction \alpha \leq q_1$, then there is $r \geq q_1, q_2$ in $\mathcal{AP}$.*

Proof. 1: Let $\mathcal{A}_i = \mathcal{A}^{q_i}$, $\underset{\sim}{\mathcal{F}}^i = \underset{\sim}{\mathcal{F}}^{q_i}$, $\mathcal{A} = \mathcal{A}_1 \cup \{\delta\}, \varepsilon = \varepsilon^{q_2}_\delta$ and $\underset{\sim}{k} = \underset{\sim}{k}^{q_2}_\delta$. In particular $\underset{\sim}{\mathcal{F}}^0 \subseteq \underset{\sim}{\mathcal{F}}^1, \underset{\sim}{\mathcal{F}}^2$, and we have to combine them into one ultrafilter $\underset{\sim}{\mathcal{F}}$ in $V^{\mathbb{P}\restriction\mathcal{A}}$. The point is to preserve 4.8(iv), that is to ensure that $\mathbb{P}\restriction\mathcal{A}$ forces the relevant family of sets (namely, $\underset{\sim}{\mathcal{F}}^1, \underset{\sim}{\mathcal{F}}^2$, and sets imposed on us by 4.8(iv)) to have the finite intersection property.

If $\mathtt{p} \in \mathbb{P}\restriction\mathcal{A}$ forces the contrary, then after extending $\mathtt{p}$ suitably we may suppose that there is a $(\mathbb{P}\restriction\mathcal{A}_1)$-name $\underset{\sim}{a}$ of a member of $\underset{\sim}{\mathcal{F}}^1$, a $(\mathbb{P}\restriction\mathcal{A}_2)$-name $\underset{\sim}{b}$ of a member of $\underset{\sim}{\mathcal{F}}^2$, and — since $\mathcal{A}_1 = \mathcal{A} \cap \delta$ — a $(\mathbb{P}\restriction\mathcal{A}_1)$-name $(\underset{\sim}{A}_n : n < \omega)$ forced by $\mathtt{p}$ to be $(\underset{\sim}{\mathcal{F}}^1, \underset{\sim}{k}, \varepsilon)$-slow (as in (iv) of 4.8) so that letting $\underset{\sim}{c} = \{n < \omega : \underset{\sim}{x}_\delta(n) \in \Gamma^{\varepsilon_\delta}_{\underset{\sim}{k}_\delta(n)} \setminus \underset{\sim}{A}_n\}$ we have:

$$\mathtt{p} \Vdash_{\mathbb{P}\restriction\mathcal{A}} \text{“}\underset{\sim}{a} \cap \underset{\sim}{b} \cap \underset{\sim}{c} = \emptyset\text{”}.$$

(i.e. we used the fact that there are three kinds of requirements of the form “a set belongs to $\mathcal{F}$ ”, each kind is closed under finite intersections).

Let $\mathtt{p}_i = \mathtt{p}\restriction\mathcal{A}_i$ for $i = 0, 1, 2$. To clarify the matter choose $\mathtt{H}^0 \subseteq \mathbb{P}\restriction\mathcal{A}_0$ generic over V so that $\mathtt{p}_0 \in \mathtt{H}^0$. Note that $\underset{\sim}{k}$ is a $(\mathbb{P}\restriction\mathcal{A}_0)$-name (4.8(iii)).

In $V[\mathtt{H}^0]$, for each $n < \omega$ let

$$\underset{\sim}{B}_n[\mathtt{H}^0] = \{v \in \Gamma^\varepsilon_{\underset{\sim}{k}(n)}[\mathtt{H}^0] :$$
$$\text{For some } \mathtt{p}_2' \in \mathbb{P}\restriction\mathcal{A}_2 \text{ with } \mathtt{p}_2' \geq \mathtt{p}_2 \text{ and } \mathtt{p}_2'\restriction\mathcal{A}_0 \in \mathtt{H}^0,$$
$$\mathtt{p}_2' \Vdash_{\mathbb{P}\restriction\mathcal{A}_2} \text{“}\underset{\sim}{x}_\delta(n) = v \text{ and } n \in \underset{\sim}{b}\text{”}\}.$$

Then $(\underset{\sim}{B}_n : n < \omega)$ is not $(\underset{\sim}{\mathcal{F}}^0\restriction\delta, \underset{\sim}{k}, \varepsilon)$-slow, since $(\underset{\sim}{B}_n : n < \omega)$ is a $\mathbb{P}\restriction\mathcal{A}_0$-name, $q_2 \in \mathcal{AP}$, and $\mathtt{p}_2 \Vdash$ “For $n \in \underset{\sim}{b}$, $\underset{\sim}{x}_\delta(n) \in \underset{\sim}{B}_n$” (and (iv) of 4.8(1)).

Also in $V[\mathtt{H}^0]$, let $\underset{\sim}{b}^+[\mathtt{H}^0] = \{n :$ for every $\mathtt{p}_0' \in \mathtt{H}^0$, $\mathtt{p}_0' \cup \mathtt{p}_2 \nVdash$ “$n \notin \underset{\sim}{b}$”$\}$. As $q_2 \in \mathcal{AP}$, we have $\underset{\sim}{b}^+ \in \underset{\sim}{\mathcal{F}}^0[\mathtt{H}^0]$. For each $n \in \underset{\sim}{b}^+[H^\delta]$ let

$$\underset{\sim}{A}^1_n[\mathtt{H}^0] =: \{v \in \Gamma^\varepsilon_{\underset{\sim}{k}(n)}[\mathtt{H}^0] : \text{ For no } \mathtt{p}_1' \geq \mathtt{p}_1 \text{ in } \mathbb{P}\restriction\mathcal{A}_1 \text{ with } \mathtt{p}_1'\restriction\mathcal{A}_0 \in \mathtt{H}^0,$$
$$\mathtt{p}_1' \Vdash_{\mathbb{P}\restriction\mathcal{A}_1} \text{“}n \in \underset{\sim}{a} \text{ and } v \notin \underset{\sim}{A}_n.\text{”}\}$$

Let $\underset{\sim}{A}^1_n[\mathtt{H}^0] = \emptyset$ if $n \notin \underset{\sim}{b}^+$.

Easily $(\underset{\sim}{A}^1_n : n < \omega)$ is $\underset{\sim}{\mathcal{F}}^0$-slow. Hence in $V[\mathtt{H}^0]$ the sequence $(\underset{\sim}{B}_n \setminus \underset{\sim}{A}^1_n : n < \omega)$ is not $(\underset{\sim}{\mathcal{F}}^0[\mathtt{H}^0])$-slow. We can compute the values of $\underset{\sim}{B}_n$ and $\underset{\sim}{A}^1_n$ in $V[\mathtt{H}^0]$. So we can find $n \in \underset{\sim}{b}^+[H^0]$ with $\underset{\sim}{B}_n \setminus \underset{\sim}{A}^1_n \neq \emptyset$, and choose $v \in \underset{\sim}{B}_n \setminus \underset{\sim}{A}^1_n$. Then there are $\mathrm{p}'_1 \in \mathbb{P}{\restriction}\mathcal{A}_1/\mathtt{H}^0$, $\mathrm{p}'_1 \geq \mathrm{p}_1$, and $\mathrm{p}'_2 \in \mathbb{P}{\restriction}\mathcal{A}_2/\mathtt{H}^0$, with $\mathrm{p}'_2 \geq \mathrm{p}_2$, so that:

$$\mathrm{p}'_1 \Vdash \text{“}n \in \underset{\sim}{a} \text{ and } v \notin \underset{\sim}{A}_n\text{”}.$$

$$\mathrm{p}'_2 \Vdash \text{“}n \in \underset{\sim}{b} \text{ and } \underset{\sim}{x}_\delta(n) = v\text{”}$$

Now $\mathrm{p} \leq \mathrm{p}'_1 \cup \mathrm{p}'_2 \in \mathbb{P}{\restriction}\mathcal{A}$ and $\mathrm{p}'_1 \cup \mathrm{p}'_2$ forces “$n \in \underset{\sim}{a} \cap \underset{\sim}{b} \cap \underset{\sim}{c}$” (over $\mathtt{H}^0$), contradicting the choice of p. This completes the proof of 4.9 (1).

2: Let $[(\mathcal{A}^{q_2} \setminus \alpha) \bigcup \{\sup \mathcal{A}^{q_2}\}] = \{\delta_i : i \leq \gamma\}$ in increasing order. Define inductively $r_i \in \mathcal{A}P$, increasing in i, with $q_2{\restriction}(\mathcal{A} \cap \delta_i) \leq r_i$, $\operatorname{dom} r_i \subseteq \delta_i$, $r_0 = q_1$; then let $r = r_\gamma$.

At successor stages $i = j+1$ we apply 4.9 (1) to $q_2{\restriction}(\mathcal{A}^{q_2} \cap \delta_j)$, r_j, $q_2{\restriction}[\mathcal{A}^{q_2} \cap (\delta_j + 1)]$.

If i is a limit of uncountable cofinality, we just take unions:

$$\mathcal{A}^{r_i} = \bigcup_{\zeta<i} \mathcal{A}^{r_\zeta}; \underset{\sim}{\mathcal{F}}^{r_i} = \bigcup_{\zeta<i} \underset{\sim}{\mathcal{F}}^{r_\zeta}; \varepsilon^{r_i} = \bigcup_{\zeta<i} \varepsilon^{r_\zeta}; \underset{\sim}{\boldsymbol{k}}^{r_i} = \bigcup_{\zeta<i} \underset{\sim}{\boldsymbol{k}}^{r_\zeta};$$

while if i is a limit of cofinality $\aleph_0$, we have actually to extend $\bigcup_{\zeta<i} \underset{\sim}{\mathcal{F}}^{r_\zeta}$ to a $\mathbb{P}{\restriction}\mathcal{A}^{r_i}$-name of an ultrafilter in $V^{\mathbb{P}{\restriction}\mathcal{A}^{r_i}}$. However, in $V^{\mathbb{P}{\restriction}\mathcal{A}^{r_i}}$, $\bigcup_{\zeta<i} \underset{\sim}{\mathcal{F}}^{r_\zeta}$ is interpreted as a filter including all cofinite subsets of ω, hence can be completed to an ultrafilter. $\square$

4.10 Claim

1. *If q_i $(i < \delta)$ is an increasing sequence of members of $\mathcal{A}P$, with $\delta < \aleph_2$, then for some $q \in \mathcal{A}P$, $q \geq q_i$ for all $i < \delta$.*
2. *If $q_1, q_2 \in \mathcal{A}P$, $\alpha < \aleph_3$, $q_2{\restriction}\alpha \leq q_1$, and $\operatorname{dom} q_1 \cap \operatorname{dom} q_2 = \operatorname{dom} q_1 \cap \alpha$, then there is $r \geq q_1, q_2$ in $\mathcal{A}P$.*

Proof. 1: We may suppose $\delta = \aleph_0$ or $\aleph_1$. Let $\mathcal{A} =: \bigcup_i \mathcal{A}^{q_i}$ be enumerated in increasing order as $\{\alpha_j : j < \gamma\}$ for the appropriate γ, and set $\alpha_\gamma = \sup \mathcal{A}$. We define an increasing sequence of members r_j of $\mathcal{A}P$ for $j \leq \gamma$ by induction on j so that:

$$\mathcal{A}^{r_j} = \{\alpha_\zeta : \zeta < j\};$$

$$q_i{\restriction}\alpha_j \leq r_j \text{ for all } i < \delta.$$

In all cases we proceed as in the proof of Claim 4.9. The only difference is that we deal with several q_i, but as they are linearly ordered there is no difficulty.

2: This is proved similarly to part (1): let $\gamma = \sup(\operatorname{dom} q_1 \cup \operatorname{dom} q_2)$. Choose by induction on $\beta \in (\operatorname{dom} q_1 \cup \operatorname{dom} q_2 \cup \{\gamma\}) \setminus \alpha$ an upper bound r_β of $q_1 \restriction \beta$ and $q_2 \restriction \beta$, increasing with β, with $\operatorname{dom} r_\beta = \beta \cap (\operatorname{dom} q_1 \cup \operatorname{dom} q_2)$. The successor step is by 4.9(i). The limit is easy too. Note: if $\operatorname{dom} q_1/E$ has only finitely many classes, when $\beta_1 \, E \, \beta_2$ iff $\bigwedge_{\gamma \in \operatorname{dom} q_2}[\gamma < \beta_1 \Leftrightarrow \gamma < \beta_2]$, then 4.9(ii) suffices. □

4.11 Proof of Theorem 4.1: The construction. We define an increasing sequence $G^\alpha \subseteq \{q \in \mathcal{AP} : \mathcal{A}^q \subseteq \alpha\}$ of $\aleph_2$-directed sets increasing in α, and a set of at most $\aleph_2$ "commitments" which G^α will meet. In particular we require that $\forall \beta < \alpha \, \exists q \in G^\alpha \; (\beta \in \mathcal{A}^q)$, and at each stage α we may make new commitments to "enter some collection of dense sets" – in set theoretic terminology – or equivalently, to "omit some type" – in model theoretic terms. We make use of $\diamondsuit_{\{\delta<\aleph_3 : \operatorname{cof}\delta=\aleph_2\}}$ to choose the commitments. The combinatorics involved in meeting the commitments are treated in [ShLH 162], and are reviewed in §A3 of the Appendix. Our summary of the construction in the present section will be less formal.

At a stage $\delta < \aleph_3$ with $\operatorname{cof} \delta = \aleph_2$, we will "guess" $\mathbb{P}$-names $\underset{\sim}{k}^1_\delta, \underset{\sim}{k}^2_\delta, \underset{\sim}{F}_\delta$, a condition $\mathbf{p}^\delta \in \mathbb{P} \restriction \delta$ and a parameter $\varepsilon_\delta \in \{1,2\}$, explained in connection with (4) below, and attempt to "kill" the possibility that $\mathbf{p}^\delta$ forces:

"$\underset{\sim}{F}_\delta : \prod_n \Gamma^1_{\underset{\sim}{k}^1_\delta(n)} \longrightarrow \prod_n \Gamma^2_{\underset{\sim}{k}^2_\delta(n)}$ induces a map which can be extended to an isomorphism:

$$\prod \Gamma^1_{\underset{\sim}{k}^1_\delta(n)}/\underset{\sim}{\mathcal{F}} \simeq \prod \Gamma^2_{\underset{\sim}{k}^2_\delta(n)}/\underset{\sim}{\mathcal{F}}\text{''}.$$

(Here we have taken $\varepsilon_\delta = 1$; otherwise the roles of 1 and 2 in this – and in all that follows – must be reversed.)

We will refer to the genericity game of [ShHL 162], as described in §A3 of the Appendix. In that game the Ghibellines can accomplish the following. For $\delta < \aleph_3$, they determine a set of compatible approximations G^δ which together will determine an ultrafilter $\mathcal{F}\restriction\delta$ on ω in $V^{\mathbb{P}\restriction\delta}$ (specifically, G^α is a subset of $\{r \in \mathcal{AP} : \operatorname{Dom} r \subseteq \alpha\}$ which is directed, increasing in α). The Guelfs set them tasks which ensure that the ultrafilter $\mathcal{F}$ which is gradually built up by the Ghibellines has all the desired properties.

Let $\mathcal{F}_0$ be a fixed nonprincipal ultrafilter on ω, in the ground model and without loss of generality there is $q \in G^0$ with $\mathcal{F}^q = \mathcal{F}_0$. For $\delta < \aleph_3$ of cofinality $\aleph_2$, let q^*_δ be an approximation $(\{\delta\}, \underset{\sim}{\mathcal{F}}^\delta, (\varepsilon_\delta), (\underset{\sim}{k}^{\varepsilon_\delta}_\delta))$, where $\underset{\sim}{\mathcal{F}}^\delta$ is the $\mathbb{P}\restriction\{\delta\}$-name of some ultrafilter on ω extending $\mathcal{F}_0$ such that

(1) $\{n : \underset{\sim}{x}_\delta(n) \in \Gamma^{\varepsilon_\delta}_{\underset{\sim}{k}^{\varepsilon_\delta}_\delta(n)}\} \in \underset{\sim}{\mathcal{F}}^\delta$;

(2) $\{n : \underset{\sim}{x}_\delta(n) \notin A_n\} \in \underset{\sim}{\mathcal{F}}^\delta$ for any $(\mathcal{F}_0, \underset{\sim}{k}_\delta^{\varepsilon_\delta}, \varepsilon_\delta)$-slow sequence (A_n) in the universe V

The Ghibellines will be required (by the Guelfs) to put q^*_δ in $G^{\delta+1}$. The Ghibellines are also obliged to make commitments of the following form, which should then be respected throughout the rest of the construction. (These commitments involve a parameter $\alpha > \delta$ to be controlled by the Ghibellines as the play progresses: of course these commitments have to satisfy density requirements.)

$(*)^\delta_{q^*, \underset{\sim}{z}^*}$:
For every $\alpha > \delta$, every $q \in G^\alpha$ with $\delta \in \operatorname{dom} q$,
every $k_\delta^{1-\varepsilon_\delta}(n)$ (really a $(\mathbb{P}{\restriction}\delta)$-name)
and every $(\mathbb{P}{\restriction}\mathcal{A}^\delta)$-name $\underset{\sim}{z}$
of a member of $\prod_n \Gamma^{1-\varepsilon_\delta}_{k_\delta(n)}$:

if $(q, \underset{\sim}{z}) \simeq (q^*, \underset{\sim}{z}^*)$ over $\delta+1$, *then* there will be some r in G^α, some $\mathrm{p}' \in \mathbb{P}{\restriction}\mathcal{A}^r$, and some $\mathbb{P}{\restriction}(\mathcal{A}^r \cap \delta)$-name $\underset{\sim}{x}$ of a member of $\prod_n \Gamma^{\varepsilon_\delta}_{\underset{\sim}{k}_\delta^{\varepsilon_\delta}(n)}$, with $r \geq q$, $\mathrm{p}' \geq \mathrm{p}^\delta$, $\underset{\sim}{F}_\delta(\underset{\sim}{x})$ is a $\mathbb{P}{\restriction}(\mathcal{A}^r \cap \delta)$-name, and:

(†)
$$\mathrm{p}' \Vdash_{\mathbb{P}\restriction\mathcal{A}^r} \text{“}\{n : \Gamma^{\varepsilon_\delta}_{k_\delta^{\varepsilon_\delta}(n)} \models R(\underset{\sim}{x}(n), \underset{\sim}{x}_\delta(n)) \iff \Gamma^{1-\varepsilon_\delta}_{k_\delta^{1-\varepsilon_\delta}(n)} \models \neg R(\underset{\sim}{F}_\delta(\underset{\sim}{x})(n), \underset{\sim}{z}(n))\} \in \underset{\sim}{\mathcal{F}}^r\text{”}$$

There is such a commitment for each $q^*, \underset{\sim}{z}^*$ with $q^*_\delta \leq q^* \in \mathcal{A}P$, $q^*{\restriction}\delta \in G^\delta$, and $\underset{\sim}{z}^*$ a $(\mathbb{P}{\restriction}\mathcal{A}^{q^*})$-name of a member of $\prod_n \Gamma^2_{\underset{\sim}{k}^2_\delta(n)}$. So apparently we are making $\aleph_3$ commitments, which is not feasible, but as we are using isomorphism types this amounts to only $2^{\aleph_1} = \aleph_2$ commitments, and this *is* feasible. This is formalized in §A3.6 in the Appendix.

These commitments can only be met when the corresponding set of approximations is dense, but on the other hand we have a stationary set δ of opportunities to meet such a commitment, and we will show that for any candidate $\underset{\sim}{F}$ for an isomorphism, either we kill it off as outlined above (by making it obvious that $\underset{\sim}{F}(\underset{\sim}{x}_\delta)$ cannot be defined), or else – after failing to do this on a stationary set – that $\underset{\sim}{F}$ must be quite special (somewhat definable) and hence even more easily dealt with, as will be seen in detail in the next few sections.

After we have obtained G^α for all α, we will let $\underset{\sim}{\mathcal{F}}^\alpha$ be $\bigcup\{\underset{\sim}{\mathcal{F}}^q : q \in G^\alpha\}$ (that is, the appropriate $(\mathbb{P}{\restriction}\alpha)$-name of a uniform ultrafilter on ω). Letting $G =: G^{\aleph_3} =: \bigcup_\alpha G^\alpha$, also $\underset{\sim}{\mathcal{F}} = \underset{\sim}{\mathcal{F}}^{\aleph_3}$ is defined.

4.12 Proof of Theorem 4.1: The heart of the matter. Now suppose toward a contradiction that after $\underset{\sim}{\mathcal{F}}$ has been constructed in this way, there are $\mathbb{P}$-names $\underset{\sim}{F}$, $\underset{\sim}{k}^1$, $\underset{\sim}{k}^2$, and a condition $\mathsf{p} \in \mathbb{P}$ such that:

(3) $$\mathsf{p} \Vdash_{\mathbb{P}} \text{“}\underset{\sim}{F} \text{ is a function from } \prod_n \Gamma^1_{\underset{\sim}{k}^1(n)} \text{ onto } \prod_n \Gamma^2_{\underset{\sim}{k}^2(n)}$$

which induces an isomorphism of the corresponding ultraproducts with respect to $\underset{\sim}{\mathcal{F}}$”.

Actually, we will want to assume in addition that p forces:

(4) $$\text{“}\{n : \|\Gamma^1_{\underset{\sim}{k}^1(n)}\| > \|\Gamma^2_{\underset{\sim}{k}^2(n)}\|\} \in \underset{\sim}{\mathcal{F}}\text{,”}$$

which could force us to increase p and to switch the roles of 1 and 2 in all that follows; this is why we have carried along a parameter ε in our definition of $\mathcal{AP}$.

We will say that a set $\mathcal{A} \subseteq \aleph_3$ is $(\underset{\sim}{F}, \underset{\sim}{k}^1, \underset{\sim}{k}^2, \mathsf{p})$-*closed* if:

i. $\underset{\sim}{k}^1, \underset{\sim}{k}^2$ are $(\mathbb{P}\restriction\mathcal{A})$-names; $\underset{\sim}{F}\restriction\mathcal{A}$ is a $(\mathbb{P}\restriction\mathcal{A})$-name;

ii. $\mathsf{p} \Vdash_{\mathbb{P}\restriction\mathcal{A}}$: “$\underset{\sim}{F}\restriction\mathcal{A}$ is a function from $\prod_n \Gamma^1_{\underset{\sim}{k}^1(n)}$ onto $\prod_n \Gamma^2_{\underset{\sim}{k}^2(n)}$ which (interpreted in $\mathbb{P}\restriction\mathcal{A}$) induces an isomorphism from $\prod_n \Gamma^1_{\underset{\sim}{k}^1(n)}/(\underset{\sim}{\mathcal{F}}\restriction\mathcal{A})$ onto $\prod_n \Gamma^2_{\underset{\sim}{k}^2(n)}/(\underset{\sim}{\mathcal{F}}\restriction\mathcal{A})$”.

iii. $\mathsf{p} \Vdash_{\mathbb{P}\restriction\mathcal{A}}$: “$\{n : \|\Gamma^1_{\underset{\sim}{k}^1(n)}\| > \|\Gamma^2_{\underset{\sim}{k}^2(n)}\|\} \in \underset{\sim}{\mathcal{F}}\restriction\mathcal{A}$.”

Properly speaking, the only actual closure condition here is clause (ii). Note that the condition in (iii) can be strengthened to:

$$\text{“}\left\{n : \|\Gamma^1_{\underset{\sim}{k}^1(n)}\| > \|\Gamma^2_{\underset{\sim}{k}^2(n)}\|^5\right\} \in \underset{\sim}{\mathcal{F}}\restriction\mathcal{A}\text{,”}$$

by the choice of the sequences (Γ^i_n) $(i = 1, 2)$.

Let C be $\{\delta < \aleph_3 : \mathrm{cof}(\delta) = \aleph_2,\ \delta \text{ is } (\underset{\sim}{F}, \underset{\sim}{k}^1, \underset{\sim}{k}^2, \mathsf{p})\text{-closed}\}$. Clearly the set C is unbounded and is closed under $\aleph_2$-limits. By our construction, for a stationary subset S_C of C we may suppose that for $\delta \in S_C$: $\underset{\sim}{F}_\delta = \underset{\sim}{F}\restriction\delta$, $\mathsf{p}^\delta = \mathsf{p}$, $\varepsilon_\delta = 1$, $\underset{\sim}{k}_\delta = \underset{\sim}{k}^1$, and that δ was $(\underset{\sim}{F}, \underset{\sim}{k}^1, \underset{\sim}{k}^2, \mathsf{p})$-closed. So $q^*_\delta \in G^{\delta+1}$, and we can find $q \in G$ such that $\underset{\sim}{z} =: \underset{\sim}{F}(\underset{\sim}{x}_\delta)$ is a $(\mathbb{P}\restriction\mathcal{A}^q)$-name, $\delta \in \mathcal{A}^q$.

At stage δ in the construction, the Ghibellines had tried to make the commitment $(*)^\delta_{q^*, \underset{\sim}{z}^*}$, with $(q^*, \underset{\sim}{z}^*) = (q, \underset{\sim}{z})$. They later failed to meet this commitment, since otherwise there would be some $r \geq q$ in G, some $\mathsf{p}' \geq \mathsf{p}$ in $\mathbb{P}\restriction\mathcal{A}^r$, and some $[\mathbb{P}\restriction(\mathcal{A}^r \cap \delta)]$-name of a member $\underset{\sim}{x}$ of $\Gamma^1_{\underset{\sim}{k}^1(n)}$, for which (†) holds:

$$\mathsf{p}' \Vdash_{\mathbb{P}\restriction\mathcal{A}^r} \text{“}\{n : [\Gamma^1_{k^1_\delta} \models R(\underset{\sim}{x}(n), \underset{\sim}{x}_\delta(n)) \iff \Gamma^2_{k^2_\delta} \models \neg R(\underset{\sim}{F}_\delta(\underset{\sim}{x})(n), \underset{\sim}{z}(n))]\} \in \mathcal{F}^r\text{”}.$$

and $\underset{\sim}{z}$ is $\underset{\sim}{F}(\underset{\sim}{x}_\delta)$. But p forced $\underset{\sim}{F}$ to induce an isomorphism, so we have a contradiction.

The failure to make the commitment $(*)^{\delta}_{q,\underset{\sim}{z}}$, implies a failure of density, which means that for some $(q', \underset{\sim}{z}') \simeq (q, \underset{\sim}{z})$ over $\delta + 1$ – and hence also for $(q, \underset{\sim}{z})$ – taking $q_0 = q{\restriction}\delta$, we will have:

(i) δ is $(\underset{\sim}{F}, \underset{\sim}{k}^1, \underset{\sim}{k}^2, \mathrm{p})$-closed.

(ii) $\mathrm{p} \in \mathbb{P}{\restriction}\mathcal{A}^{q_0}$, $\delta \in \mathcal{A}^q$, $\varepsilon^q = 1$, $\underset{\sim}{k}^q_\delta = \underset{\sim}{k}^1$, $\underset{\sim}{F}_\delta = \underset{\sim}{F}{\restriction}\delta$;

(iii) $\underset{\sim}{z}$ is a $(\mathbb{P}{\restriction}\mathcal{A}^q)$-name for a member of $\prod_n \Gamma^2_{\underset{\sim}{k}^2_\delta(n)}$;

(iv) For all $r \geq q$ in $\mathcal{A}P$ such that $r{\restriction}\delta \in G^\delta$, and $\underset{\sim}{x}$ a $(\mathbb{P}{\restriction}\mathcal{A}^{r\restriction\delta})$-name, with $\underset{\sim}{y} =: \underset{\sim}{F}(\underset{\sim}{x})$ a $(\mathbb{P}{\restriction}\mathcal{A}^{r\restriction\delta})$-name, we have:

$(*)_{\underset{\sim}{x},\underset{\sim}{y}}$ $\mathrm{p} \Vdash$ "The set $\{n : \Gamma^1_{\underset{\sim}{k}^1(n)} \models R(\underset{\sim}{x}(n), \underset{\sim}{x}_\delta(n))$ iff $\Gamma^2_{\underset{\sim}{k}^2(n)} \models R(\underset{\sim}{y}(n), \underset{\sim}{z}(n))\}$ is in $\underset{\sim}{\mathcal{F}}^r$".

(Note: another possibility of failure, $q \notin G^\alpha$, is ruled out by the choice of q).

Now we analyze the meaning of $(*)_{\underset{\sim}{x},\underset{\sim}{y}}$. Consider the following property of $(\mathbb{P}{\restriction}\delta)$-names $\underset{\sim}{x}, \underset{\sim}{y}$ for a fixed choice of $\delta \in C$, $q \in \mathcal{A}P$ with $\delta \in \mathcal{A}^q$, and $\underset{\sim}{z}$ a $(\mathbb{P}{\restriction}\mathcal{A}^q)$-name.

$(**)_{\underset{\sim}{x},\underset{\sim}{y}}$ For all $r \geq q$ in $\mathcal{A}P$ such that $r{\restriction}\delta \in G^\delta$ and $\underset{\sim}{x}$, $\underset{\sim}{y}$ are $(\mathbb{P}{\restriction}\mathcal{A}^{r\restriction\delta})$-names, $(*)_{\underset{\sim}{x},\underset{\sim}{y}}$ holds.

We explore the meaning of this property when $\underset{\sim}{y}$ is not necessarily $\underset{\sim}{F}(\underset{\sim}{x})$.

Clearly,

$(\otimes_1)$ If $\underset{\sim}{x}$ is a $(\mathbb{P}{\restriction}\delta)$-name, $\underset{\sim}{y} = \underset{\sim}{F}(\underset{\sim}{x})$, then $(**)_{\underset{\sim}{x},\underset{\sim}{y}}$.

To simplify the analysis, let $\mathtt{H}$ be generic for $\mathbb{P}{\restriction}\delta$. Let $\underset{\sim}{x}$ be a $\mathbb{P}{\restriction}\delta$-name of a real, $\mathcal{A} \subseteq \delta$. We say $\underset{\sim}{x}$ is *unrestricted* for $(\mathtt{H}, \mathcal{A}, \underset{\sim}{k}^1)$ if:

There is no $(\underset{\sim}{\mathcal{F}}{\restriction}\mathcal{A}, \underset{\sim}{k}^1)$-slow sequence $(\underset{\sim}{B}_n)_{n<\omega}$ in $V[\mathtt{H}{\restriction}\mathcal{A}]$ such that: $\{n : \underset{\sim}{x}(n) \in \Gamma^1_{\underset{\sim}{k}^1(n)} \setminus \underset{\sim}{B}_n\} \equiv \emptyset \mod \underset{\sim}{\mathcal{F}}^\delta[\mathtt{H}]$.

Observe that if $\sup \mathcal{A} < \gamma < \delta$ and $\underset{\sim}{k}_\gamma = \underset{\sim}{k}^1$, then the Cohen real $\underset{\sim}{x}_\gamma$ is forced (in $\mathbb{P}{\restriction}\delta$) to be unrestricted for $(\mathtt{H}, \mathcal{A}, \underset{\sim}{k}^1)$.

4.12A Claim. *If $\underset{\sim}{x}^1, \underset{\sim}{x}^2$ are $(\mathbb{P}{\restriction}\delta)$-names of functions in $\prod_n \Gamma^1_{\underset{\sim}{k}^1(n)}$, $\underset{\sim}{y}$ is a $(\mathbb{P}{\restriction}\delta)$-name of a member of $\prod_n \Gamma^2_{\underset{\sim}{k}^2(n)}$, and both pairs $(\underset{\sim}{x}^1, \underset{\sim}{y})$ and $(\underset{\sim}{x}^2, \underset{\sim}{y})$ satisfy the condition (**) above, then:*

(Clm)

$\mathrm{p} \Vdash_{\mathbb{P}\restriction\delta}$ "$\underset{\sim}{x}^1 = \underset{\sim}{x}^2 \mod \underset{\sim}{\mathcal{F}}{\restriction}\delta[\mathtt{H}]$ *or both are restricted for* $(\mathtt{H}, \mathcal{A}^{q_0}, \underset{\sim}{k}^1)$."

We will give the proof of this, which contains one of the main combinatorial points, in paragraph 4.13. For the present we continue with the proof of the theorem. We first record a consequence of the claim.

($\otimes_2$) If $\underset{\sim}{x}, \underset{\sim}{y}$ are $(\mathbb{P}{\restriction}\delta)$-names with $\underset{\sim}{x}$ forced by $\mathbb{P}{\restriction}\delta$ to be unrestricted for $(\mathtt{H}, \mathcal{A}^{q_0}, \underset{\sim}{k}^1)$, and the pair $(\underset{\sim}{x}, \underset{\sim}{y})$ satisfies $(**)_{\underset{\sim}{x},\underset{\sim}{y}}$, then $p \Vdash_{\mathbb{P}\restriction\delta}$ "$\underset{\sim}{F}(\underset{\sim}{x}) = \underset{\sim}{y} \bmod \underset{\sim}{\mathcal{F}}^\delta$"

Indeed, if $\mathtt{H} \subseteq \mathbb{P}{\restriction}\delta$ is generic over V, and $\underset{\sim}{F}(\underset{\sim}{x})[\mathtt{H}] = \underset{\sim}{y}_1[\mathtt{H}] \neq \underset{\sim}{y}[\mathtt{H}] \bmod \underset{\sim}{\mathcal{F}}^\delta$, then since $\underset{\sim}{F}$ is *onto* (in $V[\mathtt{H}]$, as δ is $(\underset{\sim}{F}, \underset{\sim}{k}^1, \underset{\sim}{k}^2, \mathtt{p})$-closed), there is a $(\mathbb{P}{\restriction}\delta)$-name $\underset{\sim}{x}'$ with $\underset{\sim}{F}(\underset{\sim}{x}')[\mathtt{H}] = \underset{\sim}{y}[H]$, so $\underset{\sim}{x}'[\mathtt{H}] \neq \underset{\sim}{x}[\mathtt{H}] \bmod \underset{\sim}{\mathcal{F}}^\delta$. Now $\underset{\sim}{x}, \underset{\sim}{x}', \underset{\sim}{y}$ contradict (Clm). Thus $(\underset{\sim}{F})$ holds. As $(\otimes_1) + (\otimes_2)$ holds for stationarily many δ's, it holds for $\delta = \aleph_3$ (in the natural interpretation).

In what follows, we use the statements $(\otimes_1) + (\otimes_2)$ as a kind of "definability" condition on $\underset{\sim}{F}$; but we deal with the current concrete case, rather than seeking an abstract formulation of the situation.

Let $S = \{\gamma \in S_C : \underset{\sim}{F}(\underset{\sim}{x}_\gamma)$ is (forced by p to be equal to) a $[\mathbb{P}{\restriction}(\gamma+1)]$-name $\}$. We claim that S is stationary. Let $C' \subseteq \aleph_3$ be closed unbounded, and let $\delta \in S_C$ be taken with $C' \cap S_C$ unbounded below δ. Let $q \in G$ be chosen so that $\underset{\sim}{F}(\underset{\sim}{x}_\delta)$ is a $(\mathbb{P}{\restriction}\mathcal{A}^q)$-name, let $q_0 = q{\restriction}\delta$, and $\gamma_0 = \sup \mathcal{A}^{q_0}$. It suffices to check that for $\gamma_0 < \gamma < \delta$ with $\gamma \in S_C$, we have $\gamma \in S$. So let $r_1 \in G^\delta$ be chosen so that $\underset{\sim}{y}_1 =: \underset{\sim}{F}(\underset{\sim}{x}_\gamma)$ is a $(\mathbb{P}{\restriction}\mathcal{A}^{r_1})$-name. It suffices to show that $\underset{\sim}{y}_1$ is (forced by p to be equal to) a $(\mathbb{P}{\restriction}[\mathcal{A}^{r_1} \cap (\gamma+1)])$-name. Otherwise, by a density requirement (Appendix, §A3) we can find a 1-1 order preserving function h with domain $\mathcal{A}^{r_1}$, h is the identity on $\mathcal{A}^{r_1} \cap (\gamma+1)$, $h(\min(\mathcal{A}^{r_1} \backslash (\gamma+1))) > \sup \mathcal{A}^{r_1}$, with $r_2 =: h(r_1)$ in G^δ. Let $\underset{\sim}{y}_2 = h(\underset{\sim}{y}_1)$. Then $(**)_{\underset{\sim}{x}_\gamma, \underset{\sim}{y}_i}$ holds for $i = 1, 2$, so $p \Vdash_{\mathbb{P}\restriction\delta}$ "$\underset{\sim}{y}_1 = \underset{\sim}{y}_2 \bmod \underset{\sim}{\mathcal{F}}^\delta$", but by 4.14 below we can ensure that this is not the case (by making additional commitments, cf. §A3).

Now for $\gamma \in S$ let $q_\gamma \in G^{\gamma+1}$ be chosen so that $\underset{\sim}{z}_\gamma = \underset{\sim}{F}(\underset{\sim}{x}_\gamma)$ is a $(\mathbb{P}{\restriction}\mathcal{A}^{q_\gamma})$-name, and let $\hat{\gamma} = \sup(\mathcal{A}^{q_\gamma} \cap \gamma)$. By Fodor's lemma we can shrink S so that $\hat{\gamma}$ and $\mathcal{A}_0 = \mathcal{A}^{q_\gamma} \cap \hat{\gamma}$ and $q_\gamma{\restriction}\hat{\gamma}$ are constant for $\gamma \in S$. Now choose $\delta_1 < \delta_2$ in S, and let $q_i = q_{\delta_i}$, $\mathcal{A}_i = \mathcal{A}^{q_i}$ for $i = 1, 2$, so $\mathcal{A}_1 = \mathcal{A}^{q_1} = \mathcal{A}_0 \cup \{\delta_1\}$, $\mathcal{A}_2 = \mathcal{A}^{q_2} = \mathcal{A}_0 \cup \{\delta_2\}$; also let $\mathcal{A} =: \mathcal{A}_1 \cup \mathcal{A}_2$; we now let $q_i{\restriction}\hat{\gamma}$ be called q_0. Let $\underset{\sim}{\mathcal{F}}^i = \underset{\sim}{\mathcal{F}}^{q_i}$, and set

($\underset{\sim}{d}$)

$$\underset{\sim}{d} =: \{n : \Gamma^1_{\underset{\sim}{k}^1(n)} \models R(\underset{\sim}{x}_{\delta_1}(n), \underset{\sim}{x}_{\delta_2}(n)) \iff \Gamma^2_{\underset{\sim}{k}^2(n)} \models \neg R(\underset{\sim}{z}_{\delta_1}(n), \underset{\sim}{z}_{\delta_2}(n))\}.$$

We want to find $r \in \mathcal{A}P$ with $\mathcal{A}^r = \mathcal{A}$ so that $r \geq q_1, q_2$, and $\mathtt{p} \Vdash$ "$\underset{\sim}{d} \in \underset{\sim}{\mathcal{F}}^r$". This will then mean that $\underset{\sim}{F}$ could have been "killed", after all, and will complete the argument.

Suppose this is not possible, and thus as in 4.9 (1) for some $\mathrm{p}' \geq \mathrm{p}$ in $\mathbb{P}\restriction\mathcal{A}$, if $\mathrm{p}'_i = \mathrm{p}'\restriction\mathcal{A}_i$ for $i = 0, 1, 2$, we have:

a $(\mathbb{P}\restriction\mathcal{A}_1)$-name $\underset{\sim}{a}$ of a member of $\underset{\sim}{\mathcal{F}}^1$;

a $(\mathbb{P}\restriction\mathcal{A}_2)$-name $\underset{\sim}{b}$ of a member of $\underset{\sim}{\mathcal{F}}^2$; and

a $\mathbb{P}$-name $\underset{\sim}{c} =: \{n : \underset{\sim}{x}_{\delta_2}(n) \in \Gamma^1_{\underset{\sim}{k}^1(n)} \backslash \underset{\sim}{A}_n\}$ associated with a $(\mathbb{P}\restriction\mathcal{A}_1)$-name $(\underset{\sim}{A}_n)_{n<\omega}$ of an $(\underset{\sim}{\mathcal{F}}^1, \underset{\sim}{k}^1)$-slow sequence; with

$$\mathrm{p}' \Vdash_{\mathbb{P}\restriction\mathcal{A}} \text{“}\underset{\sim}{a} \cap \underset{\sim}{b} \cap \underset{\sim}{c} \cap \underset{\sim}{d} = \emptyset\text{”}$$

We shall get a contradiction. Let $H^0 \subseteq \mathbb{P}\restriction\mathcal{A}^0$ be generic over V.

We define for every n the following $(\mathbb{P}\restriction\mathcal{A}^0)$-names:

$$\begin{aligned} \underset{\sim}{can}^1_n[H^0] = \{(u,v) \in \Gamma^1_{\underset{\sim}{k}^1(n)} \times \Gamma^2_{\underset{\sim}{k}^2(n)} :&\\ &\text{For some } \mathrm{p}''_1 \in \mathbb{P}\restriction\mathcal{A}_1 \text{ with } \mathrm{p}''_1 \geq \mathrm{p}'_1 \text{ and } \mathrm{p}''_1\restriction\mathcal{A}_0 \in \mathtt{H}^0,\\ &\mathrm{p}''_1 \Vdash_{\mathbb{P}\restriction\mathcal{A}_1/\mathtt{H}^0} \text{“}[\underset{\sim}{z}_{\delta_1}(n) = u,\ u \notin \underset{\sim}{A}_n,\ n \in \underset{\sim}{a} \text{ and } \underset{\sim}{z}_{\delta_1}(n) = v]\text{”}\} \end{aligned}$$

$$\begin{aligned} \underset{\sim}{can}^2_n[H^0] = \{(u,v) \in \Gamma^1_{\underset{\sim}{k}^1(n)} \times \Gamma^2_{k^2(n)} :&\\ &\text{For some } \mathrm{p}''_2 \in \mathbb{P}\restriction\mathcal{A}_2 \text{ with } \mathrm{p}''_2 \geq \mathrm{p}'_2 \text{ and } \mathrm{p}''_2\restriction\mathcal{A}_0 \in \mathtt{H}^0,\\ &\mathrm{p}''_2 \Vdash_{\mathbb{P}\restriction\mathcal{A}_2/\mathtt{H}^0} \text{“}[\underset{\sim}{z}_{\delta_2}(n) = u,\ n \in \underset{\sim}{b} \text{ and } \underset{\sim}{z}_{\delta_2}(n) = v]\text{”}\} \end{aligned}$$

and for $i = 1, 2$ and $u \in \Gamma^1_{\underset{\sim}{k}^1(n)}$ we let

$$\underset{\sim}{can}^i_n(u) =: \{v \in \Gamma^2_{\underset{\sim}{k}^2(n)} : (u,v) \in \underset{\sim}{can}^i_n\}$$

$$\underset{\sim}{A}^i_n =: \{u : (\exists v)(u,v) \in \underset{\sim}{can}^i_n\}$$

Now in $V[\mathtt{H}^0]$, $(\underset{\sim}{A}^i_n : n < \omega)$ is not $(\underset{\sim}{\mathcal{F}}^i, \underset{\sim}{k}^1)$-slow, and thus the set:

$$\{n : |\underset{\sim}{A}^1_n|/\|\Gamma^1_{\underset{\sim}{k}^1(n)}\|, |\underset{\sim}{A}^2_n|/\|\Gamma^1_{\underset{\sim}{k}^1(n)}\| \text{ are greater than } \|\Gamma^1_{\underset{\sim}{k}^1(n)}\|^{-1/2}\}$$

belongs to $\underset{\sim}{\mathcal{F}}^0[\mathtt{H}]$. Choose any such n, and by finite combinatorics we shall derive a contradiction. Remember that we have assumed without loss of generality that $\|\Gamma^1_{\underset{\sim}{k}^1(n)}\| > \|\Gamma^2_{\underset{\sim}{k}^2(n)}\|^5$ for a large set of n modulo $\underset{\sim}{\mathcal{F}}\restriction\mathcal{A}_0$, so wlog our n satisfies this, too. Let $g_i : \underset{\sim}{A}^i_n \longrightarrow \Gamma^2_{\underset{\sim}{k}^2(n)}$ be such that $g_i(v) \in \underset{\sim}{can}^i_n(v)$. Now $|\operatorname{range}(g_i)| \leq \|\Gamma^2_{\underset{\sim}{k}^2(n)}\|$, so there are $b_1, b_2 \in \Gamma^2_{\underset{\sim}{k}^2(n)}$ such that for $i = 1, 2$:

$$|g_i^{-1}(b_i)| \geq |\underset{\sim}{A}^i_n|/\|\Gamma^2_{\underset{\sim}{k}^2(n)}\| > \|\Gamma^1_{\underset{\sim}{k}^1(n)}\|^{1/5}.$$

Now by 4.4(ii) we find $a_i, a_i' \in g^{-1}(b_i)$ for $i = 1, 2$ with $\Gamma^1_{\underset{\sim}{k}^1(n)} \models R(a_1, a_2)$ and $\neg R(a_1', a_2')$. As either $\Gamma^1_{\underset{\sim}{k}^1(n)} \models R(b_1, b_2)$ or $\Gamma^1_{\underset{\sim}{k}^1(n)} \models \neg R(b_1, b_2)$, we can show that it is not forced by p' that $n \notin \underset{\sim}{a} \cap \underset{\sim}{b} \cap \underset{\sim}{c} \cap \underset{\sim}{d}$, a contradiction. □

4.13 Proof of the Claim 4.12A from 4.12. We first recall the situation. We had:

(i) δ is $(\underset{\sim}{F}, \underset{\sim}{k}^1, \underset{\sim}{k}^2, \mathrm{p})$-closed; $q_0 = q{\restriction}\delta$;

(ii) $\mathrm{p} \in \mathbb{P}{\restriction}\mathcal{A}^{q_0}$, $\delta \in \mathcal{A}^q$, $\varepsilon^q = 1$, $\underset{\sim}{k}^q_\delta = \underset{\sim}{k}^1$, $\underset{\sim}{F}_\delta = \underset{\sim}{F}{\restriction}\delta$;

(iii) $\underset{\sim}{z}$ is a $(\mathbb{P}{\restriction}\mathcal{A}^q)$-name for a real;

(iv) For all $r \geq q$ in $\mathcal{A}P$ such that $r{\restriction}\delta \in G^\delta$, and $\underset{\sim}{x}$ a $(\mathbb{P}{\restriction}\mathcal{A}^{r\restriction\delta})$-name, with $\underset{\sim}{y} =: \underset{\sim}{F}(\underset{\sim}{x})$ a $(\mathbb{P}{\restriction}\mathcal{A}^{r\restriction\delta})$-name, we have:
$(*)_{\underset{\sim}{x},\underset{\sim}{y}}$ $\mathrm{p} \Vdash$ "The set $\{n : \Gamma^1_{\underset{\sim}{k}^1(n)} \models R(\underset{\sim}{x}(n), \underset{\sim}{x}_\delta(n))$ iff $\Gamma^2_{\underset{\sim}{k}^2(n)} \models R(\underset{\sim}{y}(n), \underset{\sim}{z}(n))\}$ is in $\underset{\sim}{\mathcal{F}}^r$".

We defined the property $(**)_{\underset{\sim}{x},\underset{\sim}{y}}$ as follows:

$(**)_{\underset{\sim}{x},\underset{\sim}{y}}$ For all $r \geq q$ in $\mathcal{A}P$ such that $r{\restriction}\delta \in G^\delta$ and $\underset{\sim}{x}$, $\underset{\sim}{y}$ are $(\mathbb{P}{\restriction}\mathcal{A}^{r\restriction\delta})$-names, $(*)_{\underset{\sim}{x},\underset{\sim}{y}}$ holds.

Claim. *If $\underset{\sim}{x}^1, \underset{\sim}{x}^2$ are $(\mathbb{P}{\restriction}\delta)$-names of functions in $\prod_n \Gamma^1_{\underset{\sim}{k}^1(n)}$, $\underset{\sim}{y}$ is a $(\mathbb{P}{\restriction}\delta)$-name of a member of $\prod_n \Gamma^2_{\underset{\sim}{k}^2(n)}$, and both pairs $(\underset{\sim}{x}^1, \underset{\sim}{y})$ and $(\underset{\sim}{x}^2, \underset{\sim}{y})$ satisfy the condition $(**)_{\underset{\sim}{x},\underset{\sim}{y}}$ above, then:*
$\mathrm{p} \Vdash_{\mathbb{P}\restriction\delta}$ "$\underset{\sim}{x}^1 = \underset{\sim}{x}^2 \bmod \underset{\sim}{\mathcal{F}}{\restriction}\delta[\mathtt{H}]$ *or both are restricted for* $(\mathtt{H}, \mathcal{A}^{q_0}, \underset{\sim}{k}^1)$."

Proof. Suppose that $\mathrm{p} \leq \hat{\mathrm{p}} \in \mathbb{P}{\restriction}\delta$ and $\hat{\mathrm{p}}$ forces the contrary; so without loss of generality

(5) $$\hat{\mathrm{p}} \Vdash \text{"}\underset{\sim}{x}^1 \neq \underset{\sim}{x}^2 \bmod \underset{\sim}{\mathcal{F}}{\restriction}\delta[\underset{\sim}{\mathtt{H}}]\text{"};$$

(6) $$\hat{\mathrm{p}} \Vdash \text{"}\underset{\sim}{x}^1 \text{ is unrestricted for } (\mathtt{H}, \mathcal{A}^{q_0}, \underset{\sim}{k}^1)\text{."}$$

Choose any $q_1 \geq q_0$ with $q_1 \in G^\delta$ so that $\underset{\sim}{x}^1, \underset{\sim}{x}^2, \underset{\sim}{y}$ are $\mathbb{P}{\restriction}\mathcal{A}^{q_1}$-names. Now we will construct $r \geq q_1, q{\restriction}(\delta + 1)$, with r in $\mathcal{A}P$ and $\mathcal{A}^r = \mathcal{A}^{q_1} \cup \{\delta\}$, so that:

(7) $$\hat{\mathrm{p}} \Vdash \text{"}\{n : \Gamma^1_{\underset{\sim}{k}^1(n)} \models\text{'}R(\underset{\sim}{x}^1(n), \underset{\sim}{x}_\delta(n)) \iff \neg R(\underset{\sim}{x}^2(n), \underset{\sim}{x}_\delta(n))]\} \in \underset{\sim}{\mathcal{F}}^r\text{."}$$

By 4.9(2) we can also find $r' \geq r, q$, and then (7) contradicts $(**)_{\underset{\sim}{x}^1,\underset{\sim}{y}}$ & $(**)_{\underset{\sim}{x}^2,\underset{\sim}{y}}$. Thus to complete the proof of our claim, it suffices to find r.

This is the sort of problem considered in 4.9(1), with an additional set required to be in $\underset{\sim}{\mathcal{F}}{\restriction}(\mathcal{A}^{q_1} \cup \{\delta\})$. The q_0, q_1 under consideration here correspond to the q_0, q_1 of 4.9(1), and we let q_2 be $q\restriction(\delta+1)$. Following the notation of 4.9(1), set $\underset{\sim}{\mathcal{F}}^i = \underset{\sim}{\mathcal{F}}^{q_i}$, $\mathcal{A}_i = \mathcal{A}^{q_i}$ for $i = 0,1,2$, and $\mathcal{A} = \mathcal{A}_1 \cup \{\delta\} = \mathcal{A}_1 \cup \mathcal{A}_2$. We need to find $r \geq q_1, q_2$ as in 4.9(1), with (7) holding.

Suppose on the contrary that $\hat{\mathrm{p}} \leq \mathrm{p}' \in \mathbb{P}\restriction\mathcal{A}$ and p' forces "There is no $\underset{\sim}{\mathcal{F}}$ as required". Then extending p', we may suppose that we have a $\mathbb{P}\restriction\mathcal{A}_1$-name $\underset{\sim}{a}$ for a member of $\underset{\sim}{\mathcal{F}}^1$, a $\mathbb{P}\restriction\mathcal{A}_2$-name $\underset{\sim}{b}$ for a member of $\underset{\sim}{\mathcal{F}}^2$, a $\mathbb{P}\restriction\mathcal{A}_1$-name for an $(\underset{\sim}{\mathcal{F}}^1, \underset{\sim}{k}^1)$-slow sequence $(\underset{\sim}{A}_n)$ (associated with a power $d < \omega$ – cf. 4.7), such that setting:

$$\underset{\sim}{c} = \{n : \underset{\sim}{x}_\delta(n) \in \Gamma^1_{\underset{\sim}{k}^1(n)} \setminus \underset{\sim}{A}_n\}$$

$$\underset{\sim}{d} = \{n : \Gamma^1_{\underset{\sim}{k}^1(n)} \models \text{``}R(\underset{\sim}{x}^1(n), \underset{\sim}{x}_\delta(n)) \iff \neg R(\underset{\sim}{x}^2(n), \underset{\sim}{x}_\delta(n))\text{''}\}$$

we have:

$$\mathrm{p}' \Vdash_{\mathbb{P}\restriction\mathcal{A}} \text{``}\underset{\sim}{a} \cap \underset{\sim}{b} \cap \underset{\sim}{c} \cap \underset{\sim}{d} = \emptyset\text{''}$$

Let $\mathrm{p}'_i = \mathrm{p}'\restriction\mathcal{A}_i$ for $i = 0,1,2$, and take $\mathtt{H}^0 \subseteq \mathbb{P}\restriction\mathcal{A}_0$ generic over V. Without loss of generality, for some natural number d:

$$\mathrm{p}'_1 \Vdash \text{``}n \in \underset{\sim}{a} \Longrightarrow \underset{\sim}{x}^1(n) \neq \underset{\sim}{x}^2(n) \text{ and } |\underset{\sim}{A}_n| \leq \sqrt{\|\Gamma^1_{\underset{\sim}{k}^1(n)}\|} \cdot (\log \|\Gamma^1_{\underset{\sim}{k}^1(n)}\|)^d \text{ (and } \underset{\sim}{A}_n \subseteq \Gamma^1_{\underset{\sim}{k}^1(n)}).\text{''}$$

We are interested in $\underset{\sim}{B}_n[\mathtt{H}^0]$=:

$$\{v \in \Gamma^1_{\underset{\sim}{k}^1(n)} : \text{for some } \mathrm{p}''_2 \geq \mathrm{p}'_2 \text{ with } \mathrm{p}''_2\restriction\mathcal{A}_0 \in \mathtt{H}^0,\ \mathrm{p}''_2 \Vdash_{\mathbb{P}\restriction\mathcal{A}_2}\text{``}n \in \underset{\sim}{b} \text{ and } \underset{\sim}{x}_\delta(n) = v\text{''}\}$$

(which is a $(\mathbb{P}\restriction\mathcal{A}_0)$-name). Clearly the sequence $(\underset{\sim}{B}_n)$ is not $(\underset{\sim}{\mathcal{F}}^1, \underset{\sim}{k}^1)$-slow in $V[\mathtt{H}^0]$.

For each n let us also consider the set $\underset{\sim}{Y}_n[\mathtt{H}^0]$ =:

$$\{(A, v_1, v_2) : A \cup \{v_1, v_2\} \subseteq \Gamma^1_{\underset{\sim}{k}^1(n)}, v_1 \neq v_2, \text{ and for some } \mathrm{p}''_1 \text{ with } \mathrm{p}''_1 \geq \mathrm{p}'_1, \mathrm{p}''_1\restriction\mathcal{A}_0 \in \mathtt{H}^0, \mathrm{p}''_1 \Vdash \text{``}n \in \underset{\sim}{a}, \underset{\sim}{A}_n = A, \underset{\sim}{x}^1(n) = v_1, \underset{\sim}{x}^2(n) = v_2.\text{''}\}$$

For every $(A, v_1, v_2) \in \underset{\sim}{Y}_n$, we have:

$$|A| \leq \sqrt{\|\Gamma^1_{\underset{\sim}{k}^1(n)}\|} \cdot (\log \|\Gamma^1_{\underset{\sim}{k}^1(n)}\|)^d, \text{ and } v_1 \neq v_2 \tag{8}$$

As $\underset{\sim}{x}_\delta$ is unrestricted over $\mathcal{A}_0$ in $V[\mathtt{H}^0]$, for the $\underset{\sim}{\mathcal{F}}^0$-majority of n we have:

$$|\underset{\sim}{B}_n| \geq \sqrt{\|\Gamma^1_{\underset{\sim}{k}^1(n)}\|} \cdot (\log \|\Gamma^1_{\underset{\sim}{k}^1(n)}\|)^{d+2} \tag{9}$$

Now (by (6)), also for the $\underset{\sim}{\mathcal{F}}^0$ majority of n we have:

$$\underset{\sim}{C}_n =: \{v_1 \in \Gamma^1_{\underset{\sim}{k}^1(n)} : \text{There are } A, v_2 \text{ so that } (A, v_1, v_2) \in \underset{\sim}{Y}_n\} \tag{10}$$

$$\text{has at least } \sqrt{\|\Gamma^1_{\underset{\sim}{k}^1(n)}\|} \text{ members}$$

Now it will suffice to find n, $v \in \underset{\sim}{B}_n$ and $(A, v_1, v_2) \in \underset{\sim}{Y}_n$ so that

$$\Gamma^1_{\underset{\sim}{k}^1(n)} \models [R(v_1, v) \iff \neg R(v_2, v)] \,\&\, v \notin A, \tag{11}$$

as we can then choose $\mathtt{p}_1'' \in \mathbb{P}{\restriction}\mathcal{A}_1$, $\mathtt{p}_2'' \in \mathbb{P}{\restriction}\mathcal{A}_2$ with $\mathtt{p}_i'' \geq \mathtt{p}_i'$, $\mathtt{p}_i''{\restriction}\mathcal{A}_0 \in \mathtt{H}^0$ for $i = 1, 2$, so that:

$$\mathtt{p}_1'' \Vdash \text{“}n \in \underset{\sim}{a}, \underset{\sim}{A}_n = A,\ \underset{\sim}{x}^1(n) = v_1,\ \underset{\sim}{x}^2(n) = v_2\text{”};\ \mathtt{p}_2'' \Vdash \text{“}n \in \underset{\sim}{b} \text{ and } \underset{\sim}{x}_\delta(n) = v\text{”}$$

and hence $\mathtt{p}_1'' \cup \mathtt{p}_2'' \Vdash$ “$n \in \underset{\sim}{a} \cap \underset{\sim}{b} \cap \underset{\sim}{c} \cap \underset{\sim}{d}$”, a contradiction.

So it remains to find n, v and (A, v_1, v_2). For n sufficiently large satisfying (8-10), we can choose triples $t_i = (A^i, v_1^i, v_2^i) \in \underset{\sim}{Y}_n$ for $i < 5 \log \|\Gamma^1_{\underset{\sim}{k}^1(n)}\|$ with all vertices v_1^i distinct from each other and from all v_2^i. By the pseudorandomness of $\Gamma^1_{\underset{\sim}{k}^1(n)}$ (more specifically 4.4(iii)), the set

$\underset{\sim}{S} = \{v \in \Gamma^1_{\underset{\sim}{k}^1(n)} :$ For no $i < 5 \log \|\Gamma^1_{\underset{\sim}{k}^1(n)}\|$ do we have $R(v_1^i, v) \iff \neg R(v_2^i, v)\}$

has size at most $5 \log \|\Gamma^1_{\underset{\sim}{k}^1(n)}\|$. So if $\underset{\sim}{S}' =: \underset{\sim}{S} \cup \bigcup\{A^i : i < 5 \log \|\Gamma^1_{\underset{\sim}{k}^1(n)}\|\}$, then we will have: $|\underset{\sim}{S}'| \ll \sqrt{\|\Gamma^1_{\underset{\sim}{k}^1(n)}\|}(\log \|\Gamma^1_{\underset{\sim}{k}^1(n)}\|)^{d+2}$, so there is $v \in \underset{\sim}{B}_n \setminus \underset{\sim}{S}'$. Since $v \notin \underset{\sim}{S}'$, for some i (11) will hold with $(A, v_1, v_2) = (A^i, v_1^i, v_2^i)$. □

The last detail

The following was used in the proof of 4.12 (after 3.12A slightly before $(\underset{\sim}{d})$).

Claim. *Assume $q_2{\restriction}\beta \leq q_1$, $\mathcal{A}^{q_1} \subseteq \beta$. Let $q_0 = q_2{\restriction}\beta$, and write $\mathcal{A}_i$ for $\mathcal{A}^{q_i}$, $\mathcal{A} = \mathcal{A}_1 \cup \mathcal{A}_2$, and $\mathcal{F}^i$ for $\mathcal{F}^{q_i}$. Let $\mathtt{p} \in \mathbb{P}{\restriction}\mathcal{A}$ and $\mathtt{p}_i = \mathtt{p}{\restriction}\mathcal{A}_i$. Then we can find r with $\mathcal{A}^r = \mathcal{A}$ and $r \geq q_1, q_2$, so that for any $(\mathbb{P}{\restriction}\mathcal{A}_i)$-names $\underset{\sim}{y}_i$ $(i = 1, 2)$ of members of $\prod_n \Gamma^2_{\underset{\sim}{k}^2(n)}$ if:*

$$\mathtt{p}_i \Vdash_{\mathbb{P}{\restriction}\mathcal{A}_i} \text{“}\underset{\sim}{y}_i \neq \underset{\sim}{y}' \bmod \underset{\sim}{\mathcal{F}}^i\text{”}$$

for $(i = 1, 2)$ *and for all* $(\mathbb{P}{\restriction}\mathcal{A}_0)$*-names* $\underset{\sim}{y}'$*, then we have:*

$$\mathtt{p} \Vdash_{\mathbb{P}\restriction\mathcal{A}} \text{“}\underset{\sim}{y}_1 \neq \underset{\sim}{y}_2 \quad \operatorname{mod} \underset{\sim}{\mathcal{F}}^r\text{”}$$

Hence $p \Vdash_{\mathbb{P}\restriction\mathcal{A}}$ "if $\underset{\sim}{y}_i \neq \underset{\sim}{y}' \operatorname{mod} \underset{\sim}{\mathcal{F}}^i$ for $i = 1, 2$ and $\underset{\sim}{y}'$ a $(\mathbb{P}{\restriction}\mathcal{A}_0)$-name then $\underset{\sim}{y}_1 \neq \underset{\sim}{y}_2 \operatorname{mod} \underset{\sim}{\mathcal{F}}^r$".

Proof. We use induction construction. Much as in the proof of 4.9, we must deal primarily with the case in which $\mathcal{A}^{q_2} = \mathcal{A}^{q_0} \cup \{\beta\}$. Suppose toward a contradiction that $\mathtt{p} \leq \mathtt{p}' \in \mathbb{P}{\restriction}\mathcal{A}$, and with $\mathtt{p}'_i = \mathtt{p}'{\restriction}\mathcal{A}_i$ for $i = 0, 1, 2$ we have:

i. a $(\mathbb{P}{\restriction}\mathcal{A}_1)$-name $\underset{\sim}{a}$ of a member of $\underset{\sim}{\mathcal{F}}^1$;
ii. a $(\mathbb{P}{\restriction}\mathcal{A}_2)$-name $\underset{\sim}{b}$ of a member of $\underset{\sim}{\mathcal{F}}^2$;
iii. a $(\mathbb{P}{\restriction}\mathcal{A})$-name $\underset{\sim}{c} = \{n : \underset{\sim}{x}_\beta(n) \in \Gamma^2_{\underset{\sim}{k}^2(n)} \setminus \underset{\sim}{A}_n\}$ associated with a $(\mathbb{P}{\restriction}\mathcal{A}_1)$-name $(\underset{\sim}{A}_n)_{n<\omega}$ of a $(\underset{\sim}{\mathcal{F}}^1, \underset{\sim}{k}^1)$-slow sequence; and
iv. a $(\mathbb{P}{\restriction}\mathcal{A})$-name $\underset{\sim}{d} = \bigcap_{j=1}^N \underset{\sim}{d}_j$, for a finite intersection of sets of the form $\underset{\sim}{d}_j =: \{n : \underset{\sim}{y}^1_j(n) \neq \underset{\sim}{y}^2_j(n)\}$, with each $\underset{\sim}{y}^i_j$ a $\mathbb{P}{\restriction}\mathcal{A}_i$-name of a member of $\prod_n \Gamma^2_{\underset{\sim}{k}^2(n)}$, such that for each $i = 1, 2$ and $j = 1, \ldots, N$:

$\mathtt{p}_i \Vdash$ "$\underset{\sim}{y}^i_j \neq \underset{\sim}{y}' \quad \operatorname{mod} \underset{\sim}{\mathcal{F}}^i$ for any $(\mathbb{P}{\restriction}\mathcal{A}_0)$-name $\underset{\sim}{y}'$ of a member of $\prod_n \Gamma^2_{\underset{\sim}{k}^2(n)}$."

and that $\mathtt{p}' \Vdash$ "$\underset{\sim}{a} \cap \underset{\sim}{b} \cap \underset{\sim}{c} \cap \underset{\sim}{d} = \emptyset$". Let $\mathtt{H}^0$ be generic over V, $\mathtt{p}_0 \in \mathtt{H}^0$, and let us define in $V[\mathtt{H}^0]$:

$$\underset{\sim}{\Delta}^1_n[\mathtt{H}^0] =: \{(A, u_1, \ldots, u_N) : A \subseteq \Gamma^2_{\underset{\sim}{k}^2(n)}, u_1, \ldots, u_N \in \Gamma^2_{\underset{\sim}{k}^2(n)},$$

$$\text{and there is } \mathtt{p}'_1 \in \mathbb{P}{\restriction}\mathcal{A}_1,\ \mathtt{p}'_1 \geq \mathtt{p}_1,\ \mathtt{p}'_1{\restriction}\mathcal{A}_0 \in \mathtt{H}^0, \text{ and}$$

$$\mathtt{p}'_1 \Vdash_{\mathbb{P}\restriction\mathcal{A}_1} \text{“}\underset{\sim}{A}_n = A,\ \underset{\sim}{y}^1_1(n) = u_1, \ldots, \underset{\sim}{y}^1_N(n) = u_N, \text{ and } n \in \underset{\sim}{a}\text{”}.\}$$

$$\underset{\sim}{\Delta}^2_n[\mathtt{H}^0] =: \{(v_0, v_1, \ldots, v_N) : \text{ all } v_j \in \Gamma^2_{\underset{\sim}{k}^2(n)} \text{ and there is}$$

$$\mathtt{p}'_2 \in \mathbb{P}{\restriction}\mathcal{A}_2,\ \mathtt{p}' \geq \mathtt{p}_2,\ \mathtt{p}'_2{\restriction}\mathcal{A}_0 \in \mathtt{H}^0 \text{ and}$$

$$\mathtt{p}'_2 \Vdash_{\mathbb{P}\restriction\mathcal{A}_2} \text{“}\underset{\sim}{x}_\beta(n) = v_0, \underset{\sim}{y}^2_1(n) = v_1 \ldots, \underset{\sim}{y}^2_N(n) = v_N \text{ and } n \in \underset{\sim}{b}\text{”}\}$$

Without loss of generality, for some d,

$$\mathtt{p}_1 \Vdash: \text{“For } n \in \underset{\sim}{a},\ |\underset{\sim}{A}_n| \leq \sqrt{\|\Gamma^2_{\underset{\sim}{k}^2(n)}\|} \cdot (\log \|\Gamma^2_{\underset{\sim}{k}^2(n)}\|)^d.\text{”}$$

Thus:

(1) $(A, u_1, \dots, u_N) \in \underset{\sim}{\Delta}^1_n \Longrightarrow |A| \leq \sqrt{\|\Gamma^2_{\underset{\sim}{k}^2(n)}\|} \cdot (\log \|\Gamma^2_{\underset{\sim}{k}^2(n)}\|)^d.$

By the assumption on $\underset{\sim}{y}^1_1, \dots, \underset{\sim}{y}^1_N$,

(2) If $e < \omega$, $C_n \subseteq \Gamma^2_{\underset{\sim}{k}^2(n)}$, $|\underset{\sim}{C}_n| \leq e$, and $(\underset{\sim}{C}_n : n < \omega) \in V[\mathtt{H}^0]$, then

$$\{n : \text{ there is } (A, u_1, \dots, u_N) \in \underset{\sim}{\Delta}^1_n,\ u_1 k \dots, u_N \notin \underset{\sim}{C}_n \} \in \underset{\sim}{\mathcal{F}}^0.$$

Hence without loss of generality:

For $n \in \underset{\sim}{a}$, there are $(A, u^j_1, \dots, u^j_N) \in \underset{\sim}{\Delta}^1_n$, for $j \leq N+1$, with

(3) The sets $\{u^j_1, \dots, u^j_N\}$ (for $j \leq N+1$) pairwise disjoint.

As $q_2 \in \mathcal{AP}$,

(4) If $(\underset{\sim}{C}_n : n < \omega) \in V[\mathtt{H}^0]$ is $(\underset{\sim}{\mathcal{F}}^0, \underset{\sim}{k})$-slow then

$$\{n : \text{There is } (v_0, v_1, \dots, v_N) \in \underset{\sim}{\Delta}^2_n \text{ with } v_0 \notin \underset{\sim}{C}_n\} \in \underset{\sim}{\mathcal{F}}^0$$

Let $\underset{\sim}{a}^+ =: \{n : \underset{\sim}{\Delta}^1_n \neq \emptyset, \underset{\sim}{\Delta}^2_n \neq \emptyset$, moreover, $\underset{\sim}{\Delta}^1_n$ satisfies (3)$\}$ (a $\mathbb{P}\restriction\mathcal{A}^0$-name of a member of $\mathcal{F}^0$). So for $n \in \underset{\sim}{a}^+$, there are $(N+1)$-tuples $(A^{n,j}, u^{n,j}_1, \dots, u^{n,j}_N)$ for $j \leq N+1$ with the sets $\{u^{n,j}_1, \dots, u^{n,j}_N\}$ pairwise disjoint. Let $\underset{\sim}{C}_n = \bigcup_{j \leq N} A^{n,j}$ for $n \in \underset{\sim}{a}^+$, $\underset{\sim}{C}_n = \emptyset$ for $n \notin \underset{\sim}{a}^+$. So $(\underset{\sim}{C}_n)_{n<\omega} \in V[\mathtt{H}^0]$ is $(\underset{\sim}{\mathcal{F}}^0, \underset{\sim}{k}^1)$-slow, hence for some $n \in \underset{\sim}{a}^+$, there is $(v_0, v_1, \dots, v_N) \in \underset{\sim}{\Delta}^2_n$, with $v_0 \notin \underset{\sim}{C}_n$. Now for some $j \leq N+1$ we have $\bigwedge_{i=1}^N v_i \neq u^{n,j}_i$. Choose $\mathtt{p}'_2 \in \mathbb{P}\restriction\mathcal{A}_2$, $\mathtt{p}'_2 \geq \mathtt{p}_2$, with $\mathtt{p}'_2\restriction\mathcal{A}_0 \in \mathtt{H}^0$ and $\mathtt{p}'_2 \Vdash$ "$n \in \underset{\sim}{b}$, $\underset{\sim}{x}_\beta(n) = v_0, \bigwedge_{i=1}^N \underset{\sim}{y}^2_i(n) = v_i$". Choose $\mathtt{p}'_1 \in \mathbb{P}\restriction\mathcal{A}_1$, $\mathtt{p}'_1 \geq \mathtt{p}_1$, with $\mathtt{p}'_1\restriction\mathcal{A}_0 \in \mathtt{H}^0$ and $\mathtt{p}'_1 \Vdash$ "$n \in \underset{\sim}{a}$, $\underset{\sim}{A}_n = A^{n,j}$, and for all $i = 1, \dots, N$ $\underset{\sim}{y}^1_i(n) = u^{n,j}_i$." Now $\mathtt{p}'_1 \cup \mathtt{p}'_2 \Vdash$ "$n \in \underset{\sim}{a} \cap \underset{\sim}{b} \cap \underset{\sim}{c} \cap \underset{\sim}{d}$", a contradiction.

This finishes the case $\mathcal{A}_2 = \mathcal{A}_1 \cup \{\beta\}$. The general case follows as in 4.9(2). At successors we apply the case just treated. Limits of uncountable cofinality are handled by taking unions. At limits of cofinality ω we have to repeat the first argument with some variations; we do not have to worry about $\underset{\sim}{c}$, so the fact that there are several $\underset{\sim}{x}_\beta$ involved is not a problem. The problem in this case is of course to extend the union of the ultrafilters constructed so far to an ultrafilter in a slightly larger model of set theory, while retaining the main property for new names $\underset{\sim}{y}^2_i$. $\square$

Appendix. Background Material

A1. Proper and α-proper forcing

A1.1 Proper forcing

Let $\mathcal{P} = (P, \leq)$ be a partially ordered set. A cardinal λ is *$\mathcal{P}$-large* if the power set of P is in V_λ (the universe of all sets of rank less than λ). With $\mathcal{P}$ fixed and λ $\mathcal{P}$-large, let $\mathcal{V}_\lambda$ be the structure $(V_\lambda; \in, P, \leq)$.

1. For $\mathcal{M} \prec \mathcal{V}_\lambda$ and $p \in P$, p is *$\mathcal{M}$-generic* iff for each name of an ordinal $\underset{\sim}{\alpha}$ with $\underset{\sim}{\alpha} \in M$, $p \Vdash$ "$\underset{\sim}{\alpha} \in \mathcal{M}$".
2. $\mathcal{P}$ is *proper* iff for all $\mathcal{P}$-large λ and all countable elementary substructures $\mathcal{M}$ of $\mathcal{V}_\lambda$ with $\mathcal{P} \in \mathcal{M}$, each $p \in \mathcal{M}$ has an $\mathcal{M}$-generic extension in P.

A1.2 Axiom A

$\mathcal{P}$ satisfies *Axiom* A if there is a collection $\leq_n$ $(n = 1, 2, \dots)$ of partial orderings on the set P with $\leq_1$ coinciding with the given ordering $\leq$, and $\leq_{n+1}$ finer than $\leq_n$ for each n, satisfying the following two conditions:

1. If $p_1 \leq_1 p_2 \leq_2 \leq p_3 \leq_3 \dots$ then there is some $p \in P$ with $p_n \leq_n p$ for all n;
2. For all $p \in P$, any name $\underset{\sim}{\alpha}$ of an ordinal, and any n, there is a condition $q \in P$ with $p \leq_n q$, and a countable set B of ordinals, such that $q \Vdash \underset{\sim}{\alpha} \in B$.

The forcings used in §§2,3 were seen to satisfy Axiom A, and the following known result was then applied.

A1.3 Proposition. *If $\mathcal{P}$ satisfies Axiom A then $\mathcal{P}$ is proper.*

Proof. Given a countable $\mathcal{M} \prec \mathcal{V}_\lambda$ and $p \in P \cap M$, let $\underset{\sim}{\alpha}_n$ be a list of all ordinal names in $\mathcal{M}$, and use clause (2) of Axiom A to find $q_n, B_n \in \mathcal{M}$ with $q_n \in P$, B_n countable. $p \leq_1 q_1 \leq_2 q_2 \leq \dots$ and $q_n \Vdash$ "$\underset{\sim}{\alpha}_n \in B_n$. Then use clause (1) to find $q \geq$ all q_n; this q will be $\mathcal{M}$-generic. $\square$

A1.4 Countable support iteration

Our notation for iterated forcing is as follows. $\underset{\sim}{\mathcal{Q}}_\alpha$ is the name of the α-th forcing in the iteration, and $\mathcal{P}_\alpha$ is the iteration up to stage α. The sequence $\mathcal{P}_\alpha$ is called the iteration, and the $\underset{\sim}{\mathcal{Q}}_\alpha$ are called the factors. It is assumed that $\underset{\sim}{\mathcal{Q}}_\alpha$ is a $\mathcal{P}_\alpha$-name for a partially ordered set with minimum element 0, and that $\mathcal{P}_{\alpha+1}$ is $\mathcal{P}_\alpha * \underset{\sim}{\mathcal{Q}}_\alpha$.

In general it is necessary to impose some further conditions at limit ordinals. We will be concerned exclusively with countable support iteration: at a limit ordinal δ, $\mathcal{P}_\delta$ consists of δ-sequences p such that $p{\restriction}\alpha \in \mathcal{P}_\alpha$ for $\alpha < \delta$, and $\Vdash_{\mathcal{P}_\alpha} p(\alpha) = 0$ for all but countably many $\alpha < \delta$.

A1.5 Proposition. *Let $\mathcal{P}_\alpha$ be a countable support iteration of length λ with factors $\underset{\sim}{\mathcal{Q}}_\alpha$ such that for all $\alpha < \lambda$, $\Vdash_{\mathcal{P}_\alpha}$ "$\underset{\sim}{\mathcal{Q}}_\alpha$ is proper." Then $\mathcal{P}_\lambda$ is proper.*

See [Sh b, Sh f, or Jech] for the proof.

In §§2,3 we need additional iteration theorems discussed in [Sh b] in the context of ω-proper forcing. Improvements in [Sh 177] or [Sh f] make this unnecessary, but we include a discussion of the relevant terminology here. This makes our discussion compatible with the contents of [Sh b].

A1.6 α-Proper forcing

Let α be a countable ordinal. Then $\mathcal{P}$ is *α-proper* iff for every $\mathcal{P}$-large λ, every continuous increasing $\alpha+1$-sequence $(\mathcal{M}_i)_{i\leq\alpha}$ of countable elementary substructures of $\mathcal{V}_\lambda$ with $\mathcal{P} \in \mathcal{M}_0$, every $p \in P \cap M_0$ has an extension $q \in P$ which is $\mathcal{M}_i$-generic for all $i \leq \alpha$.

Axiom A implies α-properness for α countable. For example we check ω-properness. So we consider a condition p in M_0, where $(\mathcal{M}_i)_{i<\omega}$ is a sequence of suitable countable models satisfying, among other things, $\mathcal{M}_i \in \mathcal{M}_{i+1}$. There is an $\mathcal{M}_0$-generic condition p_1 above p, and we can take $p_1 \in \mathcal{M}_1$, since $\mathcal{M}_1 \prec \mathcal{V}_\lambda$. Similarly we can successively find $p_{n+1} \in P \cap M_{n+1}$ with p_{n+1} $\mathcal{M}_n$-generic, and $p_n \leq_n p_{n+1}$. A final application of Axiom A yields q above all the p_n.

Countable support iteration also preserves α-properness for each α [Sh b]. Furthermore it is proved in [Sh b, V4.3] that countable support iteration preserves the following conjunction of two properties: ω-properness and ${}^\omega\omega$-bounding. So [Sh b] contains most of the information needed in §§2,3, though we will need to add more concerning the iteration theorems below.

A2. Iteration theorems

A2.1 Fine* covering models

We recall the formalism introduced in [Sh b, Chap. VI] for proving iteration theorems. We consider collections of subtrees of ${}^{\omega>}\omega$ that cover ${}^\omega\omega$ in the sense that every function in ${}^\omega\omega$ represents a branch of one of the specified trees, and iterate forcings that do not destroy this property. Of course the precise formulation is considerably more restrictive. See discussion A2.6.

Weak covering models

A structure $(D; R)$ consisting of a set D and a binary relation R on D is called a *weak covering model* if:

1. For $x, t \in D$, $R(x,t)$ implies that t is a (nonempty) subtree of ${}^{\omega>}\omega$,

with no terminal nodes (leaves); we denote the set of branches of t by $\mathrm{Br}(t)$.

2. For every $\eta \in {}^\omega\omega$, and every $x \in \operatorname{dom} R$, there is some $t \in D$ with $R(x,t)$ and $\eta \in \mathrm{Br}(T)$. In this case, we say: (D,R) covers ${}^\omega\omega$.

$(D;R)$ should be thought of as a suitable small fragment of a universe of sets, and $R(x,t)$ is to be thought of intuitively as saying, in some manner, that the tree t has "size" at most x. In the next definition we introduce an ordering on the "sizes" and exploit more of our intutition, though certain intuitively natural axioms are omitted, as they are never needed in proofs.

Fine* covering models

A structure $\mathcal{D} = (D; R, <)$ is called a *fine* covering model* if $(D;R)$ is a weak covering model, $<$ is a partial order on $\operatorname{dom} R$ with no minimal element, and:

(1) If $x, y \in \operatorname{dom} R$ with $x < y$, then there is $z \in \operatorname{dom} R$ with $x < z < y$ (and $D \neq \emptyset$ and for every $y \in D$ there is $x < y$ in D).

(2) $x < y \,\&\, R(x,t)$-implies $R(y,t)$.

(3) In any generic extension V^* in which $(D;R)$ is a weak covering model we have:

 $(*)$ for $x < y$ (from $\operatorname{dom} R$) and $t_n \in D$ with $R(x,t_n)$ for all n there is $t \in D$ with $R(y,t)$ holding and there are indices $n_0 < n_1 < \dots$ such that: for all $\eta \in {}^\omega\omega$: if $\eta{\restriction}n_i \in \bigcup_{j\leq i} t_j$ for all i then $\eta \in \mathrm{Br}(t)$.

 $\otimes$ if $\eta \in {}^\omega\omega$, $\eta_n \in {}^\omega\omega$, $\eta_n{\restriction}n = \eta{\restriction}n$ for $n < \omega$ and $x \in \operatorname{dom} R$ then for some t, $R(x,t)$, $\eta \in \mathrm{Br}(t)$ and for infinitely many n we have $\eta_n \in \mathrm{Br}(t)$.

In particular we require $(*)$ and $\otimes$ to hold in the original universe V. Observe also that in $(3*)$ we have in particular $t_0 \subseteq t$.

Note that $(3)^+$ below implies (3).

$(3)^+$ In any generic extension V^* (of V) in which (D,R) is a weak covering model we have:

$(*)^+$ For $x < y$ and $t_n \in D$ with $R(x,t_n)$ for all n, there is $t \in D$ with $R(y,t)$ holding and there are indices $0 = n_0 < n_1 < \dots$ such that for all $\eta \in {}^\omega\omega$ if $\eta{\restriction}n_i \in \bigcup_{j\leq i} t_{n_j}$ for all i, then $\eta \in Br(t)$; we let $w = \{n_0, n_1, \dots\}$.

[Why $(3)^+ \Rightarrow (3)$? assume $(3)^+$, so let a generic extension V^* of V in which (D,R) is a weak covering model be given, so in V^*, $(*)^+$ holds. First, for $\otimes$ of (3) let η, η_n, y be given, let $x < y$; as "(D,R) is a weak covering model in V^*" for each $n < \omega$ there is $t_n \in D$ such that

$R(x, t_n) \& \eta_n \in \mathrm{Br}(t_n)$. Apply $(*)^+$ to x, y, t_n and get t which is as required there. Second, for $(*)$ of (3), let $x < y, t_n (n < \omega)$ be given. Choose inductively $y', x_n, x < x_n < y' < y$, $x_n < x_{n+1}$ (possible by condition (1)). Choose by induction on n, k_n, t_n^* such that: $t_0^* = t^*, R(x_n, t_n^*), t_n^* \subseteq t_{n+1}^*$ and $[\nu \in t_{n+1} \& \nu \restriction k_n \in t_n^* \Rightarrow \nu \in t_{n+1}^*]$. For $n =$ 0-trivial, for $n+1$ use $(*)^+$ with $\langle x_n, x_{n+1}, t_n^*, t_{n+1}, t_{n+1}, \ldots \rangle$ here standing for $\langle x, y, t_0, t_1, t_2, \ldots \rangle$ there, and we get t_{n+1}^*, w_n (for t, w there), let $k_n = \mathrm{Min}(w_n \setminus \{0\})$, easily t_n^* as required. Now apply $(*)^+$ to $\langle y', y, t_0^*, t_1^*, \ldots \rangle$ and get $t, \langle n_i : i < \omega \rangle$; thinning the n_i's we finish].

A forcing notion $\mathcal{P}$ is said to be *$\mathcal{D}$-preserving* if $\mathcal{P}$ forces: "$\mathcal{D}$ is a fine* covering model"; equivalently, $\mathcal{P}$ forces: "$(D; R)$ covers ${}^\omega\omega$." So this means that $\mathcal{P}$ does not add certain kinds of reals.

In this terminology, we can state the following general iteration theorem ([Sh 177],[Sh-f]VI§1, §2):

A2.2 Iteration theorem. *Let $\mathcal{D}$ be a fine* covering model. Let $\langle \mathcal{P}_\alpha, \underset{\sim}{\mathcal{Q}}_\beta : \alpha \leq \delta, \beta < \delta \rangle$ be a countable support iteration of proper forcing notions with each factor $\mathcal{D}$-preserving. Then $\mathcal{P}_\delta$ is $\mathcal{D}$-preserving.*

Proof. We reproduce the proof given in [Sh b, pp. 199-202], with the modifications suggested in [Sh 177]. We note that in the present exposition we have suppressed some of the terminology in [Sh b] and made other minor alterations. In particular our statement of the main theorem is slightly weaker than the one given in [Sh f]. We have also suppressed the discussion of variants of condition (3*) in the definition of fine* covering model, which occurs on pages 197-198 of [Sh b]; as a result we leave a little more to the reader.

By [Sh b, V4.4], if δ is of uncountable cofinality then there is no problem, as all new reals are added at some earlier point. So we may suppose that $\mathrm{cf}\, \delta = \aleph_0$ hence by associativity of CS iterations of proper forcing ([Sh-b], III) without loss of generality $\delta = \omega$.

We claim that $\Vdash_{\mathcal{P}_\omega}$ "$(D; R)$ covers ${}^\omega\omega$." (Note that this suffices for the proof of the iteration theorem.)

Fix $x \in \mathrm{dom}\, R$, $p \in \mathcal{P}_\omega$, $\underset{\sim}{f}$ a $\mathcal{P}_\omega$-name with $p \Vdash$ "$\underset{\sim}{f} \in {}^\omega\omega$." We need to find an extension p' of p and a tree $t \in D$ with $R(\underset{\sim}{x}, t)$ such that $p' \Vdash$ "$\underset{\sim}{f} \in \mathrm{Br}(t)$." As in the proof that countable support iteration preserves properness, we may assume without loss of generality (after increasing p) that $\underset{\sim}{f}(n)$ is a $\mathcal{P}_n$-name for all n.

By induction on n we define conditions $p^n \in \mathcal{P}_n$ and $\mathcal{P}_m$-names $\underset{\sim}{t}_{m,n}$ for $m \leq n$ with the following properties:

(1) $\Vdash_{\mathcal{P}_i}$ "$p(i) \leq p^n(i) \leq p^{n+1}(i)$" for $i < n$;

(2) If $G_m \subseteq \mathcal{P}_m$ is generic with $m \leq n$, then in $V[G_m]$ we have $(p^n(m), \dots, p^n(n-1)) \Vdash_{\mathcal{P}_n/\mathcal{P}_m}$ "$\underset{\sim}{f}(n) = \underset{\sim}{t}_{m,n}$." This is easily done; for each n, we increase p^n n times, once for each possible m. By (1) we have $p \restriction n \leq p^n \leq p^{n+1}$.

We let $\underset{\sim}{f}_m$ be the $\mathcal{P}_m$-name for an element of ${}^\omega\omega$ satisfying: $\underset{\sim}{f}_m(n) = \underset{\sim}{t}_{m,n}$ for $n \geq m$, $\underset{\sim}{f}_m(n) = \underset{\sim}{f}(n)$ for $n < m$. Then we have:

(3) $(0, \dots, 0, p^n(m)) \Vdash_{\mathcal{P}_{m+1}}$ "$\underset{\sim}{f}_m \restriction n = \underset{\sim}{f}_{m+1} \restriction n$"

(4) $\Vdash_{\mathcal{P}_n}$ "$\underset{\sim}{f} \restriction n = \underset{\sim}{f}_n \restriction n$."

Choose $x_1 < x' < x$ and then inductively $x_1 < x_2 < \dots$ with all $x_n < x'$, and choose a countable $N \prec V_\lambda$ (with λ $\mathcal{P}$-large) such that all the data $(x_n)_{n<\omega}$, $(\mathcal{P}_n, \underset{\sim}{\mathcal{Q}}_n)_{n<\omega}$, $\underset{\sim}{f}$, $(p^n)_{n<\omega}$, $(\underset{\sim}{t}_{m,n})_{m \leq n < \omega}$ lie in N. We will define conditions $q^n \in \mathcal{P}_n$ and trees $t_n \in D$ (not names!) by induction on n with $q^{n+1} \restriction n = q^n$ (hence we may write: $q^n = (q_0, q_1, \dots, q_{n-1})$) and $t_n \subseteq t_{n+1}$, satisfying the following conditions:

(A) $p \restriction n \leq q^n$;

(B) q^n is $(N, \mathcal{P}_n)$-generic;

(C) $q^n \Vdash$ "$\underset{\sim}{f}_n \in \mathrm{Br}(t_n)$";

(D) $R(x_{3n}, t_n)$;

(E) For $m < n < \omega$ we have $q^m \Vdash_{\mathcal{P}_m}$ "q_m and $p^n(m)$ are compatible in $\underset{\sim}{\mathcal{Q}}_m$".

Suppose we succeed in this endeavour. Then we can let $q = \bigcup_n q^n$. By condition (2) in A2.1 for every $n < \omega$ $R(x', t_n)$ (as $x_{3n} < x'$). Let $(n_i : i < \omega)$ be a strictly increasing sequence of natural numbers and t be as guaranteed by (*) of condition (3) of A2.1 (for $\langle t_n : n < \omega \rangle, x', x$) so $R(x, t)$ and: if $\eta \restriction n_i \subseteq \bigcup_{j \leq i} t_j$ for each $i < \omega$ then $\eta \in t$. Let $g(i) =: n_i$.

By (E) above there are conditions q'_m with $q^m \Vdash_{\mathcal{P}_m}$ "$q'_m \in \underset{\sim}{\mathcal{Q}}_m$, $q'_m \geq q_m, p^{g(m)}(m)$." Let $q' = (q'_0, q'_1, \dots)$. Then $q' \geq q \geq p$ and for $m \leq n \leq g(m)$ we will have (if we succeed in defining q_n, t_n) $q' \restriction n \Vdash_{\mathcal{P}_n}$ "$\underset{\sim}{f} \restriction n = \underset{\sim}{f}_m \restriction n$", hence:

$$q' \restriction n \Vdash_{\mathcal{P}_n} \text{"}\underset{\sim}{f} \restriction n \in \mathrm{Br}\ t_m\text{"}.$$

Now we have finished proving the existence of p', t (see before (1)) as required: $q' \Vdash$ "$\underset{\sim}{f} \in \mathrm{Br}\,(t)$", as t includes the tree: $\{\eta \in {}^{\omega >}\omega$: For all i, $\eta \restriction n_i \in \bigcup_{j \leq i} t_j\}$; and $R(x, t)$ holds. Hence we have finished proving $\Vdash_{\mathcal{P}_\omega}$ "$(D; R)$ covers ${}^\omega\omega$". So it suffices to carry out the induction.

There is no problem for $n = 0$ or 1. Assume that q^n and t_n are defined. Let $G_n \subseteq P_n$ be generic with $q_n \in G_n$. Then $\underset{\sim}{f}_{n+1}$ becomes a $\underset{\sim}{\mathcal{Q}}_n[G_n]$-name $\hat{\underset{\sim}{f}}_{n+1} = \underset{\sim}{f}_{n+1}/G_n$ for a member of ${}^\omega\omega$. As $\mathcal{P}_{n+1}$ preserves (D, R), for every $r \in \underset{\sim}{\mathcal{Q}}_n[G_n]$ and every $y \in \mathrm{dom}\, R$ there is a condition $r' \geq r$ in $\underset{\sim}{\mathcal{Q}}_n[G_n]$ such that

$$(*)\qquad r' \Vdash \text{“}\underset{\sim}{\hat{f}}_{n+1} \in \mathrm{Br}(t')\text{” for some } t' \in D \text{ with } R(y,t').$$

For each $m < \omega$, applying this to $r =: p^m(n)$, $y = x_{3n}$ we get $r' = r^n_m$, $t' = t^n_{m+1}$; we could have guaranteed $t^n_{m+1} \subseteq t^n_{m+2}$. Now choose by induction on $l < \omega$, $r^n_{m,l} \in \underset{\sim}{\mathcal{Q}}_n[G_n]$ such that: $r^n_{m,0} = r^n_m$, $r^n_{m,l} \le r^n_{m,l+1}$, $r^n_{m,l+1}$ forces a value to $\underset{\sim}{\hat{f}}_{n+1}\restriction l$. So for some $\eta^n_m \in {}^\omega\omega[G_n]$, $r^n_{m,l} \Vdash$“$\underset{\sim}{\hat{f}}_{n+1}\restriction l = \eta^n_m\restriction l$”. Note $\eta^n_m\restriction m = f_n\restriction m$. Without loss of generality, $\langle r^n_m, t^n_m, r^n_{m,\ell}, \eta^n_m : n, m, \ell < \omega\rangle$ belongs to N. Applying (3⊗) from A2.1 (to $\eta = \underset{\sim}{f}_n[G_n]$, $\eta_m = \eta^n_m$) we can find $T^I_n \in D \cap N[G_n] = D \cap N$ such that $R(x_{3n}, T^I_n)$, $\underset{\sim}{f}_n \in \mathrm{Br}(T^I_n)$ and $\eta^n_m \in \mathrm{Br}(T^I_n)$ for infinitely many $m < \omega$. Applying (3∗) from A2.1 (to $T^I_n, t^n_1, t^n_2, \ldots$ and x_{3n}, x_{3n+1}) we obtain a tree T^{II}_n. Returning to V, we have a $\mathcal{P}_n$-name $\underset{\sim}{T}$ for such a tree. For $s \in \mathcal{P}_n$, if $s \Vdash$ “$\underset{\sim}{T} = T$” for some tree T in V, let $T(s)$ be this tree. Let U be the open dense subset of $s \in \mathcal{P}_n$ for which $T(s)$ is defined. Some such function $T(\cdot)$ belongs to N, and $U \in N$. If q^n is in the generic set G_n, then some $s \in U \cap N$ is in G_n, by condition (2). Let $U \cap N = \{s_i : i < \omega\}$. Applying (3∗) there is a tree t_{n+1} satisfying:

(a) $R(x_{3n+3}, t_{n+1})$.

(b) $t_n \subseteq t_{n+1}$.

(c) for every $T \in (\mathrm{Rang}\, R) \cap N$ such that $R(x_{3n+2}, T)$ for some $k_T < \omega$ we have:

$$\nu \in T \,\&\, \nu\restriction k_T \in t_n \Rightarrow \nu \in t_{n+1}$$

We shall prove now

(d) suppose $G_n \subseteq \mathcal{P}_n$ is generic over V with $q^n \in G_n$, and $k^* < \omega$. *Then* there is q', $p^{k^*}(n) \le q' \in \underset{\sim}{\mathcal{Q}}_n[G_n] \cap N[G_n]$, such that $q' \Vdash$“$\underset{\sim}{f}_{n+1} \in \mathrm{Br}(t_{n+1})$” (though t_{n+1} is generally not in N).

Proof of (d). As $q^n \in G_n$ necessarily for some $s \in P_n \cap N$ we have $s \in G_n$ so (c) applies to T_s and $T_s = \underset{\sim}{T}^{II}_n[G_n]$ (as $T^{II}_n = \underset{\sim}{T}^{II}_n[G_n]$ is well defined and also T^I_n is well-defined and belongs to $N \cap D$ not only $N[G_n] \cap D$, as $D \subseteq V$). By the choice of T^I_n the following set is infinite

$$w = \{i < \omega : \eta^n_i \in \mathrm{Br}(T^I_n)\}$$

By the choice of t^n_{i+1}, for every $i \in w$ there exists $k_i < \omega$ such that $\eta \in t^n_{i+1} \,\&\, \eta\restriction k_i = \eta^n_i\restriction k_i \Longrightarrow \eta \in T^{II}_n$. To show (d), choose $i \in w \setminus k^*$ (exists as w is infinite, k^* will be shown to be as required in (d)).

Now $r^n_{i,k} \in N \cap \underset{\sim}{\mathcal{Q}}_n[G_n]$ is well-defined, and any q', $p^i(n) \le q' \in \underset{\sim}{\mathcal{Q}}'_n[G_n]$ which is $(N, \underset{\sim}{\mathcal{Q}}_n[G_n])$-generic is as required (note that $p^{k^*}(n) \le p^i(n)$).

We can assume without loss of generality that $\mathcal{Q}_n$ is closed under countable disjunction, so we can find $\underset{\sim}{r}_n$ compatible with $p^n(m)$ for all m such that:

$$(q_0, \ldots, q_{n-1}, q'_n) \Vdash_{\mathcal{P}_{n+1}} \text{“}\underset{\sim}{f}_{n+1} \in \operatorname{Br}(t_{n+1})\text{”}.$$

Now find $q_n \geq q'_n$ such that $(q_0, \ldots, q_{n-1}, q_n)$ is $(N, \mathcal{P}_{n+1})$-generic. This completes the induction step.

[If this infinite disjunction bothers you, define by induction on n sequences $\langle q^n_\eta : \eta \in {}^{n+1}\omega \rangle$ where $q^n_\eta \in \underset{\sim}{\mathcal{Q}}_n$ is such that for every $\eta \in {}^m\omega$ the condition $\langle q^i_{\eta \restriction (i+1)} : i < n \rangle$ is generic for N and q^n_η is above $p^{\eta(n)}(n)$.] □

A2.3 The ${}^\omega\omega$-bounding property

We leave the successor case to the reader (see A2.6(2)).

A forcing notion $\mathcal{P}$ is ${}^\omega\omega$-bounding if it forces every function in ${}^\omega\omega$ in the generic extension to be bounded by one in the ground model. In §2 we quoted the result that a countable support iteration of proper ${}^\omega\omega$-bounding forcing notions is again ${}^\omega\omega$-bounding, which is almost Theorem V.4.3 of [Sh b]. In Chapter VI, §2 of [Sh b] this result is shown to fit into the framework just given. Here D is just a single collection $\mathcal{T}$ of trees; to fit D into the general framework given previously, we would let A be any suitable partial order, $D = A \dot{\cup} \mathcal{T}$, and $R = A \times \mathcal{T}$. The set $\mathcal{T}$ will consist of all subtrees of ${}^{\omega>}\omega$ with finite ramification (as we have no measure on how small $t \in \mathcal{T}$ is, so $<, R$ are degenerate).

In a generic extension of the universe, the set $\mathcal{T}$ (as defined in the ground model) will cover ${}^\omega\omega$ if and only if every function in ${}^\omega\omega$ is dominated by one in the ground model. In fact the only relevant trees are those of the form $T_f = \{\eta \in {}^{\omega>}\omega : \eta(i) \leq f(i) \text{ for } i < \operatorname{len}\eta\}$ with f in the ground model. Thus the ${}^\omega\omega$-bounding property coincides with the property of being $\mathcal{D}$-preserving, where $\mathcal{D}$ is essentially $\mathcal{T}$, more precisely $\mathcal{D} = (A \times \mathcal{T}; R, <)$ for a suitable $R, <$ (which play no role in this degenerate case). Thus to see that the general iteration theorem applies, it suffices to check that such a $\mathcal{D}$ will be a fine* covering model. We have to check the final clause (3) of the definition of fine* covering model. In fact we will prove a strong version of $(3)^+$.

For any sequence of trees T_n in $\mathcal{T}$, there is a tree T such that for all $\eta \in {}^\omega\omega$, if $\eta \restriction i \in \bigcup_{j \leq i} T_j$ for all i, then $\eta \in \operatorname{Br}(T)$.

We will verify that this property holds in any generic extension V^* of V in which $\mathcal{D}$ covers ${}^\omega\omega$. Let $T^* = \{\eta \in {}^{\omega>}\omega : \text{for all } i \leq \operatorname{len}(\eta),\ \eta \restriction i \in \bigcup_{j \leq i} T_j\}$. If T^* is in V this will do, but since the sequence (T_n) came from a generic extension, this need not be the case. On the other hand the sequence $T^* \restriction n$ of finite trees is itself coded by a real $f \in {}^\omega\omega$, and as $\mathcal{D}$ covers ${}^{\omega}\omega$, there is a tree $T^{\bullet}$ in D which contains this code f; via a decoding, $T^{\bullet}$ can be thought of as a tree T^{o} whose nodes t are subtrees of ${}^{n\geq}\omega$ with no maximal nodes below level n, so that for any $s, t \in T^{\bullet}$ with $s \leq t$, s is the restriction of t to the level of s, and such that the sequence $T^{*}\restriction n$ actually is a branch of T^{o}. Let T be the subtree of ${}^{\omega>}\omega$ consisting of the union of all the nodes of T^{o}. Then T still has finite ramification, lies in the ground model, and contains T^{*}.

A2.4 Cosmetic changes

(a) We may want to deal just with $\mathrm{Br}(T^{*})$, where T^{*} a subtree ${}^{\omega>}\omega$ (hence downward closed): So D is a set of subtrees of T^{*}, so we can replace D by $\{\{\eta \in {}^{\omega>}\omega : \eta \in T \text{ or } (\exists \ell)[\eta\restriction\ell \in T \,\&\, \eta\restriction(\ell+1) \notin T^{*}\} : T \in D\}$.

(b) We may replace subtrees T^{*} of ${}^{\omega>}\omega$ by isomorphic trees.

(c) We may want to deal with some $(D_i; R_i, <_i)$ simultaneously; by renaming without loss of generality the D_i are pairwise disjoint, and even: $\bigwedge_{\ell=1,2} t_l \in D_{i_l} \& i_1 \neq i_2 \Longrightarrow \mathrm{Br}\, T_1 \cap \mathrm{Br}\, t_2 = \emptyset$. Then we use $(\bigcup D_i; \bigcup R_i, \bigcup <_i)$ to get the result.

(d) We may want to have $(D; R)$ (i.e. no $<$); just use $(D \cup \lesseqgtr \times D; R', <)$ where $R'(x,t)$ iff $x = (q, y)$, $q \in \lesseqgtr, y \in D, R(y,t)$, $(q_1, y_1) < (q_2, y_2)$ iff $q_1 < q_2 \,\&\, y_1 = y_2$.

A2.5 The (f, g)-bounding property

We leave the successor case to the reader (see A2.6(2)).

Let $\boldsymbol{F}$ be a family of functions in ${}^{\omega}\omega$, and $g \in {}^{\omega}\omega$ with $1 < g(n)$ for all n. We say that a forcing notion $\mathcal{P}$ has the $(\boldsymbol{F}, g)$-*bounding property* if:

(*) For any sequence $(A_k : k < \omega)$ in the ground model, with $|A_k| \in \boldsymbol{F}$ (as a function of k), and any $\underset{\sim}{\eta} \in \prod_k A_k$ in the generic extension and $\varepsilon > 0$, there is a "cover" $\mathcal{B} = (B_k : k < \omega)$ in the ground model with $B_k \subseteq A_k$, $[|B_k| > 1 \Rightarrow |B_k| < g(k)^{\varepsilon}]$ and $\underset{\sim}{\eta}(k) \in B_k$ for each k.

This notion is only of interest if $g(n) \longrightarrow \infty$ with n.

We will show that this notion is also covered by a case of the general iteration theorem of §A2.2.

Let $\mathcal{T}_{f,g}$ $[\mathcal{T}^{\varepsilon}_{f,g}]$ be the set of those subtrees T of $\bigcup_n \prod_{m<n} f(n)$ of the form $\bigcup_n \prod_{m<n} B_m$, such that $|B_n| < \max\{g(k), 2\}$ [such that $|B_n| \leq \max\{2, g(k)^{\varepsilon}\}$], where as usual $f(n)$ is thought of as the set $\{0, \dots, f(n)-1\}$. Let $\mathcal{T}_{\boldsymbol{F},g}$ be $\bigcup_{f \in \boldsymbol{F}, \varepsilon \in \lesseqgtr^{+}} \mathcal{T}^{\varepsilon}_{f,g}$. Our fine* covering model is essentially $\mathcal{T}_{\boldsymbol{F},g}$, more accurately, it is the family of $\{(\mathcal{T}_{f,g} \cup \lesseqgtr^{+}; R, <) : f \in \boldsymbol{F}\}$, where $\lessgtr^+$ is the set of positive rationals, $<$ is the order on $\lessgtr^+$, and $R(\varepsilon, t) =: \varepsilon \in \lessgtr^+ \,\&\, t \in \mathcal{T}^{\varepsilon}_{f,g}$. See A2.4(c).

Call a family $\boldsymbol{F}$ *g-closed* if it satisfies the following two closure conditions:

1. For $f \in \boldsymbol{F}$, the function $F(n) = \prod_{m<n}(f(m)+1)$ lies in $\boldsymbol{F}$;
2. For $f \in \boldsymbol{F}$, f^g is in $\boldsymbol{F}$.

If $\boldsymbol{F}$ is g-closed, $f \in \boldsymbol{F}$, and $(A_n)_{n<\infty}$ are sets with $|A_n| = f(n)$, then the function $f'(n) =$ the number of trees of the form $\prod_{m<n} B_m$ with $B_m \subseteq A_m$ and $|B_m| < g(m)$ is dominated by a function in $\boldsymbol{F}$.

Using the formalism of §A2.2, we wish to prove:

Theorem. *If $\boldsymbol{F}$ is g-closed then a countable support iteration of $(\boldsymbol{F}, g)$-bounding proper forcing notions is again an $(\boldsymbol{F}, g)$-bounding proper forcing.*

Since the $\mathcal{D}$-preserving forcing notions are the same as the $(\boldsymbol{F}, g)$-bounding ones, we need only check that $\mathcal{D}$ is a fine* covering model. Again the nontrivial condition is $(3)^+$, i.e.,

Let $f \in \boldsymbol{F}$. For any sequence of trees T_n in $\mathcal{T}_{f,g}$, $R(\varepsilon', T_n)$, $\varepsilon' < \varepsilon$ (in $\lessgtr^+$), there is a tree T in D satisfying $R(\varepsilon, T)$ and an increasing sequence n_i such that for all $\eta \in {}^{\omega}\omega$, if $\eta \restriction n_i \in \bigcup_{j \le i} T_{n_j}$ for all i, then $\eta \in \mathrm{Br}(T)$.

This must be verified in any generic extension V^* of V in which $\mathcal{D}$ covers ${}^{\omega}\omega$. Working in V, choose $(n_i)_{i<\omega}$ increasing so that $n_0 = 0$ and for $n_i \le n$ we have $\min_{n \ge n_i} g(n)^{(\varepsilon-\varepsilon')/2} > i+1$. For $n_i \le n < n_{i+1}$ set:

$$B_n = \{\eta(n) : \eta \in \cup\{T_j : n_j \le n\}.$$

(For $n < n_0$ let $B_n = \{\eta(n) : \eta \in T_0\}$.) If the sequence B_n was in the ground model, we could take $T = \bigcup_n \prod_{m<n} B_m$. Instead we have to think of the sequence B_n as a possible branch through the tree of finite sequences of subsets of $f(n)$ of size at most (say) $\max\{1, g(n)-1\}$. As $\boldsymbol{F}$ is g-closed, $\mathcal{T}_{\boldsymbol{F},g}$ contains a tree $T^\bullet$ which encodes a tree T^o of such subsets, for which the desired sequence B_n is a branch in V^*, so that the number of members of T^0 of level m is $\le g(m)^{(\varepsilon-\varepsilon')/2}$ (or is ≤ 1). Let $B^o_n = \bigcup_{b \in T^o} b(n)$. Then $B_n \subseteq B^o_n$, $\lim_{n \longrightarrow \infty} |B^o_n|/g^{\varepsilon}(n) = 0$ and $\bigcup_n \prod_{m<n}(B^o_m)$ is in V. $\square$

A2.6 Discussion

This was treated in [Sh-f,VI] [Sh-f, XVIII §3] too (the presentation in [Sh-b, VI] was inaccurate). The version chosen here goes for less generality (gaining, hopefully, in simplicity and clarity) and is usually sufficient. We consider below some of the differences.

A2.6(1) A technical difference

In the context as phrased here the preservation in the successor case of the iteration was trivial — by definition essentially. We can make the fine* covering model (in A2.1) more similar to [Sh-f, VI §1] by changing (3∗) to

$(*)'$ For $y_0 < y_1 < \dots y < x$ in $\operatorname{dom} R$ and $t_n \in D$ such that $R(y_n, t_n)$ for all n, there is $t \in D$ with $R(x,t)$ holding and indices $n_0 < n_1 < \dots$ such that $[\eta \in {}^{\omega>}\omega \;\&\; \bigwedge_i \eta{\restriction}\eta_i \in \bigcup_{j\le i} t_j \Rightarrow \eta \in t]$.

We can use this version here.

A2.6(2) Two-stage iteration

We can make the fine* covering model (in A2.1) more similar to [Sh-f, VI §1] by changing (3∗). In the context as presented here the preservation by two step iteration is trivial — by definition essentially. In [Sh-f VI, §2] we phrase our framework such that we can have: if $Q_0 \in V$ is x-preserving, $\underset{\sim}{Q}_1$ is X-preserving (over V^{Q_0}, $\underset{\sim}{Q}_1$ a Q_0-name) then $Q_0 * \underset{\sim}{Q}_1$ is x-preserving. The point is that X-preserving means $(D,R,<)^V$-preserving, i.e. $(D,R,<)$ is a definition (with a parameter in V_0). The point is that if $V_1 = V_0{}^{Q_0}$, $V_2 = V_1^{\underset{\sim}{Q}_1}$ *then* for $\eta \in ({}^{\omega}\omega)^{V_2}$ and $x \in \operatorname{dom} R$, we choose $y < x$ and $t \in D^{V_1}$ such that $\eta \in \operatorname{Br}(t)$, $R^{V_1}(y,t)$, then we look in V_0 at the tree of possible initial segments of t getting $T \in D^{V_0}$ such that $t \in \operatorname{Br}(T)$, $R^{V_0}(y,T)$. If y was chosen rightly, $\bigcup \operatorname{Br}(T)$ is as required. Here it may be advantageous to use a preservation of several $(D,R,<)$'s at once (see A2.4(c)).

A2.6(3) Several models — the real case

We may consider a (weak) (fine*) covering family of models $\langle (D_\ell, R_\ell, <_\ell) : \ell < \ell^* \rangle$ (actually a sequence) i.e. *not* that each one is a cover, but simultaneously.

(A) We say $(\overline{D}, \overline{R}) = \langle (D_\ell, R_\ell) : \ell < \ell^* \rangle$ is a weak c.f.m. if each D_ℓ is a set, R_ℓ a binary relation, $\ell^* < \omega$ and

1. $R_\ell(x,t)$ implies that t is a subtree of ${}^{\omega>}\omega$ (nonempty, no maximal models).
2. Every $\eta \in {}^{\omega}\omega$ is of kind ℓ for at least one $\ell < \ell^*$ which means: for every $x \in \operatorname{dom} R_\ell$ for some t, we have $R_\ell(x,t) \;\&\; \eta \in \operatorname{Br}(t)$.

(B) We say $(\overline{D}, \overline{R}, \overline{<})$ is a fine* c.f.m. if:

0. $(\overline{D}, \overline{R})$ is a weak family.
1. If $x \in \operatorname{dom} R_\ell \Rightarrow (\exists z) z <_\ell x$ and $\forall y <_\ell x\, \exists z (y <_\ell z <_\ell x)$ (and $D \neq \emptyset$).

2. $x <_\ell y \,\&\, R_\ell(x,t) \Rightarrow R_\ell(y,t)$.
3. For any generic extensions V^* in which $(\overline{D},\overline{R})$ is a weak c.f.m.

(∗) for every $\ell < \ell^*$ and $y <_\ell x$ (from $\operatorname{dom} R_\ell$) and $t_n \in D_\ell$ with $R_\ell(y,t_n)$ for all n there is $t \in D$ with $R_\ell(x,t)$ and there are indices $n_0 < n_1 < \dots$ such that for every $\eta \in {}^\omega\omega$: if $\eta \restriction n_i \in \bigcup_{j \le i} t_j$ for all i then $\eta \in \operatorname{Br}(t)$.

⊗ if $\ell < \ell^*$, $\eta \in {}^\omega\omega$, $\eta_n \in {}^\omega\omega$, $\eta_n \restriction n = \eta \restriction n$, $x \in \operatorname{dom} R_\ell$ and η, η_n are of kind ℓ, *then* for some t^*, $R_\ell(x,t^*)$, $\eta \in \operatorname{Br}(t^*)$ and for infinitely many $n < \omega$, $\eta_n \in \operatorname{Br}(t^*)$.

Theorem. *If $(\overline{D};\overline{R},\overline{<})$ is a fine* c.f.m., $\langle \mathcal{P}_\alpha, \underset{\sim}{\mathcal{Q}}_\beta : \alpha \le \delta, \beta < \alpha\rangle$ is a countable support iteration of proper forcing notions with each factor $(\overline{D};\overline{R},\overline{<})$-preserving. Then $\mathcal{P}_\delta$ is $(\overline{D};\overline{R},\overline{<})$-preserving.*

Proof. Similar to the previous one, with the following change. After saying that without loss of generality $\delta = \omega$ and, above p, for every n, $\underset{\sim}{f}(n)$ as a P_n-name, and choosing x_n, x', we do the following. For clarity think that our universe V is countable in the true universe or at least $\beth_3(|P_\omega|)^V$ is. We let $K = \{(n,p,G) : n < \omega, p \in P_\omega,\ G \subseteq P_n$ is generic over V and $p \restriction n \in G_n\}$. On K there is a natural order $(n,p,G) \le (n',p',G')$ if $n \le n'$, $P_\omega \models p \le p'$ and $G \subseteq G'$. Also for $(n,p,G) \in G$ and $n' \in (n,\omega)$ there are p', G' such that $(n,p,G) \le (n',p',G')$. For $(n,p,G) \in K$ let $L_{(n,p,G)} = \{g : g \in ({}^\omega\omega)^{V[G]}$ and there is an increasing sequence $\langle p_\ell : \ell < \omega\rangle$ of conditions in P_ω/G, $p \le p_0$, such that $p_\ell \Vdash \underset{\sim}{f} \restriction \ell = g \restriction \ell\}$. So:
$g \in L_{(n,p,G)} \Rightarrow \underset{\sim}{f} \restriction n = g \restriction n$
$(n,p,G) \le (n',p',G') \Rightarrow L_{(n',p',G')} \subseteq L_{(n,p,G)}$. □

Theorem. *There are ℓ_* and $(n,p,G) \in K$ such that if $(n,p,G) \le (n',p',G') \in K$ then there is $g \in L_{(n',p',G')}$ which is of the ℓ_*'th kind.*

Proof. Otherwise choose by induction (n^ℓ, p^ℓ, G^ℓ) for $\ell \le \ell^*$, in K, increasing such that: $L_{(n^{\ell+1},p^{\ell+1},G^{\ell+1})}$ has no member of the ℓ'th kind. So $L_{(n^{\ell^*},p^{\ell^*},G^{\ell^*})} = \emptyset$ contradiction.

So without loss of generality for every $(n,p,G) \in K$, $L_{(n,p,G)}$ has a member of the ℓ_*'th kind. Now we choose by induction on n, A_n, $\langle p_\eta : \eta \in {}^{n+1}\omega, \eta \restriction n \in A_n\rangle$, $\langle \underset{\sim}{f}_\eta : \eta \in A_n\rangle$, $\langle q_\eta : \eta \in A_n\rangle$, and t_n such that

(A)′ $A_n \subseteq {}^n\omega, A_0 = \{\langle\rangle\}, \eta \in A_n \Rightarrow (\exists^{\aleph_0}\ell)(\eta\char`\^\langle\ell\rangle \in A_{n+1})$ $p_\eta \in \mathcal{P}_\omega \cap N, p_{<>} = p, p_\eta \le p_{\eta\char`\^\langle\ell\rangle}, p_\eta \restriction n \le q_n$.

(B)′ q_η is $(N, \mathcal{P}_{\lg\eta})$-generic, $q_\eta \in \mathcal{P}_{\lg\eta}$ and $[\ell < \lg\eta \Rightarrow q_\eta \restriction \ell = q_{\eta\restriction\ell}]$.

(C)′ $q_\eta \Vdash$ "$\underset{\sim}{f}_\eta \in \mathrm{Br}(t_n)$ is of the ℓ_*'th kind" when $\eta \in A_n$ and $\underset{\sim}{f}_\eta$ is a $\mathcal{P}_n$-name.

(D)′ $R_{3n}(x_{3n}, t_n), t_n \subseteq t_{n+1}$.

(E)′ $p_{\eta\frown\langle\ell\rangle} \Vdash_{\mathcal{P}_w}$ "$\underset{\sim}{f}_\eta \restriction \ell = \underset{\sim}{f}_{\eta\frown\langle\ell\rangle} \restriction \ell = \underset{\sim}{f} \restriction \ell$".

This suffices, as $x_n < x'$ so $\bigwedge_n R(x', t_n)$ hence for some $\langle n_i : i < \omega\rangle$ strictly increasing and t as guaranteed by $(*)$ of (3) we find $\nu \in {}^\omega\omega$ increasing fast enough and let $q = \bigcup_{n<\omega} q_{\nu\restriction n}$. In the induction there is no problem for $n = 0, 1$. For $n+1$; first for each $\eta \in {}^{n+1}\omega$ we choose ℓ, work in $V^{\mathcal{P}_{n+1}}$ and find $\langle p_{\eta\frown\langle\ell\rangle} : \ell < \omega\rangle, \underset{\sim}{f}_\eta$, and without loss of generality they are in N. For $\eta \in {}^n\omega$ there is a $\mathcal{P}_{n+1}$-name $\underset{\sim}{t}_\eta \in N$ of a member of D, $R_\ell(x_{3n}, \underset{\sim}{t}_\eta)$, $\underset{\sim}{f}_\eta \in \mathrm{Br}(\underset{\sim}{t}_\eta)$, $(\exists^\infty \ell)\underset{\sim}{f}_{\eta\frown\langle\ell\rangle} \in \mathrm{Br}(\underset{\sim}{t}_\eta)$. Now we can replace $p_{\eta\frown\langle 1\rangle}$ by $p_{\eta\frown\langle\ell'\rangle}$, $\ell' = \mathrm{Min}\{m : m \geq \ell, \underset{\sim}{f}_{\eta\frown\langle\ell\rangle} \in \mathrm{Br}(\underset{\sim}{t}_\eta)\}$. We continue as in A2.2. Note: it is natural to use this framework e.g. for preservation of P-points. $\square$

A3. Omitting types

A3.1 Uniform partial orders

In the proof of Theorem 4.1 given in §4 we used the combinatorial principle developed in [ShLH162]. (Cf. [Sh107] for applications published earlier.) This is a combinatorial refinement of forcing with $\mathcal{AP}$ to get a $\mathbb{P}_3$-name $\underset{\sim}{\mathcal{F}}$ with the required properties in a generic extension. We now review this material.

With the cardinal λ fixed, a partially ordered set $(\mathcal{P}, <)$ is said to be *standard λ^+-uniform* if $\mathcal{P} \subseteq \lambda^+ \times \mathcal{P}_\lambda(\lambda^+)$ (we refer here to subsets of λ^+ of size strictly less than λ), satisfying the following properties (where we take e.g. $p = (\alpha, u)$ and write $\mathrm{dom}(p)$ for u):

1. If $p \leq q$ then $\mathrm{dom}\, p \subseteq \mathrm{dom}\, q$.
2. For all $p, q, r \in \mathcal{P}$ with $p, q \leq r$ there is $r' \in \mathcal{P}$ so that $p, q \leq r' \leq r$ and $\mathrm{dom}\, r' = \mathrm{dom}\, p \cup \mathrm{dom}\, q$.
3. If $(p_i)_{i<\delta}$ is an increasing sequence of length less than λ, then it has a least upper bound q, with domain $\bigcup_{i<\delta} \mathrm{dom}\, p_i$; we will write $q = \bigcup_{i<\delta} p_i$, or more succinctly: $q = p_{<\delta}$.
4. For all $p \in \mathcal{P}$ and $\alpha < \lambda^+$ there exists a $q \in \mathcal{P}$ with $q \leq p$ and $\mathrm{dom}\, q = \mathrm{dom}\, p \cap \alpha$; furthermore, there is a unique maximal such q, for which we write $q = p \restriction \alpha$.
5. For limit ordinals δ, $p \restriction \delta = \bigcup_{\alpha<\delta} p \restriction \alpha$.
6. If $(p_i)_{i<\delta}$ is an increasing sequence of length less than λ, then $(\bigcup_{i<\delta} p_i) \restriction \alpha = \bigcup_{i<\delta} (p_i \restriction \alpha)$.

7. *(Indiscernibility)* If $p = (\alpha, v) \in \mathcal{P}$ and $h : v \to v' \subseteq \lambda^+$ is an order-isomorphism onto V' then $(\alpha, v') \in \mathcal{P}$. We write $h[p] = (\alpha, h[v])$. Moreover, if $q \leq p$ then $h[q] \leq h[p]$.
8. *(Amalgamation)* For every $p, q \in \mathcal{P}$ and $\alpha < \lambda^+$, if $p{\restriction}\alpha \leq q$ and $\operatorname{dom} p \cap \operatorname{dom} q = \operatorname{dom} p \cap \alpha$, then there exists $r \in \mathcal{P}$ so that $p, q \leq r$.

It is shown in [ShHL162] that under a diamond-like hypothesis, such partial orders admit reasonably generic objects. The precise formulation is given in A3.3 below.

A3.2 Density systems

Let $\mathcal{P}$ be a standard λ^+-uniform partial order. For $\alpha < \lambda^+$, $\mathcal{P}_\alpha$ denotes the restriction of $\mathcal{P}$ to $p \in \mathcal{P}$ with domain contained in α. A subset G of $\mathcal{P}_\alpha$ is an *admissible ideal* (of $\mathcal{P}_\alpha$) if it is closed downward, is λ-directed (i.e. has upper bounds for all small subsets), and has no proper directed extension within $\mathcal{P}_\alpha$. For G an admissible ideal in $\mathcal{P}_\alpha$, $\mathcal{P}/G$ denotes the restriction of $\mathcal{P}$ to $\{p \in \mathcal{P} : p{\restriction}\alpha \in G\}$.

If G is an admissible ideal in $\mathcal{P}_\alpha$ and $\alpha < \beta < \lambda^+$, then an (α, β)-*density system* for G is a function D from pairs (u, v) in $P_\lambda(\lambda^+)$ with $u \subseteq v$ into subsets of $\mathcal{P}$ with the following properties:

(i) $D(u, v)$ is an upward-closed dense subset of $\{p \in \mathcal{P}/G : \operatorname{dom}(p) \subseteq v \cup \beta\}$;

(ii) For pairs $(u_1, v_1), (u_2, v_2)$ in the domain of D, if $u_1 \cap \beta = u_2 \cap \beta$ and $v_1 \cap \beta = v_2 \cap \beta$, and there is an order isomorphism from v_1 to v_2 carrying u_1 to u_2, then for any γ we have $(\gamma, v_1) \in D(u_1, v_1)$ iff $(\gamma, v_2) \in D(u_2, v_2)$.

An admissible ideal G' (of $\mathcal{P}_\gamma$) is said to *meet* the (α, β)-density system D for G if $\gamma \geq \alpha$, $G' \geq G$ and for each $u \in P_\lambda(\gamma)$ there is $v \in P_\lambda(\gamma)$ containing u such that G' meets $D(u, v)$.

A3.3 The genericity game

Given a standard λ^+-uniform partial order $\mathcal{P}$, the *genericity* game for $\mathcal{P}$ is a game of length λ^+ played by Guelfs and Ghibellines, with Guelfs moving first. The Ghibellines build an increasing sequence of admissible ideals meeting density systems set by the Guelfs. Consider stage α. If α is a successor, we write α^- for the predecessor of α; if α is a limit, we let $\alpha^- = \alpha$. Now at stage α for every $\beta < \alpha$ an admissible ideal G_β in some $\mathcal{P}_{\beta'}$ is given, and one can check that there is a unique admissible ideal G_{α^-} in $\mathcal{P}_{\alpha^-}$ containing $\bigcup_{\beta<\alpha} G_{\beta'}$ (remember A 3.1(5)) [Lemma 1.3, ShHL 162]. The Guelfs now supply at most λ density systems D_i over G_{α^-} for (α, β_i) and also fix an element g_α in $\mathcal{P}/G_{\alpha}^{-}$. Let α' be minimal such that $g_\alpha \in \mathcal{P}_{\alpha'}$

and $\alpha' \geq \sup \beta_i$. The Ghibellines then build an admissible ideal $G_{\alpha'}$ for $\mathcal{P}_{\alpha'}$ containing G_α^- as well as g_α, and meeting all specified density systems, or forfeit the match; they let $G_{\alpha''} = G_{\alpha'} \cap \alpha''$ when $\alpha \leq \alpha'' < \alpha'$. The main result is that the Ghibellines can win with a little combinatorial help in predicting their opponents' plans.

For notational simplicity, we assume that G_δ is an $\aleph_2$-generic ideal on $\mathcal{AP}\restriction\delta$, when $\operatorname{cf}\delta = \aleph_2$ which is true on a club in any case.

A3.4 Dl_λ

The combinatorial principle Dl_λ states that there are subsets Q_α of the power set of α for $\alpha < \lambda$ such that $|Q_\alpha| < \lambda$, and for any $A \subseteq \lambda$ the set $\{\alpha : A \cap \alpha \in Q_\alpha\}$ is stationary. This follows from $\diamondsuit_\lambda$ or inaccessibility, obviously, and Kunen showed that for successors, Dl and $\diamondsuit$ are equivalent. In addition Dl_λ implies $\lambda^{<\lambda} = \lambda$.

A3.5 A general principle

Theorem. *Assuming Dl_λ, the Ghibellines can win any standard λ^+-uniform $\mathcal{P}$-game.*

This is Theorem 1.9 of [ShHL 162].

In our application we identify $\mathcal{AP}$ with a standard $\aleph_2^+$-uniform partial order via a certain coding. We first indicate a natural coding which is not quite the right one, then repair it.

First try

An approximation $q = (\mathcal{A}, \underset{\sim}{\mathcal{F}}, \underset{\sim}{\varepsilon}, \underset{\sim}{\mathbf{k}},)$ will be identified with a pair (τ, u), where $u = \mathcal{A}$, and τ is the image of q under the canonical order-preserving map $h : \mathcal{A} \leftrightarrow \operatorname{otp}(\mathcal{A})$. One important point is that the first parameter τ comes from a fixed set T of size $2^{\aleph_1} = \aleph_2$; so if we enumerate T as $(\tau_\alpha)_{\alpha<\aleph_2}$ then we can code the pair (τ_α, u) by the pair (α, u). Under these successive identifications, $\mathcal{AP}$ becomes a standard $\aleph_2^+$-uniform partial order, as defined in §A3.1. Properties 1, 2, 4, 5, and 6 are clear, as is 7, in view of the uniformity in the iterated forcing $\mathbb{P}$, and properties 3, 8 were, in essence but not formally, stated in Claim 3.10.

The difficulty with this approach is that in this formalism, density systems cannot express nontrivial information: any generic ideal meets any density system, because for $q \leq q'$ with $\operatorname{dom} q = \operatorname{dom} q'$, we will have $q = q'$; thus $D(u, u)$ will consist of all q with $\operatorname{dom} q = u$, for any density system D.

So to recode $\mathcal{AP}$ in a way that allows nontrivial density systems to be defined, we proceed as follows.

Second try

Let $\iota : \aleph_2^+ \leftrightarrow \aleph_2^+ \times \aleph_2$ order preserving where $\aleph_2^+ \times \aleph_2$ is ordered lexicographically. Let $\pi : \aleph_2^+ \times \aleph_2 \longrightarrow \aleph_2^+$ be the projection on the first coordinate. First encode q by $\iota[q] = (\iota[\mathcal{A}], \dots)$, then encode $\iota[q]$ by $(\tau, \pi[\mathcal{A}])$, where τ is defined much as in the first try – a description of the result of collapsing q into $\operatorname{otp} \pi[\mathcal{A}] \times \aleph_2$, after which τ is encoded by an ordinal label below $\aleph_2$. The point of this is that now the domain of q is the set $\pi[\mathcal{A}]$, and q has many extensions with the same domain. After this recoding, $\mathcal{AP}$ again becomes a $\aleph_2^+$-uniform partial ordering, as before. We will need some additional notation in connection with the indiscernibility condition. It will be convenient to view $\mathcal{AP}$ simultaneously from an encoded and a decoded point of view. One should now think of $q \in \mathcal{AP}$ as a quintuple $(u, \mathcal{A}, \underset{\sim}{\mathcal{F}}, \varepsilon, \underset{\sim}{k})$ with $\mathcal{A} \subseteq u \times \aleph_2$. If $h : u \leftrightarrow v$ is an order isomorphism, and q is an approximation with domain u, we extend h to a function h_* defined on $\mathcal{A}^q$ by letting it act as the identity on the second coordinate. Then $h[q]$ is the transform of q using h_*, and has domain v.

In order to obtain *least* upper bounds for increasing sequences, it is also necessary to allow some extra elements into $\mathcal{AP}$, by adding formal least upper bounds to increasing sequences of length $< \aleph_2$.

This provides the formal background for the discussion in §3. The actual construction should be thought of as a match in the genericity game for $\mathcal{AP}$, with the various assertions as to what may be accomplished corresponding to proposals by the Guelfs to meet certain density systems. To complete the argument it remains to specify these systems and to check that they are in fact density systems.

A3.6 The major density systems

The main density systems under consideration were introduced implicitly in 4.11. Suppose that $\delta < \aleph_2$, $q \in \mathcal{AP}$ with $\delta \in \operatorname{dom} q \subseteq \aleph_2$, $q_\delta^* \le q$, and $\underset{\sim}{z}$ is a $(\mathbb{P}\restriction \operatorname{dom} q)$-name. Define a density system $D^\delta_{q,\underset{\sim}{z}}(u,v)$ for $u \subseteq v \subseteq \aleph_3$ with $|v| \le \aleph_1$ as follows. First, if $\operatorname{otp} u \le \operatorname{otp} \operatorname{dom} q$ then let $D^\delta_{q,\underset{\sim}{z}}(u,v)$ degenerate to $\mathcal{AP}\restriction v$. Now suppose that $\operatorname{otp} u > \operatorname{otp} \operatorname{dom} q$ and that $h : \operatorname{dom} q \longrightarrow u$ is an order isormorphism from $\operatorname{dom} q$ to an initial segment of u. Let $q^* = h[q]$. Call an element r of $\mathcal{AP}$ a (u,v)-*witness* if:

1. $u \subseteq \operatorname{dom} r \subseteq v$;
2. $r \ge q^*$;
3. for some $\mathrm{p} \in \mathcal{P}\restriction\mathcal{A}^r$ with $\mathrm{p} \ge \mathrm{p}^\delta$, and some $(\mathbb{P}\restriction[\mathcal{A}^r \cap \delta])$-name $\underset{\sim}{x}$, $\underset{\sim}{F}_\delta(\underset{\sim}{x})$ is a $(\mathbb{P}\restriction[\mathcal{A}^r \cap \delta])$-name; and:
4. $\mathrm{p}' \Vdash_{\mathbb{P}\restriction\mathcal{A}^r}$ "$\{n : [\Gamma^1_{k^1_\delta(n)} \models R(\underset{\sim}{x}(n), \underset{\sim}{x}_\delta(n)) \iff \Gamma^2_{k^2_\delta(n)} \models \neg R(\underset{\sim}{F}_\delta(\underset{\sim}{x})(n), \underset{\sim}{z}(n))]\} \in \mathcal{F}^r$."

Let $D^{\delta}_{q,\underset{\sim}{z}}(u,v)$ be the set of $r \in \mathcal{AP}$ with $\operatorname{dom} r = v$ such that either r is a (u,v)-witness, or else there is no (u,v)-witness $r' \geq r$.

This definition has been arranged so that $D^{\delta}_{q,\underset{\sim}{z}}(u,v)$ is trivially dense. In §4 we wrote the argument as if no default condition had been used to guarantee density, so that the nonexistence of (u,v)-witnesses is called a "failure of density". Here we adjust the terminology to fit the style of [ShHL 162].

Now we return to the situation described in 4.12. We had $\mathbb{P}$-names $\underset{\sim}{F}$, $\underset{\sim}{k}^1$, $\underset{\sim}{k}^2$, and a condition $\mathrm{p} \in \mathbb{P}$, satisfying conditions (3,4) as stated there, and we considered the set $C = \{\delta < \aleph_3 : \operatorname{cof}(\delta) = \aleph_2,\ \delta \text{ is } (\underset{\sim}{F}, \underset{\sim}{k}^1, \underset{\sim}{k}^2, \mathrm{p})\text{-closed}\}$, and a stationary set S_C on which $\underset{\sim}{F}{\restriction}\delta$, p, ε_δ, $\underset{\sim}{k}^1_\delta$ were guessed by $\diamondsuit$. Then $\underset{\sim}{z} =: \underset{\sim}{F}(\underset{\sim}{x}_\delta)$ is a $(\mathbb{P}{\restriction}\mathcal{A}^q)$-name for some $q \in G$. Let $u = \operatorname{dom} q$, $q_0 = q{\restriction}\delta$. Now we consider the following condition used in 4.12:

(iv) For all $r \geq q$ in $\mathcal{AP}$ such that $r{\restriction}\delta \in G_\delta$, and $\underset{\sim}{x}$ a $(\mathbb{P}{\restriction}\mathcal{A}^{r\restriction\delta})$-name, with $\underset{\sim}{y} =: \underset{\sim}{F}(\underset{\sim}{x})$ a $(\mathbb{P}{\restriction}\mathcal{A}^{r\restriction\delta})$-name, we have:

$(*)_{\underset{\sim}{x},\underset{\sim}{y}}$ $\mathrm{p} \Vdash$ "The set $\{n : \Gamma^1_{\underset{\sim}{k}^1(n)} \models R(\underset{\sim}{x}(n), \underset{\sim}{x}_\delta(n))$ iff $\Gamma^2_{\underset{\sim}{k}^2(n)} \models R(\underset{\sim}{y}(n), \underset{\sim}{z}(n))\}$ is in $\underset{\sim}{\mathcal{F}}^r$".

We argued in 4.12 that we could confine ourselves to the case in which (iv) holds. We now go through this more carefully. Suppose on the contrary that we have $r \geq q$ in $\mathcal{AP}$ with $r{\restriction}\delta \in G_\delta$, and a $(\mathbb{P}{\restriction}\mathcal{A}^{r\restriction\delta})$-name $\underset{\sim}{x}$, so that $\underset{\sim}{y} =: \underset{\sim}{F}(\underset{\sim}{x})$ is a $(\mathbb{P}{\restriction}\mathcal{A}^{r\restriction\delta})$-name, and a condition $\mathrm{p}' \geq \mathrm{p}$, so that

$$\mathrm{p}' \Vdash \text{"The set } \{n : \Gamma^1_{\underset{\sim}{k}^1(n)} \models R(\underset{\sim}{x}(n), \underset{\sim}{x}_\delta(n)) \text{ iff } \Gamma^2_{\underset{\sim}{k}^2(n)} \models R(\underset{\sim}{y}(n), \underset{\sim}{z}(n))\} \text{ is not in } \underset{\sim}{\mathcal{F}}^r\text{''}.$$

Let $\alpha > \sup(\operatorname{dom} r)$, $u = \{\delta\} \cup \operatorname{dom} r \cup \{\sup \operatorname{dom} r\}$. Let $q^* \in G$, $q^* \geq r{\restriction}\delta, q$, and let π collapse u to $\operatorname{otp} u$. Set $D = D^{\pi(\delta)}_{\pi[q^*],\pi[\underset{\sim}{z}]}$. Fix $v \subseteq \alpha$, and $r' \in G_\alpha \cap \mathrm{D}(u,v)$. We can copy r via an order-isomorphism inside $\alpha \times \aleph_2$, fixing $r{\restriction}\delta$, so that the result can be amalgamated with r', to yield r'', which is then a (u,v)-witness above r'. Since $r' \in \mathrm{D}(u,v)$, this means that r' is itself a (u,v)-witness in G_α. As this is all that the construction in 4.12 was supposed to achieve, this case is covered by the discussion there.

A3.7 Minor density systems

In the course of the argument in 4.12, we require two further density systems. In the course of that argument we introduced the set

$$S = \{\gamma \in S_C : \underset{\sim}{F}(\underset{\sim}{x}_\gamma) \text{ is a } [\mathbb{P}{\restriction}(\gamma+1)]\text{-name}\},$$

and argued that S is stationary. This led us to consider certain ordinals $\gamma < \delta$, with δ of cofinality $\aleph_2$, and an element $r_1 \in G_\delta$, at which point we claimed that we could produce a 1-1 order preserving function h with domain $\mathcal{A}^{r_1}$, equal to the identity on $\mathcal{A}^{r_1} \cap (\gamma+1)$, with $h(\min(\mathcal{A}^{r_1} \setminus (\gamma+1))) > \sup \mathcal{A}^{r_1}$, and $h[r_1] \in G_\delta$. More precisely, our claim was that this could be ensured by meeting suitable density systems.

For $\alpha < \aleph_2$, $q \in \mathcal{A}P{\restriction}\aleph_2$, define $\mathrm{D}_q^\alpha(u,v)$ as follows. If $(\{\alpha\} \cup \operatorname{otp} \operatorname{dom} q) \geq \operatorname{otp} u$ then let $\mathrm{D}_q^\alpha(u,v)$ degenerate. Otherwise, fix $k : (\{\alpha\} \cup \operatorname{dom} q) \longrightarrow u$ an order isomorphism onto an initial segment of u, and let $\beta = \inf(u \setminus \operatorname{range} k)$. Let $\mathrm{D}_q^\alpha(u,v)$ be the set of $r \in \mathcal{A}P$ with domain v such that $r{\restriction}v \setminus u$ contains the image of q under an order-preserving map h_0 which agrees with k below α and which carries $\inf(\mathcal{A}^q \setminus (\alpha \times \aleph_2))$ above β (i.e., above $(\beta, 0)$). The density condition corresponds to our ability to copy over part of q onto any set of unused ordinals in $(v \setminus \beta) \times \aleph_2$, recalling that $|\operatorname{dom} r| < \aleph_2$ for any $r \in \mathcal{A}P$, and then to perform an amalgamation.

For our intended application, suppose that γ, δ, r_1 are given as above, and let $u = \{\gamma\} \cup \operatorname{dom} r_1 \cup \{\sup \operatorname{dom} r_1)$. Let π be the canonical isomorphism of u with $\operatorname{otp} u$, and $\alpha = \pi(\gamma)$, $q = \pi[r_1]$. As G_δ meets D_q^α, we have $v \subseteq \delta$, and $r \in G_\delta \cap \mathrm{D}_q^\alpha(u,v)$. Then with $h = h_0 \circ \pi$, we have $h[r_1] \leq r$, and our claim is verified.

Finally, a few lines later in the course of the same argument we mentioned that the claim proved in 4.14 can be construed as the verification that certain additional density systems are in fact dense, and that accordingly we may suppose that the condition r described there lies in G.

Acknowledgments. This paper owes its existence to Annalisa Marcja's hospitality in Trento, July 1987; van den Dries' curiosity about Kim's conjecture; the willingness of Hrushovski and Cherlin to look at §4 through a dark glass; and most of all to Cherlin's insistence that this is one of the fundamental problems of model theory.

This is Number 326 in the publication list.

The author thanks the BSF, the Basic Research Fund, Israeli Academy of Sciences and the NSF for partial support of this research.

References

[AxKo] J. Ax and S. Kochen, *Diophantine problems over local rings I*, Amer. J. Math. **87** (1965), 605–630.

[Bollobas] Bollobas, *Random Graphs*, Academic Press, London-New York, 1985.

[JBSh 368] H. Judah and T.Bartoszynski and S.Shelah, *The Chicoń diagram*, Israel J. of Math.

[Gregory] J. Gregory, *Higher Souslin trees and the generalized continuum hypothesis*, J. Symb. Logic **41** (1976), 663-671.

[Jech] T. Jech, *Multiple Forcing*, Cambridge Univ. Press, Cambridge, 1986.

[Keisler] H. J. Keisler, *Ultraproducts and saturated models*, Indag. Math. **26** (1964), 178–186.

[Sh] S. Shelah, *Classification theory and the number of non-isomorphic models*, North Holland Publ. Co. Studies in Logic and the foundation of Math. **92** (1978).

[Sh b] S. Shelah, *Proper Forcing*, Lect. Notes Math. **940** (1982), Springer-Verlag, Berlin Heidelberg NY Tokyo, 496.

[Sh c] S. Shelah, *Classification theory and the number of non-isomorphic models, revised*, North Holland Publ. Co. Studies in Logic and the foundation of Math. **92** 705+xxxiv (1990).

[Sh f] S. Shelah, *Proper and Improper Forcing*, revised edition of [Sh b] (in preparation).

[Sh 13] S. Shelah, *Every two elementarily equivalent models have isomorphic ultrapowers*, Israel J. Math **10** (1971), 224–233.

[Sh 65] K. Devlin and S. Shelah, *A weak form of the diamond which follows from* $2^{\aleph_0} < 2^{\aleph_1}$, Israel J. Math. **29** (1978), 239–247.

[Sh 82] S. Shelah, *Models with second order properties III. Omitting types for $L(Q)$*, Proceedings of a workshop in Berlin, July 1977 Archiv f. Math. Logik **21** (1981), 1–11.

[Sh 107] S. Shelah, *Models with second order properties IV. A general method and eliminating diamonds*, Annals Math. Logic **25** (1983), 183–212.

[ShHL 162] S. Shelah and B. Hart and C. Laflamme, *Models with second order properties V. A general principle*, Annals Pure Appl. Logic; (to appear).

[Sh 177] S. Shelah, *More on proper forcing*, J. Symb. Logic **49** (1984), 1035–1038.

[Sh 345] S. Shelah, *Products of regular cardinals and cardinal invariant of Boolean Algebras*, Israel J. of Math **70** (1990), 129–187.

[Sh 405] S. Shelah, *Vive la differance II-refuting Kim's conjecture*, (preprint) (Oct 1989).

[ShFr 406] S. Shelah and D. Fremlin, *Pointwise compact and stable sets of measurable functions*, Journal of Symbolic Logic (submitted).

Department of Mathematics, Hebrew University, Jerusalem, ISRAEL

CODING AND RESHAPING WHEN THERE ARE NO SHARPS

Saharon Shelah and Lee J. Stanley

Abstract. Assuming $0^\sharp$ does not exist, κ is an uncountable cardinal and for all cardinals λ with $\kappa \leq \lambda < \kappa^{+\omega}$, $2^\lambda = \lambda^+$, we present a "mini-coding" between κ and $\kappa^{+\omega}$. This allows us to prove that any subset of $\kappa^{+\omega}$ can be coded into a subset, W of κ^+ which, further, "reshapes" the interval $[\kappa, \kappa^+)$, i.e., for all $\kappa < \delta < \kappa^+$, $\kappa = (\text{card } \delta)^{L[W \cap \delta]}$. We sketch two applications of this result, assuming $0^\sharp$ does not exist. First, we point out that this shows that any set can be coded by a real, via a **set** forcing. The second application involves a notion of abstract condensation, due to Woodin. Our methods can be used to show that for any cardinal μ, condensation for μ holds in a generic extension by a **set** forcing.

1. Introduction

Theorem. *Assume that $V \models ZFC +$ "$0^\sharp$ does not exist", and, in V, $\kappa \geq \aleph_2$, $Z \subseteq \kappa^{+\omega}$ and for cardinals λ with $\kappa \leq \lambda < \kappa^{+\omega}$, $2^\lambda = \lambda^+$.* **THEN** there is a cofinality preserving forcing $\mathbf{S}(\kappa) = \mathbf{S}(\kappa, Z)$ of cardinality $\kappa^{+(\omega+1)}$ such that if G is V-generic for $\mathbf{S}(\kappa)$, there is $W \subseteq \kappa^+$ such that $V[G] = V[W]$, $Z \in L[W, Z \cap \kappa]$, for all cardinals λ with $\kappa \leq \lambda < \kappa^{+\omega}$, and for all limit ordinals δ with $\kappa < \delta < \kappa^+$, $\kappa = (\text{card } \delta)^{L[W \cap \delta]}$.

Our forcing $\mathbf{S}(\kappa)$ can be thought of as a kind of Easton product between κ and $\kappa^{+\omega}$ of partial orderings which *simultaneously* perform the tasks of coding (§1.2 of [1]) and reshaping (§1.3 of [1]). Our new idea is to introduce an additional coding area used for "marking" certain ordinals. This "marking" technique is the crucial addition to the arguments of §1 of [1]. We appeal to the Covering Lemma twice: in (3.1), and again in the proof of the Proposition in (3.3). The referee has informed us that the hypothesis that $0^\sharp$ does not cannot be eliminated. Jensen first used this hypothesis in [1] to facilitate certain arguments, and then realized that *his* uses *were* eliminatable. It is not the purpose of this paper to discuss the nature of Jensen's appeals to the Covering Lemma; the interested reader may consult pp. 62, 96 and the Introduction to Chapter 8 of [1] for insight into Jensen's uses of the Covering Lemma, and how he was able to eliminate them. In [2], S. Friedman presents a rather different, more streamlined approach to avoiding such uses of Covering. It should be clear from the preceding that Jensen's appeals to the Covering Lemma are of a rather different character than ours.

To better understand the role of this "marking" technique, let us briefly recall some material from [1]. Let us first consider the possibility of coding $R \subseteq \kappa^+$ into a subset of κ, when κ is regular. In order to use almost disjoint set coding, we seem to need extra properties of the ground model, or of the set R, since, in order to carry out the *decoding* recursion across $[\kappa, \kappa^+)$ we need, e.g., an almost disjoint sequence $\vec{b} = (b_\alpha : \alpha \in [\kappa, \kappa^+))$ of cofinal subsets of κ satisfying:

$$(*) \qquad \text{for all } \theta \in [\kappa, \kappa^+),\ (b_\alpha : \alpha \leq \theta) \in L[R \cap \theta], \\ \text{and is "canonically definable" there.}$$

Such a $\vec{b}$ is called *decodable*, and it is easy to obtain a decodable $\vec{b}$ if R satisfies:

$$(**) \qquad \text{for all } \theta \in [\kappa, \kappa^+),\ (\text{card } \theta)^{L[R\cap\theta]} = \kappa.$$

If $(**)$ holds, we say that R *promptly collapses fake cardinals.*

Of course, typically $(**)$ fails, and the "reshaping" conditions of §1.3 of [1] are introduced to obtain $(**)$ in a generic extension. Our κ and R, from the previous paragraph are called γ and B in §1.3 of [1]. Unfortunately, the distributivity argument for the reshaping partial ordering given there *seems* to really require not merely that $H_{\gamma^+} = L_{\gamma^+}[B]$, but that $H_{\gamma^{++}} = L_{\gamma^{++}}[B]$, where $B \subseteq \gamma^+$. This will be the case if B is the result of coding as far as γ^+, but that is another story, which leads to Jensen's original approach to the Coding Theorem. Our appeals to the Covering Lemma focus on this point: essentially, to prove a distributivity property of the reshaping conditions. As already indicated, in Jensen's treatment, the appeals to the Covering Lemma were designed to overcome different sorts of obstacles and proved to be eliminatable.

Our approach to guaranteeing that the unions of certain increasing chains of reshaping conditions collapse the suprema of their domains is to have "marked" a cofinal sequence of small order type. Because of the need to meet certain dense sets in the course of the construction, it is too much to expect that the ordinals we **intentionally** marked are the only marked ordinals. However, what we will be able to guarantee is that they are the only members of a certain club subset which have been marked. The club will exist in a small enough inner model, thanks to the Covering Lemma. This argument is given in (3.3). We are grateful to the referee for suggesting the use of "fast clubs" in the argument of (3.3). This allowed us to streamline a more complicated argument (which also suffered from some [probably reparable] inaccuracies) in an earlier version of this paper. We use "1" to mark ordinals. To guarantee that this does not collide with requirements imposed by the "coding" part of the conditions, we set aside the limit ordinals as the only potentially marked ordinals and do not use them for coding.

1.1. Summary and organization

We now give a brief overview of the contents of this paper. In §2, we build to the definition, in (2.5), of the $\mathbf{S}(\kappa)$, along with auxiliary forcings, $\mathbf{S}_k(\kappa)$. In §3, we prove that the $\mathbf{S}(\kappa)$ are as required. The heart of the matter is (3.3), where we prove the distributivity properties of the $\mathbf{S}_k(\kappa)$. Preliminary observations are given in (3.1) and (3.2). The former shows that only increasing sequences of certain lengths are problematical. The latter is a rather routine observation about how the coding works. In the argument of (3.3), we use this in the context of forcing over $\hat{\mathcal{N}}$, a transitive set model of enough ZFC, introduced in the proof of (3.3), below. In (3.4) we put together the material of (3.1)–(3.3) to prove the Theorem. In (3.5) we make a few remarks and briefly sketch the applications mentioned in the abstract.

The partial ordering $\mathbf{S}(T,\ \lambda)$, introduced in (2.2), below, is the analogue of the reshaping partial ordering of §(1.3) of [1]. It adds a subset of λ^+, which, together with T, promptly collapses fake cardinals in $(\lambda,\ \lambda^+)$. The partial ordering $\mathbf{P}_{\kappa,\ T,\ g}$, introduced in (2.4), is a version of the coding partial ordering of §(1.2) of [1], **relative to** g. We require that $T \subseteq \kappa^+$, $g \in S(T,\ \kappa^+)$. If $p \in P_{\kappa,\ T,\ g}$, then p will have the form $(\ell(p),\ r(p))$; $\ell(p)$ is the "function part" of p and $r(p)$ is the "promise part" of p. We require that $\ell(p)$ starts to code not only T, but also g **and** that $\ell(p) \in S(T \cap \kappa,\ \kappa)$. If g were not merely a condition but generic for $\mathbf{S}(T,\ \kappa^+)$, then $\mathbf{P}_{\kappa,\ T,\ g}$ would just be the usual forcing for the almost-disjoint set coding of the "join" of T and g, with the extra requirement above, that for conditions, p, $\ell(p)$, together with $T \cap \kappa$, collapses sup dom $\ell(p)$.

Finally, the $\mathbf{S}(\kappa)$, introduced in (2.5), is the forcing which accomplishes the task of coding and reshaping, between κ and κ^+. It is defined relative to the choice of a fixed $Z \subseteq \kappa^{+\omega}$ such that $H_{\kappa^{+n}} = L_{\kappa^{+n}}[Z \cap \kappa^{+n}]$, for all $n \leq \omega$. The elements of $S(\kappa)$, are ω-sequences, $(p(n):\ n < \omega)$, where for all $n < \omega$, $p(n) = (\ell(p(n)),\ r(p(n)))$, $\ell(p(n)) \in S(Z \cap \kappa_n,\ \kappa_n)$ and $p(n) \in P_{\kappa_n,\ Z\cap\kappa_{n+1},\ \ell(p(n+1))}$. Thus, letting $\dot{G}$ be the canonical name for the generic of $\mathbf{S}(\kappa)$, letting $\dot{G}(n)$ be the canonical name for $\{\ell(p(n)) : p \in \dot{G}\}$, and letting $\dot{W}(n)$ be the canonical name for $\bigcup \dot{G}(n)$, $\mathbf{S}(\kappa)$ is a sort of Easton product of the $P_{\kappa_n,\ Z\cap\kappa_{n+1},\ \dot{W}(n+1)}$.

1.2. Notation and terminology

Our notation and terminology is intended to be standard, or have a clear meaning, e.g., *o.t.* for order type, card for cardinality. A catalogue of possible exceptions follows. When forcing, $p \leq q$ means q gives more information. Closed unbounded sets are *clubs* . The set of limit points of a set X of ordinals is denoted by X'. $A\Delta B$ is the symmetric difference of A and B, and $A\backslash B$ is the relative complement of B in A. For ordinals, $\alpha \leq \beta$, $[\alpha,\ \beta)$ is the half-open interval $\{\gamma : \alpha \leq \gamma < \beta\}$. The notation for the three other intervals are clear. It should be clear from context whether the open interval or the ordered pair is meant. OR is the class of all ordinals. For infinite cardinals, κ, H_κ is the set of all sets hereditarily of cardinality $< \kappa$, i.e. those sets x such that if t is the transitive closure of x, then card $t < \kappa$. For ordinals α, β, we write $\alpha >> \beta$ to mean that α is MUCH greater than β; the precise sense of how much greater we must take it to be is supposed to be clear from context. For models, $\mathcal{M}$, $Sk_{\mathcal{M}}$ denotes the Skolem operation in $\mathcal{M}$, where the Skolem functions are obtained in some reasonable fixed fashion. In this paper, we often suppress mention of the membership relation as a relation of a model, but it is always intended that it be one. Thus, $(M,\ A,\ \cdots)$ denotes the same model as $(M,\ \in,\ A,\ \cdots)$. All other notation is introduced as needed (we hope).

2. The Forcings

2.1 Definition. If g is a function, $\overline{g} = \{x \in \text{dom }\ g : g(x) = 1\}$.

2.2 Definition. If λ is a infinite cardinal, $T \subseteq \lambda$, then $g \in S(T,\ \lambda)$ iff there's $\delta = \delta(g) \in (\lambda,\ \lambda^+)$ such that $g : (\lambda,\ \delta) \rightarrow \{0,\ 1\}$ and for all $\alpha \in (\lambda,\ \delta]$:

$(*)_{\alpha,\ g}$ (card $\alpha)^{L[T,\ g|\alpha]} = \lambda$ (we say: g promptly collapses α).

$\mathbf{S}(T,\ \lambda) = (S(T,\ \lambda),\ \subseteq)$.

2.3 Definition. Let κ be an infinite cardinal, $T \subseteq \kappa^+$, $g \in S(T, \kappa^+)$. $\vec{b}^g = (b^g_\alpha : \alpha \in (\kappa^+, \delta(g)])$ is a sequence of almost disjoint cofinal subsets of successor ordinals $\beta \in (\kappa, \kappa^+)$ which are multiples of 3, such that for all $\alpha \in (\kappa, \delta(g)]$, $(b^g_\xi : \xi \in (\kappa^+, \alpha])$ is canonically defined in $L[T, g|\alpha]$.

2.4 Definition. With κ, T, g as in (1.5), $p = (\ell(p), r(p)) \in P_{\kappa, T, g}$ iff

(1) $\ell(p) \in S^+(T \cap \kappa, \kappa)$,

(2) if $\alpha \in (\kappa, \delta(\ell(p)))$, $\alpha = 3\alpha' + 1$, then $\ell(p)(\alpha) = 1$ iff $\alpha' \in T$. (we say: $\ell(p)$ codes T),

(3) $r(p) : \text{dom } r(p) \to \kappa^+$, $\text{dom } r(p) \in [\text{dom } g]^{<\kappa^+}$, and whenever $\alpha \in \text{dom } r(p)$, $r(p)(\alpha) \leq \xi \in b^g_\alpha \cap \delta(\ell(p))$, $\ell(p)(\xi) = g(\alpha)$,

(4) if $\alpha_1, \alpha_2 \in \text{dom } r(p)$ and $g(\alpha_1) \neq g(\alpha_2)$, then $b^g_{\alpha_1} \setminus r(p)(\alpha_1) \cap b^g_{\alpha_2} \setminus r(p)(\alpha_2) = \emptyset$.

For $p, q \in P_{\kappa, T, g}$, $p \leq q$ iff $\ell(p) \subseteq \ell(q)$, $r(p) \subseteq r(q)$; $\mathbf{P}_{\kappa, T, g} = (P_{\kappa, T, g}, \leq)$.

2.5 Definition. Let κ be an infinite cardinal. For $n \leq \omega$, let κ_n be κ^{+n}. Let $Z \subseteq \kappa_\omega$ be such that for all $n \leq \omega$, $H_{\kappa_n} = L_{\kappa_n}[Z \cap \kappa_n]$. $p \in S(\kappa, Z) = S(\kappa)$ iff dom $p = \omega$, for all $n < \omega$, $p(n) = (\ell(p(n)), r(p(n)))$, $\ell(p(n)) \in S(Z \cap \kappa_n, \kappa_n)$ and $p(n) \in P_{\kappa_n, Z\cap\kappa_{n+1}, \ell(p(n+1))}$. For $p, q \in S(\kappa)$, $p \leq q$ iff for all $n < \omega$, $\ell(p(n)) \subseteq \ell(q(n))$, $r(p(n)) \subseteq r(q(n))$. $\mathbf{S}(\kappa) = \mathbf{S}(\kappa, Z) = (S(\kappa), \leq)$.

If $k < \omega$, $S_k(\kappa) = S_k(\kappa, Z) = \{p|[k, \omega) : p \in S(\kappa)\}$; $\leq_k$ is the obvious projection of $\leq$ onto $S_k(\kappa)$. $\mathbf{S}_k(\kappa) = \mathbf{S}_k(\kappa, Z) = (S_k(\kappa), \leq_k)$.

3. The Results

Our ultimate goal in this section will be to prove that for cardinals κ with $\kappa \geq \aleph_2$, for all $k < \omega$, $\mathbf{S}_k(\kappa)$ is (κ_k, ∞)- distributive. As will be clear from what follows, by this we mean that the intersection κ_k open dense sets is dense, and not the weaker notion involving fewer than κ_k open dense sets. We denote the latter notion by $(< \kappa_k, \infty$-distributive. A useful first step will be to establish something stronger than this latter notion.

3.1 Proposition. *For all $k < \omega$, $\mathbf{S}_k(\kappa)$ is $< \kappa_k$- complete.*

Proof. Let $\theta < \kappa_k$, $(p_i : i < \theta)$ be a $\leq_k$-increasing sequence from $S_k(\kappa)$. For $i < \theta$, $k \leq n < \omega$, let $\delta_i(n) = \delta(\ell(p_i(n)))$, so, for such n, $(\delta_i(n) : i < \theta)$ is non-decreasing. Let $\delta(n) = \sup \{\delta_i(n) : i < \theta\}$. Let $\ell(p(n)) = \bigcup\{\ell(p_i(n)) : i < \theta\}$, $r(p(n)) = \bigcup\{r(p_i(n)) : i < \theta\}$, and let $p(n) = (\ell(p(n)), r(p(n)))$, for

$k \leq n < \omega$. We shall prove that $p \in S_k(\kappa)$. The only difficulty is to prove that for $k \leq n < \omega$, $(\text{card } \delta(n))^{L[Z\cap\kappa_n,\ \ell(p(n))]} = \kappa_n$. If θ is a successor ordinal or $\delta(n) = \delta_i(n)$ for some $i < \theta$, this is clear. Otherwise, $\delta(n)$ is a limit ordinal of cofinality $\leq cf\ \theta < \kappa_n$, so, by the Covering Lemma, already $(\text{cf } \delta(n))^L < \kappa_n$. But then, since $(\forall\alpha < \delta(n))(\text{card } \alpha)^{L[Z\cap\kappa_n,\ \ell(p(n))]} \leq \kappa_n$, the conclusion is clear. □

(3.2) Before proving the main lemma of the section, in (3.3), it will be helpful to simply remark (the proofs are easy, and the reader may consult [2] for an outline) that letting $\dot{G}$ be the canonical name for the generic, letting $\dot{G}(n)$ be the canonical name for $\{\ell(p(n)) : p \in \dot{G}\}$, and letting $\dot{W}(n)$ be the canonical name for $\bigcup \dot{G}(n)$, then for all $k < \omega$,

$$\Vdash_{\mathbf{S}_k(\kappa)} \text{“}(\forall k \leq n < \omega)\dot{W}(n),\ Z \cap \kappa_n \in L[Z \cap \kappa_k,\ \dot{W}(k)]\text{ ”}.$$

We shall use a variant of this fact with no further comment below, in the proof of the main lemma. We note only that by an easy density argument, it can be shown that for $k \leq n < \omega$ and $\alpha \in [\kappa_{n+1},\ \kappa_{n+2})$, there is $\eta < \kappa_{n+1}$ such that whenever $\xi \in b_\alpha^{\dot{W}(n+1)|\alpha} \setminus \eta,\ \dot{W}(n)(\xi) = 0 \Rightarrow \dot{W}(n)(\xi + 1) = \dot{W}(n+1)(\alpha)$, and that $\{\xi \in b_\alpha^{\dot{W}(n+1)|\alpha} : \dot{W}(n)(\xi) = 0\}$ is cofinal in κ_{n+1}. Thus, $\dot{W}(n+1)(\alpha)$ is read by: $\dot{W}(n+1)(\alpha) = i$ iff there is a final segment $x \subseteq b_\alpha^{\dot{W}(n+1)|\alpha}$ such that for all $\xi \in x,\ \dot{W}(n)(\xi) = i$.

(3.3) We are now ready for the main Lemma.

Lemma. *For all $k < \omega$, $\mathbf{S}_k(\kappa)$ is $(\kappa_k,\ \infty)$-distributive.*

Proof. We first note that it suffices to prove that for all $k < \omega$

> $(*)_k$: Let $p_0 \in S_k(\kappa)$, let χ be regular $\chi >> 2^{2^{\kappa_\omega}}$; let $< (*)$ be a well-ordering of H_χ in type χ; let $\mathcal{M} = (H_\chi,\ < (*),\ \{\mathbf{S}_k(\kappa)\},\ \{Z\},\ \{p_0\})$; let $\mathcal{N} \prec \mathcal{M},\ \kappa_k + 1 \subseteq |\mathcal{N}|$, card $|\mathcal{N}| = \kappa_k$. Then there is $p_0 \leq_k p^*$ which is $(\mathcal{N},\ \mathbf{S}_k(\kappa) \cap |\mathcal{N}|)$-generic.

The argument that $(*)_k$ suffices is well-known, so fix the above data. Without loss of generality, we may assume that $[|\mathcal{N}|]^{<\kappa_k} \subseteq |\mathcal{N}|$. It will often be convenient to work with the transitive collapse of $\mathcal{N}$, so let $\pi : \hat{\mathcal{N}} \to \mathcal{N}$ be the inverse of the transitive collapse map; thus, $[|\hat{\mathcal{N}}|]^{<\kappa_k} \subseteq |\hat{\mathcal{N}}|$. Let $\sigma = \pi^{-1} =$ the transitive collapse map. If $X \subseteq |\hat{\mathcal{N}}|$ and $(\hat{\mathcal{N}},\ X)$ is amenable, then we let $\pi(X) = \bigcup\{\pi(a \cap X) : a \in |\hat{\mathcal{N}}|\}$, and similarly for $\sigma(Y)$ if $(\mathcal{N},\ Y)$ is amenable. We let $\hat{\kappa}_n = \sigma(\kappa_n)$. We also let $\theta_n = \sup\ (|\mathcal{N}| \cap \kappa_n)$.

For $k < n < \omega$, note that $\hat{\kappa}_n = (\kappa_n)^{\hat{\mathcal{N}}}$, and that $\hat{\kappa}_{k+1} = \theta_{k+1}$. Note that by applying Proposition 3.1 to forcing over $\hat{\mathcal{N}}$ with $\sigma(\mathbf{S}_{k+1}(\kappa))$, we easily construct $\hat{p} \in \sigma(\mathbf{S}_{k+1}(\kappa))$ which is $(\hat{\mathcal{N}},\ \sigma(\mathbf{S}_{k+1}(\kappa)))$-generic, such that $\sigma(p_0)|[k+1,\ \omega)$ is extended by $\hat{p}$, in $\sigma(\leq_{k+1})$, such that for $k+1 \leq n < \omega$, $\hat{p}(n) \subseteq |\hat{\mathcal{N}}|$ and all proper initial segments of $\hat{p}(n)$ lie in $|\hat{\mathcal{N}}|$. In view of the discussion in (2.2), for forcing over $\hat{\mathcal{N}}$,

$$\hat{\mathcal{N}}[\hat{p}] \models \text{“}(\forall n)(k+1 \leq n < \omega \Rightarrow \hat{p}(n) \in L[\sigma(Z \cap \kappa_{k+1}),\ \hat{p}(k+1)]\text{”}.$$

Thus, $\hat{\mathcal{N}}[\hat{p}] \models$ “$\sigma(Z \cap \kappa_\omega) \in L[\sigma(Z \cap \kappa_{k+1}),\ \hat{p}(k+1)]$”.

A crucial observation is:

Proposition. $OR \cap |\hat{\mathcal{N}}| < ((\hat{\kappa}_{k+1})^+)^L$.

Proof. Let $\hat{\theta} = OR \cap |\hat{\mathcal{N}}|$, $\theta = \sup\ (OR \cap |\mathcal{N}|)$. Note that $\pi|L^{\hat{\mathcal{N}}} : L_{\hat{\theta}} \rightarrow_{\Sigma_1} L_\theta$, with critical point $\hat{\kappa}_{k+1}$. If $\hat{\theta} \geq ((\hat{\kappa}_{k+1})^+)^L$, then $0^\sharp$ exists, which proves the Proposition. $\square$

Thus, $(cf\ \hat{\kappa}_n)^L \leq (cf\ \hat{\kappa}_{k+1})^L$, for all $k+1 \leq n < \omega$. Typically, of course, $\hat{\kappa}_{k+1}$ is a (regular cardinal)^L. Let $x_{k+1} = Z \cap \hat{\kappa}_{k+1}$, $h_{k+1} = \ell(\hat{p}(k+1))$.

We shall construct in V, $\hat{p}(k)$ which is $|\hat{\mathcal{N}}|$-generic for $\mathbf{P}_{\kappa_k,\ x_{k+1},\ h_{k+1}}$, as defined in $\hat{\mathcal{N}}$. Among other properties, letting $h_k = \ell(\hat{p}(k))$, h_k will code h_{k+1}. This will be **clear** from the construction; we shall use this fact **before** showing that $(cf\ \hat{\kappa}_{k+1})^{L[Z\cap\kappa_k,\ h_k]} = \kappa_k$. This is exactly what is required to show that if we define p by letting $p(n) = (\pi(\ell(\hat{p}(n))),\ \pi(r(\hat{p}(n))))$ (recall our convention about $\pi(X)$ for $(\hat{\mathcal{N}},\ X)$ amenable), then $p \in S_k(\kappa)$ (and p is $|\mathcal{N}|$-generic for $\mathbf{S}_k(\kappa) \cap |\mathcal{N}|$).

We shall have $h_k = \ell(q_{\kappa_k})$, $r(\hat{p}(k)) = r(q_{\kappa_k})$, where $q_i = (\ell(q_i),\ r(q_i))$ and $(q_i : i \leq \kappa_k)$ is defined recursively in V, with $q_0 = \sigma(p_0(k))$. For this, in V, we let $(D_i : i < \kappa_k)$ enumerate the dense subsets, in $|\hat{\mathcal{N}}|$, of $\mathbf{P}_{\kappa_k,\ x_{k+1},\ h_{k+1}}$, as defined in $\hat{\mathcal{N}}$. For all $\theta < \kappa_k$, $(D_i : i < \theta) \in |\hat{\mathcal{N}}|$, in virtue of the closure property we have assumed for $|\hat{\mathcal{N}}|$. For all $i < \theta$, we'll have $q_i \in |\hat{\mathcal{N}}|$, so, by the same observation, for $\theta < \kappa_k$, $(q_i : i < \theta) \in |\hat{\mathcal{N}}|$.

Also, for $j < \hat{\kappa}_{k+1}$, letting $D(j)$ be the subset of $\mathbf{P}_{\kappa_k,\ x_{k+1},\ h_{k+1}}$ consisting of those r with $\delta(\ell(r)) \geq j$, as defined in $\hat{\mathcal{N}}$, clearly $D(j)$ is dense and so is among the D_i. This will guarantee that $\sup\ \{\delta(\ell(q_i)) : i < \kappa_k\} = \hat{\kappa}_{k+1}$, provided that we know that $q_{i+1} \in D_i$. This will be part of the construction and will also guarantee the genericity of p.

For $i < \kappa_k$, we'll set $\alpha_i = \delta(\ell(q_i))$. For limit $\theta \leq \kappa_k$, we let $\ell(q_\theta) = \bigcup\{\ell(q_i) : i < \theta\}$, $r(q_\theta) = \bigcup\{r(q_i) : i < \theta\}$. If $\theta < \kappa_k$, by the covering argument of the proposition of (2.1), these are always conditions, and, if $\theta < \kappa_k$, as noted above, $(q_i : i < \theta) \in |\hat{\mathcal{N}}|$, so also $q_\theta \in |\hat{\mathcal{N}}|$. So, we must define q_{i+1}, where our crucial work is done.

For each $\alpha_i \leq \alpha < \gamma < \hat{\kappa}_{k+1}$, α a limit ordinal, we define $p^{\gamma,\, \alpha,\, 1} \geq q_i$ as follows: $r(p^{\gamma,\, \alpha,\, 1}) = r(q_i)$; if $\alpha_i \leq \beta < \gamma$ and $\beta \equiv 1\ (mod\ 3)$ then $\ell(p^{\gamma,\, \alpha,\, 1})(\beta) = 0$ if $\beta' \notin Z$ & $= 1$, if $\beta' \in Z$, where β' is such that $\beta = 3\beta'+1$. If $\gamma \geq \alpha_i+\kappa_k$, we fix a subset $b \in |\hat{\mathcal{N}}| \cap L$, $b \subseteq \kappa_k$ which codes a well-ordering of κ_k in type γ, and for $\beta < \kappa_k$, we set $\ell(p^{\gamma,\, \alpha,\, 1})(\alpha_i + 3\beta + 2) = 0$ if $\beta \notin b$ & $=$ 1 if $\beta \in b$. If $\alpha_i + \kappa_k \leq \beta < \gamma$ and $\beta \equiv 2\ (mod\ 3)$, we set $\ell(p^{\gamma,\, \alpha,\, 1})(\beta) = 0$. Similarly, if $\gamma < \alpha_i + \kappa_k$, we set $\ell(p^{\gamma,\, \alpha,\, 1})(\beta) = 0$ for all $\alpha_i \leq \beta < \gamma$ such that $\beta \equiv 2\ (mod\ 3)$.

If $\alpha_i \leq \beta < \gamma$ and for some $\tau \in$ dom $r(q_i)$, $\beta \in b_\tau^{h_{k+1}} \setminus r(q_i)(\tau)$, then $\ell(p^{\gamma,\, \alpha,\, 1})(\beta) = h_{k+1}(\tau)$. Note that in virtue of (4) of (1.4), this is well-defined. For all other successor ordinals, $\alpha_i \leq \beta\gamma$ which are multiples of 3, we set $\ell(p^{\gamma,\, \alpha,\, 1})(\beta) = 0$.

Now, suppose β is a limit ordinal, $\alpha_i \leq \beta < \gamma$. We set $\ell(p^{\gamma,\, \alpha,\, 1})(\beta) = 0$, **unless** $\beta = \alpha$ & $=$ 1), if $\beta = \alpha$ (in this case, we mark α).

Then, let $p^{\gamma,\, \alpha,\, 2} \geq p^{\gamma,\, \alpha,\, 1}$ be chosen canonically in D_i. Now $(\gamma,\, \alpha) \mapsto p^{\gamma,\, \alpha,\, 2}$ is definable in $\hat{\mathcal{N}}$, and so, for each γ, in $\hat{\mathcal{N}}$, we can compute a bound, $\eta(\gamma) < \hat{\kappa}_{k+1}$, for sup {dom $\ell(p^{\gamma,\, \alpha,\, 2})$: $\alpha_i \leq \alpha < \gamma$, α a limit ordinal }, as a function of γ. Iterating η in $\hat{\mathcal{N}}$ gives us a club, E_i, of $\hat{\kappa}_{k+1}$, $E_i \in |\hat{\mathcal{N}}|$. Now, $(H_{\hat{\kappa}_{k+2}})^{\hat{\mathcal{N}}} = L_{\hat{\kappa}_{k+2}}[\sigma(Z) \cap \hat{\kappa}_{k+2}]$, so all clubs of $\hat{\kappa}_{k+1}$ which lie in $|\hat{\mathcal{N}}|$, and, in particular, E_i, lie in $L[\sigma(Z) \cap \hat{\kappa}_{k+2}]$. Already in L, card $\hat{\kappa}_{k+2} =$ card $\hat{\kappa}_{k+1}$. So, in $L[\sigma(Z) \cap \hat{\kappa}_{k+2}]$ there is $\theta < (\hat{\kappa}_{k+1})^+$ such that all clubs of $\hat{\kappa}_{k+1}$ which lie in $|\hat{\mathcal{N}}|$, in fact, lie in $L_\theta[\sigma(Z) \cap \hat{\kappa}_{k+2}]$. This, however, readily gives us that unless (card $\hat{\kappa}_{k+1})^{L[\sigma(Z)\cap\hat{\kappa}_{k+2}]} = \kappa_k$ (and in this case, there is no problem in proving that q_{κ_k} is a condition), there is a club C of $\hat{\kappa}_{k+1}$, $C \in L[\sigma(Z) \cap \hat{\kappa}_{k+2}]$, such that C grows faster than any club of $\hat{\kappa}_{k+1}$ which lies in $|\hat{\mathcal{N}}|$. In particular, C grows faster than E_i, so that for sufficiently large $\gamma < \hat{\kappa}_{k+1}$, all E_i-intervals above γ miss C. In V, fix $C^* \subseteq C$, *o.t.* $C^* = \kappa_k$, C^* a club of $\hat{\kappa}_{k+1}$.

The idea of the above is that in constructing $p^{\gamma,\, \alpha,\, 1}$, we have "marked" α and our hope is that in passing from $p^{\gamma,\, \alpha,\, 1}$ to $p^{\gamma,\, \alpha,\, 2}$, we have not inadvertently "marked" anything else. While this is too much to hope for, in general, we shall be able to get that we have not marked anything else **in C**, provided we choose γ sufficiently large so that every interval of E_i, above γ, misses C. So, GOOD's winning strategy, finally, to go from q_i to q_{i+1}, is to take γ to be the least ordinal $> \alpha_i$, $\gamma \in C$ which, as above, is sufficiently large that the interval $[\gamma,\, \eta(\gamma)) \cap C = \emptyset$, **and such that there is** $\alpha^* \in [\alpha_{\mathbf{i}},\, \gamma) \cap \mathbf{C}^*$ and then to take $q_{i+1} = p^{\gamma,\, \alpha^*,\, 2}$. Thus, GOOD has "marked" a member of C^* and nothing else in C, while obtaining $q_{i+1} \in D_i$.

Now, since, as remarked above, we know from the construction that h_k

codes h_{k+1}, in $L[Z \cap \kappa_k,\ h_k]$, we can recover $\sigma(Z) \cap \hat{\kappa}_{k+2}$, and therefore C. But then, by the construction, we have that $\{\alpha \in C : (h_k(\alpha),\ h_k(\alpha+1)) = (1,\ 1)\}$ is a cofinal subset of C^*. Thus, as required, $(cf\ \hat{\kappa}_{k+1})^{L[Z\cap\kappa_k,\ h_k]} = \kappa_k$. This completes the proof. $\square$

(3.4) Taken together, (3.1)–(3.3) give us the following Lemma, which, in turn, gives us the Theorem of the Introduction:

Lemma. *Forcing with* $\mathbf{S}(\kappa)$ *preserves cofinalities, GCH, and if* G *is* V*-generic for* $\mathbf{S}(\kappa)$*, then, in* $V[G]$ *there is* $W \subseteq \kappa^+$ *such that* $V[G] = V[W]$*,* $Z \in L[W,\ Z \cap \kappa]$ *and for all* $n \leq \omega$*,* $H_{\kappa_n} = L_{\kappa_n}[W]$ *and for* $\kappa < \alpha < \kappa^+$*,* $(\text{card}\ \alpha)^{L[W\cap\alpha]} = \kappa$*.*

Proof. Of course $W = \bigcup\{\overline{\ell(p(0))} : p \in \dot{G}\}$. It is a routine generalization of arguments from Chapter 1 of [1] to see that for all k, there is $\mathbf{Q}_k \in V^{\mathbf{S}_k(\kappa)}$ such that $\mathbf{S}(\kappa) \cong \mathbf{S}_k(\kappa) * \mathbf{Q}_k$, and $\Vdash_{\mathbf{S}_k(\kappa)}$ "$\mathbf{Q}_k$ is $\kappa_{k+1}-$ c.c. and card $Q_k = \kappa_{k+1}$". Further, for $k = 0$, (2.3) gives us that $\mathbf{S}(\kappa)$ is $(\kappa,\ \infty)$-distributive and clearly card $S(\kappa) = \kappa_\omega^+$. Thus, preservation of GCH is clear, as is the preservation of all cardinals except possibly κ_ω^+. The argument here is routine: if this failed, then letting $\gamma = (cf\ \kappa_\omega^+)^{V^{\mathbf{S}(\kappa)}}$, for some $0 < k < \omega$, $\gamma = \kappa_k$. But then, since $(cf\ \kappa_\omega^+)^{V^{\mathbf{S}_k(\kappa)}} > \kappa_k$, forcing with $\mathbf{Q}_k$ over $V^{\mathbf{S}_k(\kappa)}$ would have to collapse a cardinal $\geq \kappa_{k+1}$ which is impossible. $\square$

3.5 Remarks and applications

(1) If we start from an arbitrary $Z' \subseteq \kappa_\omega$, we can, of course, code Z' by first coding Z' into a Z, as above (e.g., by coding Z' into Z on odd ordinals), and then proceeding as above.

(2) In work in progress, we are attempting to develop a combinatorial approach to coding the universe by a real (when $0^\sharp$ does not exist). Part of our approach is to use the Easton product of the $\mathbf{S}(\kappa)$, for $\kappa = \aleph_2$, or κ a limit cardinal, as a preliminary forcing, to simplify the universe before doing the main coding.

(3) Several people have observed that the $\mathbf{S}(\kappa)$ afford a method of coding any **set** of ordinals using a **set** forcing over models of GCH where $0^\sharp$ does not exist. This can be done as follows. Let $X \subseteq \lambda$, and assume, without loss of generality, that $\lambda \geq \aleph_2$. Code X into a $Z \subseteq \lambda^{+\omega}$, where Z has the properties assumed above. Then, force

with $\mathbf{S}(\lambda)$ to get W, as above. Finally, since W reshapes the interval (λ, λ^+), we can continue to code W down to a real, using one of the usual methods of coding by a real.

(4) Woodin has introduced the following abstract notion of condensation. $A \subseteq \delta$ **has condensation** iff there's an algebra, $\mathcal{A} \in V$ with underlying set δ, such that for any generic extension V' of V:

(*) if $X \subseteq \delta$ and X is the underlying set of a subalgebra of $\mathcal{A}$, and $\pi : (\mathcal{A}^*, A^*) \to (\mathcal{A}|X, A \cap X)$, where π is the inverse of the transitive collapse map, then $A^* \in V$.

δ has condensation iff for all $A \subseteq \delta$, A has condensation. This notion has been investigated by Woodin's student, D. Law, in his dissertation [3], and by Woodin himself.

S. Friedman has observed that using (3), above, it can be shown that for any cardinal μ, we can force condensation for μ via a set forcing. We omit the proof, except to say that according to Friedman, this is not a routine consequence of the usual sort of condensation for $L[r]$, but rather involves a closer look at the coding apparatus.

Acknowledgments. The research of the first author was partially supported by the NSF, the Basic Research Fund of the Israel Academy of Science and the Mathematical Sciences Research Institute. The research of the second author was partially supported by NSF grant DMS-8806536 and the Mathematical Sciences Research Institute, and, for the preparation of the final version of this paper, by a grant from the Reidler Foundation.

This paper is Number 294 in the first author's publication list.

We are both grateful to the administration and staff of the Mathematical Sciences Research Institute for their hospitality during portions of 1989–90.

REFERENCES

1. A. Beller, R. Jensen and P. Welch, *Coding the Universe*, London Mathematical Society, Lecture Notes Series, vol. 47, Cambridge University Press, Cambridge, 1982.
2. S. Friedman, *A guide to 'Coding the universe' by Beller, Jensen, Welch*, J. Symbolic Logic **50** (1985), 1002–1019.
3. D. Law, *Doctoral dissertation, California Institute of Technology* (to appear).

DEPARTMENT OF MATHEMATICS, HEBREW UNIVERSITY, JERUSALEM, ISRAEL

DEPARTMENT OF MATHEMATICS, LEHIGH UNIVERSITY, BETHLEHEM PA 18015

Mathematical Sciences Research Institute Publications

(continued)

Volume 20	Dazord and Weinstein (eds.): Symplectic Geometry, Groupoids, and Integrable Systems
Volume 21	Moschovakis (ed.): Logic from Computer Science
Volume 22	Ratiu (ed.): The Geometry of Hamiltonian Systems
Volume 23	Baumslag and Miller (eds.): Algorithms and Classification in Combinatorial Group Theory
Volume 24	Montgomery and Small (eds.): Noncommutative Rings
Volume 25	Akbulut and King: Topology of Real Algebraic Sets
Volume 26	Judah, Just, and Woodin (eds.): Set Theory of the Continuum